The Foundations of Computability Theory

Borut Robič

The Foundations
of Computability Theory

Second Edition

Borut Robič
Faculty of Computer and Information Science
University of Ljubljana
Ljubljana, Slovenia

ISBN 978-3-662-62423-4   ISBN 978-3-662-62421-0   (eBook)
https://doi.org/10.1007/978-3-662-62421-0

This Springer imprint is published by the registered company Springer-Verlag GmbH, DE part of Springer Nature.
The registered company address is: Heidelberger Platz 3, 14197 Berlin, Germany

To Marinka, Gregor and Rebeka Hana

I still think that the best way to learn a new idea is to see its history, to see why someone was forced to go through the painful and wonderful process of giving birth to a new idea. ...Otherwise, it is impossible to guess how anyone could have discovered or invented it.

— *Gregory J. Chaitin*, Meta Maths, The Quest for Omega

Preface

Context

The paradoxes discovered in Cantor's set theory sometime around 1900 began a crisis that shook the foundations of mathematics. In order to reconstruct mathematics, freed from all paradoxes, Hilbert introduced a promising program with formal systems as the central idea. Though the program was unexpectedly brought to a close in 1931 by Gödel's famous theorems, it bequeathed burning questions: "What is computing? What is computable? What is an algorithm? Can every problem be algorithmically solved?" This led to *Computability Theory*, which was born in the mid-1930s, when these questions were resolved by the seminal works of Church, Gödel, Kleene, Post, and Turing. In addition to contributing to some of the greatest advances of twentieth-century mathematics, their ideas laid the foundations for the practical development of a universal computer in the 1940s as well as the discovery of a number of algorithmically unsolvable problems in different areas of science. New questions, such as "Are unsolvable problems equally difficult? If not, how can we compare their difficulty?" initiated new research topics of *Computability Theory*, which in turn delivered many important concepts and theorems. The application of these is central to the multidisciplinary research of *Computability Theory*.

Aims

Monographs in *Theoretical Computer Science* usually strive to present as much of the subject as possible. To achieve this, they present the subject in a definition–theorem–proof style and, when appropriate, merge and intertwine different related themes, such as computability, computational complexity, automata theory, and formal-language theory. This approach, however, often blurs historical circumstances, reasons, and the motivation that led to important goals, concepts, methods, and theorems of the subject.

My aim is to compensate for this. Since the fundamental ideas of theoretical computer science were either motivated by historical circumstances in the field or developed by pure logical reasoning, I describe *Computability Theory*, a part of *Theoretical Computer Science*, from this point of view. Specifically, I describe the difficulties that arose in mathematical logic, the attempts to recover from them, and how these attempts led to the birth of *Computability Theory* and later influenced it. Although some of these attempts fell short of their primary goals, they put forward crucial questions about computation and led to the fundamental concepts of *Computability Theory*. These in turn logically led to still new questions, and so on. By describing this evolution I want to give the reader a deeper understanding of the foundations of this beautiful theory. The challenge in writing this book was therefore to keep it accessible by describing the historical and logical development while at the same time introducing as many modern topics as needed to start the research. Thus, I will be happy if the book makes good reading before one tackles more advanced literature on *Computability Theory*.

Contents

There are three parts in this book.

Part I (Chaps. 1–4) *Chapter 1* is introductory: it discusses the *intuitive* comprehension of the concept of the algorithm. This comprehension was already provided by Euclid and sufficed since 300 B.C.E. or so. In the next three chapters I explain how the need for a rigorous, *mathematical definition* of the concepts of the algorithm, computation, and computability was born. *Chapter 2* describes the events taking place in mathematics around 1900, when *paradoxes* were discovered. The circumstances that led to the paradoxes and consequently to the foundational crisis in mathematics are explained. The ideas of the three main schools of recovery—*intuitionism*, *logicism*, and *formalism*—that attempted to reconstruct mathematics are described. *Chapter 3* delves into formalism. This school gathered the ideas and results of other schools in the concept of the *formal axiomatic system*. Three particular systems that played crucial roles in events are described; these are the formal axiomatic systems of *logic*, *arithmetic*, and *set theory*. *Chapter 4* presents *Hilbert's Program*, a promising formalistic attempt that would use formal axiomatic systems to eliminate all the paradoxes from mathematics. It is explained how *Hilbert's Program* was unexpectedly shattered by Gödel's *Incompleteness Theorems*, which state, in effect, that not every truth can be proved (in a formal system).

Part II (Chaps. 5–9) *Hilbert's Program* left open a question about the existence of a particular algorithm, the algorithm that would solve the *Entscheidungsproblem*. Since this algorithm might not exist, it was necessary to formalize the concept of the algorithm—only then would a proof of the non-existence of the algorithm be possible. Therefore, *Chapter 5* discusses the fundamental questions: "What is an algorithm? What is computation? What does it mean when we say that a function or problem is computable?" It is explained how these intuitive, informal concepts were formally defined in the form of the *Computability* (*Church–Turing*) *Thesis* by different yet equivalent *models of computation*, such as µ-recursive functions and (general) recursive functions, λ-calculus, the Turing machine, the Post machine, and Markov algorithms. ***Chapter 6*** focuses on the *Turing machine*, which most convincingly formalized the intuitive concepts of computation. Several equivalent variants of the Turing machine are described. Three basic uses of the Turing machine are presented: function computation, set generation, and set recognition. The existence of the *universal Turing machine* is proved and its impact on the creation and development of *general-purpose computers* is described. The equivalence of the Turing machine and the RAM model of computation is proved. In ***Chapter 7***, the first basic yet important theorems are deduced. These include the relations between *decidable*, *semi-decidable*, and *undecidable sets* (i.e., decision problems), the *Padding Lemma*, the *Parameter* (i.e., *s-m-n*) *Theorem*, and the *Recursion Theorem*. The latter two are also discussed in view of the recursive procedure calls in the modern general-purpose computer. ***Chapter 8*** is devoted to *incomputability*. It uncovers a surprising fact that, in effect, not everything that is defined can be computed (on a usual model of computation). Specifically, the chapter shows that not every computational problem can be solved by a computer. First, the incomputability of the *Halting Problem* is proved. To show that this is not just a unique event, a list of selected incomputable problems from various fields of science is given. Then, in ***Chapter 9***, methods of proving the incomputability of problems are explained; in particular, proving methods by *diagonalization*, *reduction*, the *Recursion Theorem*, and *Rice's Theorem* are explained.

   Part III (Chaps. 10–15)   In this part attention turns to *relative computability*. I tried to keep the chapters "bite-sized" by focusing in each on a single issue only. ***Chapter 10*** introduces the concepts of the oracle and the *oracle Turing machine*, describes how computation with such an external help would run, and discusses how oracles could be replaced in the real world by actual databases or networks of computers. ***Chapter 11*** formalizes the intuitive notion of the "*degree of unsolvability*" of a problem. To do this, it first introduces the concept of *Turing reduction*, the most general reduction between computational problems, and then the concept of *Turing degree*, which formalizes the notion of the degree of unsolvability. This formalization makes it possible to define, in ***Chapter 12***, an operator called the *Turing jump* and, by applying it, to construct a *hierarchy* of infinitely many Turing degrees. Thus, a surprising fact is discovered that for every unsolvable problem there is a more difficult unsolvable problem; there is no most difficult unsolvable problem. ***Chapter 13*** expands on this intriguing fact. It first introduces a view of the *class of all Turing degrees* as a mathematical structure. This eases expression of relations between the degrees. Then several properties of this class are proved, revealing the highly complex structure of the class. ***Chapter 14*** introduces *computably enumerable* (*c.e.*) *Turing degrees*. It then presents *Post's Problem*, posing whether there exist c.e. degrees other than **0** below the degree **0'**. Then the *priority method*, discovered and used by Friedberg and Muchnik to solve *Post's Problem*, is described. ***Chapter 15*** introduces the *arithmetical hierarchy*, which gives another, arithmetical view of the degrees of unsolvability. Finally, ***Chapter 16*** lists some suggestions for further reading.

Approach

The main lines of the approach are:

- *Presentation levels.* I use two levels of presentation, the fast track and detours. The *fast track* is a *fil rouge* through the book and gives a bird's-eye view of *Computability Theory*. It can be read independently of the *detours*. These contain detailed proofs, more demanding themes, additional historical facts, and further details, all of which can safely be skipped while reading on the fast track. The two levels differ visually: detours are written in small font and are put into *Boxes* (between gray bars, with broken lower bar), so they can easily be skipped or skimmed on first reading. Proofs are given on both levels whenever they are difficult or long.

- *Clarity.* Whenever possible I give the motivation and an explanation of the circumstances that led to new goals, concepts, methods, or theorems. For example, I explicitly point out with **NB** (nota bene) marks those situations and achievements that had important impact on further development in the field. Sometimes **NB** marks introduce conventions that are used in the rest of the book. New notions are introduced when they are naturally needed. Although I rigorously deduce theorems, I try to make proofs as intelligible as possible; this I do by commenting on tricky inferences and avoiding excessive formalism. I give intuitive, informal explanations of the concepts, methods, and theorems. Figures are given whenever this can add to the clarity of the text.

- *Contemporary terminology.* I use the recently suggested terminology and describe the reasons for it in the Bibliographic Notes; thus, I use partial computable (p.c.) functions (instead of partial recursive (p.r.) functions); computable functions (instead of recursive functions); computably enumerable (c.e.) sets (instead of recursively enumerable (r.e.) sets); and computable sets (instead of recursive sets).

- *Historical account.* I give an extended historical account of the mathematical and logical roots of *Computability Theory*.

- *Turing machine.* After describing different competing models of computation, I adopt the Turing machine as *the* model of computation and build on it. I neither formally prove the equivalence of these models, nor do I teach how to program Turing machines; I believe that all of this would take excessive space and add little to the understanding of *Computability Theory*. I do, however, rigorously prove the equivalence of the Turing machine and the RAM model, as the latter so closely abstracts real-life, general-purpose computers.

- *Unrestricted computing resources.* I decouple *Automata Theory* and *Formal-Language Theory* from *Computability Theory*. This enables me to consider general models of computation (i.e., models with unlimited resources) and hence focus freely on the question "What can be computed?" In this way, I believe, *Computability Theory* can be seen more clearly and it can serve as a natural basis for the development of *Computational Complexity Theory* in its study of "What can be computed efficiently?" Although I don't delve into *Computational Complexity Theory*, I do indicate the points where *Computational Complexity Theory* would take over.

- **Shortcuts to relative computability.** I introduce oracles in the usual way, after explaining classical computability. Readers eager to enter relative computability might want to start with Part II and continue on the fast track.

- **Practical consequences and applications.** I describe the applications of concepts and theorems, whenever I am aware of them.

Finally, in describing *Computability Theory* I do not try to be comprehensive. Rather, I view the book as a first step towards more advanced texts on *Computability Theory*, or as an introductory text to *Computational Complexity Theory*.

Audience

This book is written at a level appropriate for undergraduate or beginning graduate students in computer science or mathematics. It can also be used by anyone pursuing research at the intersection of theoretical computer science on the one hand and physics, biology, linguistics, or analytic philosophy on the other.

The only necessary prerequisite is some exposure to elementary logic. However, it would be helpful if the reader has had undergraduate-level courses in set theory and introductory modern algebra. All that is needed for the book is presented in Appendix A, which the reader can use to fill in the gaps in his or her knowledge.

Teaching

There are several courses one can teach from this book. A course offering the *minimum* of *Computability Theory* might cover (omitting boxes) Chaps. 5, 6, 7; Sects. 8.1, 8.2, 8.4; and Chap. 9. Such a course might be continued with a course on *Complexity Theory*. An *introductory* course on *Computability Theory* might cover Parts I and II (omitting most boxes of Part I). A beginning *graduate*-level course on *Computability Theory* might cover all three parts (with all the details in boxes). A course offering a *shortcut* (some 60 pages) to *Relative Computability* (Chaps. 10 to 15) might cover Sect. 5.3; Sects. 6.1.1, 6.2.1, 6.2.2; Sects. 6.3, 7.1, 7.2, 7.3; Sects. 7.4.1, 7.4.2, 7.4.3; Sects. 8.1, 8.2, 9.1, 9.2; and then Chaps. 10 through 15.

PowerPoint slides covering all three parts of the text are maintained and available at:

`http://lalg.fri.uni-lj.si/fct`

Origin

This book grew out of two activities: (1) the courses in *Computability and Computational Complexity Theory* that I have been teaching at the University of Ljubljana, and (2) my intention to write a textbook for a course on algorithms that I also teach.

When I started working on (2) I wanted to explain the $\mathcal{O}$-notation in a satisfactory way, so I planned an introductory chapter that would cover the basics of *Computational Complexity Theory*. But to explain the latter in a satisfactory way, the basics of *Computability Theory* had to be given first. So, I started writing on computability. But the story repeated once again and I found myself describing the *Mathematical Logic* of the twentieth century. This regression was due to (i) my awareness that, in the development of mathematical sciences, there was always some *reason* for introducing a new notion, concept, method, or goal, and (ii) my belief that the text should describe such reasons in order to present the subject as clearly as possible. Of course, many historical events and logical facts were important in this respect, so the chapter on *Computability Theory* continued to grow.

At the same time, I was aware that students of *Computability and Computational Complexity Theory* often have difficulty in grasping the meaning and importance of certain themes, as well as in linking up the concepts and theorems as a whole. It was obvious that before a new concept, method, or goal was introduced, the student should be given a historical or purely logical *motivation* for such a step. In addition, giving a *bird's-eye view* of the theory developed up to the last milestone also proved to be extremely advantageous.

These observations coincided with my wishes about the chapter on *Computability Theory*. So the project continued in this direction until the "chapter" grew into a text on *The Foundations of Computability Theory*, which is in front of you.

Acknowledgments

I would like to express my sincere thanks to all the people who read all or parts of the manuscript and suggested improvements, or helped me in any other way. I benefited from the comments of my colleagues *Uroš Čibej* and *Jurij Mihelič*. In particular, *Marko Petkovšek* (University of Ljubljana, Faculty of Mathematics and Physics, Department of Mathematics), and *Danilo Šuster* (University of Maribor, Faculty of Arts, Department of Philosophy) meticulously read the manuscript and suggested many improvements. The text has benefited enormously from their assistance. Although errors may remain, these are entirely my responsibility.

Many thanks go to my colleague *Boštjan Slivnik*, who skilfully helped me on several occasions to deal with TEX and its fonts. I have used drafts of this text in courses on *Computability and Computational Complexity Theory* that are given to students of computer science by our faculty, and to students of computer science and mathematics in courses organized in collaboration with the Faculty of Mathematics and Physics, University of Ljubljana. For their comments I particularly thank the students *Žiga Emeršič, Urša Krevs, Danijel Mišanović, Rok Resnik, Blaž Sovdat, Tadej Vodopivec,* and *Marko Živec.* For helpful linguistic suggestions, discussions on the pitfalls of English, and careful proofreading I thank *Paul McGuiness.*

I have made every reasonable effort to get permissions for inclusion of photos of the scientists whose contributions to the development of *Computability Theory* are described in the book. It turned out that most of the photos are already in the public domain. Here, *Wikimedia* makes praiseworthy efforts in collecting them; so does the online *MacTutor History of Mathematics Archive* at the University of St Andrews, Scotland. They were both very helpful and I am thankful to them. For the other photos I owe substantial thanks to the *Archives of the Mathematisches Forschungsinstitut Oberwolfach*, Germany, *King's College Library*, Cambridge, UK, and the *Los Alamos National Laboratory Archives*, USA. I have no doubt that photos make this serious text more pleasant. The following figures are courtesy of Wikimedia: Figs. 1.3, 1.4, 1.5, 1.7, 2.5, 2.6, 2.7, 2.9, 3.1, 3.6, 3.9, 5.4, and 5.11. The following figures are courtesy of the MacTutor History of Mathematics archive: Figs. 1.6, 2.4, 2.8, 4.8, 5.5, and 5.8. The following figures are courtesy of the King's College Library, Cambridge: Figs. 5.6 (AMT/K/7/9), 6.1 (AMT/K/7/14). The following figures are courtesy of the Archives of the Mathematisches Forschungsinstitut Oberwolfach: Figs. 2.2, 2.10, 3.5, 3.8, 4.5, 5.2, and 5.3. Figure 3.7 is courtesy of Los Alamos National Laboratory Archives, USA. (Unless otherwise indicated, this information has been authored by an employee or employees of the Los Alamos National Security, LLC (LANS), operator of the Los Alamos National Laboratory under Contract No. DE-AC52-06NA25396 with the U.S. Department of Energy. The U.S. Government has rights to use, reproduce, and distribute this information. The public may copy and use this information without charge, provided that this Notice and any statement of authorship are reproduced on all copies. Neither the Government nor LANS makes any warranty, express or implied, or assumes any liability or responsibility for the use of this information.)

I also thank the staff at Springer for all their help with the preparation of this book. In particular, I thank *Ronan Nugent*, my editor at Springer in Heidelberg, for his advice and kind support over the past few years. Finally, I thank the anonymous reviewers for their many valuable suggestions.

Ljubljana, January 2015   *Borut Robič*

Preface to the Second Edition

This is a completely revised edition, with about ninety pages of new material. In particular:

1. To improve the clarity of exposition, some terminological inconsistencies and notational redundancies have been removed. Thus all partial computable functions are now uniformly denoted by (possibly indexed) φ, instead of using ψ for the functions induced by particular Turing machines.

2. Various kinds of typos, minor errors, and aesthetic as well as grammatical flows have been corrected.

3. An entirely new Section 3.1.2 on **The Notion of Truth** has been added in Chapter 3. The section describes Alfred Tarski's definition of the notion of truth in formal languages and his attempts to formulate a similar definition for natural languages. Since the new section represents a natural bridge between the notion of the formal axiomatic system and the notion of its model, it has been inserted between the old sections on formal axiomatic systems and their interpretations.

4. Another major change is in Chapter 5, in Section 5.2.3 on models of computation, where the discussion of the **Post Machine** has been completely rewritten.

5. To comply with the up-to-date terminology, the recursive functions (as defined by Gödel and Kleene) have been renamed to μ-**recursive functions**. In this way, general recursive functions (as defined by Gödel and Herbrand) can simply be called **recursive functions**.

6. An entirely new Chapter 16 **Computability (Church-Turing) Thesis Revisited** has been added. The chapter is a systematic and detailed account of the origins, evolution, and meaning of this thesis.

7. Accordingly, some sections with **Bibliographic Notes** have been augmented.

8. Some sections containing **Problems** have been extended with new problems. Where required, definitions introducing the key notions and comments on these notions have been added.

9. A **Glossary** relating to computability theory has been added to help the reader.

10. Finally, **References** have been expanded by ninety new bibliographic entries.

Acknowledgments for the Second Edition

I am grateful to *Benjamin Wells* (University of San Francisco), who meticulously read the first edition of the book and suggested so many improvements. The text has greatly benefited from his comments.

I am grateful to *Ronan Nugent*, Senior Editor at Springer, for his advice and kind support during the preparation of this edition. I thank the anonymous copyeditor for many valuable comments and the staff at Springer for their help with the preparation of this edition.

Ljubljana, May 2020   *Borut Robič*

Contents

Contents

Part I
THE ROOTS OF COMPUTABILITY THEORY

In this part we will describe the events taking place in mathematics at the beginning of the twentieth century. Paradoxes discovered in Cantor's set theory around 1900 started a crisis in the foundations of mathematics. To reconstruct mathematics free from all paradoxes, Hilbert introduced a promising program with formal systems as the central idea. Though an end was unexpectedly put to the program in 1931 by Gödel's famous theorems, it bequeathed burning questions: What is computing? What is computable? What is an algorithm? Can every problem be solved algorithmically? These were the questions that led to the birth of *Computability Theory*.

Chapter 1
Introduction

*A recipe is a set of instructions describing how to prepare
something. A culinary recipe for a dish consists of the required
ingredients and their quantities, equipment and environment
needed to prepare the dish, an ordered list of preparation steps,
and the texture and flavor of the dish.*

Abstract The central notions in this book are those of the algorithm and computation, not a particular algorithm for a particular problem or a particular computation, but the algorithm and computation in general. The first algorithms were discovered by the ancient Greeks. Faced with a specific problem, they asked for a set of instructions whose execution in the prescribed order would eventually provide the solution to the problem. This view of the algorithm sufficed since the fourth century B.C.E., which meant there was no need to ask questions about algorithms and computation in general.

1.1 Algorithms and Computation

In this section we will describe how the concept of the algorithm was traditionally intuitively understood. We will briefly review the historical landmarks connected with the concept of the algorithm up to the beginning of the twentieth century.

1.1.1 The Intuitive Concept of the Algorithm and Computation

Every computational problem is associated with two sets: a set $\mathcal{A}$, which consists of all the possible input data to the problem, and a set $\mathcal{B}$, which consists of all the possible solutions. For example, consider the problem

"Find the greatest common divisor of two positive natural numbers."

After we have chosen the input data (e.g., 420 and 252), the solution to the problem is defined (84). Thus, we can think of the problem as a function $f : \mathcal{A} \to \mathcal{B}$, which maps the input data to the corresponding solution to the problem (see Fig. 1.1).

R. Robič, *The Foundations of Computability Theory*, https://doi.org/10.1007/978-3-662-62421-0_1

Fig. 1.1 A problem is viewed as a function mapping the input data to the corresponding solution

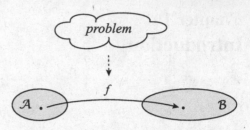

What does f look like? How do we compute its value $f(x)$ for a given x? The way we do this is described and governed by the associated *algorithm* (see Fig. 1.2).

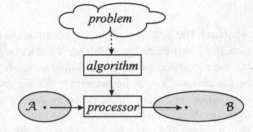

Fig. 1.2 The algorithm directs the processor in order to compute the solution to the problem

Definition 1.1. (Algorithm Intuitively) An **algorithm** for solving a problem is a finite set of instructions that lead the processor, in a finite number of steps, from the input data of the problem to the corresponding solution.

This was an informal definition of the algorithm. As such, it may raise questions, so we give some additional explanations. An algorithm is a *recipe*, a finite list of instructions that tell how to solve a problem. The *processor* of an algorithm may be a human, or a mechanical, electronic, or any other device, capable of mechanically following, interpreting, and executing instructions, while using no self-reflection or external help. The *instructions* must be simple enough so that the processor can execute them, and they have to be unambiguous, so that every next step of the execution is precisely defined. A *computation* is a sequence of such steps. The *input data* must be reasonable, in the sense that they are associated with solutions, so that the algorithm can bring the processor to a solution.

Generally, there exist many algorithms for solving a computational problem. However, they differ because they are based on different ideas or different design methods.

Example 1.1. (Euclid's Algorithm) An algorithm for finding the greatest common divisor of two numbers was described around 300 B.C.E. by Euclid[1] in his work *Elements*. The algorithm is:

Divide the larger number by the other.
Then keep dividing the last divisor by the last remainder—unless the last remainder is 0.
In that case the solution is the last nonzero remainder.[2]

For the input data 420 and 252, the computation is:

$$
\begin{aligned}
420 : 252 &= 1 + \text{remainder } 168; \\
252 : 168 &= 1 + \text{remainder } \ 84; \\
168 : 84 &= 2 + \text{remainder } \ \ 0; \quad \text{and the solution is 84.}
\end{aligned}
$$

Fig. 1.3 Euclid
(Courtesy: See Preface)

Observe that there exists another algorithm for this problem that is based on a different idea:

Factorize both numbers
and multiply the common factors.
The solution is this product.

For the input data 420 and 252, the computation is:

$$
\begin{aligned}
420 &= 2^2 \times 3 \times 5 \times 7 \\
252 &= 2^2 \times 3^2 \times \ 7 \\
&\quad 2^2 \times 3 \ \ \times \ 7 = 84; \quad \text{and the solution is 84.}
\end{aligned}
$$

□

---

[1] Euclid, 325–265 B.C.E., Greek mathematician, lived in Alexandria, now in Egypt.

[2] This algorithm was probably not discovered by Euclid. The algorithm was probably known by Eudoxus of Cnidus (408–355 B.C.E.), but it may even pre-date Eudoxus.

1.1.2 Algorithms and Computations Before the Twentieth Century

Euclid's algorithm is one of the first known nontrivial algorithms designed by a human. Unfortunately, the clumsy Roman number system, which was used in the ancient Mediterranean world, hindered computation with large numbers and, consequently, the application of such algorithms. Only after the positional decimal number system was discovered between the first and fourth centuries in India could large numbers be written succinctly. This enabled the Persian mathematician al-Khwarizmi[3] to describe in the year 825 algorithms for computing with such numbers, and in 830 algorithms for solving linear and quadratic equations. His name is the origin of the word *algorithm*.

Fig. 1.4 al-Khwarizmi
(Courtesy: See Preface)

In the seventeenth century, the first attempts were made to *mechanize* the algorithmic solving of *particular* computational problems of interest. In 1623, Schickard[4] tried to construct a machine capable of executing the operations + and − on natural numbers. Ten years later, a similar machine was successfully constructed by Pascal.[5] But Leibniz[6] saw further into the future. From 1666 he was considering a universal language (Latin *lingua characteristica universalis*) that would be capable of describing any notion from mathematics or the other sciences. His intention was to associate basic notions with natural numbers in such a way that the application of arithmetic operations on these numbers would return more complex numbers that would represent new, more complex notions. Leibniz also considered a universal computing machine (Latin *calculus ratiocinator*) capable of computing with such numbers. In 1671 he even constructed a machine that was able to carry out the operations $+, -, \times, \div$. In short, Leibniz's intention was to replace certain forms of human reflection (such as thinking, inferring, proving) with mechanical and mechanized arithmetic.

---

[3] Muhammad ibn Musa al-Khwarizmi, 780–850, Persian mathematician, astronomer, geographer; lived in Baghdad.

[4] Wilhelm Schickard, 1592–1635, German astronomer and mathematician.

[5] Blaise Pascal, 1623–1662, French mathematician, physicist, and philosopher.

[6] Gottfried Wilhelm Leibniz, 1646–1716, German philosopher and mathematician.

In 1834, a century and a half later, new and brilliant ideas led Babbage[7] to his design of a new, conceptually different computing machine. This machine, called the *analytical engine*, would be capable of executing *arbitrary programs* (i.e., algorithms written in the appropriate way) and hence of solving *arbitrary computational problems*. This led Babbage to believe that every computational problem is mechanically, and hence algorithmically, solvable.

Fig. 1.5 Blaise Pascal (Courtesy: See Preface)   **Fig. 1.6** Gottfried Leibniz (Courtesy: See Preface)   **Fig. 1.7** Charles Babbage (Courtesy: See Preface)

Nowadays, Leibniz's and Babbage's ideas remind us of several concepts of contemporary *Computability Theory* that we will look at in the following chapters. Such concepts are Gödel's arithmetization of formal axiomatic systems, the universal computing machine, computability, and the *Computability Thesis*. So Leibniz's and Babbage's ideas were much before their time. But they were too early to have any practical impact on mankind's comprehension of computation.

As a result, the concept of the algorithm remained firmly at the intuitive level, defined only by common sense as in Definition 1.1. It took many events for the concept of the algorithm to be rigorously defined and, as a result, for *Computability Theory* to be born. These events are described in the next chapter.

1.2 Chapter Summary

The algorithm was traditionally intuitively understood as a recipe, i.e., a finite list of directives written in some language that tells us how to solve a problem mechanically. In other words, the algorithm is a precisely described routine procedure that can be applied and systematically followed through to a solution of a problem. Because there was no need to define formally the concept of the algorithm, it remained firmly at the intuitive, informal level.

---

[7] Charles Babbage, 1791–1871, British mathematician.

Chapter 2
The Foundational Crisis of Mathematics

> *A paradox is a situation that involves two or more facts or qualities that contradict each other.*

Abstract The need for a formal definition of the concept of algorithm was made clear during the first few decades of the twentieth century as a result of events taking place in mathematics. At the beginning of the century, Cantor's naive set theory was born. This theory was very promising because it offered a common foundation to all the fields of mathematics. However, it treated infinity incautiously and boldly. This called for a response, which soon came in the form of logical paradoxes. Because Cantor's set theory was unable to eliminate them—or at least bring them under control—formal logic was engaged. As a result, three schools of mathematical thought—intuitionism, logicism, and formalism—contributed important ideas and tools that enabled an exact and concise mathematical expression and brought rigor to mathematical research.

2.1 Crisis in Set Theory

In this section we will describe the axiomatic method that was used to develop mathematics since its beginnings. We will also describe how Cantor applied the axiomatic method to develop his set theory. Finally, we will explain how paradoxes revealed themselves in this theory.

2.1.1 Axiomatic Systems

The basic method used to acquire new knowledge in mathematics and similar disciplines is the *axiomatic method*. Euclid was probably the first to use it when he was developing his geometry.

R. Robič, *The Foundations of Computability Theory*, https://doi.org/10.1007/978-3-662-62421-0_2

9

Axiomatic Method

When using the axiomatic method, we start our treatment of the field of interest by carefully selecting a few *basic notions* and making a few basic statements,[1] called *axioms* (see Fig. 2.1). An axiom is a statement that asserts either that a basic notion has a certain property, or that a certain relation holds between certain basic notions. We do not try to define the basic notions, nor do we try to prove the axioms. The basic notions and axioms form our *initial theory* of the field.

We then start *developing* the theory, i.e., extending the initial knowledge of the field. We do this systematically. This means that we must *define* every new notion in a clear and precise way, using only basic or previously defined notions. It also means that we must try to *prove* every new proposition,[2] i.e., *deduce*[3] it only from axioms or previously proven statements. Here, a *proof* is a finite sequence of mental steps, i.e., inferences,[4] that end in the realization that the proposition is a logical consequence of the axioms and previously proven statements. A provable proposition is called a *theorem* of the theory. The process of proving is informal, *content-dependent* in the sense that each conclusion must undoubtedly follow from the *meaning* of its premises.

Informally, the development of a theory is a process of discovering (i.e., deducing) new theorems—in Nagel and Newman's words, as Columbus discovered America—and defining new notions in order to facilitate this process. (This is the *Platonic* view of mathematics; see Box 2.1.) We say that axioms and basic notions constitute an *axiomatic system*.

Fig. 2.1 A theory has axioms, theorems, and unprovable statements

---

Box 2.1 (Platonism).

According to the *Platonic view* mathematics *does not create* new objects but, instead, *discovers already existing* objects. These exist in the non-material world of abstract *Ideas*, which is accessible only to our intellect. For example, the idea of the number 2 exists *per se*, capturing the state of "twoness", i.e., the state of any gathering of anything and something else—and nothing else. In the

---

[1] A *statement* is something that we say or write that makes sense and is either true or false.

[2] A *proposition* is a statement for which a proof is either required or provided.

[3] A *deduction* is the process of reaching a conclusion about something because of other things (called premises) that we know to be true.

[4] An *inference* is a conclusion that we draw about something by using information that we already have about it. It is also the process of coming to a conclusion.

material world, *Ideas* present themselves in terms of imperfect copies. For example, the *Idea* of a triangle is presented by various copies, such as the figures $\triangle, \triangledown, \triangleleft, \triangleright$, and love triangles too. It can take considerable instinct to discover an *Idea*, the comprehension of which is consistent with the sensation of its copies in the material world. The agreement of these two is the criterion for deciding whether the *Idea* really exists.

Evident Axiomatic Systems

From the time of Euclid to the mid-nineteenth century it was required that axioms be statements that are in perfect agreement with human experience in the particular field of interest. The validity of such axioms was beyond any doubt, because they were clearly confirmed by reality. Thus, no proofs of axioms were required. Such axioms are *evident*. Euclidean elementary geometry is an example of an evident axiomatic system, because it talks of points, lines, and planes, which are evident idealizations of the corresponding real-world objects.

However, in the nineteenth century serious doubts arose as to whether evident axiomatic systems are always appropriate. This is because it was found that experience and intuition may be misleading. (For an example of such a situation see Box 2.2.) This led to the concept of the *hypothetical* axiomatic system.

Box 2.2 (Euclid's Fifth Axiom).

In two-dimensional geometry a line parallel to a given line L is a line that does not intersect with L. Euclid's fifth axiom, also called the *Parallel Postulate*, states that at most one parallel can be drawn through any point not on L. (In fact, Euclid postulated this axiom in a different but equivalent form.)

To Euclid and other ancients the fifth axiom seemed less obvious than the other four of Euclid's axioms. This is because a parallel line can be viewed as a line segment that never intersects with L, even if it is extended indefinitely. The fifth axiom thus implicitly speaks about a certain occurrence in arbitrarily removed regions of the plane, that is, that the segment and L will never meet. However, since Aristotle the ancients were well aware that one has to be careful when dealing with infinity. For example, they were already familiar with the notion of asymptote, a line that approaches a given curve but meets the curve only at infinity.

To avoid the vagueness and controversy of Euclid's fifth axiom, they undertook to deduce it from Euclid's other four axioms; these caused no disputes. However, all attempts were unsuccessful until 1868, when Beltrami[5] proved that Euclid's fifth axiom *cannot be deduced* from the other four axioms. In other words, Euclid's fifth axiom is *independent* of Euclid's other four axioms.

NB *The importance of Beltrami's discovery is that it does not belong to geometry, but rather to the science about geometry, and, more generally, to metamathematics, the science about mathematics. About fifty years later, metamathematics would come to the fore more explicitly and play a key role in the events that led to a rigorous definition of the notion of the algorithm.*

Since the eleventh century, Persian and Italian mathematicians had tried to prove Euclid's fifth axiom indirectly. They tried to refute all of its alternatives. These stated either that there are no parallels, or that there are several different parallels through a given point. When they considered

---

[5] Eugenio Beltrami, 1835–1900, Italian mathematician.

these alternatives they unknowingly discovered *non-Euclidean geometries*, such as elliptic and hyperbolic geometry. But they cast them away as having no evidentness in reality. According to the usual experience they viewed reality as a space where only Euclidean geometry can rule.

In the nineteenth century, Lobachevsky,[6] Bolyai,[7] and Riemann[8] thought of these geometries as true alternatives. They showed that if Euclid's fifth axiom is replaced by a different axiom, then a different non-Euclidean geometry is obtained. In addition, there exist in reality examples, also called *models*, of such non-Euclidean geometries. For instance, Riemann replaced Euclid's fifth axiom with the axiom that states that there is no parallel to a given line L through a given point. The resulting geometry is called *elliptic* and is modeled by a sphere. In contrast to this, Bolyai and Lobachevsky selected the axiom that allows several parallels to L to pass through a given point. The resulting *hyperbolic* geometry holds, for example, on the surface of a saddle.

NB *These discoveries shook the traditional standpoint that axioms should be obvious and clearly agree with reality. It became clear that intuition and experience may be misleading.*

Hypothetical Axiomatic Systems

After the realization that instinct and experience can be delusive, mathematics gradually took a more abstract view of its research subjects. No longer was it interested in the (potentially slippery) *nature* of the basic notions used in an axiomatic system. For example, arithmetic was no longer concerned with the question of *what* a natural number really is, and geometry was no longer interested in *what* a point, a line, and a plane really are. Instead, mathematics focused on the *properties* of and *relations* between the basic notions, which could be defined without specifying what the basic notions are in reality. A basic notion can be *any* object that fulfills all the conditions given by the axioms.

Thus the role of the axioms has changed; now an axiom is only a *hypothesis*, a speculative statement about the basic notions taken to hold, although nothing is said about the true nature and existence of such basic notions. Such an axiomatic system is called *hypothetical*. For example, by using nine axioms Peano in 1889 described properties and relations typical of natural numbers without explicitly defining a natural number. Similarly, in 1899 Hilbert developed elementary geometry, where no explicit definition of a point, line, and plane is given; instead, these are defined implicitly, only as possible objects that satisfy the postulated axioms.

Because the nature of basic notions lost its importance, also the requirement for the evidentness of axioms as well as their verifiability in reality was abandoned. The obvious link between the subject of mathematical treatment and reality vanished. Instead of axiomatic evidentness the fertility of axioms came to the fore, i.e., the number of theorems deduced, their expressiveness, and their influence. The reasonableness and applicability of the theory developed was evaluated by the importance of successful *interpretations*, i.e., applications of the theory to various domains of reality. Depending on this, the theory was either accepted, corrected, or cast off.

---

[6] Nikolai Ivanovich Lobachevsky, 1792–1856, Russian mathematician and geometer.

[7] János Bolyai, 1802–1860, Hungarian mathematician.

[8] Georg Friedrich Bernhard Riemann, 1826–1866, German mathematician.

NB *This freedom, which arose from the hypothetical axiomatic system, enabled scientists to make attempts that eventually bred important new areas of mathematics. Set theory is such an example.*[9]

2.1.2 Cantor's Naive Set Theory

A theory with a hypothetical axiomatic system that will play a special role in what follows was the *naive set theory* founded by Cantor.[10] Let us take a quick look at this theory.

Fig. 2.2 Georg Cantor
(Courtesy: See Preface)

Basic Notions, Concepts, and Axioms

In 1895 Cantor defined the concept of a set as follows.

Definition 2.1. (Cantor's Set) A **set** is any collection of definite, distinguishable objects of our intuition or of our intellect to be conceived as a whole (i.e., regarded as a single unity).

Thus, an object can be any thing or any notion, such as a number, a pizza, or even another set. If an object x is in a set S, we say that x is a *member* of S and write $x \in S$. When x is not in S, it is not a member of S, so $x \notin S$. Given an object x and a set S, either $x \in S$ or $x \notin S$—there is no third choice. This is known as the *Law of Excluded Middle*.[11]

Cantor did not develop his theory from explicitly written axioms. However, later analyses of his work revealed that he used three principles in the same fashion as axioms. For this reason we call these principles the **Axioms of Extensionality**, **Abstraction**, and **Choice**. Let us describe them.

---

[9] Another example of a theory with a hypothetical axiomatic system is group theory

[10] Georg Cantor, 1845–1918, German mathematician.

[11] The law states that for any logical statement, either that statement is true, or its negation is—there is no third possibility (Latin *tertium non datur*).

Axiom 2.1 (Extensionality). *A set is completely determined by its members.*

Thus a set is completely described if we list all of its members (by convention between braces "{" and "}"). For instance, $\{\diamond, \triangleleft, \circ\}$ is a set whose members are $\diamond, \triangleleft, \circ$, while one of the three members of the set $\{\diamond, \{\triangleleft, \triangleright\}, \circ\}$ is itself a set. When a set has many members, say a thousand, it may be impractical to list all of them; instead, we may describe the set perfectly by stating the *characteristic property* of its members. Thus a set of objects with the property P is written as $\{x \mid x$ has the property $P\}$ or as $\{x \mid P(x)\}$. For instance, $\{x \mid x$ is a natural number $\land\ 1 \leqslant x \leqslant 1{,}000\}$.

What property can P be? Cantor's liberal-minded standpoint in this matter is summed up in the second axiom:

Axiom 2.2 (Abstraction). *Every property determines a set.*

If there is no object with a given property, the set is *empty*, that is, $\{\}$. Due to the *Axiom of Extensionality* there is only one empty set; we denote it by $\emptyset$.

Cantor's third principle is summed up in the third axiom:

Axiom 2.3 (Choice). *Given any set $\mathcal{F}$ of nonempty pairwise disjoint sets, there is a set that contains exactly one member of each set in $\mathcal{F}$.*

We see that the set and the membership relation $\in$ are such basic notions that Cantor defined them informally, in a descriptive way. Having done this he used them to rigorously define other notions in a true axiomatic manner. For example, he defined the relations $=$ and $\subseteq$ on sets. Specifically, two sets $\mathcal{A}$ and $\mathcal{B}$ are *equal* (i.e., $\mathcal{A} = \mathcal{B}$) if they have the same objects as members. A set $\mathcal{A}$ is a *subset* of a set $\mathcal{B}$ (i.e., $\mathcal{A} \subseteq \mathcal{B}$) if every member of $\mathcal{A}$ is also a member of $\mathcal{B}$. Cantor also defined the operations $^{-}, \cup, \cap, -, 2^{\cdot}$ that construct new sets from existing ones. For example, if $\mathcal{A}$ and $\mathcal{B}$ are two sets, then also the *complement* $\overline{\mathcal{A}}$, the *union* $\mathcal{A} \cup \mathcal{B}$, the *intersection* $\mathcal{A} \cap \mathcal{B}$, the *difference* $\mathcal{A} - \mathcal{B}$, and the *power set* $2^{\mathcal{A}}$ are sets.

Applications

Cantor's set theory very quickly found applications in different fields of mathematics. For example, Kuratowski[12] used sets to define the *ordered pair* (x,y), i.e., a set of two elements with one being the first and the other the second in some order. The definition is $(x,y) \stackrel{\text{def}}{=} \{\{x\}, \{x,y\}\}$. (The ordering of $\{x\}$ and $\{x,y\}$ is implicitly imposed by the relation $\subseteq$, since $\{x\} \subseteq \{x,y\}$ but not vice versa.) Two ordered pairs are equal if they have equal first elements and equal second elements. Now the *Cartesian product* $\mathcal{A} \times \mathcal{B}$ could be defined as the set of all ordered pairs (a,b), where $a \in \mathcal{A}$ and $b \in \mathcal{B}$. The sets $\mathcal{A}$ and $\mathcal{B}$ need not be distinct. In this case, $\mathcal{A}^2$ was used to denote $\mathcal{A} \times \mathcal{A}$ and, in general, $\mathcal{A}^n \stackrel{\text{def}}{=} \mathcal{A}^{n-1} \times \mathcal{A}$, where $\mathcal{A}^1 = \mathcal{A}$.

---

[12] Kazimierz Kuratowski, 1896–1980, Polish mathematician and logician.

Many other important notions and concepts that were in common use although informally defined were at last rigorously defined in terms of set theory, e.g., the concepts of function and natural number. For example, a *function* $f : \mathcal{A} \to \mathcal{B}$ is a set of ordered pairs (a, b), where $a \in \mathcal{A}$ and $b = f(a) \in \mathcal{B}$, and there are no two ordered pairs with equal first components and different second components. Based on this, set-theoretic definitions of injective, surjective, and bijective functions were easily made. For example, a bijective function is a function $f : \mathcal{A} \to \mathcal{B}$ whose set of ordered pairs contains, for each $b \in \mathcal{B}$, at least one ordered pair with the second component b (surjectivity), and there are no two ordered pairs having different first components and equal second components (injectivity).

Von Neumann used sets to construct *natural numbers* as follows. Consider the number 2. We may imagine that it represents the state of "twoness", i.e., the gathering of one element and one more different element—and nothing else. Since the set $\{0, 1\}$ is an example of such a gathering, we may define $2 \overset{\text{def}}{=} \{0, 1\}$. Similarly, if we imagine 3 to represent "threeness", we may define $3 \overset{\text{def}}{=} \{0, 1, 2\}$. Continuing in this way, we arrive at the general definition $n \overset{\text{def}}{=} \{0, 1, 2, \ldots, n-1\}$. So a natural number can be defined as the *set* of all of its predecessors. What about the number 0? Since 0 has no natural predecessors, the corresponding set is empty. Hence the definition $0 \overset{\text{def}}{=} \emptyset$. We now see that natural numbers can be constructed from $\emptyset$ as follows: $0 \overset{\text{def}}{=} \emptyset$; $1 \overset{\text{def}}{=} \{\emptyset\}$; $2 \overset{\text{def}}{=} \{\emptyset, \{\emptyset\}\}$; $3 \overset{\text{def}}{=} \{\emptyset, \{\emptyset\}, \{\emptyset, \{\emptyset\}\}\}$; $\ldots$; $n+1 \overset{\text{def}}{=} n \cup \{n\}$; $\ldots$ Based on this, other definitions and constructions followed (e.g., of rational and real numbers).

NB *Cantor's set theory offered a simple and unified approach to all fields of mathematics. As such it promised to become the foundation of all mathematics.*

But Cantor's set theory also brought new, quite surprising discoveries about the so-called cardinal and ordinal numbers. As we will see, these discoveries resulted from Cantor's *Axiom of Abstraction* and his view of infinity. Let us go into details.

Cardinal Numbers

Intuitively, two sets have the same "size" if they contain the same number of elements. Without any counting of their members we can assert that two sets are *equinumerous*, i.e., of the same "size", *if* there is a bijective function mapping one set onto the other. This function pairs every member of one set with exactly one member of the other set, and vice versa. Such sets are said to have the same *cardinality*. For example, the sets $\{\diamond, \lhd, \circ\}$ and $\{a, b, c\}$ have the same cardinality because $\{(\diamond, a), (\lhd, b), (\circ, c)\}$ is a bijective function. In this example, each of the sets has cardinality ("size") 3, a *natural* number. We denote the cardinality of a set $\mathcal{S}$ by $|\mathcal{S}|$.

Is the cardinality always a natural number? Cantor's *Axiom of Abstraction* guarantees that the set $\mathcal{S}_P = \{x \mid P(x)\}$ exists for *any* given property P. Hence, it exists also for a P that is shared by infinitely many objects. For example, if we put

$P \equiv$ "is natural number," we get the set of *all* natural numbers. This set is not only an interesting and useful mathematical object, but (according to Cantor) it also *exists* as a perfectly defined and accomplished unity. Usually, it is denoted by $\mathbb{N}$. It is obvious that the cardinality of $\mathbb{N}$ cannot be a natural number because any such number would be too small. Thus Cantor was forced to introduce a new kind of number and designate it with some new symbol not used for natural numbers. He denoted this number by $\aleph_0$ (read *aleph zero* [13]). Cardinality of sets can thus be described by the numbers that Cantor called *cardinal numbers*. A *cardinal number* (or *cardinal* for short) can either be *finite* (in that case it is natural) or *transfinite*, depending on whether it measures the size of a finite or infinite set. For example, $\aleph_0$ is a transfinite cardinal that describes the size of the set $\mathbb{N}$ as well as the size of any other infinite set whose members can all be listed in a sequence.

Does every infinite set have the cardinality $\aleph_0$? Cantor discovered that this is not so. He proved (see Box 2.3) that the cardinality of a set S is strictly less than the cardinality of its power set 2^S—*even when S is infinite!* Consequently, there are larger and larger infinite sets whose cardinalities are larger and larger transfinite cardinals—and this never ends. He denoted these transfinite cardinals by $\aleph_1, \aleph_2, \ldots$. Thus, there is no largest cardinal.

Cantor also discovered (using diagonalization, a method he invented; see Sect. 9.1) that there are more real numbers than natural ones, i.e., $\aleph_0 < c$, where c denotes the cardinality of $\mathbb{R}$, the set of real numbers. (For the proof see Example 9.1 on p. 207.) He also proved that $c = 2^{\aleph_0}$. But where is c relative to $\aleph_0, \aleph_1, \aleph_2, \ldots$? Cantor conjectured that $c = \aleph_1$, that is, $2^{\aleph_0} = \aleph_1$. This would mean that there is no other transfinite cardinal between $\aleph_0$ and c and consequently there is no infinite set larger than $\mathbb{N}$ and smaller than $\mathbb{R}$. Yet, no one succeeded in proving or disproving this conjecture, until Gödel and Cohen finally proved that *neither* can be done (see Box 4.3 on p. 65). Cantor's conjecture is now known as the *Continuum Hypothesis*.

Box 2.3 (Proof of Cantor's Theorem).

Cantor's Theorem states: $|S| < |2^S|$, *for every set S.*

Proof. (a) First, we prove that $|S| \leq |2^S|$. To do this, we show that S is equinumerous to a subset of 2^S. Consider the function $f : S \to 2^S$ defined by $f : x \mapsto \{x\}$. This is a bijection from S onto $\{\{x\} | x \in S\}$, which is a subset of 2^S. (b) Second, we prove that $|S| \neq |2^S|$. To do this, we show that there is no bijection from S onto 2^S. So let $g : S \to 2^S$ be an *arbitrary* function. Then g cannot be surjective (and hence, neither is it bijective). To see this, let $\mathcal{N}$ be a subset of S defined by $\mathcal{N} = \{x \in S | x \notin g(x)\}$. Of course, $\mathcal{N} \in 2^S$. But $\mathcal{N}$ is not a g-image of any member of S. *Suppose* it were. Then there would be an $m \in S$ such that $g(m) = \mathcal{N}$. Where would be m relative to $\mathcal{N}$? If $m \in \mathcal{N}$, then $m \notin g(m)$ (by definition of $\mathcal{N}$), and hence $m \notin \mathcal{N}$ (as $g(m) = \mathcal{N}$)! Conversely, if $m \notin \mathcal{N}$, then $m \notin g(m)$ (as $g(m) = \mathcal{N}$), and hence $m \in \mathcal{N}$ (by definition of $\mathcal{N}$)! This is a contradiction. We conclude that g is not a surjection, and therefore neither is it a bijection. Since g was an arbitrary function, we conclude that there is no bijection from S onto 2^S.   □

---

[13] $\aleph$ is the first symbol of the Hebrew alphabet.

Ordinal Numbers

We have seen that one can introduce order into a set of *two* elements. This can easily be done with other finite and infinite sets, and it can be done in many different ways. Of special importance to Cantor was the so-called *well-ordering*, because this is the way natural numbers are ordered with the usual relation $\leqslant$. For example, each of the sets $\{0,1,2\}$ and $\mathbb{N}$ is well-ordered with $\leqslant$, that is, $0 < 1 < 2$ and $0 < 1 < 2 < 3 < \cdots$. (Here $<$ is the strict order corresponding to $\leqslant$.) But well-ordering can also be found in other sets and for relations other than the usual $\leqslant$. When two well-ordered sets differ only in the naming of their elements or relations, we say that they are *similar*.

Cantor's aim was to *classify* all the well-ordered sets according to their similarity. In doing so he first noticed that the usual well-ordering of the set $\{0,1,2,\ldots,n\}$, $n \in \mathbb{N}$, can be represented by a single natural number $n+1$. (We can see this if we construct $n+1$ from $\emptyset$, as von Neumann did.) For example, the number 3 represents the ordering $0 < 1 < 2$ of the set $\{0,1,2\}$. But the usual well-ordering of the set $\mathbb{N}$ cannot be described by a natural number, as any such number is too small. Once again a new kind of a "number" was required and a new symbol for it was needed. Cantor denoted this number by ω and called it the *ordinal number*.

Well-ordering of a set can thus be described by the *ordinal number*, or *ordinal* for short. An ordinal is either *finite* (in which case it is natural) or *transfinite*, depending on whether it represents the well-ordering of a finite or infinite set. For example, ω is the transfinite ordinal that describes the usual well-ordering in $\mathbb{N}$. Of course, in order to use ordinals in classifying well-ordered sets, Cantor required that two well-ordered sets have the same ordinal *iff* they are similar. (See details in Box 2.4.) Then, once again, he proved that there are larger and larger transfinite ordinals describing larger and larger well-ordered infinite sets—and this never ends. There is no largest ordinal.

NB *With his set theory, Cantor boldly entered a curious and wild world of infinities.*

2.1.3 Logical Paradoxes

Unfortunately, the great leaps forward made by Cantor's set theory called for a response. This came around 1900 when logical paradoxes were suddenly discovered in this theory. A *paradox* (or *contradiction*) is an unacceptable conclusion derived by apparently acceptable reasoning from apparently acceptable premises (see Fig. 2.3).

Burali-Forti's Paradox. The first logical paradox was discovered in 1897 by Burali-Forti.[14] He showed that in Cantor's set theory there exists a well-ordered set Ω whose ordinal number is *larger than itself*. But this is a contradiction. (See details in Box 2.4.)

---

[14] Cesare Burali-Forti, 1861–1931, Italian mathematician.

Cantor's Paradox. A similar paradox was discovered by Cantor himself in 1899. Although he proved that, for any set $\mathcal{S}$, the cardinality of the power set $2^{\mathcal{S}}$ is strictly larger than the cardinality of $\mathcal{S}$, he was forced to admit that this cannot be true of *the set $\mathcal{U}$ of all sets*. Namely, the existence of $\mathcal{U}$ was guaranteed by the *Axiom of Abstraction*, just by defining $\mathcal{U} = \{x \mid x = x\}$. But if the cardinality of $\mathcal{U}$ is less than the cardinality of $2^{\mathcal{U}}$, which also exists, then $\mathcal{U}$ is not the largest set (which $\mathcal{U}$ is supposed to be since it is the set of *all* sets). This is a contradiction.

Russell's Paradox. The third paradox was found in 1901 by Russell.[15] He found that in Cantor's set theory there exists a set $\mathcal{R}$ that *both is and is not* a member of itself. How? Firstly, the set $\mathcal{R}$ defined by

$$\mathcal{R} = \{\mathcal{S} \mid \mathcal{S} \text{ is a set } \textit{and } \mathcal{S} \text{ does not contain itself as a member}\}$$

must exist because of the *Axiom of Abstraction*. Secondly, the *Law of Excluded Middle* guarantees that $\mathcal{R}$ either contains itself as a member (i.e., $\mathcal{R} \in \mathcal{R}$), or does not contain itself as a member (i.e., $\mathcal{R} \notin \mathcal{R}$). But then, using the definition of $\mathcal{R}$, each of the two alternatives implies the other, that is, $\mathcal{R} \in \mathcal{R} \iff \mathcal{R} \notin \mathcal{R}$. Hence, each of the two is both a true and a false statement in Cantor's set theory.

Fig. 2.3 A paradox is an unacceptable statement or situation because it defies reason; for example, because it is (or at least seems to be) both true and false

Why Do We Fear Paradoxes?

Suppose that a theory contains a logical statement such that both the statement and its negation can be deduced. Then it can be shown (see Sect. 4.1.1) that *any* other statement of the theory can be deduced as well. So in this theory everything is deducible! This, however, is not as good as it may seem at first glance. Since deduction is a means of discovering truth (i.e., what is deduced is accepted as true) we see that in such a theory every statement is true. But a theory in which everything is true has no cognitive value and is of no use. Such a theory must be cast off.

---

[15] Bertrand Russell, 1872–1970, British mathematician, logician, and philosopher.

Box 2.4 (Burali-Forti's Paradox).

A set S is *well-ordered* by a relation $\prec$ if the following hold: 1) $a \nprec a$; 2) $a \neq b \Rightarrow a \prec b \vee b \prec a$; and 3) every nonempty $\mathcal{X} \subseteq S$ has $m \in \mathcal{X}$ such that $m \prec x$ for every other $x \in \mathcal{X}$. For example, $\mathbb{N}$ is well-ordered with the usual relation $<$ on natural numbers. Well-ordering is a special case of the so-called *linear ordering*, i.e., a well-ordered set is also linearly ordered. For example, $\mathbb{Z}$, the set of integers, is linearly ordered by the usual relation $<$.

Suppose we do not want to distinguish between two linearly ordered sets that differ only in the naming of their elements or relations. We want to consider such sets as being similar, because they obviously share the same "type of order".

Let us define the notion "type of order" precisely. Let two sets $\mathcal{A}$ and $\mathcal{B}$ be linearly ordered with relations $\prec_{\mathcal{A}}$ and $\prec_{\mathcal{B}}$, respectively. We say that $\mathcal{A}$ and $\mathcal{B}$ are *similar* if there is a bijection $f : \mathcal{A} \to \mathcal{B}$ such that $a \prec_{\mathcal{A}} b \Longleftrightarrow f(a) \prec_{\mathcal{B}} f(b)$. The function f renames the elements of $\mathcal{A}$ to the elements of $\mathcal{B}$ while respecting both relations. We can easily prove that similarity is an equivalence relation between linearly ordered sets. So we can define the *order type* to be an equivalence class of similar, linearly ordered sets. Informally, an order type is the feature shared by all linearly ordered sets that differ only in the naming of their elements and relations.

Having defined the order types we might want to compare them. Unfortunately, they may not be comparable. It can be shown, however, that *order types of well-ordered sets* are themselves *linearly ordered* by some relation $\prec_o$. (Actually, $\prec_o$ is the usual set-membership relation $\in$.) Because such order types are ordered in a similar way to integers, we call them *ordinal numbers* (or *ordinals* for short). Hence, the definition: An *ordinal* is an equivalence class of similar well-ordered sets. For example, sets similar to $\{0, 1, \ldots, n\}$ have the same ordinal; we denote it by the natural number $n + 1$. This cannot be done with sets similar to $\mathbb{N}$, so we use ω to denote their ordinal.

For each ordinal α there is exactly one ordinal $\alpha' \stackrel{\text{def}}{=} \alpha \cup \{\alpha\}$ that is the $\prec_o$-successor of α. (We also denote α' by $\alpha + 1$.) It follows that there is no $\prec_o$-largest ordinal.

This is where Burali-Forti entered. He proved that Cantor's set theory allows the construction of a set Ω of *all* the ordinals. He also showed that such an Ω leads to a paradox. Namely, Ω would not only be linearly ordered by $\prec_o$, but also well-ordered by $\prec_o$. As such, Ω would be associated with the corresponding ordinal, say α_Ω. But where would α_Ω be relative to Ω? Since Ω is the set of *all* the ordinals, it must be that $\alpha_\Omega \in \Omega$. On the other hand, α_Ω must be $\prec_o$-larger than any member of Ω, and therefore larger than itself.

2.2 Schools of Recovery

In this section we will describe the three main schools of mathematical thought that significantly contributed to the struggle against paradoxes in mathematics. These are *intuitionism*, *logicism*, and *formalism*. We will show how their discoveries were synthesized in the concept of a formal axiomatic system and then in a clear awareness that a higher, metamathematical language is needed to investigate such systems.

2.2.1 Slowdown and Revision

The discovery of the paradoxical sets Ω, $\mathcal{U}$, and $\mathcal{R}$ was shocking, because Cantor's set theory was supposed to become a firm foundation for all other fields of mathematics and should, therefore, have been free of paradoxes. But the simplicity of Russell's Paradox, which used only two basic notions of set and membership relation, revealed that paradoxes originated deep in Cantor's theory, in the very definition of the concept of a set. It was this definition of a set and the unrestricted use of the *Axiom of Abstraction* that allowed the existence of the sets Ω, $\mathcal{U}$, and $\mathcal{R}$ that, in the end, caused and revealed paradoxical situations. So it was clear that objects like Ω, $\mathcal{U}$, and $\mathcal{R}$ should *not* be recognized as existing sets.

Therefore, Cantor's *naive* definition of the concept of a set (Sect. 2.1.2) should be restricted somehow. But this was easier said than done. Namely, Cantor's definition of a set was so natural and of such common sense that it was far from clear how to restrict it and, at the same time, retain all the sound parts of the theory. If a set is not what Cantor thought about, then what was it? And what was it not?

This once again triggered a critical reflection about the basic concepts, notions, principles, methods, and tools of set theory and logic, which might be sources of paradoxes. The aim was to make the necessary corrections to them, so that they, as a whole, would again act as a foundation for the development of mathematics and other axiomatic areas of science, but this time *safe from all paradoxes*. It turned out that no universally accepted resolutions could be expected. The critiques and proposals went in several directions, of which the three mainstream directions were called **intuitionism**, **logicism**, and **formalism**. Because they all contributed to future events, we briefly review them.

2.2.2 Intuitionism

Intuitionism argued for greater mathematical rigor in the process of proving and it advocated a non-Platonic view that the existence of a mathematical object is closely connected to the existence of its mental construction.

Fig. 2.4 Jan Brouwer
(Courtesy: See Preface)

The school was initiated by Brouwer[16] and then further developed by his student Heyting.[17] Brouwer was critical of the way in which Cantor's mathematics viewed the existence of *infinite* sets, and of the way in which mathematics was using the *Law of Excluded Middle*. He proposed a thorough change of this view as well as severe restrictions on the use of the law. Specifically, unlike Cantor, who considered infinite sets as *actualities*, i.e., accomplished objects, intuitionism advocated the classical point of view that infinite sets are no more than *potentialities*, i.e., objects that are *always* under construction, making it possible to construct as many members as needed, but *never all*. This view called for a change in the way that the *existence* of objects in infinite sets should be proven: an object is recognized as a member of an infinite set *if and only if* the object has been *constructed* or the existence of such a construction is beyond doubt. We give the details in Box 2.5.

Using these principles, intuitionism reconstructed several parts of classical mathematics and showed that such intuitionistic mathematics is free of all *known* paradoxes. Unfortunately, the price for this was rather high: large parts of mathematics had to be cast off, because it seemed impossible to reconstruct them according to intuitionistic principles. In addition, in the reconstructed mathematics, surprising changes occurred; for example, every (constructed) function is continuous.

Not surprisingly, it turned out that only a few researchers were willing to make such radical sacrifices.

NB *Nevertheless, the intuitionistic demand for mathematical rigor survived and was partially taken into account in the events to follow.*

Box 2.5 (Intuitionism).

This school argued for greater mathematical rigor in several ways.

View of Infinity. Since Aristotle, mathematics understood infinity only as the potentiality (i.e., possibility), never as the actuality (i.e., accomplishment). For instance, it is true that natural numbers $0, 1, 2, \ldots$ continue endlessly, yet up to any natural number there are only finitely many of them, and when we say that what remains is infinite we only mean that the rest, although growing ever larger, remains never accomplished. So, in the classical view infinity is by nature never accomplished, never actual. In contrast, Cantor's view of infinity was different, indeed radically Platonic: *"Any set, regardless of its size, is as much real as its members are real,"* he boldly advocated. To Cantor the set $\{0, 1, 2, \ldots\}$ was an actual, accomplished mathematical object.

Intuitionism returned to the classical view of infinity as potentiality. According to this view, using an appropriate procedure, we can find in an infinite set as many members as we wish, but never all of them. To treat infinite sets as actual, accomplished unities, is wrong, said intuitionists, and may lead to paradoxes (as shown by Russell and others).

But there are also differences between classical mathematics and intuitionism. (We use symbols $\exists$ for "exists"; : for "such that"; $\equiv$ for "is"; $\neg$ for "not"; $\vee$ for "or"; $\forall$ for "for all"; see Appendix B.)

---

[16] Luitzen Egbertus Jan Brouwer, 1881–1966, Dutch mathematician and philosopher.

[17] Arend Heyting, 1898–1980, Dutch mathematician and logician.

Existence of Objects. Intuitionism treats the existence of mathematical objects differently from classical mathematics. In classical mathematics, mathematical objects exist *per se*, as Platonic ideas (see p. 10). Consequently, statements about mathematical objects are either true or false. Intuitionism does not accept this view. Instead, it advocates that the only things that exist *per se* are mental, mathematical constructions, while the existence of an object that has *not* been constructed remains *dubious*. For intuitionism, *to exist is the same as to be constructed.*

For instance, in classical mathematics, given a set S and a property P sensible for the members of S, we are always allowed to *indirectly* prove that there exists a member of S with the property P, i.e., that the statement $\exists x \in S : P(x)$ is true. To do this, we first make the hypothesis $H \equiv \neg \exists x \in S : P(x)$, stating that such a member *does not* exist. Then we try to deduce from H a contradiction. If we succeed in this, we conclude that H is false. Now comes the critical step: since classical mathematics fully accepts the *Law of Excluded Middle*, there can be no other alternative but to conclude that $\neg H \equiv \exists x \in S : P(x)$ is true, i.e., that such a member of S *exists*. But note that, generally, we have no idea about this member, or how to find it.

Intuitionism does not accept such an indirect proof of existence when the set S is *infinite*. Indeed, it rejects any proof of existence that neither constructs the alleged object, nor describes how to construct it at least in principle.

Use of Logic. The intuitionistic point of view was also reflected in the use of logic. For example, because of the *Law of Excluded Middle*, classical mathematics takes for granted that, for *any* statement F, either F or $\neg F$ is true. Hence, the statement $F \vee \neg F$ is *a priori* true, even though we may never determine the truth-values of F and $\neg F$. Intuitionism, in contrast, treats the truth-values of F and $\neg F$ as *dubious*, until they are actually determined in some indisputable way.

To explain the reasons for such caution, let S be a set, P a property sensible for the members of S, and F the statement $\forall x \in S : P(x)$. So F conjectures that every member of S has the property P. How can we indisputably determine whether or not F is true? Can we always do this?

First, we can try to prove in one sweep that *all* the members of S have the property P. (We can use various techniques, such as mathematical induction.) If it turns out that we are unable to construct a proof that works for every member of S, it might be that F is *false*. However, it might also be that F is *true*, where $P(x)$ holds for every $x \in S$, but for a *different* reason in each case. That is, our inability to construct a one-sweep proof might be due to the lack of a recognizable pattern, i.e., a common reason for which different members of S share the property P. In this case, we can neither prove F (because the "proof" would be infinitely long) nor refute it (because F is true).

We must therefore resort to some other method to settle the conjecture F. If S is *finite*, we can in principle check, for each $x \in S$ individually, whether or not $P(x)$ holds. When the checking is finished, we know either that F is true, or that it is false. But what if S is *infinite*? We can still do the checking, but we must be aware of the following. We may check as many members of S as we like, say 10^{18}, and find that each of them has the property P—but, generally, there is *no* way of knowing whether, for a member yet to be checked, P holds or not. So we keep checking in the hope that such a member will be reached soon. But, if in truth F is true, the checking will continue indefinitely, and we will never find out whether F is true or false. (By the way, this is the present situation with *Goldbach's Conjecture*; see Box 5.4 on p. 100.)

Finally, we may try to prove $F \equiv \forall x \in S : P(x)$ by contradiction. As usually, we assume the converse, i.e., that $\neg \forall x \in S : P(x)$ is true. In classical mathematics, where the equivalence $\neg \forall x \in S : P(x) \iff \exists x \in S : \neg P(x)$ holds for arbitrary S, we would try to deduce a contradiction from the more promising right-hand side of the equivalence. In intuitionism, however, the equivalence does not *a priori* hold; namely, if S is infinite, the statement $\exists x \in S : \neg P(x)$ is *dubious* until we have *constructed* an $x \in S$ for which $\neg P(x)$ holds. (As we have seen above, this may not be easy.) As long as $\exists x \in S : \neg P(x)$ is dubious, it cannot be used to deduce a contradiction, and our proving by contradiction is stalled.

So in some situations the truth-value of a statement F cannot be indisputably determined.

2.2.3 Logicism

Logicism aimed to found mathematics on pure logic. As a side-effect it developed the notation by which mathematics was at last given concise and precise expression. The main contributions to this school were made by Boole, Frege, Peano, Russell, and Whitehead.

Fig. 2.5 George Boole (Courtesy: See Preface)

Fig. 2.6 Gottlob Frege (Courtesy: See Preface)

Fig. 2.7 Giuseppe Peano (Courtesy: See Preface)

Boole

In the middle of the nineteenth century scientists noticed that, from Aristotle onward, logical deduction had been using various *self-evident* rules of inference that, surprisingly, had never been rigorously analyzed and written down.

Boole[18] was among the first to become aware of the pitfalls of this. He embarked on the question of how to express logical statements by means of algebraic expressions containing the operations "and", "or", and "not", and then algebraically manipulate these expressions to pursue logical deduction. He described his discoveries in the book *The Laws of Thought* (1854) and thus founded *algebraic logic*. His logic was further developed by Peirce[19] and others in the early twentieth century to become what is now known as *Propositional Calculus* **P** (see Appendix A). Since then a more precise and clear expression of logical statements has been possible.

Frege and Peano

Frege[20] was aiming even higher. His goal was to show that arithmetic can be deduced from pure logic. In particular, he planned to define number-theoretic notions (i.e., numbers, relations, and operations on numbers) by pure logical notions, and to deduce arithmetical axioms from logical axioms.

---

[18] George Boole, 1815–1864, English mathematician and philosopher.

[19] Charles Sanders Peirce,1839–1914, American philosopher, logician, mathematician, and scientist.

[20] Friedrich Ludwig Gottlob Frege, 1848–1925, German mathematician, logician, and philosopher.

Like Boole, Frege was well aware that a natural language, such as German, has structural, rhetorical, psychological, and other characteristics that often blur the meaning of its own statements and, consequently, the argumentation of the deductions. This required the introduction of a new, formal notation by which mathematics and logic could be given concise and precise expression. In particular, such a notation should be able to isolate all the important logical principles of inference while throwing off the lumber of natural language. In other words, the notation should be able to support purely logical deduction. So, in 1879, Frege proposed his *Begriffs-schrift*, a "conceptual notation", capable of giving mathematics and logic better expression. Begriffsschrift was based on an alphabet of *symbols*, from which mathematical and logical expressions were constructed using *rules of construction*. An important innovation of Frege was that these rules directed the purely *mechanical manipulation* of symbols, without appealing to intuition or to the (possible) meaning of symbols. In addition, Frege introduced *quantified variables* and thus laid the foundations of the *First-Order Logic* (which we will describe later). The inferences were described diagrammatically, so they were in this respect somewhat unusual. Nevertheless, Begriffsschrift was capable of precisely and concisely representing the inferences that involved arbitrary mathematical statements.

At the same time, Peano[21] developed another symbolic language for expressing mathematical statements. He used innovative logical symbols (e.g., $\in, \Rightarrow$) in order to distinguish between logical and other operations. In 1895, he published a book *Fomulario Mathematico* where he expressed fundamental theorems of mathematics in his symbolic language. Peano's notation proved to be more practical than Frege's notation and after having gone through further development is in common use today.

In short, among Frege's and Peano's contributions to logic were the analysis of logical concepts, the foundation of the *First-Order Logic* **L** (see Appendix A, p. 364), and the introduction of a standard formal notation.

Russell and Whitehead

Russell's goal was even more ambitious than Frege's. He wanted to deduce *all* mathematics from logic. Namely, at the end of the nineteenth century it had already been shown that many concepts of algebra and analysis can be defined by means of number-theoretic notions, which, in turn, can be defined with purely logical notions.

To avoid his own paradox, Russell invented the *Theory of Types*. There are three requirements in this theory: 1) A *hierarchy of types* must be established. A *type* can be a member of any well-ordered set, e.g., a natural number. 2) Each mathematical object must be assigned to a type. 3) Each mathematical object must be constructed exclusively from objects of lower types in the hierarchy. As a result, the set $\mathcal{U}$ of all sets cannot exist in this theory (because $\mathcal{U} \in \mathcal{U}$), and neither does Russell's Paradox (for if $\mathcal{R}$ existed, we would have $\mathcal{R} \notin \mathcal{R}$ because of its type, and consequently $\mathcal{R} \in \mathcal{R}$ due to its definition: a contradiction). Similarly, Ω would not exist (as $\Omega \in \Omega$).

---

[21] Giuseppe Peano, 1858–1932, Italian mathematician.

These ideas were described in the 1910–13 book *Principia Mathematica* (*PM*) by Whitehead[22] and Russell. Using symbolic notation based on Peano's work, they developed from logical and three additional axioms the theory of sets and cardinal, ordinal, and real numbers, while avoiding all *known* paradoxes. The deductions were long, even cumbersome, yet many shared the opinion that the remaining fields of mathematics could also be deduced (at least in principle).

Fig. 2.8 Alfred Whitehead
(Courtesy: See Preface)

Fig. 2.9 Bertrand Russell
(Courtesy: See Preface)

Did *Principia Mathematica* put an end to the crisis in mathematics? Not really. There were imperfections in *PM*. First of all, there was a kind of aesthetic flaw in the set of *PM*'s axioms, because in addition to logical axioms there were three axioms not recognized as purely logical. One of these was Cantor's *Axiom of Choice*. More importantly, it remained unclear as to whether *PM* is *consistent*, i.e., it avoids, besides all *known* paradoxes, also *all the other* paradoxes that may still be hidden in various fields of mathematics, patiently awaiting their discovery. This question became known as the *Consistency Problem* of *PM*. In addition, it was not clear whether *PM* was *complete*, i.e., whether exactly all true statements are provable within *PM*. This was the *Completeness Problem* of *PM*. Consequently, *PM* was not widely accepted.[23]

NB *Nevertheless, PM was all-important for future events, because it finally developed 1) a symbolic language for the concise and precise expression of mathematical statements from an arbitrary field of mathematics; and 2) a concise formulation of all the rules of inference used in the deduction of mathematical theorems. In addition, PM led to a clear formulation of the problems of consistency and completeness of a particular axiom system, the PM.*[24]

---

[22] Alfred North Whitehead, 1861–1947, British mathematician and philosopher.

[23] In addition, it would soon turn out that Russell's Paradox, as well as other paradoxes stemming from Cantor's liberal *Axiom of Abstraction*, can be eliminated just by a *two-level hierarchy of sets and classes*, instead of the complicated infinite hierarchy of types. (See Box 3.7 on p. 50.)

[24] As we will see in Chap. 4, the two problems were later solved in general by Gödel.

The concepts and tools developed by intuitionism and logicism were used by *formalism*, the third of the schools that attempted to resolve the crisis in mathematics.

2.2.4 Formalism

Formalism could not accept the radical measures suggested by intuitionism. It wished to keep all classical mathematics. After all, classical mathematics had been proving its immense usefulness from the very beginning.

To achieve this, formalism focused on a radically different aspect of human mathematical activity. Instead of being the *meaning* (i.e., semantics, contents) of mathematical expressions and inferences, the subject of the formalists' research was their *structure* (i.e., syntax, form). Formalism focused on the formal-language formulation of human mathematical activities and their results, as well as on the relations between these formulations.

This school was initiated by Hilbert[25] and then developed in close collaboration with Ackermann,[26] Bernays,[27] and others.

Fig. 2.10 David Hilbert
(Courtesy: See Preface)

Syntax vs. Semantics

Hilbert became fully aware that it is sensible to draw a distinction between syntactic notions (i.e., notions referring to the structure of mathematical expressions) and semantic notions (i.e., notions referring to the meaning of mathematical expressions). For instance, the interpretation of a theory is a semantic notion. Recall that interpretation gives a meaning to a theory developed in a hypothetical axiomatic system (see p. 12). To describe the interpretation one needs to describe its domain, which is, mathematically, a set. But the concept of a set was not clear at that time. So Hilbert advocated a focus on syntactic notions, as the research of these seemed to require

---

[25] David Hilbert, 1862–1943, German mathematician.

[26] Wilhelm Friedrich Ackermann, 1896–1962, German mathematician.

[27] Paul Isaac Bernays, 1888–1977, Swiss mathematician.

only the non-problematic parts of mathematics, that is, basic logic and some basic arithmetic.

Let us describe these ideas in greater detail. Because it had been seen that mathematical concepts, such as that of the set, may be vague, also inference incorporating such concepts may be false, and may, eventually, lead to paradoxes. On the other hand, mathematical notions are always expressed in the *words* of some language, either natural, such as English, or symbolic, designed just for this purpose. A word is a finite sequence of *symbols* from some finite alphabet. Formalism noticed that every symbol is perfectly clear *per se*, that is, a symbol is comprehended as soon as it is recognized as a discrete part of the reality, without any further intuitive or logical analysis. This comprehension of symbols is independent of their intended meaning, which might previously be associated with them (such as the operation of addition with the symbol "+"). So why not comprehend words in that manner as well? One should only ignore the intended meaning of the word at hand and comprehend and treat it simply as a finite sequence of symbols. *Expressions*, i.e., sequences of words, could also be treated in the same fashion and, finally, *sequences of expressions* too.

After the banishment of meaning from language constructs, one would be free to focus on their structure (syntax). But why do that? The reason is that one could found mathematical inference on a clear and precise structure (syntax) of language constructs, instead of on their (sometimes) unclear meaning (semantics). The syntax is always clear, provided it is rigorously and precisely defined (as was the case with logicism). As a result, a proof (deduction) would simply be a finite sequence of language constructs (expressions), built according to a finite number of rules. The gain would be improved control over the process of deduction and, finally, the elimination of paradoxes.

Formal Axiomatic Systems

In order to implement these ideas, formalism invented *formal axiomatic systems*. Each such system offers 1) a rigorously defined *symbolic language*; 2) a set of *rules of construction*, i.e., syntactic rules that are used to build well-formed expressions, called *formulas*, of the language; and 3) a set of *rules of inference* that are used to build well-formed sequences of formulas, called *derivations* or *formal proofs*. Each formula or derivation is viewed and treated exclusively as a finite sequence of symbols of the language. Hence, though each formula has a definite structure, no meaning is to be seen or searched for in it. Some of the formulas are distinguished as *axioms*. Given a finite set of formulas, one may *infer* a new formula by applying a rule of inference. Formulas that can be derived by a finite sequence of inferences from axioms only are called *theorems*. Axioms, theorems, and other formulas make up the *theory* belonging to the formal axiomatic system at hand. A detailed discussion of formal axiomatic systems and their theories will appear in the next chapter.

Interpretation

Let us stress that formalists were aware of the fact that there was a limit to neglecting the semantics. After all, their ultimate goal was to establish conditions for the development of sound, safely *applicable* theories. They were aware that each theory, developed in a formal axiomatic system, should eventually be given some meaning; otherwise it would be of no use. In other words, the theory should be *interpreted*. Informally, an interpretation of a theory in a field of interest maps formulas of the theory into statements about (some) objects of the field. We will discuss interpretation again shortly.

NB *Formalism cast out the issues of meaning from the development of a theory, and shifted them to a later interpretation. What were the expected benefits of this? Such a theory could clearly show the syntactic properties of its expressions and expose various relations between these properties. Laid bare, the whole theory could be examined by* metamathematics *and subjected to its judgment.*

Metamathematics

When a theory is developed in a formal axiomatic system, the only things that can be examined within or about it are its *expressions*, the *syntactic properties* of expressions, and the *relations* between them. All these are unambiguously determined by the formal system (i.e., its language and rules of construction and inference). Thus, syntactic aspects of the theory can be systematically analyzed without the interference of semantic issues. Only now can one raise well-defined questions *about the theory* and propose answers to such questions.

Fig. 2.11 A statement about the theory belongs to its metatheory

But questions and statements about the theory are no longer part of the theory. Instead, they belong to the higher "theory about the theory," which is called a *metatheory*, or, more generally, *metamathematics*.[28] Thus, the subject matter of a metatheory is some other theory.

---

[28] meta- (Greek μετά) = after, beyond, about

Metamathematical statements are formulated in a *metalanguage*. This is a natural language or a fragment of it that is appropriately augmented by special symbols and other objects. We will say more about metalanguage and two such symbols, $\vdash$ and $\models$, shortly.

The proving of metamathematical statements is still necessary; however, it is not formal, in contrast to proving within the formal system. Instead, the usual (i.e., semantic, informal) proving is used, where each inference in a metamathematical proof must be grounded in the *meaning* of its premises. Of course, premises are metamathematical statements, so they can refer only to syntactic aspects of the theory.

In addition, to avoid any doubts that might arise because of the use of infinity, only *finite* objects and techniques are allowed in metamathematical proofs. Such a cautious and indisputable way of reasoning is called *finitism*.

Goals of Formalism

Formalism harbored hopes that the analysis of formal systems would provide answers to many important metamathematical questions about the theories of interest. Specifically, these were the two well-known questions concerning mathematics developed in *Principia Mathematica* (see p. 25):

* The *Consistency Problem* of *PM* $\equiv$ "Is the math developed in *PM* consistent?"
* The *Completeness Problem* of *PM* $\equiv$ "Is the math developed in *PM* complete?"

But the ultimate goals of formalism were even more ambitious. Specifically, formalists intended to:

1. develop *all* mathematics in *one* formal axiomatic system;
2. *prove* that such mathematics is free of *all* known and unknown paradoxes.

2.3 Chapter Summary

The axiomatic method was used to develop mathematics since its beginnings. The evident axiomatic system required that basic notions and axioms be clearly confirmed by reality. Since it was found that human experience and intuition may be misleading, the hypothetical axiomatic system was introduced. Here, axioms are only hypotheses whose fertility is more important than their link to reality. Such axiomatic systems offered more freedom in the search for interesting and useful theories. This approach was taken by Cantor when he developed his *Set Theory*. Because Cantor treated the existence of infinite sets naively, this resulted in several paradoxes in his theory.

Intuitionism, logicism, and formalism were three schools that reflected critically on the mathematical and logical notions and concepts that might be the cause of paradoxes.

Intuitionism advocated for greater rigor in the process of proving and for the non-Platonic view that the existence of mathematical objects is closely connected to the existence of their mental constructions. Intuitionism reconstructed several parts of classical mathematics that were free of all known paradoxes. But, at the same time, large parts of mathematics had to be cast off, as it seemed impossible to reconstruct them in the intuitionistic manner. Few researchers were willing to make such a sacrifice.

Logicism, the second school, developed a formal notation by which mathematics was given concise and precise expression. It also bore *Principia Mathematica*, a book that finally developed a symbolic language of mathematics and concisely formulated its rules of inference. In addition, it brought an awareness of the importance of the problems of consistency and completeness of axiomatic theories.

The third school, formalism, built on the ideas and tools developed by intuitionism and logicism, and aspired to retain all mathematics. Formalism acknowledged that the syntax and semantics of mathematical expressions should be clearly separated and dealt with in succession. It introduced the concept of the formal axiomatic system, i.e., an environment for the mechanical, syntax-oriented development of a theory. In addition, it introduced a clear distinction between a theory and its metatheory.

Chapter 3
Formalism

The form of something is its shape and structure. Something that is done in a formal way has a very ordered, organized method and style. Formalism is a style in which great attention is paid to the form rather than to the contents of things.

Abstract The great ideas and tools that intuitionism and logicism discovered in solving the crisis in mathematics were gathered by formalism in the concept of the formal axiomatic system. Later, formal axiomatic systems led to seminal discoveries about axiomatic theories and mathematics in general. Particularly important to us is the fact that formal axiomatic systems also gave rise to the need for a deeper understanding of the concepts of algorithm and computation. To appreciate this need, we devote this chapter to the understanding of formal axiomatic systems in general and describe those particular formal axiomatic systems that played a crucial role in the events to follow.

3.1 Formal Axiomatic Systems and Theories

In this section we will describe what a formal axiomatic system is and how a theory is developed in such a system. We will then show how meaning, and consequently a possible application, is given to a formally developed theory. Finally, we will describe several formal axiomatic systems and their theories that played important roles in the development of the notions of algorithm and computation.

3.1.1 What Is a Formal Axiomatic System?

A formal axiomatic system (for short *f.a.s.*) **F** is determined by three entities: a symbolic language, a set of axioms, and a set of rules of inference.

Symbolic Language

The basic building blocks of the symbolic language are *symbols*. There are a countable (potentially infinite) number of them and they constitute the *alphabet* of the language. In the alphabet there are *individual-constant symbols* (e.g., a, b, c), *individual-variable symbols* (e.g., x, y, z), *function symbols* (e.g., f, g, h), *predicate symbols* (e.g., P, Q, R), *logical connectives* (e.g., $\lor$, $\land$, $\Rightarrow$, $\Leftrightarrow$, $\lnot$), *quantification symbols* (e.g., $\forall$, $\exists$), and *punctuation marks* (e.g., comma ","; colon ":"; parentheses "(", ")"). Usually, there is the *equality symbol* (i.e., $=$). In some cases, certain function symbols will be designated as *function-variable symbols* and certain predicate symbols will be designated as *predicate-variable symbols*. (The reasons for this naming of the symbols will become clear when we discuss their interpretation.)

From symbols one constructs larger building blocks of the language, i.e., symbols are combined into arbitrary finite sequences called *words*. Some of these are called terms and are inductively defined by the following *rule of construction*: A *term* is either an individual-constant symbol or an individual-variable symbol, or it is a word $f(t_1, t_2, \ldots, t_k)$, where t_i are terms and f is a k-ary function symbol.

A formula is defined inductively by another syntactical *rule of construction*: A *formula* is an expression $P(t_1, t_2, \ldots, t_k)$, where t_i are terms and P is a k-ary predicate symbol; or it is an expression $t_1 = t_2$, where t_1, t_2 are terms; or it is one of the expressions $F \lor G$, $F \land G$, $F \Rightarrow G$, $F \Leftrightarrow G$, $\lnot F$, $\forall \tau F$, $\exists \tau F$, where F and G are formulas and τ is a variable symbol.

The symbols $\forall$ and $\exists$ are called the *universal* and *existential quantification symbol*, respectively, while $\forall \tau$ and $\exists \tau$, where τ is a variable symbol, are called the *universal* and *existential quantifier*, respectively. If τ immediately following $\forall$ or $\exists$ can only be an individual-variable symbol, then the symbolic language is said to be of the *first order*. If, however, τ can be a function-variable symbol or a predicate-variable symbol, then the language is of the *second order*.

If $\forall \tau F$ and $\exists \tau F$ are formulas, F is called the *scope* of $\forall \tau$ and $\exists \tau$, respectively. An *occurrence* of a variable symbol σ is *bound* in a formula G iff either σ is the variable of $\forall$ or $\exists$ in G, or it is within the scope of $\forall \sigma$ or $\exists \sigma$ in G. Otherwise, the occurrence is said to be *free* in G. We say that the variable symbol σ is bound (free) in G iff σ has a bound (free) occurrence in G. A formula with at least one free variable symbol is said to be *open*. A formula with no free variable symbols is said to be *closed*; a closed formula is also called a *sentence*. If F is a formula and t is a term, then t is said to be *free for* x *in* F iff no occurrences of x in F lie within the scope of any quantifier $\forall y$, where y is a variable in t.

Notice that the construction of the building blocks of the language is governed exclusively by rules of construction that are syntactic by nature. Consequently, neither intended nor possible meanings of the building blocks interfere in their construction.

Example 3.1. (Term, Formula, Sentence) The symbols a and x are terms. If f and g are function symbols, then $f(x)$ and $g(a, f(x))$ are both terms. If P and Q are predicate symbols, then $P(a, x)$ and $Q(a, x, f(x))$ are formulas; so is $P(a, x) \lor Q(a, x, f(x))$. The formula $\forall x \exists y P(x, y)$ is a sentence because its individual-variable symbols x and y are bound by $\forall$ and $\exists$, respectively. The formula $\forall x \exists y R(x, y, z)$ is open because the individual-variable symbol z is free in R. If h is a function-variable symbol, then the formula $\forall h P(a, h(a))$ belongs to a second-order language. $\square$

Axioms

Axioms are selected formulas. If there is a procedure that can decide whether a formula is an axiom, then the set of axioms is said to be *computable* and the theory developed in the f.a.s. is said to be *computably axiomatizable*, or just *axiomatizable*. There are *logical* and *proper* (i.e., non-logical) axioms. Logical axioms are present in every f.a.s., while proper axioms vary from system to system. As we will see shortly, logical axioms are intended to epitomize the principles of pure logical reflection, while proper axioms condense other special basic notions and facts.

Rules of Inference

A *rule of inference*, say $\mathscr{R}$, specifies the conditions in which, given a set $\mathcal{P}$ of formulas, called the *premises* of $\mathscr{R}$, one is allowed to *derive* another formula, say F. The formula F is called the *conclusion* drawn from premises $\mathcal{P}$ by the rule of inference $\mathscr{R}$. We also say that F *directly follows from* premises $\mathcal{P}$ by the rule $\mathscr{R}$ and write

$$\mathcal{P} \overset{\mathscr{R}}{\vdash} F. \qquad \text{(Rule of inf. } \mathscr{R})$$

Two usual rules of inference are Modus Ponens and Generalization. *Modus Ponens* (*MP*) says: "If G and G $\Rightarrow$ F are premises, then the conclusion F directly follows." That is: "If G is asserted to hold, and G implies F, then also F holds." In short:

$$G, G \Rightarrow F \overset{MP}{\vdash} F. \qquad \text{(Modus Ponens)}$$

Generalization (*Gen*) says: "If F(x) is a premise, then the conclusion $\forall x F(x)$ directly follows." That is: "If F holds for an unspecified x, then F holds for every x." In short:

$$F(x) \overset{Gen}{\vdash} \forall x F(x). \qquad \text{(Generalization)}$$

Example 3.2. (Inference) Greg now plays guitar. If Greg plays guitar, Becky sings. So (by *MP*), she now sings. Becky likes ice cream. So (by *Gen*), she likes vanilla, lemon, ... ice cream. □

Development of the Theory

When the symbolic language, axioms, and rules of construction are fixed, the *development* of the *theory* belonging to the defined formal axiomatic system **F** can start.[1]

---

[1] In the *Platonic view*, as soon as **F** is defined, also the theory belonging to **F** is perfectly defined: it consists of *all* those propositions that are *provable* within **F**—regardless of whether or not they have actually been proved. In this sense, the theory is "static" and defined from its very birth. Adopting the Platonic view allows us to identify an f.a.s. and its theory and denote both by **F**. The development of the theory **F** is by discovering new (existing) theorems, i.e., finding their proofs. In contrast, the *constructivist view* takes the theory to consist of *all* propositions that have been *proved* within **F**. At its birth, the theory only contains **F**'s axioms, and then grows by absorbing new propositions as they are proved. At any stage of its development, the theory is denoted by th(**F**).

During the development, the following strict rules must be obeyed. Firstly, each new notion must be defined by the basic notions or previously defined notions. Secondly, each new proposition must be *derived* (i.e., *formally proved*) before it is named a *theorem*. Here, a *derivation* (i.e., *formal proof*) of a formula F in the theory **F** is a finite sequence of formulas such that 1) the last formula of the sequence is F, and 2) each formula of the sequence is either an axiom of **F**, or it directly follows from some of the preceding formulas of the sequence by one of the rules of inference of **F**. Such a formula F is called a *theorem* of the theory **F**. That F is a theorem of **F** we denote by

$$\vdash_{\mathbf{F}} F.$$

Example 3.3. (Derivation) Let us derive a simple formula $t = t$ in the formal axiomatic system **A**, called the *Formal Arithmetic*. (We will describe **A** in detail on p. 46.) In the derivation below, the left column contains enumerated formulas that appear in the derivation, and the right column explains, for each formula, how the formula was inferred. At this point, the symbols $\oplus$ and 0 should not be given any meaning although they resemble the usual symbols for addition and the number zero.

1. $t \oplus 0 = t$	Axiom 5 of **A** with x substituted by t.
2. $t \oplus 0 = t \Rightarrow (t \oplus 0 = t \Rightarrow t = t)$	Axiom 1 of **A** with x, y, z subst. by $t \oplus 0, t, t$, resp.
3. $t \oplus 0 = t \Rightarrow t = t$	*Modus Ponens* of premises 1. and 2.
4. $t = t$	*Modus Ponens* of premises 1. and 3.

Derivations are finite sequences of formulas, which are in turn finite sequences of symbols. So, formally, a derivation is a finite sequence of symbols. For example, the above derivation is the sequence

$$1.t \oplus 0 = t, 2.t \oplus 0 = t \Rightarrow (t \oplus 0 = t \Rightarrow t = t) \overset{MP}{\underset{1,2}{\vdash}} 3.t \oplus 0 = t \Rightarrow t = t \overset{MP}{\underset{1,3}{\vdash}} 4.t = t$$

□

It seems that the development of a theory **F** in a formal axiomatic system is nothing more than a meaningless manipulation of symbols that runs according to a given set of strict rules. It is a kind of a game of symbols regulated by certain rules. In other words, the development of the theory **F** is strictly formal, so **F** is a rather strange-looking theory. Yet the reasons for this are rather meaningful:

It is in principle much easier to maintain and check the validity of a proof
that is being constructed in a formal axiomatic system
than a proof being constructed in a non-formal axiomatic system
where validity of inferences is decided by creative human thought.

This is because in a formal axiomatic system deduction is more of a mechanical process, so it is less prone to human errors. Such a deduction, now called a derivation, can be checked in a purely combinatorial way by checking whether the formal rules of construction and inference have been obeyed. In contrast, the situation in a non-formal axiomatic system is quite different: There, a proof is a mental process that involves the meaning of the constituents of the proof and the relations between them. As such the proof is vulnerable and prone to errors because of man's subjective judgment whether an inference is valid.

3.1.2 The Notion of Truth

In the previous chapter, we stated on several occasions that something was true or false. In phrasing our statement, we used the words "true" and "false" in the way customary in natural language such as the whole of English. Because of this, we assumed without hesitation that the reader would intuitively comprehend the statement and thus need no further explanation. Nevertheless, we did touch on an important notion, the notion of *truth*. Since truth will explicitly or implicitly play a significant role in the rest of the book, we devote this section to shedding some light on it. In this we will follow Tarski (see Bibliographic Notes to Chapter 4).

Elusiveness of the Notion of Truth

The prevailing usage of the words "true" and "truth" originates in the *classical* conception of truth. This conception is epitomized by the following saying from Aristotle's *Metaphysics*:

> To say of *what is* that *it is not*, or of *what is not* that *it is*, is false,
> while to say of *what is* that *it is*, and of *what is not* that *it is not*, is true.

Based on this, most early modern philosophers of the 1800s professed to accept as a definition of the notion of truth something like the following:

$$\textit{Truth} \text{ is agreement of thought with its object.} \qquad (1a)$$
$$\textit{Truth} \text{ is correspondence to reality.} \qquad (1b)$$

However, no final definition was accepted because philosophers' opinions differed about the *location* of the objects and, hence, about the *nature* of agreement. Specifically, should our thought agree with (i) objects located in an external world; or (ii) objects located in our mind along with our thought; or (iii) the interaction between our mind and the external world? A similar situation emerged in analytic philosophy, the mainstream Anglophone philosophy since the beginning of the 1900s, which incorporated mathematical precision and argumentative clarity. Analytic philosophers differed in their views of what "correspondence" and "reality" mean.

NB *All in all, the definition of the notion of truth proved to be elusive.*

It is therefore not surprising that this elusiveness—added to the well-known perplexities of infinity, deceptiveness of intuition, and the recently discovered paradoxes—made mathematicians of the 1900s *suspicious* of the notion of truth, thus avoiding its use in their endeavor. For example, formalism avoided dealing with truth by focusing on mechanical symbol manipulation (see previous page and Sect. 2.2.4) and postponing the matters of truth to later stages (see Sect. 3.1.3).

Tarski on the Notion of Truth

In contrast, Tarski[2] foresaw important applications of the notion of truth in mathematics. He embarked on a project to rehabilitate it.

Fig. 3.1 Alfred Tarski
(Courtesy: See Preface)

Tarski insisted that the answer to the question "What is truth?" should fulfill the following two natural requirements: it should give (i) a *definition* of the notion of truth; and (ii) a *description of usage* of the notion that would agree with its prevailing usage in natural language. Accordingly, Tarski first focused on natural languages. See Box 3.1 for some of his discoveries.

Box 3.1 (Definition of Truth for Natural Languages).

Tarski started with the following question: "What things bear truth or falsity?" The obvious answer was that it is *sentences* that are the bearers of truth because it is a sentence that must be written or spoken if we want to express something true or false—assuming that the sentence is meaningful.[3]

Material Adequacy of the Definition of Truth

Tarski then founded his research into the notion of truth on the following principle:

$$\text{Saying something is equivalent to saying it is true.} \tag{2}$$

So, saying that snow is white is equivalent to saying that *the sentence* "Snow is white." *is true*:

$$\text{"Snow is white." } is\ true \quad iff \quad \text{snow is white.} \tag{2a}$$

Note that the equivalence (2a) defines the meaning of the word "true" only when "true" refers to the particular sentence "Snow is white." Thus, (2a) is just a *partial definition* of the word "true".

But we can construct more partial definitions of the word "true" by taking other sentences, e.g., "Blood is red.", " Coal is black.", "Grass is green.", "One and one is two.", and so on. The obtained partial definitions share the same form

$$\text{"\_\_\_\_" } is\ true \quad iff \quad \_\_\_\_. \tag{3}$$

---

[2] Alfred Tarski (born Teitelbaum), 1901–1983, Polish-American mathematician and logician.

[3] Philosophically, things are more intricate. When a sentence is written or spoken, an orthographic or phonological *pattern* is produced. Now, is it the sentence or its pattern that (i) bears meaning; or (ii) may bear more than one meaning; or (iii) whose meaning may depend on features of context?

In (3), the same sentence goes into each blank ____; thus, "Blood is red." *is true iff* blood is red, and "Coal is black." *is true iff* coal is black. Note the difference between "____" and ____ in (3): while ____ on the right-hand side is a sentence that says something that is or is not actually the case in the world, the quoted blank "____" on the left-hand side is a *reference* to this sentence, i.e., it indicates that the sentence is *mentioned* (*referred to*) in the left-hand side. So, to say something *about* a sentence, we quote the sentence. Since "____" and ____ are of different characters, the partial definitions obtained from (3) are not circular.

Tarski wanted to advance from partial definitions of the word "true" to a *general* one. Clearly, such a definition would conform with the prevailing conception of the notion of truth *iff* it included *every* partial definition of the word "true" of the form (3). Such a general definition (if there is one) of the notion of truth would be called *materially adequate*.

Definition of Truth for Natural Languages

Thus, Tarski was led to the following question: "Can we construct a materially adequate definition of the notion of truth?" But he eventually realized that if we want to materially adequately define the notion of truth for *natural language*, several problems arise. *Firstly*, the number of partial definitions of the form (3) is enormous, possibly infinite, so we are forced to admit that any realistic definition of the notion of truth composed of partial definitions is itself doomed to be partial. *Secondly*, sentences may contain the very key words "true" ("false", "truth", "falsity") that we wish to define; take, for example, the sentence "Truth exists." *Thirdly*, sentences may express assertions about themselves; this is because references, used to mention sentences, belong to natural language, so a sentence may contain a reference to itself. Such sentences are said to be *self-referential*. These sentences are *not a priori* controversial: take, for example, the sentence "This sentence contains five words." But if self-reference is combined with other features of the language, they may become such: take, for example, the sentence "This sentence is false." We leave it to the reader to infer the consequences of each of the premises (i) the sentence is true, and (ii) the sentence is false. Either premise leads to its negation in spite of using intuitively certain forms of reasoning. In other words, if this sentence bears truth or falsity, it bears both of them, contradicting the *Law of Excluded Middle* (see p. 13). Because of this, the sentence is called the *Liar Paradox*.

So, natural language contains a sentence which seems to be both true and false. Unless we can overcome this paradox, we cannot develop a consistent theory of truth; and without this, we cannot satisfactorily understand the relation between our thought and language and the world around us.

Object Language vs. Metalanguage

Tarski became aware of the following: (i) The freedom and power of expression of natural languages enable the construction of paradoxical sentences; and (ii) The notion of truth for natural language should not be discussed in that language only.

Because of this, he advocated a clear distinction between the language $\mathcal{L}$ *for which* we want to define the notion of truth, and the language $\overline{\mathcal{L}}$ *in which* the definition will be formulated and its implications discussed. He called $\mathcal{L}$ the *object language* and $\overline{\mathcal{L}}$ he called the *metalanguage* of $\mathcal{L}$. Tarski showed that paradoxes are inevitable if both $\mathcal{L}$ and $\overline{\mathcal{L}}$ are the whole of a natural language.

Consequently, the metalanguage $\overline{\mathcal{L}}$ must be sufficiently rich to enable the discussion of the object language $\mathcal{L}$. We may expect that $\overline{\mathcal{L}}$ will contain $\mathcal{L}$ and also a means to refer to the objects of $\mathcal{L}$. In addition, there may be some special objects in $\overline{\mathcal{L}}$.

Specifically, so far we have been writing as if both $\mathcal{L}$ and $\overline{\mathcal{L}}$ were the whole of English. If, however, we make a distinction between $\mathcal{L}$ and $\overline{\mathcal{L}}$, then every instance of (3), such as (2a), becomes a sentence of $\overline{\mathcal{L}}$ (because it says something about a sentence of $\mathcal{L}$). Now, since ____ on the right-hand side of (3) represents a sentence of $\mathcal{L}$, it follows that $\overline{\mathcal{L}}$ must contain $\mathcal{L}$. Next, the "____" on the left-hand side of (3) indicates that $\overline{\mathcal{L}}$ must contain a referencing mechanism, such as quotation, so that sentences of $\mathcal{L}$ can be mentioned in $\overline{\mathcal{L}}$. Finally, "*is true*" in (3) suggests that $\overline{\mathcal{L}}$ must contain a special unary predicate, say *Is_a_true_sentence_of_$\mathcal{L}$*, which is usually called the *truth predicate*.

Tarski's Definition of Truth for Formalized Languages

Difficulties in defining the notion of truth for natural languages $\mathcal{L}$ led Tarski to restrict his ambitions and adapt his program as follows:

- We should define the notion of truth for a *fragment* $\mathcal{F} \subseteq \mathcal{L}$ of natural language; and
- Although we expect that focusing on $\mathcal{F}$ will restrict the applicability of the definition of truth, we should try to keep the classical concept of truth essentially intact.

Which fragment $\mathcal{F}$ of natural language should we focus on? Pragmatics suggests that it should be a fragment—call it the *restricted language*—that will serve for the purposes of science in general, i.e., the whole realm of intellectual inquiry.

But can we precisely define the notion of truth for such a restricted language? Tarski discovered that the answer is *yes, if* the restricted language $\mathcal{F}$ satisfies the following four conditions:

1) its full vocabulary is available;

2) its rules of construction are precisely (formally) formulated;

3) its rules of construction refer exclusively to the form of expressions; and

4) the meaning (truth or falsity) of an expression depends exclusively on its form.

Restricted languages $\mathcal{F}$ that fulfill these conditions are said to be *formalized*.

Clearly, *symbolic languages* of formal axiomatic systems (see Sect. 3.1.1) are formalized. But also other fragments of natural languages can be formalized, though in a less strict and abstract manner.

In summary, formalized languages are adequate for expression in logical and mathematical theories while admitting the definition of the notion of truth.

But to fully realize the latter, Tarski already showed (see Box 3.1) that one more condition must be fulfilled:

5) there must be a clear distinction between the formalized (object) language $\mathcal{F}$ and its metalanguage $\overline{\mathcal{F}}$.

Then the notion of truth for $\mathcal{F}$ *can* be defined.

Box 3.2 shows how truth is defined for $\mathcal{F}$ which is the *Propositional Calculus* **P**.

Box 3.2 (Definition of Truth for Propositional Calculus).

Let us define the notion of truth for the *Propositional Calculus* **P**. We will do this in three stages:

1. We define the alphabet and expressions over it. The alphabet contains symbols for individual constants $(a, b, c, \ldots)$, individual variables $(x, y, z, \ldots)$, and logical connectives $(\neg, \vee, \wedge, \Rightarrow, \Leftrightarrow)$. We then define an *expression* of **P** to be a finite sequence of these symbols.

   But not every expression will be regarded as meaningful; e.g., $\vee ax\wedge$ is such an expression.

2. We define the *syntax* of **P**. Our intention is to syntactically define the meaningful expressions as those constructed by certain rules of construction. The definition will also be a (syntactic) criterion for deciding whether or not an expression is meaningful. The definition is *inductive*:

   a. First, we need some *a priori* meaningful building blocks. The intention is that, in the third stage, these will denote true or false elementary mathematical assertions. So we now define: An *atomic sentence s* is the expression consisting of a *single* constant or variable symbol.
   b. Next, we inductively define (general) sentences as follows: A *sentence R* is either an atomic sentence s or a logical connective $\neg, \vee, \wedge, \Rightarrow, \Leftrightarrow$ combined with one or two sentences: $\neg S, S \vee P, S \wedge P, S \Rightarrow P$, or $S \Leftrightarrow P$, where S and P are sentences.
   c. Finally, we make certain that nothing else is a sentence: Sentences of **P** are *exactly* the expressions generated by the rules a and b.

   We will need this: *For every sentence R there is a unique way to break it up into its components.* (The proof of this proposition is left to the reader.)

3. We define the *semantics* of **P**, that is, we assign to each sentence R a *truth-value* (meaning). We do this by an *inductively defined* function v:

   a. First, we define a function $v_0 : s \mapsto v_0(s) \in \{\text{true}, \text{false}\}$, called the *atomic truth assignment*, that assigns to each atomic sentence s a truth-value.
   b. Next, we extend v_0 to a function $v : R \mapsto v(R) \in \{\text{true}, \text{false}\}$, called the *truth assignment*, that assigns to each general sentence R a truth-value while *respecting the rules of construction* defined in previous stage. To achieve this, the following must hold for any sentences R, S, P and any atomic sentence s:

$$\text{if} \quad R = s \quad\quad \text{then} \quad v(R) = v_0(s) \in \{\text{true}, \text{false}\};$$

$$\text{if} \quad R = \neg S \quad\quad \text{then} \quad v(R) = v(\neg S) \quad = \begin{cases} \text{true}, & \text{if } v(S) = \text{false}; \\ \text{false}, & \text{otherwise}; \end{cases}$$

$$\text{if} \quad R = S \vee P \quad \text{then} \quad v(R) = v(S \vee P) \quad = \begin{cases} \text{true}, & \text{if } v(S) = \text{true or } v(P) = \text{true}; \\ \text{false}, & \text{otherwise}; \end{cases}$$

$$\text{if} \quad R = S \wedge P \quad \text{then} \quad v(R) = v(S \wedge P) \quad = \begin{cases} \text{true}, & \text{if } v(S) = \text{true and } v(P) = \text{true}; \\ \text{false}, & \text{otherwise}; \end{cases}$$

$$\text{if} \quad R = S \Rightarrow P \quad \text{then} \quad v(R) = v(S \Rightarrow P) = \begin{cases} \text{true}, & \text{if } v(S) = \text{false or } v(P) = \text{true}; \\ \text{false}, & \text{otherwise}; \end{cases}$$

$$\text{if} \quad R = S \Leftrightarrow P \quad \text{then} \quad v(R) = v(S \Leftrightarrow P) = \begin{cases} \text{true}, & \text{if } v(S) = v(P); \\ \text{false}, & \text{otherwise}. \end{cases}$$

Since every sentence R can be uniquely broken up into atomic sentences s, every v_0 has a *unique* extension to v, and $v(R)$ depends only on the values $v(s) = v_0(s)$ for those s that occur in R.

3.1.3 Interpretations and Models

As we have seen in Sect. 3.1.1, all the aspects of meaning have been expelled from
the development of a theory and postponed to a later stage. We have now come to
the point where the potential applications of the developed theory can be searched
for. So, now the question is: "How does a formally developed theory get connected
to actuality and, at last, gain meaning?"

On one hand, the first ideas of the *intended meaning* may be involved in the very
first stage of establishing a formal axiomatic system, that is, in setting up its symbolic language and choosing its proper axioms. This usually happens when a formal
system is established in order to be used in an investigation of a particular, concrete
field of interest. Some of the concrete fields of interest that we will discuss in the
next sections are concerned with logical statements, natural numbers, and sets. A
chosen field of interest is called the field of the *intended* (or *standard*) *interpretation*
of the formal axiomatic system (and of its theory as well). It is, therefore, reasonable to choose an alphabet and rules of construction in such a way that the resulting
language is capable of a precise and comfortable description of any situation in this
field. (For example, in Example 3.3 (p. 34) we used symbols $\oplus$ and $=$ with the obvious intention that these symbols will later be interpreted as the addition and equality
in the set $\mathbb{N}$ of natural numbers.) In addition, together with axioms and rules of inference, the language should facilitate the analysis of the situation. Of course, when
the development of the theory starts, the role of the intended meaning diminishes
and syntactic issues come to the fore.

On the other hand, one may just as well define a formal axiomatic system and
develop its theory irrespective of any particular field of interest. One just mechanically develops the theory through the disciplined use of formal rules of construction
and inference.

In any case, eventually some meaning must be assigned to the theory if the theory
is to be applied somewhere. How is this done?

Interpretation of a Theory

To assign a particular meaning to a theory $\mathbf{F}$, one has to *interpret* $\mathbf{F}$ in a particular *mathematical structure*[4] $\mathscr{S} = (\mathcal{D}om, \cdot)$. Informally, this means that one has
to choose a particular class $\mathcal{D}om$ and particular functions and relations defined on
$\mathcal{D}om$, and define, for every closed formula of $\mathbf{F}$, how the formula is to be understood

---

[4] Informally, a mathematical structure is a class $\mathcal{D}om$ endowed with additional mathematical objects, such as functions and relations defined on $\mathcal{D}om$, and certain designated elements.
For example, groups, rings, vector spaces, and partially ordered sets are structures. Formally, a
mathematical structure is an ordered set $\mathscr{S} = (\mathcal{D}om, R_0, \ldots, R_k, f_0, \ldots, f_m, e_0, \ldots, e_n)$, where $\mathcal{D}om$
is a class, $R_0, \ldots, R_k$ are relations on $\mathcal{D}om$, $f_0, \ldots, f_m$ are functions from Cartesian powers of $\mathcal{D}om$
into $\mathcal{D}om$, and $e_0, \ldots, e_n$ are designated elements of $\mathcal{D}om$. For example, $(\mathbb{N}, =, +, *, 0, 1)$, the ring
of natural numbers, is a structure. For brevity we used the *dot* to stand for all the particular func-
tions, relations, and designated elements to be considered on $\mathcal{D}om$.

as a statement about the members, functions, and relations of $\mathcal{D}om$. More formally, the *interpretation* of a theory $\mathbf{F}$ is an ordered pair $(\iota, \mathscr{S})$, where ι is a mapping $\iota : \mathbf{F} \to \mathscr{S}$ that assigns to each symbol, term, and formula of $\mathbf{F}$ its "meaning" in $\mathcal{D}om$. (See Fig. 3.2.) The "meaning" can be an element of $\mathcal{D}om$, or a function or relation defined on $\mathcal{D}om$. The class $\mathcal{D}om$ is called the *domain of the interpretation*. Usually, the mapping ι is defined inductively in accordance with $\mathbf{F}$'s rules of construction. Further details about the definition of the interpretation are in Box 3.3.

Fig. 3.2 When a theory $\mathbf{F}$ is interpreted in a structure $\mathscr{S} = (\mathcal{D}om, \cdot)$, the mapping ι assigns to each formula $F \in \mathbf{F}$ a formula $\iota(F) \in \mathscr{S}$

Box 3.3 (Interpretation).

How does the interpretation ι assign meaning to a theory $\mathbf{F}$ in a structure $\mathscr{S} = (\mathcal{D}om, \cdot)$?
A *symbol* of $\mathbf{F}$ gets its meaning as follows:

- an individual-constant symbol c is mapped to an element of the domain: $\iota(c) \in \mathcal{D}om$;
- a k-ary function symbol f is mapped to a k-ary function defined on $\mathcal{D}om$: $\iota(f) : \mathcal{D}om^k \to \mathcal{D}om$;
- a k-ary predicate symbol P is mapped to a k-ary relation defined on $\mathcal{D}om$: $\iota(P) \subseteq \mathcal{D}om^k$;
- logical connectives get their usual meanings ($\vee$ "or"; $\wedge$ "and"; $\Rightarrow$ "implies"; $\Leftrightarrow$ "iff"; $\neg$ "not");
- quantification symbols get their usual meanings ($\forall$ "for all"; $\exists$ "exists");
- punctuation marks get their usual meanings (comma, colon, parentheses);
- the equality symbol $=$ always gets its usual meaning (i.e., the equality relation).

The meaning of a *term* of $\mathbf{F}$ goes with its construction:

- a term that is an individual-constant symbol c gets the same meaning as the symbol: $\iota(c) \in \mathcal{D}om$;
- a term $f(t_1, t_2, \ldots, t_k)$ is mapped to $\iota(f)(\iota(t_1), \iota(t_2), \ldots, \iota(t_k))$; this is an element of $\mathcal{D}om$.

A *formula* of $\mathbf{F}$, too, gets its meaning inductively:

- a formula $P(t_1, t_2, \ldots, t_k)$ is mapped to $\iota(P)(\iota(t_1), \iota(t_2), \ldots, \iota(t_k))$; this is a statement that is true *iff* the elements $\iota(t_i)$ of $\mathcal{D}om$ are related by the relation $\iota(P)$;
- a formula $F \vee G$ is mapped to the statement $\iota(F) \vee \iota(G)$. This statement is true *iff* at least one of the statements $\iota(F), \iota(G)$ is true;
- the formulas $F \wedge G$, $F \Rightarrow G$, $F \Leftrightarrow G$, and $\neg F$ are mapped to statements $\iota(F) \wedge \iota(G)$, $\iota(F) \Rightarrow \iota(G)$, $\iota(F) \Leftrightarrow \iota(G)$, and $\neg\iota(F)$, respectively. The statements are true according to the well-known rules of the *Propositional Calculus* $\mathbf{P}$ (see Box. 3.2 on p. 39 and Appendix A, p. 363).

Finally, let $F(x)$ be a *formula* of $\mathbf{F}$ with a *free* individual-variable symbol x. Then:

- the formula $\forall x F(x)$ is mapped to the statement $\iota(\forall x F(x))$, which is true *iff* the statement $\iota(F(x))$ is true for every $\iota(x) \in \mathcal{D}om$;
- the formula $\exists x F(x)$ is mapped to the statement $\iota(\exists x F(x))$, which is true *iff* the statement $\iota(F(x))$ is true for at least one $\iota(x) \in \mathcal{D}om$.

Satisfiability and Validity

Note that *free* variable symbols have not been assigned exact meanings under the interpretation $(\iota, \mathscr{S})$. Instead, we only know that a free individual-variable symbol represents *any element* of the domain $\mathcal{D}om$. Similarly, a free function-variable symbol and a free predicate-variable symbol, when they exist, represent *any function* and *any relation* on the domain $\mathcal{D}om$, respectively.

Consequently, free variable symbols still await someone to fix their meanings. This can usually be done in many ways and, in general, fixing the meanings of free variable symbols affects the truth-value of the formula. Let us explain this in detail.

If a formula F is *closed* (has no free variable symbols), then its interpretation $\iota(F)$ is a statement about the state of affairs in $\mathcal{D}om$. According to the classical conception of truth (see Sect. 3.1.2), the statement $\iota(F)$ is either *true* or *false*, depending on its conformity with the existing situation in $\mathcal{D}om$. Such a formula is $P(a, b)$ in Fig. 3.3.

Fig. 3.3 A theory **F** is interpreted in the structure $(\mathbb{N}, =, <, +, *, 0, 1)$. The mapping ι assigns to the predicate symbol P the usual relation $<$ on $\mathbb{N}$, and to individual-constant symbols a and b natural numbers 3 and 5, respectively. $P(a, b)$ is true as it conforms with reality. Since the individual-variable symbol x is free in $P(a, x)$, the mapping ι assigns *no* meaning (i.e., no number) to x. Since $P(a, x)$ says nothing about any specific situation in $\mathbb{N}$, it is neither true nor false

If, however, a formula F is *open*, then it contains free variable symbols. (For example, $P(a, x)$ in Fig. 3.3 is an open formula.) Since the interpretation $(\iota, \mathscr{S})$ did not assign meanings to these variable symbols, $\iota(F)$ says nothing definite about the situation in $\mathcal{D}om$. Thus $\iota(F)$ is neither true nor false at this point and, indeed, is not yet a statement about the state of affairs in $\mathcal{D}om$. However, as soon as all the free variable symbols *are* assigned meanings, $\iota(F)$ becomes either a true or a false statement about $\mathcal{D}om$. (Clearly, free individual-variable symbols are assigned particular elements of $\mathcal{D}om$, while free function-variable symbols and free predicate-variable symbols are assigned particular functions and relations on $\mathcal{D}om$, respectively.)

Later, we can *reassign* meanings to one or more free variable symbols of the formula. The change in the meanings of the free variable symbols generally *affects* the truth-value of the statement $\iota(F)$. Regarding this we emphasize two special cases:

- if $\iota(F)$ is true for *at least one* assignment of meanings to its free variable symbols, then we say that F is *satisfiable under the interpretation* $(\iota, \mathscr{S})$;
- if $\iota(F)$ is true for *every* assignment of meanings to its free variable symbols, then we say that F is *valid under the interpretation* $(\iota, \mathscr{S})$ and designate this by

$$\underset{\models}{\overset{(\iota, \mathscr{S})}{}} F.$$

Logical Validity

If a formal axiomatic system has been established to investigate a particular field of interest, then there is an obvious interpretation of its theory; this is the *intended interpretation*. (If the intended interpretation becomes usual, it is called *standard*.) For example, the formal axiomatic system **A** (which will be described shortly) was defined to formalize arithmetic, so the intended interpretation of **A** uses the structure $(\mathbb{N}, =, +, *, 0, 1)$. However, given a formal axiomatic system **F**, there may be several different interpretations $(\iota, \mathscr{S})$ of its theory **F**, each of which differs in ι or $\mathscr{S}$. With regard to this, of particular importance are those formulas of **F** that are valid under *every* interpretation of **F**. Such formulas are said to be *logically valid*. (See Fig. 3.4.)

Fig. 3.4 A formula F that is valid under every interpretation $(\iota, \mathscr{S})$ is said to be logically valid

The logical validity of a formula depends only on its structure and the general properties of functions, relations, and quantifications; it is independent of any interpretation. We denote that F is a logically valid formula by

$$\models F.$$

Observe that the *logical axioms* of **F** must be logically valid. This should not be surprising because logical axioms are meant to epitomize the principles of pure logical thought, and such principles *should be* (and are) independent of the current field of man's interest, i.e., the field of interpretation. Thus they must remain valid, irrespective of the interpretation.

Model of a Theory

What about the other kind of axiom: *proper axioms*? These are meant to abstract specific basic notions and facts typical of the current field of interest (which is typically the intended interpretation of the theory). This leads us to the concept of a *model* of a theory. Given a theory **F**, it is natural to be interested only in interpretations $(\iota, \mathscr{S})$ under which *all the proper axioms* of **F** are valid. (Otherwise, **F** would be of no use under the interpretation.) Under such interpretations *all* the axioms of **F** are valid (as logical axioms are already logically valid). Each such interpretation $(\iota, \mathscr{S})$ is called a *model* of the theory **F**. Intuitively, a model of a theory is any field of interest that the theory sensibly formalizes.

Example 3.4. (Model of a Theory) The set of natural numbers with usual operations is a model of *Peano's Arithmetic* (pp. 12 and 46). A sphere is a model of elliptic geometry (Box 2.2, p. 11). So is a geoid, the shape of Earth. And our Universe is a model of *General Relativity Theory*.   □

A theory may have several models. When a *formula* F of a theory **F** is valid in *every* model of **F**, the formula is said to be *valid in the theory* **F**. This is denoted by

$$\models_{\mathbf{F}} F.$$

A formula F valid in **F** represents a certain mathematical *Truth* expressible in the f.a.s. **F**. We will say that such an F represents a *Truth* in **F**. Each axiom of **F** represents a *Truth* in **F**.

We prefer theories with large power of expression. After we have designed a formal axiomatic system **F** and developed (some of) its theory **F**, we are interested in the existence and kinds of its models, i.e., fields that are sensibly formalized by **F**. We are confronted with questions such as "Does **F** have models? If so, how many are there? What are the differences between them? What is their applicability?" It is not very important that a model be a part of the *real, actual* world; instead, it suffices that the model behave as a *possible* part of the world. When does that happen? It turns out that the theory **F** has to be *consistent*, i.e., it must not allow a derivation of two contradictory theorems (and, hence, has no paradoxes). But the consistency of a theory is a semantic notion, so it must be dealt with within the corresponding metatheory (metamathematics). We will return to the question of consistency soon.

3.2 Formalization of Logic, Arithmetic, and Set Theory

Some of the formally developed theories and their models that played an important role in solving the crisis in the foundations of mathematics were concerned with the following fields of interest: the structure and use of logical statements, the arithmetic of natural numbers, and the construction and use of sets. The corresponding first-order formal axiomatic systems and theories are called the *First-Order Logic* **L**, the *Formal Arithmetic* **A**, and the two *Axiomatic set theories* **NBG** and **ZFC**. In this section we will get acquainted with each of them.

Formalization of Logic

It was clear that in order to develop *any* theory in a logically unassailable way, the corresponding formal axiomatic system must offer all the necessary logical principles and tools (i.e., logical symbols, logical axioms, rules of inference). This called for serious reflection on all the principles of pure logical reasoning, which should result in a formal list of all of them. Fortunately, this was done by Boole, Frege, Peano, Whitehead, Russell and other logicists (see Sect. 2.2.3). Formalism was able to gather all the undisputed logical principles in a formal axiomatic system called *First-Order Logic* (with equality) and denoted by **L**.

- **First-Order Logic[5] L** (with equality). The alphabet of the language of **L** has a potentially infinite number of individual-variable symbols $x, y, \ldots$; the equality symbol $=$; logical connectives $\neg$, $\Rightarrow$, $\forall$; and the usual punctuation marks. Using the given logical connectives one may define additional logical connectives, such as $\wedge$, $\vee$, $\Leftrightarrow$, $\exists$. The symbolic language is of the first order. The terms and formulas are defined in the usual way; e.g., $\forall x \forall y (x = y \Rightarrow y = x)$ is a formula. Instead of explicit logical axioms, of which there are infinitely many, there are *axiom schemas* that describe all of them (see Box 3.4). The two *equality axioms* are recognized as logical. There are no proper axioms. Because of this, **L** is said to be a *pure logic theory*. The rules of inference are *Modus Ponens* and *Generalization*.

Box 3.4 (Logical Axioms of L).

First-Order Logic **L** (with equality) has *axiom schemas*.[6] We use these to build concrete logical axioms. Thus, if F, G, H stand for arbitrary formulas, the following are logical axioms:

1)	$F \Rightarrow (G \Rightarrow F)$	2) $\quad (F \Rightarrow (G \Rightarrow H)) \Rightarrow ((F \Rightarrow G) \Rightarrow (F \Rightarrow H))$
3)	$(\neg G \Rightarrow \neg F) \Rightarrow ((\neg G \Rightarrow F) \Rightarrow G)$	4) $\quad \forall x F(x) \Rightarrow F(t)$
5)	$\forall x (F \Rightarrow G) \Rightarrow (F \Rightarrow \forall x G)$	6) $\quad \forall x (x = x)$
7)	$x = y \Rightarrow (F(x, x) \Rightarrow F(x, y))$	

The schemas 1–3 are also axioms of the *Propositional Calculus* **P** (see Appendix A, p. 363). In **4**, t is a term free for x in $F(x)$. In **5**, x is not free in F. In **7**, $F(x, y)$ arises from $F(x, x)$ by replacing some or all of the free occurrences of x by y (these occurrences are free for y).

First-Order Formal Axiomatic Systems and Theories

Many other formal axiomatic systems are *extensions* of the First-Order Logic **L** (with equality). Each of them contains, besides everything that **L** has, additional *proper symbols* (i.e., constant symbols $a, b, c, \ldots$; function symbols $f, g, h, \ldots$; and predicate symbols $P, Q, R, \ldots$) and *proper axioms* (i.e., axioms that are inspired by the intended interpretation). The rules of inference are *Modus Ponens* and *Generalization*. Most often these systems use first-order language. In such cases we call them *first-order formal axiomatic systems*, and their theories we call *first-order theories*.

Especially important to us will be three first-order theories: *Formal Arithmetic* **A**, which formalizes the arithmetic of natural numbers, and the two *Axiomatic set theories* **ZFC** and **NBG**, which formalize set theory in two different ways.

Formalization of Arithmetic

It was clear that in order to formally develop any nontrivial mathematical theory, natural numbers had to be taken into account. This is because natural numbers play

---

[5] Also called *First-Order Predicate Calculus*.

[6] An *axiom schema* is a formula in the metalanguage of an f.a.s., in which one or more metalinguistic variables, called *schematic variables*, appear. These variables stand for *any* formula (which may be required to satisfy certain conditions) of the f.a.s.

a key role in the construction of other kinds of numbers (e.g., integer, rational, algebraic, real), and consequently in the development of any nontrivial mathematical theory (e.g., algebra, analysis). Fortunately, the grounding for this had already been laid; the properties of natural numbers had been described in 1889 by Peano's nine axioms (as mentioned on p. 12). The formal axiomatic system describing arithmetic is called the *Formal Arithmetic*[7] and is denoted by **A**.

- **Formal Arithmetic A.** The alphabet of the language of **A** contains all the symbols of **L** and, in addition, the following *proper symbols*: the individual-constant symbol 0, the unary function symbol $'$, and the binary function symbols $\oplus$ and $\odot$. Usually, but not necessarily, one can define symbols that are abbreviations for other symbols, e.g., $\oslash, \oslash, \ominus, \ominus$. The terms and formulas are constructed as usual. For instance, x' is a term, $x \oplus 0 = x$ is an open formula, and $\forall x \forall y \forall z (x \odot (y \oplus z) = (x \odot y) \oplus (x \odot z))$ is a closed formula (i.e., sentence). The symbolic language is of the first order. In addition to logical axioms (actually axiom schemas), which were inherited from **L**, *Formal Arithmetic* **A** has nine *proper axioms* (see Box 3.5).[8] The proper axioms summarize the characteristic properties of natural numbers as discovered by Peano. There are no additional rules of inference besides *Modus Ponens* and *Generalization*, inherited from **L**.

  The standard model of the theory **A** is $(\iota, (\mathbb{N}, =, +, *, 0, 1))$, with the domain $\mathbb{N}$ being the set of natural numbers and ι, the interpretation that assigns meanings to formulas in the usual way. For example, the meaning of the individual-constant symbol 0 is the natural number 0; that is, $\iota(0) = 0$. An individual-variable symbol x means any natural number, that is, $\iota(x) \in \mathbb{N}$. The meaning of the function symbol $'$ is the successor function, that is, $\iota(x') = \iota(x) + 1$. Hence, $0'$ means the natural number 1, $(0')'$ means the number 2, etc. The binary function symbols $\oplus$ and $\odot$ are interpreted, as expected, as the addition and multiplication of natural numbers. Each closed formula is mapped by ι to a statement about natural numbers, which is either true or false. The formula $x \oplus 0 = x$ is open, because x is free in it. In the standard model, the formula means: "Adding 0 to a natural number gives the same number." Since this is true for every assignment of a natural number to x, the formula is valid under the standard interpretation of **A**.

Box 3.5 (Proper Axioms of A).

1)	$\forall x \forall y \forall z (x = y \Rightarrow (x = z \Rightarrow y = z))$	**2)**	$\forall x \forall y (x = y \Rightarrow x' = y')$
3)	$\forall x (0 \neq x')$	**4)**	$\forall x \forall y (x' = y' \Rightarrow x = y)$
5)	$\forall x (x \oplus 0 = x)$	**6)**	$\forall x \forall y (x \oplus y' = (x \oplus y)')$
7)	$\forall x (x \odot 0 = 0)$	**8)**	$\forall x \forall y (x \odot y' = (x \odot y) \oplus x)$
9)	$F(0) \wedge \forall x (F(x) \Rightarrow F(x')) \Rightarrow \forall x F(x)$, for any formula F with free x (but see also Box 3.8)		

Standard interpretation of axioms: **2)** Equal natural numbers have equal successors. **3)** 0 is not a successor of any natural number. **4)** If the successors of two natural numbers are equal, then the numbers are equal. **5)** Adding 0 to a natural number gives the same number. **6)** This axiom

---

[7] Also called *Peano Arithmetic* and denoted by **PA**.

[8] The proper axioms listed in Box 3.5 are not the same as those originally proposed by Peano.

describes how to add a successor of a natural number. **7)** Multiplying a natural number by 0 gives 0. **8)** This axiom describes how to multiply by the successor of a natural number. **9)** This is the *Axiom of Mathematical Induction*. It postulates the following principle. Let $F(x)$ be a relation on $\mathbb{N}$ with a free variable x. (We say that F is a *property*.) If $F(0)$ is true, and, if for any natural n, $F(n)$ implies $F(n+1)$, then $F(x)$ is true for all natural numbers x. (See also Box 3.8.)

Formalization of Set Theory

Recall that according to Cantor's naive set theory, the set $\mathcal{S}_P = \{x \mid P(x)\}$ exists for the *arbitrary* property P. So, if we set $P \equiv$ "is a set," there is a set $\mathcal{S}_P = \mathcal{U}$ of all sets. But Russell deduced that then there exists a set which, paradoxically, at the same time is and is not a member of itself (see Sect. 2.1.3). Hence, $\mathcal{U} = \mathcal{S}_P$, the set of all sets, is a paradoxical object as well. The obvious conclusion was that the object $\mathcal{U} = \mathcal{S}_P$ should not exist as a set. As a result, it was necessary to reconsider carefully when, for a given property P, the object $\mathcal{S}_P = \{x \mid x \text{ has the property } P\}$ has the status of a set and when it does not. Which definitions of sets are to be allowed and which are not? It was clear that if $\mathcal{U}$ had existed, it would have been a huge set. So, was it the colossal size of $\mathcal{U}$ that led to Russell's Paradox? Should we allow only those properties P that define objects $\mathcal{S}_P$ of reasonable size? But what is a "reasonable" size of a set? In light of the way mathematics should deal with large objects (sets), two views arose and led to two axiomatic set theories:

- **Axiomatic Set Theory ZFC.** The first view was advocated by Zermelo,[9] Fraenkel,[10] and Skolem.[11] Their plan was to

  find axioms that will ensure the existence of all sets needed in mathematics, and that will, at the same time, prevent the construction of too-large sets.

Fig. 3.5 Ernst Zermelo (Courtesy: See Preface)

Fig. 3.6 Abraham Fraenkel (Courtesy: See Preface)

---

[9] Ernst Friedrich Ferdinand Zermelo, 1871–1953, German mathematician.

[10] Abraham Halevi Fraenkel, 1891–1965, German (later Israeli) mathematician.

[11] Thoralf Albert Skolem, 1887–1963, Norwegian mathematician.

Based on this idea they gradually, during 1908–30, defined a formal axiomatic system **ZF**. In this system one can derive all the important theorems of Cantor's naive set theory, while avoiding all the *known* logical paradoxes. The developed theory is called *Zermelo-Fraenkel axiomatic set theory*. Today, this is a standard set theory. When the *Axiom of Choice* is added to its proper axioms, the theory is denoted by **ZFC**. (For details about the proper axioms of **ZFC**, see Box 3.6.)

• **Axiomatic Set Theory NBG.** The second view was less conservative. It was advocated by von Neumann,[12] Bernays, and Gödel.[13] Their belief was that

> *paradoxes do not follow from the existence of too-large sets,*
> *but from allowing every (large) set to be a member of some other set.*

Fig. 3.7 John von Neumann (Courtesy: See Preface)   **Fig. 3.8** Paul Bernays (Courtesy: See Preface)   **Fig. 3.9** Kurt Gödel (Courtesy: See Preface)

During 1925–40 they gradually defined a formal axiomatic system **NBG**. In addition to the two usual basic notions (i.e., the set and the relation $\in$), **NBG** introduced one more basic notion, *class*, which is a generalization of the notion of set. The advantage of this formal axiomatic system is that there are only a finite number of axioms (because there are no axiom schemas). The theory developed in this system is called *von Neumann-Bernays-Gödel's set theory*. (For further details see Box 3.7.)

What is the relationship between **ZF** and **NBG**? It was found that whatever can be proved in **ZF** can also be proved in **NBG**. The opposite holds only for the formulas of **NBG** that are also formulas of **ZF**. (This is because the notion of class is unknown to **ZF**.) Because of this, **NBG** is said to be a *conservative extension of* the theory **ZF**. Theorem 3.1 condenses all of this.

Theorem 3.1. *For any formula* F *of* **ZF** *it holds that* $\underset{\text{NBG}}{\vdash} F$ *iff* $\underset{\text{ZF}}{\vdash} F$.

---

[12] John von Neumann, 1903–1957, Hungarian–American mathematician.

[13] Kurt Gödel, 1906–1978, Austrian–American logician, mathematician, and philosopher.

Let us note that the notion of a class is often used in **ZF** too, but it is not formally defined. For example, while an object $\{w \mid w \in z \land P(w)\}$ is surely a set when z is a set (by the *Separation Schema*), an object $\{w \mid P(w)\}$ is called a class for safety's sake (as it may not exist as a set). Thus, one can talk about "the class of all sets" knowing that "the set of all sets" does not exist.

Box 3.6 (ZFC, Zermelo-Fraenkel Axiomatic Set Theory).

Formal Axiomatic System **ZFC**. The alphabet has symbols from **L** and two proper symbols: an individual-constant symbol $\emptyset$ and a binary relation symbol $\in$. The terms and formulas are as usual. The language is of the first order. The rules of inference are *Modus Ponens* and *Generalization*. In addition to the logical axioms of **L**, there are nine proper axioms:

1) $\forall x \forall y (\forall w (w \in x \Leftrightarrow w \in y) \Rightarrow x = y)$ *(Axiom of Extensionality)*

2) $\forall y \exists z \forall w (w \in z \Leftrightarrow w \in y \land P(w))$, for any P with no free z.   *(Separation Schema)*

3) $\forall x \forall y \exists z \forall w (w \in z \Leftrightarrow w = x \lor w = y)$ *(Axiom of Pair)*

4) $\forall y \exists z \forall w (w \in z \Leftrightarrow \exists x (w \in x \land x \in y))$ *(Axiom of Union)*

5) $\forall u \forall v \forall w (f(u,v) \land f(u,w) \Rightarrow v = w) \Rightarrow \forall y \exists z \forall w (w \in z \Leftrightarrow \exists x (x \in y \land f(x,w)))$,
    for any f with no free z.   *(Substitution Schema)*

6) $\exists z (\emptyset \in z \land \forall x \in z : x \cup \{x\} \in z)$ *(Axiom of Infinity)*

7) $\forall y \exists z \forall w (w \in z \Leftrightarrow w \subseteq y)$ *(Axiom of Power Set)*

8) $\forall y \neq \emptyset \; \exists x \in y : x \cap y = \emptyset$ *(Axiom of Regularity)*

9) $\forall y (\forall x \in y : x \neq \emptyset \Rightarrow \exists f \in (\cup y)^y \; \forall x \in y : f(x) \in x)$ *(Axiom of Choice)*

Standard Interpretation. The domain of the standard interpretation was described by von Neumann, so we denote it by $\mathcal{V}$. Von Neumann insisted that $\mathcal{V}$ contains *exactly all the sets*. Thus, if x is in $\mathcal{V}$, then x is a set. Now, if an element of x had not itself been a set, it would not have been in $\mathcal{V}$, and this would have led to trouble. To avoid this, von Neumann required that each element of a set be itself a set. Such sets are called *hereditary*. Thus, $\mathcal{V}$ contains exactly all the hereditary sets. It might appear that there are useful sets, such as $\{0,1,2\}$, that are not hereditary. We will see shortly that this is not so.

Remarks. In the following we describe, for each proper axiom, the motivation for adding it to the axioms of **ZF**, its meaning, and its consequences. When interpreting a proper axiom, bear in mind that the individual-variable symbols mean hereditary sets.

1)   *Axiom of Extensionality: A set is completely determined by its elements.*
     Consequences: Two sets with the same elements are equal. $\mathcal{V}$ contains exactly hereditary sets. There is at most one empty set, $\emptyset$. Motivation for axiom 2: Does $\emptyset$ exist?

2)   *Separation Schema: The set $z = \{w \mid w \in y \land P(w)\}$ exists.*
     Comment: y is any set and P is any property defined by a formula of **L**. Consequences: $\emptyset$ is a set. (Proof: $P(w) \equiv \neg(w = w)$). For any sets $\mathcal{A}$ and $\mathcal{B}$, also $\mathcal{A} \cap \mathcal{B} \overset{\text{def}}{=} \{w \mid w \in \mathcal{A} \land w \in \mathcal{B}\}$ and $\mathcal{A} \setminus \mathcal{B} \overset{\text{def}}{=} \{w \mid w \in \mathcal{A} \land w \notin \mathcal{B}\}$ are sets. Motivation for axiom 3: We need more sets.

3)   *Axiom of Pair: For any x and y there is a set z containing exactly x and y.*
     Consequences: Ordered pair $(x,y) \overset{\text{def}}{=} \{\{x\}, \{x,y\}\}$ is a set. $\{\emptyset\}$ is a set (as $\{\emptyset\} = \{\emptyset, \emptyset\}$); so is $\{\{\emptyset\}\}$ (as $\{\{\emptyset\}\} = \{\{\emptyset\}, \{\emptyset\}\}$); and so on. Defining $0 \overset{\text{def}}{=} \emptyset$, $1 \overset{\text{def}}{=} \{\emptyset\}$, and $2 \overset{\text{def}}{=} \{\emptyset, \{\emptyset\}\}$, we obtain the numbers $0, 1, 2$ and count to two. Motivation for axiom 4: We cannot define larger numbers in this way, because we cannot construct sets with more than two members. So we need more sets.

4)   *Axiom of Union: For any family y of sets x there is a set z that is the union of the sets x.*
     Consequences: Now we can construct sets with more than two elements and define any
     natural number, e.g., $3 \overset{\text{def}}{=} 2 \cup \{2\} = \{\emptyset, \{\emptyset\}, \{\emptyset, \{\emptyset\}\}\}$, using the definition $n + 1 \overset{\text{def}}{=} n \cup \{n\}$.
     The definition is applicable on every $n \in \mathbb{N}$, where $\mathbb{N}$ denotes the collection of all natural
     numbers. Motivation: Is $\mathbb{N}$ a set? We will postulate this in the *Axiom of Infinity* (see below).
     However, it turns out that $\mathbb{N}$ alone does not allow for the development of a full theory of
     ordinals and for the use of transfinite induction. This is why we first introduce axiom 5.

5)   *Substitution Schema: If the domain y of a function f is a set, then its range z is also a set.*
     Comment: In the schema, $f(x,y)$ denotes a function $x \mapsto y$. Consequences: There exist certain
     well-ordered sets, i.e., ordinal numbers.

6)   *Axiom of Infinity: There is an inductive infinite set z.*
     Comment: A set z is defined to be *inductive* if $\emptyset \in z \wedge \forall x(x \in z \Rightarrow x \cup \{x\} \in z)$. A set z is
     defined to be *infinite* if it is equipollent to a proper subset of z. Consequence: $\mathbb{N}$ is a set.
     Motivation for axiom 7: Some sets still cannot be constructed (e.g., the power set of a set).

7)   *Axiom of Power Set: For any set y there is a set z containing all the subsets w of y as members.*
     Motivation: The axioms of pair, union, and power set allow for the construction of larger sets
     from smaller ones. Thus, a set exists if it can be constructed only from $\emptyset$ and $\mathbb{N}$, which are
     the only sets whose existence was postulated by axioms. What about "irregular" sets, such
     as Russell's $\mathcal{R}$? At this point we can still construct a set x where $x \in y \in x$ for some set y. We
     should confine constructions so that only "regular" (i.e., reasonable) sets will exist. Axiom 8
     takes care of this.

8)   *Axiom of Regularity: Any nonempty set y contains an element x such that x and y have no
     common elements.*
     Consequence: There can be no set x such that $x \in y \in x$ for some set y (else, we would have
     $x \in \{x,y\} \cap y$ and $y \in \{x,y\} \cap x$, implying that $\{x,y\}$ would not contain an element sharing
     no elements with $\{x,y\}$, in contradiction with axiom 8). This prevents Russell's Paradox.

9)   *Axiom of Choice: For any family y of nonempty sets x there is a function f that assigns to
     each member x of y a member of x.*
     Comment: $(\cup y)^y$ denotes the set of all functions from y to the union $\cup y$ of all elements of y.

Box 3.7 (NBG, von Neumann-Bernays-Gödel's Axiomatic Set Theory).

Basic Ideas. In addition to the two usual basic notions of a set and a membership relation $\in$, there
is also the notion of a *class*. Each set is also a class, but some classes are not sets. Classes that are
not sets are called *proper classes*. A characteristic of a proper class is that it is not a member of
any class (and hence, of any set). The intention of such a definition of a class is now clear: proper
classes should represent collections that are too large to be sets, and non-proper classes (that is,
sets) should represent all the reasonably large sets that are used in mathematics.

Drawing a distinction between sets and proper classes enables us to prevent paradoxes. Let us
see how this works on Russell's Paradox. Define the class $\mathcal{R} = \{\mathcal{S} \mid \mathcal{S} \text{ is a set } \wedge \mathcal{S} \notin \mathcal{S}\}$. Like
every class, $\mathcal{R}$ either is or is not a member of itself. Let us see whether $\mathcal{R}$, even as a class, gives
rise to Russell's Paradox $\mathcal{R} \in \mathcal{R} \Leftrightarrow \mathcal{R} \notin \mathcal{R}$. If $\mathcal{R} \in \mathcal{R}$, then $\mathcal{R}$ is a set that is a member of itself,
and hence $\mathcal{R} \notin \mathcal{R}$. This is a contradiction. Assume now that $\mathcal{R} \notin \mathcal{R}$. This means that $\mathcal{R}$ is not a set
or $\mathcal{R} \in \mathcal{R}$. The latter alternative is impossible because of the assumption, which leaves us the first
alternative: $\mathcal{R}$ is not a set. Therefore, $\mathcal{R}$ is a *proper class*. We have seen that this deduction, which
in Cantor's naive set theory led to Russell's Paradox, now luckily ends up with the conclusion that
$\mathcal{R}$ is a proper class. In a similar way Buralli-Forti's and Cantor's paradoxes are eliminated. So how
was this system formally defined?

Formal Axiomatic System **NBG**. There are many similarities with **ZF**. The alphabet has all the symbols of **L** and two proper constant symbols $\emptyset$ and $\in$. The terms and formulas are built as usual. The symbolic language is of the first order. The rules of inference are *Modus Ponens* and *Generalization*. In addition to the logical axioms inherited from **L**, there are proper axioms. How were these selected?

Recall that Cantor's *Axiom of Abstraction* postulated that "Every property P defines a set S_P." As we have seen, the authors of **ZF** limited the properties P so that S_P are reasonable and not too large. In contrast, the authors of **NBG** argued as follows:

> If we demanded that S_P be a *class*, then S_P might not be a set (but be a proper class). Hence, the fear of too-large *sets* might become superfluous. But then, could P again be an *arbitrary* property? Could we declare the following generalization of the *Axiom of Abstraction*: "Every property P defines a *class* S_P"?

It turned out that such an axiom is bad, for it would allow $\mathcal{R}' = \{S \mid S \text{ is a class} \wedge S \notin S\}$ to be a class, and this class would again lead to Russell's Paradox $\mathcal{R}' \in \mathcal{R}' \Leftrightarrow \mathcal{R}' \notin \mathcal{R}'$. Thus, more caution was needed in order to generalize the *Axiom of Abstraction*. The result of the search is the following **Axiom of Class Existence**: *Every property P of sets defines a class*. Hence, a class cannot be determined by a property of proper classes, but only by a property of sets. This finally leads to an informal definition of a class:

> *A class is a collection of sets that have in common a property P*: $\{S \mid S \text{ is a set} \wedge P(S)\}$.

Of course, the sets S must exist in the first place. This is ensured by other axioms (as in **ZF**). For this **NBG** has three groups of proper axioms.

The first group initially consisted only of the *Axiom of Class Existence* to establish the notion of a class. This axiom is actually an axiom schema, because it represents an infinite number of axioms, one for each property P of sets. It turned out that the schema can be replaced by only eight axioms! These axioms now constitute the first group. In the second group is the following **Axiom of Extensionality**: *Two classes are equal if they have the same elements*. The third group consists of axioms that, similarly to **ZF**, postulate the existence of *sets* obtained either *ex nihilo* (such as $\emptyset$ and $\mathbb{N}$) or by a construction from existing ones.

Second-Order Formal Axiomatic Systems and Theories

It turned out that certain properties of mathematical objects cannot be defined in a first-order symbolic language. So, the basic notions and axioms referring to such properties cannot be stated in these languages. Consequently, there are no first-order theories about such objects. In such cases, it often turns out that the quantifiers $\forall$ and $\exists$ should be applicable to function-variable symbols and/or predicate-variable symbols, something that is not allowed in first-order symbolic languages. For instance, first-order languages do not enable us to define the *completeness* of the set $\mathbb{R}$ of real numbers, or the concepts of *torsion group* and *mathematical induction* (see Box 3.8 for further details). If, however, the action of quantifiers is expanded to function-variable or predicate-variable symbols, we obtain a second-order symbolic language, a second-order formal axiomatic system, and a second-order theory.

Unfortunately, second-order theories are not as useful as they seem. This is because they lack some important properties that are characteristic of first-order theories. For example, for any theory it is important to know whether the theory has models, how many there are, what their properties are and what the relations between them are. Such questions are dealt with in *model theory*. Two of its important theorems are the *Compactness Theorem*[14] and the "downward" *Löwenheim-Skolem Theorem*[15]. But, in general, the theorems do not hold in second-order theories. Thus, second-order languages and theories may be more powerful in their expression, but they are less amenable to a metamathematical treatment.

As we will see, the deficiencies of second-order theories had no opportunity to manifest themselves and influence formalism, because it was not long before a disappointment about first-order theories came as a result of Gödel's *Incompleteness Theorems*.

Box 3.8 (Expression of First-Order Languages).

We give examples of where a second-order language is needed.

- *Mathematical Induction.* The *Axiom of Mathematical Induction* is usually written as the first-order defining formula $F(0) \wedge \forall i (F(i) \Rightarrow F(i+1)) \Longrightarrow \forall n F(n)$, to which an *explanation* is added stating that $F(x)$ can be *any* formula with x as a free variable; see Box 3.5. (The formula $F(x)$ describes a property of x.) But observe that the defining formula is actually an axiom schema; only after F has been substituted with an actual formula is a particular axiom (i.e., one of infinitely many) obtained. In order to write in a symbolic language that the principle of mathematical induction holds *for any formula F*, we have to add $\forall F$ to the defining formula. This gives us a *second-order* formula $\forall F (F(0) \wedge \forall i (F(i) \Rightarrow F(i+1)) \Longrightarrow \forall n F(n))$. First-order symbolic language is too weak to describe completely the principle of mathematical induction.

- *Completeness of $\mathbb{R}$.* A fundamental property of the set $\mathbb{R}$ of real numbers is *completeness*: every nonempty subset of $\mathbb{R}$ that is bounded above has a least upper bound. How can we express this in a symbolic language? Let us start generally. Let $\mathbb{R}$ be the domain of an interpretation, and B an arbitrary property of the sets $S \subseteq \mathbb{R}$. We can describe the fact that B holds for every set $S \subseteq \mathbb{R}$ by the formula $\forall S : B(S)$. But this formula does not belong to any first- or second-order language, because $\forall$ refers to *sets* of elements of the domain. By viewing subsets of $\mathbb{R}$ as sets $S_P = \{x \in \mathbb{R} \mid P(x)\}$, where P are predicates, the formula transforms into the *second-order* formula $\forall P : B(S_P)$, because $\forall$ now binds the predicate-variable P. If we fix the property B to $B(S_P) = $ "if S_P is bounded above then it has an l.u.b.," we obtain the second-order formula $\forall P (\exists b \forall x (P(x) \Rightarrow x \leqslant b) \Longrightarrow \exists \ell \forall u (\forall x (P(x) \Rightarrow x \leqslant u) \Leftrightarrow \ell \leqslant u))$ stating that $\mathbb{R}$ is complete. That is: For every P, if S_P is bounded above by b, then S_P has an l.u.b. (an ℓ which is $\leqslant$ than any upper bound u of S_P).

- *Torsion Groups.* Let $(G, \cdot)$ be a group with unit e. We say that G is a *torsion* group if for every $a \in G$ there is an $n \geqslant 1$ such that $a^n = e$. How can we define the property $P(G) \equiv$ "*G* is a *torsion group*"? Let the domain of interpretation be $(G, \cdot)$. Let us try the seemingly obvious: $P(G) \equiv \forall a \in G \exists n \geqslant 1 \, a^n = e$. Notice that the interpretation of this formula is a proposition that also considers natural numbers—but these are *not* in the domain of interpretation. To fix that we might define $P(G) \equiv \forall a \in G \, (a = e \vee a \cdot a = e \vee a \cdot a \cdot a = e \vee \ldots)$ and thus avoid

---

[14] *Compactness Theorem:* A first-order theory has a model if every finite part of the theory does.

[15] *Löwenheim-Skolem Theorem:* If a theory has a model, then it has a countable model. The generalization is called the "upward" *Löwenheim-Skolem Theorem* and states: If a first-order theory has an infinite model, then for every infinite cardinal κ it has a model of size κ. Since such a theory is unable to pin down the cardinality of its infinite models, it cannot have exactly one infinite model (up to isomorphism).

mentioning natural numbers. However, this formula is no longer finite and, hence, not in a first-order symbolic language. In any case, it turns out that there is no finite set of first-order formulas whose models are precisely the torsion groups.

3.3 Chapter Summary

A formal axiomatic system is determined by a symbolic language, a set of axioms, and a set of rules of inference. The axioms are logical or proper. All of this is the initial theory associated with the formal axiomatic system. The theory is then systematically extended into a larger theory. The development is formal: new notions must be defined from existing ones, and propositions must be formally proved, i.e., derived from axioms or previously proved formulas. Each proved formula is a theorem of the theory. Such a syntax-oriented and rigorous development of the theory is a mechanical process and because of this better protected from man's fallibility.

The notion of truth is so fundamental that philosophers have been trying to capture it in a satisfying definition since Aristotle. Even today, there are debates about the possibility of achieving such a definition for natural language. However, Tarski demonstrated that this is possible for suitably formalized fragments of natural language. For pragmatic reasons we usually choose such fragments of natural language that allow scientific discourse.

At any stage of the development, the theory can be interpreted in a chosen field of interest, called the domain of interpretation. The interpretation defines how a formula must be understood as a statement about the elements of the domain. Each interpretation of a theory under which all the axioms of the theory are valid is called a model of the theory. A theory may have several different models. A model is not necessarily a part of the real world.

Formal axiomatic systems both protected the development of theories from man's fallibility and preserved the freedom given by the hypothetical axiomatic system. The three particular fields of mathematics whose formal axiomatic systems and theories played a crucial role in the events that followed are logic, arithmetic, and Cantor's set theory. The corresponding formal axiomatic systems are *First-Order Logic* **L**, *Formal Arithmetic* **A**, and the two *Axiomatic Set Theories* **ZF** and **NBG**.

Chapter 4
Hilbert's Attempt at Recovery

*If something is consistent, no part of it contradicts or conflicts
with any other part. If something is complete, it contains all the
parts that it should contain. If something is decidable, we can
establish the fact of the matter after considering the facts.*

Abstract *Hilbert's Program* was a promising formalistic attempt to recover mathematics. It would use formal axiomatic systems to put mathematics on a sound footing and eliminate all the paradoxes. Unfortunately, the program was severely shaken
by Gödel's astonishing and far-reaching discoveries about the general properties of
formal axiomatic systems and their theories. Thus Hilbert's attempt fell short of
formalists' expectations. Nevertheless, although shattered, the program left open an
important question about the existence of a certain algorithm—a question that was
to lead to the birth of *Computability Theory*.

4.1 Hilbert's Program

In this section we will describe *Hilbert's Program*. In order to understand the goals
of the program, we will first define the fundamental metamathematical problems of
formal axiomatic systems and their theories. Then we will describe the goals of the
program and Hilbert's intentions that influenced the program.

4.1.1 Fundamental Problems of the Foundations of Mathematics

The rigor and syntactic orientation of formal axiomatic systems not only protected
them from man's fallibility, but also enabled a precise definition and investigation of
various metamathematical problems, i.e., questions about their theories. Naturally,
these questions were closely linked to the burning question of protecting mathematics from paradoxes. They are called the *problems of the foundations of mathematics*.

Of special importance to the history of the notion of algorithm and *Computability
Theory* will be the following four problems of the foundations of mathematics:

© The Author(s), under exclusive license to Springer-Verlag GmbH, DE,
part of Springer Nature 2020
R. Robič, *The Foundations of Computability Theory*, https://doi.org/10.1007/978-3-662-62421-0_4

1. **Consistency Problem.** Let **F** be a first-order theory. (So we can use the logical connective ¬.) Suppose that there is a closed formula F in **F** such that both F and ¬F are derivable in **F**. Then the contradictory formula F ∧ ¬F immediately follows! We say that such a theory is *inconsistent*. But we can readily show that in an inconsistent theory *any* formula of the theory can be derived! (See Box 4.1.)

Box 4.1 (Derivation in an Inconsistent Theory).

Suppose that formulas F and ¬F are derivable in **F** and let A be an *arbitrary* formula in **F**. Then we have the following derivation of A:

1. F	Supposition.
2. ¬F	Supposition.
3. F ⇒ (¬A ⇒ F)	Ax. 1 of **L** (Box 3.4) with ¬A instead of G (i.e.,¬A/G)
4. ¬A ⇒ F	From 1. and 3. by *MP*.
5. ¬F ⇒ (¬A ⇒ ¬F)	Ax. 1 of **L** (Box 3.4) with ¬F/F and ¬A/G.
6. ¬A ⇒ ¬F	From 2. and 5. by *MP*.
7. (¬A ⇒ ¬F) ⇒ ((¬A ⇒ F) ⇒ A)	Ax. 3 of **L** (Box 3.4) with A/G.
8. (¬A ⇒ F) ⇒ A	From 6. and 7. by *MP*.
9. A	From 4. and 8. by *MP*.

An inconsistent theory has no cognitive value. This is why we seek consistent theories (Fig. 4.1). So the following metamathematical question is important:

*Consistency Problem: "Is a theory **F** consistent?"*

For example, in 1921, Post proved that the *Propositional Calculus* **P** is consistent.

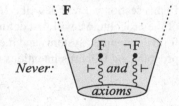

Fig. 4.1 In a consistent theory **F** for no formula F both F and ¬F are derivable in **F**

2. **Syntactic Completeness Problem.** Let **F** be a consistent first-order theory and F an arbitrary closed formula of **F**. Since **F** is consistent, F and ¬F are not both derivable in **F**. But, what if neither F nor ¬F is derivable in **F**? In such a case we say that F is *independent* of **F** (as it is neither provable nor refutable in **F**). This situation is undesirable. We prefer that *at least* one of F and ¬F be derivable in **F**. When this *is* the case for *every* closed formula of **F**, we say that **F** is *syntactically complete* (see Fig. 4.2). Thus, in a consistent and syntactically complete theory every closed formula is *either* provable *or* refutable. Informally, there are no "holes" in such a theory, i.e., no closed formulas independent of the theory. So, the next metamathematical question is important:

*Syntactic Completeness Problem: "Is a theory **F** syntactically complete?"*

The answer tells us whether **F** guarantees that, for no formula of **F**, the search for either a proof or a refutation of the formula is *a priori* doomed to fail. For instance, we know that the *Propositional Calculus* **P** and *First-Order Logic* **L** are *not* syntactically complete.

Fig. 4.2 In a syntactically complete theory **F** it holds, for every formula F, that F or ¬F is derivable in **F**

Always:

3. **Decidability Problem.** Let **F** be a consistent and syntactically complete first-order theory, and F an arbitrary formula of **F**. The derivation of F or ¬F may be highly intricate. Consequently, the search for a formal proof or refutation of F is inevitably dependent on our ingenuity. As long as F is neither proved nor refuted, we can be sure that this is because of our lack of ingenuity (because **F** is syntactically complete). Now suppose that there existed an *algorithm*—called a *decision procedure*—capable of answering—in *finite* time, and for *any* formula F of **F**—the question "Is F derivable in **F**?" Such a decision procedure would be considered *effective* because it could *decide*, for any formula, whether or not it is a theorem of **F**. When such a decision procedure exists, we say that the theory **F** is *decidable*. (See Fig. 4.3.) So, the metamathematical question, called the

*Decidability Problem: "Is a theory **F** decidable?"*

is important because the answer tells us whether **F** allows for a systematic (i.e., mechanical, algorithmic) and effective search of formal proofs. In a decidable theory we can, at least *in principle*, develop the theory without investing our ingenuity and creativity. For instance, the *Propositional Calculus* **P** is known to be a decidable theory; the corresponding decision procedure uses the well-known *truth-tables* and was discovered in 1921 by Post.[1]

Fig. 4.3 In a decidable theory **F** there is a decision procedure (algorithm) that tells, for arbitrary formula F, whether or not F is derivable in **F**

Decision procedure for
*"Is F derivable in **F**?"*

---

[1] Emil Leon Post, 1897–1954, American mathematician, born in Poland.

4. **Semantic Completeness Problem.** Let **F** be a consistent first-order theory. When interpreting **F**, we are particularly interested in the formulas that are *valid in the theory* **F**, i.e., formulas that are valid in *every* model of **F**. Such formulas represent *Truths* in **F** (see p. 44). Now, we know that all the axioms of **F**, both logical and proper, represent *Truths* in **F**. If, in addition, the rules of inference of **F** preserve the property "to represent a *Truth* in **F**", then also every theorem of **F** represents a *Truth* in **F**. When this is the case, we say that **F** is *sound*. Informally, in a sound theory we cannot deduce something that is not a *Truth*, so the theory may have cognitive value. Specifically, it can be proved that *Modus Ponens* and *Generalization* preserve the *Truth*-ness of formulas. So we can assume that the theories we are interested in are sound.

To summarize, a theory **F** is *sound* when the following holds: If a formula F is a theorem of **F**, then F represents a *Truth* in **F**; in short

$$If \vdash_F F \ then \models_F F. \qquad\qquad \text{(\textbf{F} is sound)}$$

However, the opposite may not hold: A sound theory **F** may contain a formula that represents a *Truth* in **F**, yet the formula is *not* derivable in **F**. This situation can arise when **F** lacks some axiom(s).

Of course, we would prefer a sound theory whose axioms suffice for deriving *every* *Truth*-representing formula in the theory. When this *is* the case, the theory is said to be *semantically* complete. Thus, a theory **F** is *semantically complete* when the following holds: A formula F is a theorem of **F** *if and only if* F represents a *Truth* in **F**; in short

$$\vdash_F F \ if \ and \ only \ if \models_F F. \qquad\qquad \text{(\textbf{F} is semantically complete)}$$

The metamathematical question

Semantic Completeness Problem: "Is a theory **F** *semantically complete?"*

is of the greatest importance because the answer tells us whether the syntactic property "to be a theorem of **F**" coincides with the semantic property "to represent a *Truth* in **F**" (see Fig. 4.4). That *Propositional Calculus* **P** and *First-Order Logic* **L** are semantically complete theories was proved by Post (1921) and Gödel (1930), respectively. (The latter is known as *Gödel's Completeness Theorem*.)

NB *If there is an interpretation* $(\iota, \mathscr{S})$ *of* **F** *and a formula* $G \in$ **F** *such that* G *is valid under* $(\iota, \mathscr{S})$ *but not a theorem of* **F***, then* **F** *is not semantically complete. This will happen in Sect. 4.2.3 in the standard model of* **A***,* $(\mathbb{N}, =, +, *, 0, 1)$.

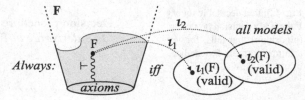

Fig. 4.4 In a semantically complete theory **F** a formula F is derivable *iff* F is valid in **F** (i.e., valid in every model of **F**)

4.1.2 Hilbert's Program

Let us now return to the foundational crisis of mathematics at the beginning of the twentieth century. During 1920–28, Hilbert gradually formed a list of goals—called *Hilbert's program*—that should be attained in order to base the whole of mathematics on new foundations that would prevent paradoxes.

Fig. 4.5 David Hilbert
(Courtesy: See Preface)

Hilbert's Program consisted of the following goals:

A. find an f.a.s. **M** having a computable set of axioms and capable of deriving all the theorems of mathematics;
B. prove that the theory **M** is semantically complete;
C. prove that the theory **M** is consistent;
D. construct an algorithm that is a decision procedure for the theory **M**.

Note that the goal D asks for a *constructive* proof that **M** is decidable, i.e., a proof by exhibiting a decision procedure for **M**. Let us denote this procedure by D_{Entsch} since Hilbert called the *Decidability Problem* for **M** the *Entscheidungsproblem*.

Intention

Hilbert's intention was that, having attained the goals A, B, C, D, every mathematical statement would be mechanically verifiable. How could they be verified? We should first write the statement as a sentence, i.e., a closed formula F of **M** (hence goal A). How would we find out whether the statement represented by F is a mathematical *Truth*? If **M** were a semantically complete theory (hence goal B), we would be sure that F represents a *Truth* in **M** *iff* F is a theorem of **M**. Therefore, we could focus on syntactic issues only. If **M** were a consistent theory (hence goal C), the formulas F and ¬F could not both be theorems. Finally, we would apply the decision procedure D_{Entsch} (hence goal D) to find out which of F and ¬F is a theorem of **M**.

NB *Hilbert expected that* **M** *would be syntactically complete.*

What is more, by using the decision procedure D_{Entsch}, mathematical statements could be algorithmically, that is, mechanically classified into 𝒯*ruths* and non-𝒯*ruths*. There would be no need for human ingenuity in mathematical research; one would just systematically generate mathematical statements (i.e., sentences), check them by D_{Entsch}, and collect only those that are 𝒯*ruths*.[2]

Finitism

Hilbert expected that the consistency and semantic completeness of **M** could be proved only by analyzing the *syntactic* properties of **M** and its formulas. To avoid deceptive intuition, he defined the kind of reasoning, called *finitism*, one should preferably use in such an analysis. Here, Hilbert approached the intuitionist view of infinity. In particular, proofs of the goals B and C should be *finitist* in the sense that they should use finite objects and methods that are constructive, at least in principle. For instance, the analysis should avoid actual infinite sets, the use of the *Law of Excluded Middle* in certain existence proofs, and the use of *transfinite induction*. (We will informally describe what transfinite induction is in Box 4.7 on p. 68.)

4.2 The Fate of Hilbert's Program

After Hilbert proposed his program, researchers started investigating how to attain the goals A, B, C, and D. While the research into the formalization of mathematics (goal A) and the decidability of mathematics (goal D) seemed to be promising, it took only a few years before Gödel discovered astonishing and far-reaching facts about the semantic completeness (goal B) and consistency (goal C) of formally developed mathematics. In this section we will give a detailed explanation of how this happened.

4.2.1 Formalization of Mathematics: Formal Axiomatic System M

So, what should the sought-for formal axiomatic system **M** look like? Preferably it would be a first-order or, if necessary, second-order formal axiomatic system. Probably it would contain one of the formal axiomatic systems **ZFC** or **NBG** in order to introduce sets. Perhaps it would additionally contain some other formal axiomatic systems that formalize other fields of mathematics (analysis, for example). Despite these open questions, it was widely believed that **M** should inevitably contain the following two formal axiomatic systems (see Fig. 4.6):

---

[2] Today, one would use a computer to perform these tasks. Of course, when Hilbert proposed his program, there were no such devices, so everything was a burden on the human processor.

1. *First-Order Logic* **L**. This would bring to **M** all the tools needed for the logically unassailable development of the theory **M**, that is, all mathematics. The trust in **L** was complete after the consistency of **L** had been proved with finitist methods, and after Gödel and Herbrand had proved the semantic completeness of **L** —Herbrand even with finitist methods.

2. *Formal Arithmetic* **A**. This would bring natural numbers to **M**. Since natural numbers play a key role in the construction of other kinds of numbers (e.g., rational, irrational, real, complex), they are indispensable in **M**.

Fig. 4.6 Mathematics as a theory **M** belonging to the formal axiomatic system **M**

4.2.2 Decidability of M: Entscheidungsproblem

Recall that the goal of the *Entscheidungsproblem* was: Construct an algorithm D_{Entsch} that will, for any formula F of **M**, decide whether F is derivable in **M**; in short, whether $\vdash_{\mathbf{M}} F$. (See Fig. 4.7).

Hopes that there was such a D_{Entsch} were raised by the syntactic orientation of formal axiomatic systems and their theories, and, specifically, by their view of a derivation (formal proof) as a finite sequence of language constructs built according to a finite number of syntactic rules. At first sight, the search for a derivation of a formula F could proceed, at least in principle, as follows:

> *systematically generate finite sequences of symbols of* **M**, *and*
> *for each newly generated sequence*
> *check whether the sequence is a proof of* F *in* **M**;
> *if so, then answer* YES *and halt.*

If, in truth, F were derivable in **M**, and if a few reasonable assumptions held (see Box 4.2), then the procedure would find a formal proof of F. However, if in truth F were not derivable in **M**, the procedure would never halt, because it would keep generating and checking candidate sequences. But notice that if a newly generated sequence is not a proof of F, it may still be a proof of ¬F. So we check this possibility too. We obtain the following improved procedure:

>   *systematically generate finite sequences of symbols of* **M**, *and*
>   *for each newly generated sequence*
>   *check whether the sequence is a proof of* F *in* **M**;
>   *if so, then answer* YES *and halt*
>   *else check whether the sequence is a proof of* ¬F *in* **M**;
>   *if so, then answer* NO *and halt.*

Assuming that either F or ¬F is provable in **M**, the procedure always halts. In Hilbert's time there was wide belief that **M** would be syntactically complete.

Fig. 4.7 D_{Entsch} answers in finite time with YES or NO the question *"Is* F *a theorem of* **M**?"

Box 4.2 (Recognition of Derivations).

Is a sequence of symbols a derivation of F in **M**? Since the sequence is finite, it can only be composed of *finitely* many formulas of **M**. There are also *finitely* many rules of inference in **M** that can connect these formulas in a syntactically correct derivation of F. Assuming that we can find out in finite time whether a formula (contained in the sequence) directly follows from a finite set of premises (contained in the sequence) by a rule of inference of **M**, we can decide in finite time whether the sequence is a derivation of F in **M**. To do this, we must systematically check a finite number of possible triplets (formula, set of premises, rule).

Here, an assumption is needed. Since a premise can also be an axiom of **M**, we must assume that there is a procedure capable of deciding in finite time whether a formula is an axiom of **M**. Today, we say that such a set of axioms is *computable* (and the corresponding theory **M** is *computably axiomatizable*). Hence goal A of *Hilbert's Program*.

If the theory **M** were *consistent*, then at most one of F and ¬F would be derivable in **M**. If, in addition, **M** were *syntactically complete*, then at least one of F and ¬F would be derivable in **M**. Consequently, for an arbitrary F of such an **M**, the procedure would halt in a finite time and answer either YES (i.e., F is a theorem of **M**) or NO (i.e., ¬F is a theorem of **M**). So, **M** would be *decidable* and the above procedure would be the decision procedure D_{Entsch}. Thus we discovered the following relationship:

>   *if*   **M** is consistent *and* **M** is syntactically complete
>   *then*   there is a decision procedure for **M**, i.e., **M** is decidable.

NB *These questions caused the birth of Computability Theory, which took over the research connected with goal D of Hilbert's Program, and finally provided answers to these and many other questions.*

We will return to questions about the decision procedure in the next chapter. Later we will describe how the new theory solved the *Entscheidungsproblem* (see Theorem 9.2 on p. 217). For the present, we continue describing what happened to the other two all-important goals (B and C) of *Hilbert's Program*.

4.2.3 Completeness of **M**: Gödel's First Incompleteness Theorem

So, how successful was proving the semantic completeness of **M** (goal B)? In 1931, hopes of finding such a proof were dashed by 25-year-old Gödel. He proved the following metamathematical theorem.

Fig. 4.8 Kurt Gödel
(Courtesy: See Preface)

Theorem 4.1. (First Incompleteness Theorem) *If the Formal Arithmetic* **A** *is consistent, then it is semantically incomplete.*

Informally, the *First Incompleteness Theorem* tells us that if **A** is a consistent theory, then it is not capable of proving *all 𝒯ruths* about natural numbers; there are statements about natural numbers that are true, but are unprovable within **A**.

Gödel proved this theorem by constructing an independent formula G in **A**, i.e., a formula G such that neither G nor ¬G is derivable in **A**. In addition, he proved that G represents a *𝒯ruth* about natural numbers, i.e., that the interpretation of G is true in the standard model $(\mathbb{N}, =, +, *, 0, 1)$ of **A**. (This is the place to recall **NB** on p. 58.)

What is more, he proved that even if G were added to the proper axioms of **A**, the theory **A′** belonging to the extended formal axiomatic system would still be semantically incomplete—now because of some other formula G′ independent of **A′** yet true in the standard model. Finally, he proved the following generalization: *Any* consistent extension of the set of axioms of **A** gives a semantically incomplete theory. (For a more detailed explanation of Gödel's proof, see Box 4.4 on p. 65.)

Informally, the generalization tells us that *no* consistent theory that includes **A** is capable of proving *all Truths* about the natural numbers; there will *always* be statements about the natural numbers that are true, yet unprovable within that theory.

4.2.4 Consequences of the First Incompleteness Theorem

Gödel's discovery revealed unexpected limitations of the axiomatic method and seriously undermined *Hilbert's Program*. Since **M** was supposed to be a consistent extension of **A**, it would inevitably be semantically incomplete! This means that the mathematics developed as a formal theory **M** would be like a "Swiss cheese full of holes" (Fig. 4.9), with some of the mathematical *Truths* dwelling in the holes, inaccessible to usual mathematical reasoning (i.e., logical deduction in **M**).

Fig. 4.9 Mathematics developed in the formal axiomatic system **M** would not be semantically complete

The independent formulas G, G′, … of the proof of the *First Incompleteness Theorem* are constructed in such a shrewd way that they express their own undecidability and, at the same time, represent *Truths* in **M** (see **NB** on p. 58 and Box 4.4).

But in other holes of **M** sit independent formulas that tell us nothing about their own validity in **M**. When such a formula is brought to light and its independence of the theory is uncovered, we may declare that either the formula or its negation represents a *Truth* in **M** (mathematics). In either case our choice does not affect the consistency of the theory **M**. However, the choice may not be easy, because either choice may have a model that behaves as a possible part of mathematics. An example of this situation is the formula that represents (i.e., describes) the *Continuum Hypothesis*, which we encountered on p. 16. (See more about this in Box 4.3.)

Remark. Nevertheless, all of this still does not mean that such *Truths* will *never* be recognized. Note that the *First Incompleteness Theorem* only says that for recognizing such *Truths* the *axiomatic method* is too weak. So there still remains a possibility that such *Truths* will be proven (recognized) with some *other* methods surpassing the axiomatic method in its proving capability. Such methods might use non-finitist tools or any other tools yet to be discovered.

Box 4.3 (Undecidability of the *Continuum Hypothesis*).

Intuitively, the *Continuum Hypothesis* (CH) conjectures that there is no set with more elements than natural numbers and fewer than real numbers. In short, there is no cardinal between $\aleph_0$ and c.

In 1940, Gödel proved the following metamathematical theorem: *If* **ZFC** *is consistent, then* CH *cannot be refuted in* **ZFC**. Then, in 1963, Cohen[3] proved: *If* **ZFC** *is consistent, then* CH *cannot be proved in* **ZFC**. Thus, CH is independent of **ZFC**. Gödel's and Cohen's proofs show that neither CH nor ¬CH is a $\mathscr{T}ruth$ in **ZFC**. Discussions about whether or not to add CH to **ZFC** (and hence mathematics) still continue.

There is a similar situation with the generalization of CH. Let α be any ordinal and $\aleph_\alpha$ and $\aleph_{\alpha+1}$ the cardinalities of a set and its power set, respectively. The *Generalized Continuum Hypothesis* (GCH) conjectures: There is no cardinal between $\aleph_\alpha$ and $\aleph_{\alpha+1}$, i.e., $2^{\aleph_\alpha} = \aleph_{\alpha+1}$.

Box 4.4 (Proof of the *First Incompleteness Theorem*).

The theorem states: *Every axiomatizable consistent theory that includes* **A** *is incomplete.* That is: Every axiomatizable theory that is consistent and sufficiently strong has countably infinitely many formulas that are true statements about natural numbers but are not derivable in the theory.

How did Gödel prove this? His first breakthrough was the idea to *transform metamathematical statements about the theory* **A** *into formulas of the very theory* **A**. In this way, each statement *about* **A** would become a formula *in* **A**, and therefore accessible to a formal treatment *within* **A**. In particular, the metamathematical statement $\mathscr{G}$ saying that "a given formula of **A** is not provable in **A**" would be transformed into a formula of **A**. Gödel's second breakthrough was *the construction of this formula and the use of it in proving its own undecidability.* The main steps of the proof are:

1. *Arithmetization of* **A**. A *syntactic object* of **A** is a symbol, a term, a formula, or any finite sequence of formulas (e.g., a formal proof).

   First, Gödel showed that with every syntactic object X one can associate a precisely defined natural number $\gamma(X)$—today called the *Gödel number* of X. Different syntactic objects have different Gödel numbers, but there are natural numbers that are not Gödel numbers. The computation of $\gamma(X)$ is straightforward. Also straightforward is testing to see whether a number is a Gödel number and, if so, constructing the syntactic object from it. (See Problems on p. 71.)

   Second, Gödel showed that with every *syntactic relation* (defined on the set of all syntactic objects) there is associated a precisely defined *numerical relation* (defined on $\mathbb{N}$). In particular, with the syntactic relation $D(X, Y) \equiv$ "X is a derivation of formula Y" is associated a numerical relation $D \subseteq \mathbb{N}^2$ such that $D(X, Y)$ *iff* $D(\gamma(X), \gamma(Y))$. All this enabled Gödel to describe **A** only with natural numbers and numerical relations. We say that he *arithmetized*[4] **A**.

2. *Arithmetization of Metatheory* $\overline{A}$. Gödel then arithmetized the metatheory $\overline{A}$, i.e., the theory about **A**. A metatheoretical proposition in $\overline{A}$ is a statement $\mathscr{F}$ that (in natural language and using special symbols like ⊢ and ⊨) states something about the syntactic objects and syntactic relations of **A**. Since Gödel was already able to substitute these with natural numbers and numerical relations, he could translate $\mathscr{F}$ into a statement referring only to natural numbers and numerical relations. But notice that such a statement belongs to the theory **A** and is, therefore, representable by a formula of **A**! We see that Gödel was now able to transform metatheoretical statements of $\overline{A}$ into formulas of **A**.

---

[3] Paul Cohen, 1934–2007, American mathematician.

[4] Recall (p. 6) that Leibniz had a similar idea, though he aimed to arithmetize man's reflection.

3. *Gödel's Formula.* Now Gödel could do the following: 1) he could transform any metamathematical statement $\mathscr{F}$ about formulas of **A** into a formula F about natural numbers; 2) since natural numbers can represent syntactic objects, he could interpret F as a statement about syntactic objects; and 3) since formulas themselves are syntactic objects, he could make it so that *some formula is a statement about itself*. How did he do that?

Let $\mathscr{G}(\mathtt{H})$ be a metamathematical statement defined by $\mathscr{G}(\mathtt{H}) \equiv$ "Formula H is not provable in **A**." To $\mathscr{G}(\mathtt{H})$ corresponds a formula in **A**; let us denote it by $\mathtt{G}(h)$, where $h = \gamma(\mathtt{H})$. The formula $\mathtt{G}(h)$ states the same as $\mathscr{G}(\mathtt{H})$, but in number-theoretic vocabulary. Specifically, $\mathtt{G}(h)$ states: "The formula with Gödel number h is not provable in **A**."

Now, also $\mathtt{G}(h)$ is a formula, so it has a Gödel number, say g. What happens if we take $h := g$ in $\mathtt{G}(h)$? The result is $\mathtt{G}(g)$, *a formula that asserts about itself that it is not provable in* **A**. Clever, huh? To improve readability, we will from now on write G instead of $\mathtt{G}(g)$.

4. *Incompleteness of* **A**. Then, Gödel proved: G *is provable in* **A** *iff* ¬G *is provable in* **A**. (In the proof of this he used so-called ω-consistency; later, Rosser showed that usual consistency suffices.) Now suppose that **A** is consistent. So, G and ¬G are not both provable. Then it remains that *neither* is provable. Hence, **A** *is syntactically incomplete*. Because there is no proof of G, we see that what G asserts is in fact *true*. Thus, G is true in the standard model $(\mathbb{N}, =, +, *, 0, 1)$ of **A**, yet it is not provable in **A**. In other words, **A** is *semantically incomplete* (see **NB** on p. 58).

5. *Incompleteness of Axiomatic Extensions of* **A**. The situation is even worse. Because G represents a $\mathscr{T}ruth$ about natural numbers, it seems reasonable to admit it to the set of axioms of **A**, hoping that the extended formal axiomatic system will result in a better theory $\mathbf{A}^{(1)}$. Indeed, $\mathbf{A}^{(1)}$ *is* consistent (assuming **A** is). But again, there is a formula $\mathtt{G}^{(1)}$ of $\mathbf{A}^{(1)}$ (not equivalent to G) that is independent of $\mathbf{A}^{(1)}$. So, $\mathbf{A}^{(1)}$ is syntactically incomplete. What is more, $\mathtt{G}^{(1)}$ is true in the standard model $(\mathbb{N}, =, +, *, 0, 1)$. Hence, $\mathbf{A}^{(1)}$ is semantically incomplete.

If we insist and add $\mathtt{G}^{(1)}$ to the axioms of $\mathbf{A}^{(1)}$, we get a consistent yet semantically incomplete theory $\mathbf{A}^{(2)}$ containing an independent formula $\mathtt{G}^{(2)}$ (not equivalent to any of G, $\mathtt{G}^{(1)}$) that is true in $(\mathbb{N}, =, +, *, 0, 1)$. Gödel proved that we can continue in this way indefinitely, but each extension $\mathbf{A}^{(i)}$ will yield a consistent and semantically incomplete theory (because of some formula $\mathtt{G}^{(i)}$, which is not equivalent to any of the formulas $\mathtt{G}, \mathtt{G}^{(1)}, \ldots, \mathtt{G}^{(i-1)}$). □

4.2.5 Consistency of **M**: Gödel's Second Incompleteness Theorem

What about the consistency of the would-be theory **M** (goal C)? Hilbert believed that it would suffice to prove the consistency of *Formal Arithmetic* **A** only. Then, the consistency of other formalized fields of mathematics (due to their construction from **A**) and, finally, the consistency of all formalized mathematics **M** would follow. Thus, the proof of the consistency of **M** would be *relative* to **A**. But this also means that, eventually, the consistency of **A** should be proved with *its own means* and *within* **A** *alone*. In other words, the proof should be constructed without the use of other fields of **M**—except for the *First-Order Logic* **L**—because, at that time, their consistency (being relative) would not be established beyond any doubt. *Formal Arithmetic* **A** should demonstrate its own consistency! We say that the proof of the consistency of **A** should be *absolute*. A method that tried to prove the consistency of **A** is described in Box 4.5.

Box 4.5 (Absolute Proof of Consistency).

We have seen in Box 4.1 (p. 56) that for any first-order theory **F** the following holds: *If there is a formula* F *such that both* F *and* ¬F *are provable in* **F***, then* arbitrary *formula* A *of* **F** *is provable in* **F**. In an inconsistent system everything is provable. Now we see: The consistency of **F** would be proved if we found a formula B of **F** that is *not* provable in **F**.

The question now is, how do we find such a formula B? We can use the following method. Let *P* be any property of the formulas of **F** such that 1) *P* is shared by all the axioms of **F**, and 2) *P* is preserved by the rules of inference of **F** (i.e., if *P* holds for the premises of a rule, it does so for the conclusion). Obviously, theorems of **F** have the property *P*. Now, if we find in **F** a formula B that *does not* have the property *P*, then B is not a theorem of **F** and, consequently, **F** is consistent. Obviously, we must construct such a property *P* that will facilitate the search for B.

Using this method the consistency of *Propositional Calculus* **P** was proved as well as the consistency of *Presburger Arithmetic* (i.e., arithmetic where addition is the only operation).

But in 1931 Gödel also buried hopes that an absolute proof of the consistency of **A** would be found. He proved:

Theorem 4.2. (Second Incompleteness Theorem) *If the Formal Arithmetic* **A** *is consistent, then this cannot be proved in* **A**.

The proof of the theorem is described in Box 4.6.

In other words, **A** cannot demonstrate its own consistency.

Box 4.6 (Proof of the *Second Incompleteness Theorem*).

The theorem says: *If* **A** *is consistent, then we cannot prove this using only the means of* **A**.

In proving this theorem Gödel used parts of the proof of his first theorem. Let 𝒞 be the following metamathematical statement: 𝒞 ≡ "**A** is consistent." This statement too is associated with a formula of **A**—denote it by C—that says the same thing as 𝒞, but by using number-theoretic vocabulary. Gödel then proved that the formula C ⇒ G is provable in **A**. Now, if C were provable in **A**, then (by *Modus Ponens*) also G would be provable in **A**. But in his *First Incompleteness Theorem* Gödel proved that G is *not* provable in **A** (assuming **A** is consistent). Hence, also C is not provable in **A** (if **A** is consistent).   □

4.2.6 Consequences of the Second Incompleteness Theorem

Gödel's discovery revealed that proving the consistency of the *Formal Arithmetic* **A** would require means that are more complex—and therefore *less transparent*—than those available in **A**. Of course, a less transparent object or tool may also be more disputable and more controversial, at least in view of Hilbert's finitist recommendations.

In any case, in 1936, Gentzen[5] proved the consistency of **A** by using *transfinite induction* in addition to usual finitist tools. (See the description of transfinite induction in Box 4.7.) Following Gentzen, several other non-finitist consistency proofs of **A** were found. Finally, the belief was accepted that arithmetic **A** is in fact consistent, and that Hilbert's finitist methods may sometimes be too strict.

Did these non-finitist consistency proofs of **A** enable researchers to prove (relative to **A**, as expected by Hilbert) the consistency of other formalized fields of mathematics and, ultimately, of all mathematics **M**? Unfortunately, no. Namely, there is a generalization of the *Second Incompleteness Theorem* stating: *If a consistent theory* **F** *contains* **A**, *then the consistency of* **F** *cannot be proved within* **F**. Of course, this also holds when **F** := **M**, the would-be f.a.s. for all mathematics. This was the second heavy blow to *Hilbert's Program*.

To prove the consistency of all mathematics, one is forced to use external means (non-finitist, metamathematical, or others yet to be discovered), which may be disputable in view of the finitist philosophy of mathematics. But fortunately, the *Second Incompleteness Theorem* does not imply that the formally developed mathematics **M** would be *in*consistent. It only tells us that the chances of proving the consistency of such a mathematics in *Hilbert's way* are null.[6]

Box 4.7 (Transfinite Induction).

This is a method of proving introduced by Cantor. Let us first recall mathematical induction. If $a_0, a_1, \ldots$ is a sequence of objects (e.g., real numbers) and P is a property sensible of objects a_i, then to prove that *every* element of the sequence has this property, we use mathematical induction as follows: 1) we must prove that $P(a_0)$ holds, and 2) we must prove that $P(a_n) \Rightarrow P(a_{n+1})$ holds for an arbitrary natural number n. In other words, if P holds for an element a_n, then it holds for its *immediate successor* a_{n+1}.

To make the description of transfinite induction more intuitive, let us take a sequence $a_0, a_1, \ldots$, where a_i are real numbers and there is no index after which all the elements are equal. Suppose that the sequence converges and a^* is the limit. (If we take, for example, $a_n = \frac{n}{n+1}$, we have $a^* = 1$.) The limit a^* is not a member of the sequence, because $a^* \neq a_n$ for every *natural* number n. But we can consider a^* to be an *infinite (in order)* element of the sequence, that is, the element that comes

---

[5] Gerhard Karl Erich Gentzen, 1909–1945, German mathematician and logician.

[6] Even if the mathematics is inconsistent, there are attempts to overcome this. A recent approach to accommodate inconsistency of a theory in a sensible manner is *paraconsistent logic* of Priest.[7] The approach challenges the classical result from Box 4.1 (p. 56) that from contradictory premises *anything* can be inferred. "Mathematics is not the same as its foundations," advocate the researchers of paraconsistency, "so contradictions may not necessarily affect all of 'practical' mathematics." They have shown that in certain theories, called *paraconsistent*, contradictions may be allowed to arise, but they need not infect the whole theory. In particular, a paraconsistent axiomatic set theory has been developed that includes cardinals and ordinals and is capable of supporting the core of mathematics. Further developments of different fields of mathematics, including arithmetic, in paraconsistent logics are well underway.

[7] Graham Priest, 1948, English-Australian analytic philosopher and logician.

after every element a_n, where $n \in \mathbb{N}$. Since ω is the smallest ordinal that is larger than any natural number (see p. 17), we can write $a^* = a_\omega$. The sequence can now be extended by a_ω and denoted

$$a_0, a_1, \ldots ; a_\omega .$$

Now, what if we wanted to prove that the property P holds for every element of this *extended* sequence? It is obvious that mathematical induction cannot possibly succeed, because it does not allow us to infer $P(a_\omega)$. The reason is that a_ω is not an immediate successor of any a_n, $n \in \mathbb{N}$, so we cannot prove $P(a_n) \Rightarrow P(a_\omega)$ for any natural n.

An ordinal that is neither 0 nor the immediate successor of another ordinal is called a *limit ordinal*. There are infinitely many limit ordinals, with ω being the smallest of them. Mathematical induction fails at each limit ordinal. *Transfinite induction* remedies that.

Principle of Transfinite Induction: Let $\langle S, \preccurlyeq \rangle$ be a well-ordered set and P a property sensible for its elements. Then P holds for *every* element of S if the following condition is met:

- P holds for $y \in S$ if P holds for *every* $x \in S$ such that $x \prec y$.

Transfinite induction is a generalization of mathematical induction. It can be used to prove that a given property P holds for all ordinals (or all elements of a well-ordered set; see Appendix A). Normally it is used as follows:

1. *Suppose* that P does not hold for all ordinals.
2. Therefore, there is the smallest ordinal, say α, for which we have $\neg P(\alpha)$.
3. Then we try to deduce a contradiction.
4. If we succeed, we conclude: P holds for every ordinal.

4.3 Legacy of Hilbert's Program

The ideas of Whitehead and Russell put forward in their *Principia Mathematica* (Sect. 2.2.3) proved to be unrealistic. Mathematics cannot be founded on logic only.

Also, *Hilbert's Program* (Sect. 4.1.2) failed. The mechanical, syntax-directed development of mathematics within the framework of formal axiomatic systems may be safe from paradoxes, yet this safety does not come for free. The mathematics developed in this way suffers from semantic incompleteness and the lack of a possibility of proving its consistency. All this makes Hilbert's ultimate idea of the mechanical development of mathematics questionable.

Aspiration and Inspiration

Consequently, it seems that research in mathematics cannot avoid human inspiration, ingenuity, and intuition (deceptive though that can be). See Fig. 4.10. *A fortiori*, Leibniz's idea (see p. 6) of replacing human reflection by mechanical and mechanized arithmetic is just an illusion. Mathematics and other axiomatic sciences selfishly guard their *Truths*; they admit to these *Truths* only humans who, in addition to demonstrating a strong aspiration for knowledge, demonstrate sufficient inspiration and ingenuity.

Fig. 4.10 Research cannot avoid inspiration, ingenuity, and intuition

4.4 Chapter Summary

Hilbert proposed a promising recovery program to put mathematics on a sound footing and eliminate all the paradoxes. To do this, the program would use formal axiomatic systems and their theories.

More specifically, Hilbert aimed to define a formal axiomatic system **M** such that the theory **M** developed in it would contain the whole of mathematics. The theory **M** would also comply with several fundamental requirements: It would be consistent, semantically complete, and decidable. In addition, Hilbert required that a decision procedure for **M** should be devised, i.e., an *algorithm* should be constructed capable of deciding, for any formula of **M**, whether the formula represents a mathematical *Truth*.

It was soon realized that **M** must contain at least *First-Order Logic* **L** and *Formal Arithmetic* **A**. This, however, enabled Gödel to discover that there can be no such **M**! In particular, in his *First Incompleteness Theorem*, Gödel proved that if **A**, the *Formal Arithmetic*, is consistent, then it is also semantically incomplete. What is more, any consistent formal axiomatic theory that includes **A** is semantically incomplete. In his *Second Incompleteness Theorem*, Gödel proved that if **A** is consistent, this cannot be proved in **A**. Moreover, if a consistent formal axiomatic theory contains **A**, then the consistency of the theory cannot be proved in the theory.

Although Gödel's discovery shattered *Hilbert's Program, the problem of finding an algorithm that is a decision procedure for a given theory remained topical.*

Problems

Definition 4.1. (Gödel numbering) The arithmetization of the *Formal Arithmetic* **A** associates each syntactic object X of **A** with the corresponding **Gödel number** $\gamma(X) \in \mathbb{N}$. The function γ can be defined in the following way:

- the following symbols are associated with Gödel numbers as decribed:
  - *logical connectives*, "¬",1; "∨", 2; "⇒",3; (that is, $\gamma(\neg) = 1$; $\gamma(\vee) = 2$; $\gamma(\Rightarrow) = 3$)
  - the *quantification symbol*, "∀",4;
  - the *equality symbol*, "=",5;
  - the *individual-constant symbol*, "0",6;
  - the unary function symbol for the *successor function*, "′",7;
  - *punctuation marks*, "(",8; ")",9; ",",10.

- *individual-variable symbols* are associated with increasing prime numbers greater than 10; for example, x,11; y,13; z,17; and so on.

- *predicate symbols* are associated with squares of increasing prime numbers greater than 10; for example, P,11^2; Q,13^2; R,17^2; and so on.

- a *formula* F, viewed as a sequence $F = s_1 s_2 \ldots s_k$ of symbols, is associated with the number $\gamma(F) = p_1^{\gamma(s_1)} p_2^{\gamma(s_2)} \ldots p_k^{\gamma(s_k)}$, where p_i is the ith prime number. For example, the Gödel number of the axiom $\forall x \forall y (x = y \Rightarrow x' = y')$ is $2^4 3^{11} 5^4 7^{13} 11^8 13^{11} 17^5 19^{13} 23^3 29^{11} 31^7 37^5 41^{13} 43^7 47^9$.

- a *formal proof (derivation)* D, viewed as a sequence $D = F_1, F_2, \ldots, F_n$ of formulas, is associated with the Gödel number $\gamma(D) = p_1^{\gamma(F_1)} p_2^{\gamma(F_2)} \ldots p_n^{\gamma(F_n)}$, where p_i is the ith prime number.

Remark. Gödel's original arithmetization was more succinct: "0",1; "′",3; "¬",5; "∨",7; "∀",9; "(",11; ")",13. The symbols such as $\wedge, \Rightarrow, =, \exists$ are only abbreviations and can be represented by the previous ones. Individual-variable symbols were associated with prime numbers greater than 13, and predicate symbols with squares of these prime numbers. Gödel also showed that, for $k \geqslant 2$, k-ary function and predicate symbols can be represented by the previous symbols.

4.1. Prove: The function $\gamma : \Sigma^* \to \mathbb{N}$ is injective. (*Remark.* Σ denotes the set of symbols of **A**.)
[*Hint.* Use the *Fundamental Theorem of Arithmetic*.]

4.2. Let $n \in \mathbb{N}$. Describe how we can decide whether or not there exists an $F \in \mathbf{A}$ such that $n = \gamma(F)$.

4.3. Let $n = \gamma(F)$ for some $F \in \mathbf{A}$. Describe how we can reconstruct $F = \gamma^{-1}(n)$.

4.4. Informally describe how we can decide, for any sequence $F_1, F_2, \ldots, F_n$ of formulas of **A**, whether or not the sequence is a derivation of F_n in **A**.
[*Hint.* See Box 4.2 on p. 62.]

Bibliographic Notes

The following excellent monographs and articles were consulted and are fine sources for the subjects covered in Part I of this book:

- A general treatment of the *axiomatic method* is Tarski [258]. Here, after an introduction to mathematical logic, the use of the axiomatic method in the development of theories is described. Examples of such theories are given (e.g., elementary number theory), and several metatheoretical questions are defined. A shorter exposition of the axiomatic method is in Tarski [256, Sect. The Notion of Proof]. The axiomatization of geometry is given in Hilbert [103].
- Introductions to *mathematical logic* are Robbin [200], Mendelson [155], and Stoll [252], and recently Hinman [106], Rautenberg [196] and Smullyan [239]. In-depth and more demanding are Kleene [125], Shoenfield [216], Manin [145], and recently Ebbinghaus et al. [65]. A different perspective on the *First-Order Logic* is presented in Smullyan [238]. *Second-Order* languages and theories are discussed in Ebbinghaus et al. [65], Boolos and Jeffrey [22], and Boolos et al. [21]. In Slovenian, an in-depth treatise on mathematical logic is Prijatelj [193]. I combined this with Prijatelj [192].
- The relation between mathematics and *metamathematics* is described in depth in Kleene [124].
- Burali-Forti described his paradox in [27]. Russell described his paradox to Frege in a letter [207]. English translations of both can be found in van Heijenoort [268]; here, also other English translations of *seminal publications* between 1879 and 1931 can be found. Explanations of a range of paradoxes can be found in Priest [190] and Sainsbury [211].
- Initial *intuitionistic ideas* about the development of mathematics can be found in Brouwer [26]. See also Kleene [124]. A more in-depth description of intuitionism is Heyting [101]. But see also a historical account of the development of intuitionistic reasoning in Moschovakis [159].
- The main ideas of *logicism* were published in Boole [20], Frege [74], Peano [174], and Whitehead and Russell [276]. Russell's achievements and impact on modern science and philosophy are described in Irvine [114].
- *Formalism* and Hilbert's finitist program were finalized and stated in Hilbert and Ackermann [104]. Here, first-order formal axiomatic systems were introduced. The need for an absolute proof of the consistency of *Formal Arithmetic* was explained in Hilbert [102]. A description of the basic ideas of formalism is in Kleene [124]. For a historical account of the development of Hilbert's view of mathematical proof, his achievements, and their impact, see Sieg [227, 230]. Putnam [194] explains the connection between the *axiomatizability* of a theory, the existence of an algorithm for *"effective" production* of theorems in a single infinite sequence, and the nonexistence of a *decision procedure* for telling whether or not a formula will eventually occur in this sequence.
- The definition of *truth* in formalized languages was given by Tarski [254, 257]. See also Tarski's shorter exposition in [256]. Although deemed impossible by Tarski, Kripke [130] shows how a language can consistently contain its own truth predicate (here Fitting [72] may help); but there is still debate on this. For an introductory survey of philosophical inquiries into truth see Burgess and Burgess [29]. Truth can be taken as the basic notion of a hypothetical axiomatic system; Halbach [94] examines the most important axiomatizations of truth.
- Basics of *model theory* can be found in all of the above monographs on mathematical logic. For a first study of model theory see Hodges [109]; for even more in-depth treatments of models see Hodges [110] and Chang and Keisler [33]. A concise account of model theory is Tent and Ziegler [260]. For a shorter exposition of the interplay between truth and formal proof see Tarski [256, Sect. The Relationship of Truth and Proof]. Vaught [269] surveys Tarski's work in model theory.
- An in-depth discussion of *second-order* formal axiomatic systems and their theories can be found in Ebbinghaus et al. [65]. Logicism, intuitionism, and formalism are extensively described in George and Velleman [78].
- The truth-table method and the proofs that *Propositional Calculus* **P** is a consistent, decidable, and semantically complete theory appeared in Post [180]. See also Tarski [258]. The semantic completeness of *First-Order Logic* **L** was proved by Gödel [80].

- Introductions to *axiomatic set theory* ZFC are Halmos [95], Suppes [253], and Goldrei [88]. For an advanced-level treatment of ZFC, see Levy [141], Winfried and Weese [277], and recently Kunen [133] and Jech [117]. About the axiomatic set theory NBG, see Mendelson [155]. An in-depth historical account of alternative set theories such as NBG is Holmes et al. [111]. The rich history of set theory from Cantor to Cohen is described in Kanamori [120]. That the *Continuum Hypothesis* is independent of ZFC is shown in Cohen [39], and that it is independent of NBG is shown in Smullyan and Fitting [240]. A historical account of the problems of continuum is Steprāns [249]. For an in-depth explanation of the methods used to construct independence proofs in ZFC, see Kunen [132]. The *Axiom of Choice* is extensively analyzed and discussed in Jech [116] and Moore [158].
- The roots of Hilbert's foundational work are systematically traced to the radical transformation of mathematics in Sieg [230].
- Several formal theories were proved to be *decidable* or *undecidable*. Here, Tarski played an important role. In 1930, he proved—by constructing a decision procedure—the decidability of real closed fields and, consequently, of the first-order theory of $\mathbb{R}$ (under $+$ and $*$) [255]. This was a remarkable and surprising result (see van den Dries [267]) because it was already known, due to Church [37] and J. Robinson [201], that the first-order theory of natural numbers $\mathbb{N}$ (under $+$ and $*$) and the first-order theory of rational numbers $\mathbb{Q}$ (under $+$ and $\times$) were both *un*decidable. The undecidability proofs for many other theories, which were obtained between 1938 and 1952 by Tarski, A. Mostowski, and R.M. Robinson, appeared in [259]. See also the surveys of Tarski's discoveries on decidable theories in [62] and on undecidable theories in [153].
- Gödel's *Incompleteness Theorems* appeared in [81]. For in-depth yet user-friendly explanations of both theorems, see Nagel and Newman [168], Smullyan [237], Gödel [84] and Smith [236]. An in-depth discussion of Gödel's *Incompleteness Theorems* as well as his *Completeness Theorem* from the historical perspective is van Atten and Kennedy [266]. Gödel's impact on modern mathematics and logic is presented in Baaz et al. [12].
- Kleene [124] gives a brief account of the use of *transfinite induction* in Gentzen's proof of the consistency of *Formal Arithmetic*.
- The ideas of *paraconsistent logic* are described in Priest [191].
- Many concepts of modern mathematics that we use throughout this monograph are nicely discussed in Stewart [250]. Mathematical logic until 1977 is extensively described in Barwise [14]. For an extensive and up-to-date treatment of set theory, see Jech [117]. The *philosophical aspects* of mathematical logic, set theory, and mathematics in general can be found in Benacerraf and Putnam [16], Machover [143], George and Velleman [78], and Potter [188]. Two expositions of the interplay between mathematical logic and computability are Epstein and Cornielli [68] and Chaitin [32].

Part II
CLASSICAL COMPUTABILITY THEORY

Our intuitive understanding of the concept of the algorithm, which perfectly sufficed for millennia, proved to be insufficient as soon as the non-existence of a certain algorithm had to be proven. This triggered the search for a model of computation, a formal characterization of the concept of the algorithm. In this part we will describe different competing models of computation. The models are equivalent, so we will adopt the Turing machine as the most appropriate one. We will then describe the Turing machine in greater detail. The existence of the universal Turing machine will be proven and its impact on the creation and development of the general-purpose computer will be explained. Then, several basic yet crucial theorems of *Computability Theory* will be deduced. Finally, the existence of incomputable problems will be proven, a list of such problems from practice will be given, and several methods for proving the incomputability of problems will be explained.

Chapter 5
The Quest for a Formalization

A model of a system or process is a theoretical description that can help you understand how the system or process works.

Abstract The difficulties that arose at the beginning of the twentieth century shook the foundations of mathematics and led to several fundamental questions: "What is an algorithm? What is computation? What does it mean when we say that a function or problem is computable?" Because of *Hilbert's Program*, intuitive answers to these questions no longer sufficed. As a result, a search for appropriate definitions of these fundamental concepts followed. In the 1930s it was discovered— miraculously, as Gödel put it—that all these notions can be formalized, i.e., mathematically defined; indeed, they were formalized in several completely different yet equivalent ways. After this, they finally became amenable to mathematical analysis and could be rigorously treated and used. This opened the door to the seminal results of the 1930s that marked the beginning of *Computability Theory*.

5.1 What Is an Algorithm and What Do We Mean by Computation?

We have seen (Sect. 4.2.2) that *if* **M** is consistent and syntactically complete *then* there exists a decision procedure for **M**. But the hopes for a complete and unquestionably consistent **M** were shattered by Gödel's Theorems. Does this necessarily mean that there cannot exist a decision procedure for **M**? Did this put an end to research on the *Entscheidungsproblem*? In this case, no. Namely, the proofs of Gödel's Theorems only used logical notions and methods; they did not involve loose notions of algorithm and computation. Specifically, the *First Incompleteness Theorem* revealed that there are mathematical *Truths* that cannot be *derived* in **M**. But it was not obvious that there could be no other way of *recognizing* every mathematical *Truth*. (After all, Gödel himself was able to determine that his formulas $G, G^{(1)}, \ldots$ are true although undecidable.) Because of this, the *Decidability Problem* for **M** and, in particular, the *Entscheidungsproblem* with its quest for a decision procedure (algorithm) D_{Entsch}, kept researchers' interest. Yet, the *Entscheidungsproblem* proved to be much harder than expected (see p. 61). What is an algorithm, anyway?

R. Robič, *The Foundations of Computability Theory*, https://doi.org/10.1007/978-3-662-62421-0_5

It became clear that the problem could not be solved unless the intuitive, loose definition of the concept of the algorithm was replaced by a formal, rigorous definition.

5.1.1 Intuition and Dilemmas

So, let us return to the intuitive definition of the algorithm (Definition 1.1 on p. 4), which was at Hilbert's disposal: An algorithm for solving a given problem is a recipe consisting of a finite number of instructions that, if strictly followed, leads to the solution of the problem. When Hilbert set the goal *"Find an algorithm that is a decision procedure for* **M***,"* he, in essence, asked us to *conceive* an appropriate recipe (functioning, at least in principle, as a decision procedure for **M**) by using only common sense, logical inference, knowledge, experience, and intuition (subject only to finitist restrictions). In addition, the recipe was to come with an idea of how it would be executed (at least in principle). We see that the notions of the algorithm and its execution were entirely intuitive.

But there were many questions about such an understanding of the algorithm and its execution. What would be the *kind* of basic instructions used to compose algorithms? In particular, would they execute in a discrete or a continuous way? Would their execution and results be predictable (i.e., deterministic) or probabilistic (i.e., dependent on random events)? These questions were relevant in view of the discoveries being made in physics at the time.[1]

Which instructions should be *basic*? Should there be only *finitely* many of them? Would they suffice for composing *any* algorithm of interest? If there were *infinitely* many basic instructions, would that not demand a processor of unlimited capability? Would that be realistic? But if the processor were of limited capability, could it be *universal*, i.e., capable of executing *any* algorithm of interest?[2]

Then there were more down-to-earth questions. How should the *processor* be constructed in order to execute the algorithms? Where would the processor keep the algorithms and where would the input data be? Should data be of arbitrary or limited size? Should storage be limited or unlimited? Where and how would basic

---

[1] *Is nature discrete or continuous?* At the beginning of the twentieth century it was discovered that energy exchange in nature seems to be *continuous* at the macroscopic level, but is *discrete* at the microscopic level. Specifically, energy exchange between matter and waves of frequency v is only possible in discrete portions (called quanta) of sizes nhv, $n = 1, 2, 3 \ldots$, where h is Planck's constant. Notice that some energy must be consumed during the instruction execution.

Is nature predictable or random? Nature at the macroscopic level (i.e., nature dealt with by classical physics) is predictable. That is, each event has a cause, and when an event seems to be random, it is only because we lack knowledge of its causes. Such randomness of nature is said to be *subjective*. In contrast, there is *objective* randomness in microscopic nature (i.e., nature dealt with by quantum physics). Here, an event may be entirely unpredictable, having no cause until it happens. Only a probability of occurrence can be associated with the event.

So, how do all these quantum phenomena impact instruction execution? (Only recently have quantum algorithms appeared; they strive to use these phenomena in problem solving.)

[2] Recall that such universality of a processor was Babbage's goal nearly a century ago (p. 6).

instructions execute? Where would the processor keep the intermediate and final results? In addition, there were questions how to deal with *algorithm–processor* pairs. For example, should it be possible to encode these pairs with natural numbers in order to enable their rigorous, or even metatheoretical, treatment?[3] Should the execution time (e.g., number of steps) of the algorithm be easily derivable from the description of the algorithm, the input data, and the processor?

5.1.2 The Need for Formalization

The "big" problem was this: How does one answer the question "Is there an algorithm that solves a given problem?" when it is not clear *what* an algorithm is?

Fig. 5.1 To decide whether something exists we must first understand what it should be

To prove that there *is* an algorithm that solves the problem, it sufficed to construct *some* candidate recipe and show that the recipe meets all the conditions (i.e., the recipe has a finite number of instructions, which are reasonably difficult, and can be mechanically followed and executed by any processor, be it human or machine, leading it, in finite time, to the solution of the problem). The loose, intuitive understanding of the concept of the algorithm was no obstacle for such a constructive existence proof.

In contrast, proving that there is *no* algorithm for the problem was a much bigger challenge. A non-existence proof should reject *every* possible recipe by showing that it does not meet all the conditions necessary for an algorithm to solve the problem. However, to accomplish such a proof, a *characterization* of the concept of the algorithm was needed. In other words, a property had to be found such that *all algorithms* and *algorithms only* have this property. Such a property would then be characteristic of algorithms. In addition, a precise and rigorous definition of the processor, i.e., the environment capable of executing algorithms, had to be found. Only then would the necessary condition for proving the non-existence of an algorithm be fulfilled. Namely, having the concept of the algorithm characterized, one would be in a position to systematically (i.e., with mathematical methods) eliminate all the

---

[3] This idea was inspired by *Gödel numbers*, introduced in his *First Incompleteness Theorem*.

infinitely many possible recipes by showing that none of them could possibly fulfill the conditions necessary for an algorithm to solve the problem.

A definition that formally characterizes the basic notions of algorithmic computation (i.e., the algorithm and its environment) is called a **model of computation**.

NB *From now until p. 99 and then on p. 104, we will use quotation marks to refer to the then-intuitive understanding of the notions of the algorithm, computation, and computable function. Thus, "algorithm", "computation", and "computable" function. So Hilbert asked to construct an "algorithm" D_{Entsch} that would answer the question $\underset{M}{\vdash}?F$ for the arbitrary formula $F \in M$.*

5.2 Models of Computation

In this section we will describe the search for an appropriate model of computation. The search started in 1930. The goal was to find a model of computation that would characterize the notions of "algorithm" and "computation". Different ideas arose from the following question:

What could a model of computation take as an example?

On the one hand, it was obvious that man is capable of complex "algorithmic computation", yet there was scarcely any idea how he does this. On the other hand, while the operation of mechanical machines of the time was well understood, it was far from complex, human-like "algorithmic computation".

As a result, three attempts were made: modeling the computation after *functions*, after *humans*, and after *languages*. Each direction proposed important models of computation. In this section we will describe them in detail.

5.2.1 Modeling After Functions

The first direction focused on the question

What does it mean when we say that we "compute" the value of a function $f : A \to B$, or when we say that the function is "computable"?

To get to an answer, it was useful to confine the discussion to functions that were as simple as possible. It seemed that such functions were the *total numerical functions* $f : \mathbb{N}^k \to \mathbb{N}$, where $k \geqslant 1$. If f is such a function, then for *any* k-tuple $(x_1, \ldots, x_k)$ of natural numbers, there is a unique natural number called the value of f at $(x_1, \ldots, x_k)$ and denoted by $f(x_1, \ldots, x_k)$. (Later, in Sects. 5.3.3 and 5.3.4, we will see that the restriction to *total* functions had to be relaxed and the discussion extended to *partial* (i.e., total *and non*-total) functions.)

After this, the search for a definition of "computable" total numerical functions began. It was obvious that any such definition should fulfill two requirements:

1. *Completeness Requirement:* The definition should include *all* the "computable" total numerical functions, *and nothing else.*
2. *Effectiveness Requirement:* The definition should make evident, for each such function f, an *effective procedure* for calculating the value $f(x_1, \ldots, x_k)$.

An **effective procedure** was to be a finite set S of instructions such that
   a. each instruction in S is *exact* and expressed by a finite number of symbols;
   b. S can, in practice and in principle, be carried out by a *human*;
   c. S can be carried out *mechanically*, without insight, intuition, or ingenuity;
   d. if carried out without error, S yields the *desired* result in *finitely* many steps.

Thus, "computable" functions were usually called **effectively calculable**.

The first requirement asked for a *characterization* of the intuitively computable total numerical functions. Only if the second requirement was fulfilled would the defined functions be considered *algorithmically* computable. Notice that although the notion of the effective procedure is a refinement of the intuitive notion of the algorithm (see Definition 1.1 on p. 4), it is still an intuitive, informal notion. Of course, an algorithm for f would be an effective procedure disclosed by f's definition.

Definitions were proposed by Gödel and Kleene (μ-recursive functions), Herbrand and Gödel (general recursive functions), and Church (λ-definable functions).

μ-Recursive Functions

In the proof of his Second Theorem, Gödel introduced numerical functions, the construction of which resembled the derivations of theorems in formal axiomatic systems and their theories. More precisely, Gödel fixed three simple *initial functions*, $\zeta : \mathbb{N} \to \mathbb{N}$, $\sigma : \mathbb{N} \to \mathbb{N}$, and $\pi_i^k : \mathbb{N}^k \to \mathbb{N}$ (called the *zero*, *successor*, and *projection* function, respectively), and two *rules of construction* (called *composition* and *primitive recursion*) for constructing new functions from the initial and previously constructed ones.[4] (There are more details in Box 5.1.)

The functions constructed from ζ, σ, and π by finitely many applications of composition and primitive recursion are total and said to be *primitive recursive*. Although Gödel's intention was to use them in proving his *Second Incompleteness Theorem*, they displayed a property much desired at that time. Namely, the construction of a primitive recursive function is also an effective procedure for computing its values. So, a construction of such a function seemed to be the formal counterpart of the "algorithm", and Gödel's definition of primitive recursive functions seemed to be the wished-for definition of the "computable" total numerical functions.

However, Ackermann and others found the total numerical functions called the *Ackermann functions* (see p. 108), which were intuitively computable but not primitive recursive. So Gödel's definition did not meet the *Completeness Requirement*.

---

[4] The resemblance between function construction and theorem derivation is obvious: Initial functions correspond to axioms, and rules of construction correspond to rules of inference.

Fig. 5.2 Kurt Gödel   **Fig. 5.3** Stephen Kleene
(Courtesy: See Preface)   (Courtesy: See Preface)

This deficiency was eliminated in 1936 by Kleene.[5] He added to Gödel's definition a third rule of construction, called the μ-*operation*, which seeks indefinitely through the series of natural numbers for one satisfying a primitive recursive relation. (See Box 5.1.) The functions that can be constructed from ζ, σ, and π by finitely many applications of composition, primitive recursion, *and* the μ-operation are said to be μ-**recursive**.

NB *Kleene assumed that the μ-operation would be applied to construct only total functions, although it could also return* non-*total functions (i.e., those that are undefined for some arguments).*

The class of μ-recursive total functions proved to contain any conceivable intuitively computable total numerical function. So, Gödel-Kleene's definition became a plausible formalization of a "computable" numerical function. Consequently, construction of a μ-recursive function became a plausible formalization of the notion of the "algorithm". All this was gathered in the following model of computation.

Model of Computation (Gödel-Kleene's Characterization):

- *An "algorithm"* is a construction of a μ-recursive function.
- *A "computation"* is a calculation of a value of a μ-recursive function that proceeds according to the construction of the function.
- *A "computable" function* is a μ-recursive total function.

Box 5.1 (μ-Recursive Functions).

Informally, a function is said to be μ-*recursive* if either it is an initial function or it has been constructed from initial or previously constructed functions by a finite application of the three rules of construction. *Remark*: For brevity, we will write in this box $\vec{n}$ instead of $n_1, \ldots, n_k$.

The *initial functions* are:

a. $\zeta(n) = 0$, for every natural n (*Zero* function)
b. $\sigma(n) = n + 1$, for every natural n (*Successor* function)
c. $\pi_i^k(\vec{n}) = n_i$, for arbitrary $\vec{n}$ and $1 \leqslant i \leqslant k$ (*Projection* (or *identity*) function)

---

[5] Stephen Cole Kleene, 1909–1994, American mathematician.

The *rules of construction* are:

1. *Composition.* Let the given functions be $g : \mathbb{N}^m \to \mathbb{N}$ and $h_i : \mathbb{N}^k \to \mathbb{N}$, for $i = 1, \ldots, m$.
   Then the function $f : \mathbb{N}^k \to \mathbb{N}$ defined by

   $$f(\overrightarrow{n}) \stackrel{\text{def}}{=} g(h_1(\overrightarrow{n}), \ldots, h_m(\overrightarrow{n}))$$

   is said to be constructed by *composition* of the functions g and h_i, $i = 1, \ldots, m$.

2. *Primitive Recursion.* Let the given functions be $g : \mathbb{N}^k \to \mathbb{N}$ and $h : \mathbb{N}^{k+2} \to \mathbb{N}$.
   Then the function $f : \mathbb{N}^{k+1} \to \mathbb{N}$ defined by

   $$f(\overrightarrow{n}, 0) \stackrel{\text{def}}{=} g(\overrightarrow{n})$$
   $$f(\overrightarrow{n}, m+1) \stackrel{\text{def}}{=} h(\overrightarrow{n}, m, f(\overrightarrow{n}, m)), \text{ for } m \geqslant 0$$

   is said to be constructed by *primitive recursion* from the functions g and h.

3. μ-*Operation (Unbounded Search).* Let the given function be $g : \mathbb{N}^{k+1} \to \mathbb{N}$.
   Then the function $f : \mathbb{N}^k \to \mathbb{N}$ defined by

   $$f(\overrightarrow{n}) \stackrel{\text{def}}{=} \mu x g(\overrightarrow{n}, x)$$

   is said to be constructed by the μ-*operation* from the function g. Here, the μ-operation is defined as follows: $\mu x g(\overrightarrow{n}, x) \stackrel{\text{def}}{=}$ least $x \in \mathbb{N}$ such that $g(\overrightarrow{n}, x) = 1 \wedge g(\overrightarrow{n}, z)$ is defined for $z = 0, \ldots, x$.
   NB $\mu x g(\overrightarrow{n}, x)$ may be undefined, e.g., when g is such that $g(\overrightarrow{n}, x) \neq 1$ for every $x \in \mathbb{N}$.

The *construction* of a μ-recursive function f is a finite sequence $f_1, f_2, \ldots, f_\ell$, where $f_\ell = f$ and each f_i is either one of the initial functions ζ, σ, π, or is constructed by one of the rules 1,2,3 from its predecessors in the sequence. Taken formally, the construction is a finite sequence of symbols. But there is also a practical side of construction: After fixing the values of the input data, we can mechanically and effectively calculate the value of f simply by following its construction and calculating the values of the intermediate functions. A construction is an algorithm.

Example 5.1. (Addition) Let us construct the function $\text{sum}(n_1, n_2) \stackrel{\text{def}}{=} n_1 + n_2$. We apply the following idea: To calculate $\text{sum}(n_1, n_2)$ we first calculate $\text{sum}(n_1, n_2 - 1)$ and then its successor. The computation involves primitive recursion, which terminates when $\text{sum}(n_1, 0)$ should be calculated. In the latter case, the sum is just the first summand, n_1. In the list below, the left-hand column contains initial and constructed functions needed to implement the idea, and the right-hand column explains, for each function, why it is there or how it was constructed. The function sum is f_5.

1. $\pi_1^1(x)$ to extract its argument and make it available for use
2. $\pi_3^3(x, y, z)$ to introduce the third variable, which will eventually be the result $x + y$
3. $\sigma(x)$ to increment its argument
4. $f_4(x, y, z)$ to increment the third argument; constructed by composition of 3. and 2.
5. $f_5(x, y)$ to compute $x + y$; constructed by primitive recursion from 1. and 4.

Now, the construction of sum is a sequence of functions and information about the rules applied:
$f_1 = \pi_1^1(n_1)$; $f_2 = \pi_3^3(n_1, n_2, n_3)$; $f_3 = \sigma(n_1)$; $f_4(n_1, n_2, n_3)$[rule 1, f_3, f_2]; $f_5(n_1, n_2)$[rule 2, f_1, f_4].
The function f_5 is constructed by primitive recursion from functions f_1 and f_4, so we have:
$f_5(n_1, n_2) \stackrel{5.}{=} f_4(n_1, n_2 - 1, f_5(n_1, n_2 - 1)) \stackrel{4.}{=} f_5(n_1, n_2 - 1) + 1 \stackrel{5.}{=} \ldots \stackrel{4.}{=} f_5(n_1, 0) + n_2 \stackrel{1.}{=} \pi_1^1(n_1) + n_2 = n_1 + n_2$. Given n_1 and n_2, say $n_1 = 2$ and $n_2 = 3$, we can calculate $f(2, 3)$ by following the construction and calculating the values of functions: $f_1 = \pi_1^1(n_1) = 2$; $f_2 = \pi_3^3(n_1, n_2, n_3) = n_3$; $f_3 = \sigma(n_1) = 3$; $f_4(n_1, n_2, n_3) = \sigma(\pi_3^3(n_1, n_2, n_3)) = \sigma(n_3) = n_3 + 1$; $f_5(n_1, n_2) = f_5(2, 3) = f_4(2, 2, f_5(2, 2)) = f_5(2, 2) + 1 = f_4(2, 1, f_5(2, 1)) + 1 = f_5(2, 1) + 2 = f_4(2, 0, f_5(2, 0)) + 2 = f_5(2, 0) + 3 = 2 + 3 = 5$. □

(General) Recursive Functions

In 1931, the then 23-year-old Herbrand[6] investigated how to define the total numerical functions $f : \mathbb{N}^k \to \mathbb{N}$ using systems of equations. Before suffering a fatal accident while mountain climbing in the French Alps, he explained his ideas in a letter to Gödel. We cite Herbrand (with the function names changed):

> If f denotes an unknown function and $g_1, \ldots, g_k$ are known functions, and if the g's and f are substituted in one another in the most general fashions and certain pairs of the resulting expressions are equated, then, if the resulting set of functional equations has one and only one solution for f, f is a recursive function.

Fig. 5.4 Jacques Herbrand
(Courtesy: See Preface)

Gödel noticed that Herbrand did not make clear what the rules for computing the values of such an f would be. He also noted that such rules would be the same for all the functions f defined in Herbrand's way. Thus, Gödel improved on Herbrand's idea in two steps. First, he added two conditions to Herbrand's idea:

- A system of equations must be in *standard* form, where f is only allowed to be on the left-hand side of the equations, and it must appear as

$$f(g_i(\ldots), \ldots, g_j(\ldots)) = \ldots$$

(Note: This is not required for $g_1, \ldots, g_k$, so these may be defined by recursion.)
- A system of equations must guarantee that f is *well defined* (i.e., single-valued).

Let us denote by $\mathscr{E}(f)$ a system of equations that fulfills the two conditions and defines a function $f : \mathbb{N}^k \to \mathbb{N}$. Second, Gödel started to search for the rules by which $\mathscr{E}(f)$ is used to compute the values of f. In 1934, he realized that there are only two such rules:

- In an equation, all occurrences of a variable can be *substituted* by the same number (i.e., the value of the variable).
- In an equation, an occurrence of a function can be *replaced* by its value.

Thus Gödel (*i*) specified the form of equations and required that (*ii*) exactly one equation $f(\ldots) = \ldots$ is deducible by substitution and replacement in $\mathscr{E}(f)$, and (*iii*) the equations giving the values of $g_1, \ldots, g_k$ are defined in a similar way.

A function $f : \mathbb{N}^k \to \mathbb{N}$ for which there exists a system $\mathscr{E}(f)$ Gödel called **general recursive**. Today we call it **recursive**.

---

[6] Jacques Herbrand, 1908–1931, French logician and mathematician.

It seemed that any conceivable intuitively computable total numerical function could be defined and effectively computed in a mechanical fashion by some system of equations of this kind. The *Completeness Requirement* and *Effectiveness Requirement* of such a definition of "computable" functions seemed to be satisfied. So, Herbrand and Gödel's ideas were merged in the following model of computation.

Model of Computation (Herbrand-Gödel's Characterization):

- *An "algorithm"* is a system of equations $\mathscr{E}(f)$ for some function f.
- *A "computation"* is a calculation of a value of a (general) recursive function f that proceeds according to $\mathscr{E}(f)$ and the rules 1, 2.
- *A "computable" function* is a (general) recursive total function.

λ-definable functions

We start with two examples that give us the necessary motivation.

Example 5.2. (Motivation) What is the value of the term $(5-3)*(6+4)$? To get an answer, we first rewrite the term in the prefix form, $*(-(5,3),+(6,4))$, and getting rid of parentheses we obtain $*-5\,3+6\,4$. This sequence of symbols *implicitly represents the result* (the number 20) *by describing a recipe for calculating it*. Let us call this sequence the *initial term*. Now we can compute the result by a series of *reductions* (elementary transformations) of the initial term: $*-5\,3+6\,4 \rightarrow *2+6\,4 \rightarrow *2\,10 \rightarrow 20$. For example, in the first reduction we *applied* $-$ to 5 and 3 and then replaced the calculated *subterm* $-\,5\,3$ by its value 2. Note that there is a different series of reductions that ends with the same result: $*-5\,3+6\,4 \rightarrow *-5\,3\,10 \rightarrow *2\,10 \rightarrow 20$. $\square$

Example 5.3. (Motivation) Usually, a function $f : \mathbb{N} \to \mathbb{N}$ is defined by the equation $f(x) = [\ldots x \ldots]$, where the right-hand side is an expression containing x. Alternatively, we might define f by the expression $f = \lambda x.[\ldots x \ldots]$, where λx would by convention indicate that x is a variable in $[\ldots x \ldots]$. Then, instead of writing $f(x), x = a$, we could write $\lambda x.[\ldots x \ldots]a$, knowing that a should be substituted for each occurrence of x in $[\ldots x \ldots]$. For example, instead of writing $f(x) = x^y$ and $g(y) = x^y$, which are two different functions, we would write $\lambda x.x^y$ and $\lambda y.x^y$, respectively. Instead of writing $f(x), x=3$ and $g(y), y=5$ we would write $(\lambda x.x^y)3$ and $(\lambda y.x^y)5$, which would result in 3^y and x^5, respectively. Looking now at x^y as a function of two variables, we would indicate this by $\lambda xy.x^y$. Its value at $x = 3, y = 5$ would then be $((\lambda xy.x^y)3)5 = (\lambda y.3^y)5 = 3^5$. $\square$

During 1931–1933, Church[7] proposed, based on similar ideas, a model of computation called the λ-*calculus*. We briefly describe it. (See Box 5.2 for the details.)

Let f be a function and $a_1, \ldots, a_n$ its arguments. Each a_i can be a number or another function with its own arguments. Thus, functions can nest within other functions. Church proposed a way of describing f and $a_1, \ldots, a_n$ as a finite sequence of symbols that *implicitly represents the value* $f(a_1, \ldots, a_n)$ *by describing a recipe for calculating it*. He called this sequence the *initial λ-term*. The result $f(a_1, \ldots, a_n)$ is

---

[7] Alonzo Church, 1903–1995, American mathematician and logician.

computed by a systematic transformation of the initial λ-term into a *final* λ-term that *explicitly* represents the value $f(a_1,\ldots,a_n)$. The transformation is a series of elementary transformations, called *reductions*. Informally, a reduction of a λ-term applies one of its functions, say g, to g's arguments, say $b_1,\ldots,b_m$, and replaces the λ-terms representing g and $b_1,\ldots,b_m$ with the λ-term representing $g(b_1,\ldots,b_m)$.

Fig. 5.5 Alonzo Church
(Courtesy: See Preface)

Generally, there are several different transformations of a λ-term. However, Church and Rosser[8] proved that the final λ-term is practically independent of the order in which the reductions are made; specifically, the final λ-term, *when it exists*, is defined up to the renaming of its variables.

Church called functions that can be defined as λ-terms, λ-**definable**. It seemed that any conceivable intuitively computable total numerical function was λ-definable and could effectively be calculated in a mechanical manner. So this definition of "computable" numerical functions seemed to fulfill the *Completeness* and *Effectiveness Requirements*. Thus, Church proposed the following model of computation.

Model of Computation (Church's Characterization):

- *An "algorithm"* is a λ-term.
- *A "computation"* is a transformation of an initial λ-term into a final one.
- *A "computable" function* is a λ-definable total function.

Box 5.2 (λ-calculus).

Let $f,g,x,y,z,\ldots$ be variables. Well-formed expressions will be called λ-terms.

A λ-*term* is a well-formed expression defined inductively as follows:

a.   a variable is a λ-term (called an *atom*);
b.   if M is a λ-term and x a variable, then $(\lambda x. M)$ is a λ-term (built from M by *abstraction*);
c.   if M and N are λ-terms, then (MN) is a λ-term (called the *application of M to N*).

---

[8] John Barkley Rosser, 1907–1989, American mathematician and logician.

Remarks. Informally, abstraction exposes a variable, say x, that occurs (or even doesn't occur) in a λ-term M and "elevates" M to a *function* of x. This function is denoted by $\lambda x. M$ and we say that x is now *bound* in M by λx. A variable that is not bound in M is said to be *free* in M. A λ-term can be interpreted as a function or an argument of another function. Thus, in general, the λ-term (MN) indicates that the λ-term M (which is interpreted as a function) can be *applied* to the λ-term N (which is interpreted as an argument of M). We also say that M can *act* on N. By convention, the application is a left-associative operation, so (MNP) means $((MN)P)$. However, we can override the convention by using parentheses. The outer parentheses are often omitted, so MN means (MN). Any λ-term of the form $(\lambda x. M)N$ is called a β-*redex* (for reasons to become known shortly). A λ-term may contain zero or more β-redexes.

λ-terms can be transformed into other λ-terms. A transformation is a series of one-step transformations called β-*reductions* and denoted by $\rightarrow_\beta$. There are two rules to do a β-reduction:

1. α-*conversion* (denoted $\mapsto_\alpha$) renames a bound variable in a λ-term;
2. β-*contraction* (denoted $\mapsto_\beta$) transforms a β-redex $(\lambda x. M)N$ into a λ-term obtained from M by substituting N for every bound occurrence of x in M. Stated formally: $(\lambda x. M)N \equiv M[x := N]$.

Remarks. Intuitively, a β-contraction is an actual *application* (i.e., *acting*) of a function to its arguments. However, before a β-contraction is started, we must apply all the necessary α-conversions to M to avoid unintended effects of the β-contraction, such as unintended binding of N's free variables in M. When a λ-term contains no β-redexes, it cannot further be β-reduced. In this case the term is said to be a β-*normal form* (β-*nf*). Intuitively, such a λ-term contains no function to apply.

A *computation* is a transformation of an initial λ-term t_0 with a sequence of β-reductions, that is

$$t_0 \rightarrow_\beta t_1 \rightarrow_\beta t_2 \rightarrow_\beta \cdots$$

If t_i is a member of this sequence, we say that t_0 is β-reducible to t_i and denote this by $t_0 \xrightarrow{*}_\beta t_i$. The computation terminates *if* and when some β-nf t_n is reached. This λ-term is said to be *final*.

A non-final λ-term may have several β-redexes. Each of them is a candidate for the next β-contraction. Since the selection of this usually affects the subsequent computation, we see that, in general, there exist different computations starting in t_0. Hence the questions: "Which of the possible computations is the 'right' one? Which of the possibly different final λ-terms is the 'right' result?" Fortunately, Church and Rosser proved that the order of β-reductions does not matter that much. Specifically, the final λ-term—when it exists—is defined up to α-conversion (i.e., up to the renaming of its bound variables). In other words: If different computations terminate, they return practically equal results. This is the essence of the following theorem.

Theorem. (Church-Rosser) If $t_0 \xrightarrow{*}_\beta U$ and $t_0 \xrightarrow{*}_\beta V$, then there is W such that $U \xrightarrow{*}_\beta W$ and $V \xrightarrow{*}_\beta W$.

Since we may *chose* a terminating computation, we can *standardize computations* by fixing a rule for selecting β-redexes. For example, we may *always immediately reduce the leftmost β-redex*. So the initial term and the fixed rule make up a determinate *algorithm* for carrying out the computation.

Natural numbers are represented by (rather unintuitive) β-nfs c_i, called *Church's numerals*:

$$\begin{aligned}
c_0 &\equiv \lambda f x. f^0 x &\equiv& \lambda f.(\lambda x. x) \\
c_1 &\equiv \lambda f x. f^1 x &\equiv& \lambda f.(\lambda x. fx) \\
c_2 &\equiv \lambda f x. f^2 x &\equiv& \lambda f.(\lambda x. f(fx)) \\
&\vdots \\
c_n &\equiv \lambda f x. f^n x &\equiv& \lambda f.(\lambda x. f(\ldots f(fx) \ldots))
\end{aligned}$$

A function $f : \mathbb{N}^k \to \mathbb{N}$ is λ-*definable* if there is a λ-term F such that

$$\text{if } f(n_1, \ldots, n_k) = \begin{cases} m & \text{then } F c_{n_1} \ldots c_{n_k} \xrightarrow{*}_\beta c_m; \\ \text{undefined} & \text{then } F c_{n_1} \ldots c_{n_k} \text{ has no } \beta\text{-nf.} \end{cases}$$

Example 5.4. (Addition) The function $\text{sum}(n_1, n_2) = n_1 + n_2$ can be defined in λ-calculus by a λ-term $S \equiv \lambda abfx. af(bfx)$. To prove this, we must check that $Sc_{n_1}c_{n_2} \xrightarrow{*}_\beta c_{n_1+n_2}$ for any n_1, n_2. So we compute: $Sc_{n_1}c_{n_2} \equiv (Sc_{n_1})c_{n_2} \equiv ((\lambda abfx. af(bfx))c_{n_1})c_{n_2} \rightarrow_\beta (\lambda bfx. c_{n_1}f(bfx))c_{n_2} \rightarrow_\beta$
$\rightarrow_\beta \lambda fx. c_{n_1}f(c_{n_2}fx) \equiv \lambda fx. (\lambda fx. f^{n_1}x)f((\lambda fx. f^{n_2}x)fx) \rightarrow_\beta \lambda fx. (\lambda x. f^{n_1}x)((\lambda fx. f^{n_2}x)fx) \rightarrow_\beta$
$\rightarrow_\beta \lambda fx. (\lambda x. f^{n_1}x)((\lambda x. f^{n_2}x)x) \rightarrow_\beta \lambda fx. (\lambda x f^{n_1}x)(f^{n_2}x) \rightarrow_\beta \lambda fx. f^{n_1}f^{n_2}x \equiv \lambda fx. f^{n_1+n_2}x \equiv c_{n_1+n_2}$ $\square$

5.2.2 Modeling After Humans

The second direction in the search for an appropriate model of computation was an attempt to model computation after humans. The idea was to abstract man's activity when he mechanically solves computational problems. A seminal proposal of this kind was made by Turing,[9] whose model of computation was inspired both by human "computation" and a real mechanical device (for details see Sect. 16.3.6). In what follows we give an informal description of the model and postpone rigorous treatment of it to later sections (see Sect. 6.1).

Turing Machine

Turing took the quest for a definition of mechanical computation at face value. It has been suggested that he was inspired by his mother's typewriter. In 1936, the then 24-year-old conceived his own model of computation, which he called the *a-machine* (automatic machine) and also the *logical computing machine*. Today, the model is called the **Turing machine** (or *TM* for short).

Fig. 5.6 Alan Turing
(Courtesy: See Preface)

The Turing machine consists of several components (see Fig. 5.7):

1. a *control unit* (corresponding to the human brain);
2. a potentially infinite *tape* divided into equally sized *cells* (corresponding to the paper used during human "computation");
3. a *window* that can move over any cell and makes it accessible to the control unit (corresponding to the human eye *and* the hand with a pen).

---

[9] Alan Mathison Turing, 1912–1954, British mathematician, logician, and computer scientist.

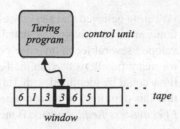

Fig. 5.7 Turing machine
(*a*-machine)

The control unit is always in some *state* from a finite set of states. Two of the states are the *initial* and the *final* state. There is also a *program* in the control unit (corresponding to the "algorithm" that a human uses when "computing" the solution of a given problem). We call it the **Turing program**. Different Turing machines have different Turing programs.

Before the machine is started, the following must be done:

1. an *input word* (i.e., the input data written in an alphabet Σ) is written on the tape;
2. the window is shifted to the beginning of the input word;
3. the control unit is set to the initial state.

From then on, the machine operates independently, in a purely mechanical fashion, step by step, as directed by its Turing program. At each step, the machine reads the symbol from the cell under the window into the control unit and, based on this symbol and the current state of the control unit, does the following:

1. writes a symbol to the cell under the window (while deleting the old symbol);
2. moves the window to one of the neighboring cells or leaves the window as it is;
3. changes the state of the control unit.

The machine *halts* if the control unit enters the final state or if its Turing program has no instruction for the next step.

How can Turing machines calculate the values of numerical functions? The general idea is to associate a Turing machine T with extensional mapping of T's inputs to T's outputs. More specifically, take any Turing machine T and any $k \in \mathbb{N}$ and then *define* a function $f_T^{(k)}$ as follows:

If the input word to T represents natural numbers $n_1, \ldots, n_k \in \mathbb{N}$, and T halts, and after halting the tape contents make up a word over Σ that represents a natural number, say m, then let m be the value of the function $f_T^{(k)}$ at $n_1, \ldots, n_k$; that is,

$$f_T^{(k)}(n_1, \ldots, n_k) \overset{\text{def}}{=} m.$$

We can view $f_T^{(k)}$ as a function that maps k words over Σ into a word over Σ. The values of $f_T^{(k)}$ can be mechanically computed simply by executing the Turing program of T. Today, we say that the function $f_T^{(k)}$ is **Turing-computable**. Actually, *any* function f for which there exists a Turing machine T such that $f = f_T^{(k)}$, for some k, is said to be Turing-computable.

Turing believed that any conceivable intuitively computable total numerical function can be effectively calculated with some Turing machine. To show this, he developed several techniques for constructing Turing machines and found machines for many functions used in mathematics. Thus, it seemed that "computable" functions could be identified with Turing-computable total functions. In sum, it seemed that this definition of a "computable" function would fulfill the *Completeness* and *Effectiveness Requirements*. Hence, the following model of computation.

Model of Computation (Turing's Characterization):

- *An "algorithm"* is a Turing program.
- *A "computation"* is an execution of a Turing program on a Turing machine.
- *A "computable" function* is a Turing-computable total function.

Box 5.3 (Memorizing During Computation).

It seems that there is an important difference between a human and a Turing machine. A human can see only a finite portion of the scribbles under his or her nose, yet is able to remember some of the previously read ones and use them in the "computation". It seems that the Turing machine's control unit—lacking explicit storage—does not allow for this.

However, this is not true. We will see later that the control unit can *simulate* finite storage. The basic idea is that the control unit memorizes the symbol that has been read from the tape by changing to a state that corresponds to the symbol. This enables the Turing machine to take into account the memorized symbol during its operation. In order to implement this idea, the states and instructions of the Turing machine have to be appropriately defined and interpreted during the execution. (Further details will be given in Sects. 6.1.2 and 6.1.3.)

Example 5.5. (Addition) That sum(n_1, n_2) is Turing-computable is shown in Example 6.1(p.115).□

5.2.3 Modeling After Languages

The third direction in the search for an appropriate model of computation focused on modeling after languages. The idea was to view human mathematical activity as the transformation of a sequence of words (description of a problem) into another sequence of words (description of the solution of the problem), where the transformation proceeds according to certain rules. Thus, the "computation" was viewed as a sequence of steps that gradually transform a given input expression into an output expression. The rules that govern the transformation are called **productions**. Each production describes how a current expression should be partitioned into fixed and variable sub-expressions and, if the partition is possible, how the sub-expressions must be changed and reordered to get the next expression. So, productions are of the form

$$\alpha_0 x_1 \alpha_1 x_2 \ldots \alpha_{n-1} x_n \alpha_n \rightarrow \beta_0 x_{i_1} \beta_1 x_{i_2} \ldots \beta_{m-1} x_{i_m} \beta_m$$

where α_i, β_j are fixed sub-expressions and x_k are variables whose values are arbitrary sub-expressions of the current expression. Then, any finite set of productions is called a **grammar**. Based on these ideas, Post and Markov proposed two models of computation, called the *Post machine* and the *Markov algorithms*.

The Post Machine

In the 1920s, Post investigated the decidability of logic theories. He viewed the derivations in theories as man's language activity and, for this reason, developed his *canonical systems*. (We will return to these in Sect. 6.3.2.) The research on canonical systems forced Post to invent an abstract machine, now called the **Post machine**.

Fig. 5.8 Emil Post
(Courtesy: See Preface)

The Post machine over the alphabet Σ is founded on a pair (G, Q), where G is a directed graph and Q is a queue. The intention is that Q will receive input data (encoded in Σ) and G will operate on Q, eventually leaving the encoded result in it.

We set $\Sigma = \{0, 1\}$ since symbols of larger alphabets can be encoded in Σ; and we introduce a symbol # to mark, when necessary, distinguished places in Q's contents.

To implement the intention, we must define the basic instructions and how they will be executed in G. Let $V = \{v_1, \ldots, v_n\}$ be the set of G's vertices, for some $n \in \mathbb{N}$, and let $A \subseteq V \times V$ be the set of G's arcs. Each vertex $v_i \in V$ contains an instruction, generically denoted by $\text{INST}(v_i)$, and each arc $(v_i, v_j) \in A$ can pass a signal from v_i to v_j. By definition, v_1 has no incoming arcs; this is the *initial vertex*. Some, potentially none, of the other vertices have no outgoing arcs; these are the *final vertices*. Each of the remaining vertices has at least one incoming arc and, depending on its instruction, either one or four outgoing arcs.

The instructions are of four kinds. The first kind, the START instruction, is contained in (and only in) the initial vertex v_1. This instruction executes as follows: (1) it copies a given *input word* from G's environment into the queue Q and (2) it *triggers* the instruction $\text{INST}(v_j)$ by sending a signal via the arc $(v_1, v_j) \in A$. Of the second kind are the instructions ACCEPT and REJECT. These are contained in (and only in) the final vertices. When an accept (or reject) instruction is executed, the input word is recognized as *accepted* (or *rejected*); thereafter the computation *halts*. The third kind of instruction is ENQ(z), where $z \in \Sigma \cup \{\#\}$. When executed in a vertex v_i, the instruction (1) attaches the symbol z to the end of Q, i.e., changes $Q = (x_1 \ldots x_m)$ to $Q = (x_1 \ldots x_m z)$, and (2) triggers INST(v_j) via $(v_i, v_j) \in A$. The fourth kind of instruction is DEQ(p,q,r,s), where $2 \leqslant p,q,r,s \leqslant n$. When executed in v_i, it (1) checks Q for emptiness and in this case triggers INST(v_p) via $(v_i, v_p) \in A$; otherwise, it removes the head of Q, i.e., changes $Q = (x_1 x_2 \ldots x_m)$ to $Q = (x_2 \ldots x_m)$, and (2) triggers INST(v_q), INST(v_r), or INST(v_s) via the corresponding arc, if the removed head was 0, 1, or #, respectively.

The graph G augmented with the above instructions and rules of execution is called the **Post program.**[10] (See Fig. 5.9.)

Fig. 5.9 A Post program

To develop the above ideas into a model of computation, the Post program must be supplemented by a supporting environment. The result is a structural view of the Post machine (see Fig. 5.10) that bears some resemblance to the Turing machine.

Fig. 5.10 Post machine

But there are differences too. The Post machine has a *control unit* that can store a Post program and execute its instructions; a potentially infinite *read-only tape* divided into equally sized *cells*; a *window* that can move to the right over any cell thus making it accessible to the control unit for reading; a *queue* that can store the input word (with input data), or the current word (with the intermediate results), or the final word (with the final result).

---

[10] In the late twentieth century, a similar model of computation, the *data-flow graph*, was used. Here, vertices contain usual instructions such as arithmetical, logic, I/O, and jumps; arcs pass intermediate results that act as triggering signals; the instruction in a vertex is triggered as soon as the last awaited operand appears on an incoming arc; and several instructions can be triggered simultaneously.

Before the machine is started, the following must be done: An input word (i.e., input data encoded in the alphabet Σ) is written on the tape; the window is positioned over the first symbol of the input word; and the queue is emptied.

The computation starts by triggering the instruction START of the Post program. From then on, the machine operates independently, in a purely mechanical fashion, step by step, as directed by its Post program. At each step, the previously triggered instruction executes and triggers another instruction. The machine *halts* if and when one of the instructions ACCEPT or REJECT has executed.

During the computation the contents of the queue gradually transform from the input word to a final word, the result of the computation. Note that the computation may never halt (when none of the final vertices is ever reached).

Can Post machines compute the values of functions, say numerical functions? Take any Post machine P and any $k \in \mathbb{N}$, and *define* a function $f_P^{(k)}$ as follows:

If the input word to P represents natural numbers $n_1, \ldots, n_k$, and P halts, and after halting the queue contains a word over Σ that represents a natural number, say m, then let m be the value of the function $f_P^{(k)}$ at $n_1, \ldots, n_k$; that is

$$f_P^{(k)}(n_1, \ldots, n_k) \overset{\text{def}}{=} m.$$

We can view $f_P^{(k)}$ as a function that maps k words over Σ into a word over Σ. The values of $f_P^{(k)}$ can be mechanically computed by P. In general, we say that a function f is **Post-computable** if there is a Post machine P such that $f = f_P^{(k)}$, for some k. This brings us to the following model of computation.

Model of Computation (Post's Characterization):

- *An "algorithm"* is a Post program.
- *A "computation"* is an execution of a Post program on a Post machine.
- *A "computable" function* is a Post-computable total function.

Markov Algorithms

Later, in the Soviet Union, similar reasoning was applied by Markov.[11] In 1951, he described a model of computation that is now called the **Markov algorithm**.

A Markov algorithm is a finite sequence M of *productions*

---

[11] Andrey Andreyevič Markov, Jr., 1903–1979, Russian mathematician.

$$\alpha_1 \to \beta_1$$
$$\alpha_2 \to \beta_2$$
$$\vdots$$
$$\alpha_n \to \beta_n$$

where α_i, β_i are words over a given alphabet Σ. The sequence M is also called the *grammar*. A production $\alpha_i \to \beta_i$ is said to be *applicable* to a word w if α_i is a subword of w. If such a production is actually applied to w, it transforms w so that it replaces the leftmost occurrence of α_i in w with β_i.

An execution of a Markov algorithm is a sequence of steps that gradually transform a given *input word* via a sequence of *intermediate words* into some *output word*. At each step, the last intermediate word is transformed by the first applicable production of M. Some productions are said to be *final*. If the last applied production was final, or if there was no production to apply, then the execution halts and the last intermediate word is the *output word*.

Fig. 5.11 Andrey Markov
(Courtesy: See Preface)

Markov algorithms can be used to calculate the values of numerical functions. Let M be any Markov algorithm and $k \in \mathbb{N}$. Then *define* a function $f_M^{(k)}$ as follows:

If the input word to M represents natural numbers $n_1, \ldots, n_k$, and M halts, and after halting the output word represents a natural number, say m, then let m be the value of the function $f_M^{(k)}$ at $n_1, \ldots, n_k$; that is

$$f_M^{(k)}(n_1, \ldots, n_k) \stackrel{\text{def}}{=} m.$$

We can view $f_M^{(k)}$ as a function from $(\Sigma^*)^k$ to Σ^* whose values can be mechanically computed by M. In general, we say that a function f is **Markov-computable** if there is a Markov algorithm M such that $f = f_M^{(k)}$, for some k.

In sum, we can define the following model of computation.

Model of Computation (Markov's Characterization):

- *An "algorithm"* is a Markov algorithm (grammar).
- *A "computation"* is an execution of a Markov algorithm.
- *A "computable" function* is a Markov-computable total function.

5.2.4 Reasonable Models of Computation

Although the described models of computation are completely different, they share an important property: They are *reasonable*. Clearly, we cannot prove this because reasonableness is not a formal notion. But we can give some facts in support.

First we introduce new notions. An *instance* of a model of computation is called the *abstract computing machine*. Examples of such a machine M are:

- a particular construction of a μ-recursive function (with the rules for using the construction to calculate the function values);
- a particular system of equations $\mathscr{E}(f)$ (with substitution and replacement rules);
- a particular λ-term (with α-conversion and β-reduction rules);
- a particular Turing program (in a Turing machine);
- a particular Post program (in a Post machine); and
- a particular Markov grammar (with the rules of production application).

As the computation on an abstract computing machine M goes on, the status (e.g., contents, location, value, state) of each component of M may change. At any step of the computation, the statuses of the relevant components of M make up the so-called *internal configuration* of M at that step. Informally, this is a snapshot of M at a particular point of its computation.

Now we return to the reasonableness of the models. For any abstract computing machine M belonging to any of the proposed models of computation, it holds that

1. M is of *limited capability*. This is because:

   a. it has finitely many different basic instructions;
   b. each basic instruction takes at least one step to be executed;
   c. each basic instruction has a finite effect, i.e., the instruction causes a limited change in the current internal configuration of M.

2. There is a *finite-size description of M*, called the *code of M* and denoted by $\langle M \rangle$.
3. The code $\langle M \rangle$ is *effective* in the sense that, given an internal configuration of M, the code $\langle M \rangle$ enables us to "algorithmically construct" the internal configuration of M after the execution of an instruction of M.
4. M is *unrestricted*, i.e., it has *potentially infinite resources* (e.g., time and space).

The proposed models of computation are *reasonable* from the standpoint of *Computability Theory*, which is concerned with the questions "What is an algorithm? What is computation? What can be computed?" Firstly, this is because these models do not offer unreasonable computational power (see item 1 above). Secondly, the answers to the questions are not influenced by any limitation of the computing resources except that a finite computation must use only a finite amount of each of the resources (see item 4). Hence, any problem that cannot be computed even with unlimited resources will remain such if the available resources become limited.[12]

---

[12] In *Computational Complexity Theory*, a large special part of *Computability Theory*, the requirements (1c) and (4) are stiffer: In (1c) the effect of each instruction must be *reasonably large* (not just finite); and in (4) *the resources are limited*, so their consumption is highly important.

5.3 Computability (Church-Turing) Thesis

The diversity of the proposed models of computation posed the obvious question: "Which model (if any) is the *right* one?" Since each of the models fulfilled the *Effectiveness Requirement*, the issue of effectiveness was replaced by the question "Which model is the most natural and appealing?" Of course, the opinions were subjective and sometimes different. As with the *Completeness Requirement*, it was not obvious whether the requirement was *truly* fulfilled by *any* of the models. In this section we will describe how the *Computability Thesis*, which in effect states that *each* of the proposed models fulfills this requirement, was born. (A systematic and detailed account of the origins and evolution of the thesis is given in Chap. 16.)

5.3.1 History of the Thesis

Church. By 1934, Kleene managed to prove that every conceivable intuitively computable total numerical function was λ-definable. For this reason, in the spring of 1934, Church conjectured that "computable" numerical functions are exactly λ-definable total functions. In other words, Church suspected that the intuitive concepts of algorithm and computation are appropriately formalized by his model of computation. He stated this in the following thesis:

Church Thesis. *"algorithm"* $\longleftrightarrow$ λ-term

Church presented the thesis to Gödel, the authority in mathematical logic of the time, but he rejected it as unsatisfactory. Why? At that time, Gödel was reflecting on the relation between intuitively computable functions and (general) recursive functions. He suspected that the latter might be a formalization of the former, yet he was well aware of the fact that such an equivalence could not possibly be proved, because it would equate two concepts of which one is formal (i.e., precisely defined) and the other is informal (i.e., intuitive). In his opinion, researchers needed to continue analyzing the intuitive concepts of algorithm and computation. Only after their intrinsic components and properties were better understood would it make sense to propose a thesis of this kind.

Shortly after, Kleene, Church, and Rosser proved the equivalence between the λ-definable functions and the (general) recursive functions in the sense that every λ-definable total function is (general) recursive total, and vice versa. Since λ-calculus (being somewhat unnatural) was not well accepted in the research community, Church restated his thesis in the terminology of the equivalent (general) recursive functions and, in 1936, finally published it.

This, however, did not convince Gödel (and Post). In their opinion, the equivalence of the two models still did not indicate that they fulfilled the *Completeness Requirement*, i.e., that they fully captured the intuitive concepts of algorithm and computation.

Turing. Independently, Turing pursued his research in England. He found that every conceivable intuitively computable total numerical function was Turing-computable. He also proved that the class of Turing-computable functions remains the same under many generalizations of the Turing machine. So, he suspected that "computable" functions are exactly Turing-computable total functions. That is, he suspected that the intuitive concepts of algorithm and computation are appropriately formalized by his model of computation. In 1936, he published his thesis.

> **Turing Thesis.** *"algorithm"* ⟷ *Turing program*

Gödel accepted the Turing machine as the model of computation that convincingly formalizes the concepts of "algorithm" and "computation". He was convinced[13] by the simplicity and generality of the Turing machine, its mechanical working, its resemblance to human activity when solving computational problems, Turing's reasoning and analysis of intuitively computable functions, and his argumentation that such functions are exactly Turing-computable total functions.

In 1937, Turing also proved that Church's and his model are equivalent in the sense that what can be computed by one can also be computed by the other. That is,

$$\lambda\text{-}definable \iff Turing\text{-}computable$$

He proved this by showing that each of the two models can simulate the basic instructions of the other.

Because of all of this, the remaining key researchers accepted the Turing machine as the most appropriate model of computation. Soon, the Turing machine met general approbation.

5.3.2 The Thesis

As the two theses were equivalent, they were merged into the *Church-Turing Thesis*. More recently, this has also been given the neutral name *Computability Thesis* (CT).

> **Computability Thesis.** *"algorithm"* ⟷ *Turing program (or equivalent model)*

Gradually, it was proved that other models of computation are equivalent to the Turing machine or some other equivalent model. This *confluence of ideas*, the equivalence of diverse models, strengthened belief in the *Computability Thesis*.[14]

---

[13] If the reader has doubts, he or she may compare Examples 5.1 (p.83), 5.4 (p.88), and 6.1 (p.115).

[14] Interestingly, a similar situation arose in physics ten years before. To explain the consequences of the quantization of energy and the unpredictability of events in microscopic nature, two theories were independently developed: matrix mechanics (Werner Heisenberg, 1925) and wave mechanics (Erwin Schrödinger, 1926). Though the two theories were completely different, Schrödinger proved that they are equivalent in the sense that physically they mean the same. This and their capability of accurately explaining and predicting physical phenomena strengthened belief in the quantum explanation of microscopic nature.

If the *Computability Thesis* is in truth correct, then, in principle, this cannot be proved. (One should give rigorous arguments that an informal notion is equivalent to another, formal one—but this was the very goal of the formalization described above.) In contrast, if the thesis is wrong, then there may be a proof of this. (One should conceive an intuitively computable function and prove that the function is not Turing-computable.) However, until now no one has found such a proof. Consequently, most researchers believe that the thesis holds.[15]

The *Computability Thesis* proclaimed the following formalization of intuitive basic concepts of computing:

Formalization. *Basic intuitive concepts of computing are formalized as follows:*

$$\text{"algorithm"} \longleftrightarrow \textit{Turing program}$$

$$\text{"computation"} \longleftrightarrow \textit{execution of a Turing program on a Turing machine}$$

$$\text{"computable" function} \longleftrightarrow \textit{Turing-computable total function}$$

NB *The Computability Thesis established a bridge between the intuitive concepts of "algorithm", "computation", and "computability" on the one hand, and their formal counterparts defined by models of computation on the other. In this way it finally opened the door to a mathematical treatment of these intuitive concepts.*

The situation after the acceptance of the *Computability Thesis* is shown in Fig. 5.12.

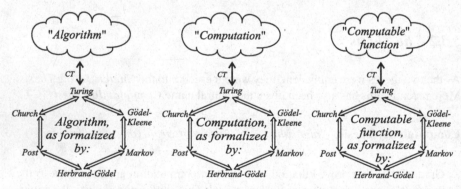

Fig. 5.12 Equivalent formalizations of intuitive concepts. By the *Computability Thesis* (CT), they also adequately capture the intuitive concepts

The reader will find an extended treatise of the *Computability Thesis* in Chap. 16.

---

[15] To this day, several new models of computation have been proposed and proved to be equivalent to the Turing machine. Some of the notable ones are the *register machine*, which, in various variants such as RAM and RASP, epitomizes modern computers (we will describe RAM in Sect. 6.2.7); the *cellular automaton* (von Neumann, Conway, Wolfram), which is inspired by the development of artificial organisms; and *DNA computing* (Adleman), which uses DNA molecules to compute.

At last we can refine Definition 1.1 and Fig. 1.2 (p. 4) to Definition 5.1 and Fig. 5.13.

Fig. 5.13 A Turing program directs a Turing machine to compute the solution to the problem

Definition 5.1. (Algorithm Formally) The **algorithm** for solving a problem is a Turing program that leads a Turing machine from the input data of the problem to the corresponding solution.

NB *Since the concepts of "algorithm" and "computation" are now formalized, we no longer need to use quotation marks to distinguish between their intuitive and formal meanings. In contrast, with the concept of "computable" function we will be able to do this only after we have clarified which functions (total or non-total too) we must talk about. This is the subject of the next two subsections.*

5.3.3 Difficulties with Total Functions

Recall that in the search for a definition of "computable" functions some researchers pragmatically focused on *total* numerical functions (see Sect. 5.2.1). These are the functions $f : \mathbb{N}^k \to \mathbb{N}$ that map *every* k-tuple of natural numbers into a natural number. However, it became evident that the restriction to total functions was too severe. There were several reasons for this.

1. **The Notion of Totality.** Focusing on total functions tacitly assumed that we can decide, for arbitrary f, whether f is a total function. This would be necessary, for example, after each application of the μ-operation in μ-recursive function construction (see Box 5.1, p. 82).

   But finding out whether f is total may not be a finite process; in the extreme case, we must check individually, for each k-tuple in $\mathbb{N}^k$, whether f is defined. (Only later, in Sect. 9.4, will we be able to prove this. For starters, see Box 5.4 for a function which is potentially of this kind.)

   Unfortunately, this meant that the formalization of the concept of "computable" function had a serious weak point: it was founded on a notion (totality) that is disputable in view of the finitist philosophy of mathematics.

Box 5.4 (Goldbach's Conjecture).

In 1742, Goldbach[16] proposed the following conjecture G:

$G \equiv$ *"Every even integer greater than 2 is the sum of two primes."*

For example, $4 = 2+2$; $6 = 3+3$; $8 = 3+5$; $10 = 3+7$. But Goldbach did not prove G. What is more, in spite of many attempts of a number of prominent mathematicians, to this date, the conjecture remains an open problem. In particular, no pattern was found such that, for every natural n, $4 + 2n = p(n) + q(n)$ and $p(n)$, $q(n)$ are primes. (See discussion in Box 2.5, p. 21.)

Let us now define the Goldbach function $g : \mathbb{N} \to \mathbb{N}$ as follows:

$$g(n) \overset{\text{def}}{=} \begin{cases} 1, & \text{if } 4 + 2n \text{ is the sum of two primes;} \\ \text{undefined}, & \text{otherwise.} \end{cases}$$

Is g total? This question is equivalent to the question of whether G holds. Yet, all attempts to answer either of the questions have failed. What is worse, there is a possibility that G is one of the undecidable *Truths* of arithmetic, whose existence was proved by Gödel (see Sect. 4.2.3).

Does this mean that, lacking any pattern $4 + 2n = p(n) + q(n)$, the only way of finding out whether g is total is by checking, for each natural n individually, whether $4 + 2n$ is the sum of two primes? Indeed, in 2012, G was verified by computers for n up to $4 \cdot 10^{18}$. However, if g is in truth total, this is not a finite process, and will never return the answer "g is total." In sum: For certain functions we might not be able to decide whether or not they are total.

2. **Diagonalization.** An even more serious consequence of the restriction to total functions was discovered by a method called *diagonalization*.

By the *Computability Thesis*—as stated on p. 98—every intuitively computable function is μ-recursive total (see p. 82). But is it really so? It was soon noticed that there are *countably* infinitely many μ-recursive functions (as many as their constructions), whereas it was known that there are *uncountably* many numerical functions (see Appendix A, p. 368). So there are numerical functions that are not μ-recursive—and *a fortiori* not μ-recursive total. But the question was: Is there an *intuitively computable* numerical function that is not μ-recursive total? If there is, how do we find it when all intuitively computable functions that researchers have managed to conceive have turned out to be μ-recursive total?

Success came with a method called *diagonalization*. We will describe this method in depth later, in Sect. 9.1, but we can still briefly describe the idea of the construction. Using diagonalization, a certain numerical function g was defined in a somewhat unusual way (typical of diagonalization), but it was still evident how one could effectively calculate its values. So, g was intuitively computable. But the definition of g was such that it implied a contradiction *if g was supposed to be μ-recursive and total*. Hence, g could not be both μ-recursive and total!

It turned out that to prevent the contradiction it sufficed to omit only the supposition that g was total. So, g was both μ-recursive *non*-total and intuitively computable! This was an indication that the search for a formalization of intuitively computable functions should be extended to all functions, total and *non*-total.

---

[16] Christian Goldbach, 1690–1764, German mathematician.

Let us describe what has been said in more detail:

a. Constructions of μ-recursive total functions are finite sequences of symbols over a finite alphabet. The set of all such constructions is well-ordered in *shortlex order*.[17] Thus, given arbitrary $n \in \mathbb{N}$, the nth construction is precisely defined (in shortlex order). Let f_n denote the μ-recursive total function whose construction is nth in this order.

b. Define a function $g : \mathbb{N}^{k+1} \to \mathbb{N}$ as follows:

$$g(n, a_1, \ldots, a_k) \overset{\text{def}}{=} f_n(a_1, \ldots, a_k) + 1.$$

c. The function g is *intuitively computable!* Namely, the intuitive algorithm for calculating its values for arbitrary $(a_1, \ldots, a_k) \in \mathbb{N}^k$ is straightforward:
  • find the nth construction; then
  • use the construction to calculate $f_n(a_1, \ldots, a_k)$; and then
  • add 1 to the result.

d. Is the function g μ-*recursive and total?* Suppose it is. Then, there is a construction of g, so g is one of the functions $f_1, f_2, \ldots$, i.e., $f_m = g$ for some $m \in \mathbb{N}$.

e. Let us now focus on the value $g(m, m, \ldots, m)$. The definition of g gives $g(m, m, \ldots, m) = f_m(m, \ldots, m) + 1$. On the other hand, we have (by d above) $f_m(m, \ldots, m) = g(m, m, \ldots, m)$. From these two equations we obtain

$$g(m, m, \ldots, m) = g(m, m, \ldots, m) + 1$$

and then $0 = 1$, a contradiction. Thus, the supposition that g is μ-recursive and total implies a contradiction. In sum, g *cannot be both* μ-*recursive and total*.

The conclusion is inescapable: There are *intuitively computable* numerical functions that are *not both* μ-*recursive and total!* At first, researchers thought that the definition of μ-recursive functions (see Box 5.1, p. 82) should be extended by additional initial functions and/or rules of construction. But they realized that this would not prevent the contradiction: *Any extended definition* of μ-recursive functions that would be used to construct only *total* μ-recursive functions would lead in the same way as above to the function g and the contradiction $0 = 1$.

Did this refute the *Computability Thesis*? Fortunately not so; the thesis only had to be slightly adapted. Namely, it was noticed that no contradiction would have occurred if the construction of *all* (total and non-total) μ-recursive functions had been allowed. Why? If $f_1, f_2, \ldots$ was the sequence of *all* μ-recursive functions, then $g = f_m$ could be non-total. Since the value $g(m, m, \ldots, m)$ could be undefined, the equation $g(m, m, \ldots, m) = g(m, m, \ldots, m) + 1$ would not inevitably be contradictory, as undefined plus one is still undefined.

---

[17] Sort the constructions by increasing length, and those of the same length lexicographically.

3. **Experience.** Certain well-known numerical functions were not total. For example, rem : $\mathbb{N}^2 \to \mathbb{N}$, defined by rem$(m,n)$ = "remainder from dividing m by n." This function is not total because it is not defined for pairs $(m,n) \in \mathbb{N} \times \{0\}$. Nevertheless, rem could be treated as intuitively computable because of the following algorithm: If $n > 0$, then compute rem(m,n) and halt; otherwise, return a warning (e.g., $-1 \notin \mathbb{N}$). Notice that the algorithm returns the value whenever rem is defined; otherwise, it warns that rem is not defined.

4. **Computable Total Extensions.** We can view the algorithm in the previous paragraph as computing the values of some other, but total function rem*, defined by

$$\text{rem}^*(m,n) \stackrel{\text{def}}{=} \begin{cases} \text{rem}(m,n) & \text{if } (m,n) \in \mathbb{N} \times (\mathbb{N} - \{0\}); \\ -1 & \text{otherwise.} \end{cases}$$

The function rem* has the same values as rem, whenever rem is defined. We say that rem* is a *total extension* of rem to $\mathbb{N} \times \mathbb{N}$. Now, if every intuitively computable partial function had a "computable" total extension, then there would be no need to consider partial functions in order to study "computability". However, we will see that this is not the case: There are intuitively computable partial functions that have no "computable" total extensions. This is yet another reason for the introduction of *partial* functions in the study of "computable" functions.

5.3.4 Generalization to Partial Functions

The difficulties described in the previous section indicated that the definition of a "computable" function should be founded on *partial* functions instead of only total functions. So, let us recall the basic facts about partial functions and introduce some useful notation. We will be using Greek letters to denote partial functions, e.g., φ.

Definition 5.2. (Partial Function) We say that $\varphi : \mathcal{A} \to \mathcal{B}$ is a **partial function** if φ *may be* undefined for some elements of $\mathcal{A}$. (See Fig. 5.14.) If φ is defined for $a \in \mathcal{A}$, we write

$$\varphi(a)\downarrow;$$

otherwise we write $\varphi(a)\uparrow$. The set of all the elements of $\mathcal{A}$ for which φ is defined is the **domain** of φ, denoted by

$$\text{dom}(\varphi).$$

Hence dom$(\varphi) = \{a \in \mathcal{A} \mid \varphi(a)\downarrow\}$. Thus, for partial φ we have dom$(\varphi) \subseteq \mathcal{A}$. In the special case when dom$(\varphi) = \mathcal{A}$, we omit the adjective partial and say that φ is a **total** function (or just a function). When it is clear that a function is total, we denote it by a Latin letter, e.g., f, g.

The expression

$$\varphi(a)\!\downarrow = b$$

says that φ is defined for $a \in \mathcal{A}$ and its value is b. The set of all elements of $\mathcal{B}$ that are φ-images of elements of $\mathcal{A}$ is the **range** of φ, denoted by

$$\mathrm{rng}(\varphi).$$

Hence $\mathrm{rng}(\varphi) = \{b \in \mathcal{B} \mid \exists a \in \mathcal{A} : \varphi(a)\!\downarrow = b\}$. The function φ is **surjective** if $\mathrm{rng}(\varphi) = \mathcal{B}$, and it is **injective** if different elements of $\mathrm{dom}(\varphi)$ are mapped into different elements of $\mathrm{rng}(\varphi)$. Partial functions $\varphi : \mathcal{A} \to \mathcal{B}$ and $\psi : \mathcal{A} \to \mathcal{B}$ are said to be **equal**, and denoted by

$$\varphi \simeq \psi$$

if they have the same domains and the same values; that is, for every $x \in \mathcal{A}$ it holds that $\varphi(x)\!\downarrow \Longleftrightarrow \psi(x)\!\downarrow$ and $\varphi(x)\!\downarrow \Longrightarrow \varphi(x) = \psi(x)$.

Fig. 5.14 Partial function $\varphi : \mathcal{A} \to \mathcal{B}$ is defined on the set $\mathrm{dom}(\varphi) \subseteq \mathcal{A}$

The improved definition of an intuitively computable function (the definition that would consider non-total functions in addition to total ones) was expected to retain the intuitive appeal of the previous definition (see p. 98). Fortunately, this was not difficult to achieve. One had only to take into account all the possible outcomes of the calculating function's values for arguments where the function is *not* defined.

So let $\varphi : \mathcal{A} \to \mathcal{B}$ be a partial function, $a \in \mathcal{A}$, and suppose that $\varphi(a)\!\uparrow$. There are two possible outcomes when an attempt is made to calculate $\varphi(a)$:

1. the computation halts and returns a nonsensical result (not belonging to $\mathrm{rng}(\varphi)$);
2. the computation does not halt.

In the first outcome, the nonsensical result is also the signal (i.e., warning) that φ is not defined for a. The second outcome is the trying one: Neither do we receive the result nor do we receive the warning that $\varphi(a)\!\uparrow$. As long as the computation goes on, we can only wait and hope that it will soon come—not knowing that all is in vain (Fig. 5.15). The situation is even worse: We will prove later, in Sect. 8.2, that there is no general way to find out whether all is in vain. In other words, we can never find out that we are victims of the second outcome.

Fig. 5.15 What does one do if a computation on a Turing machine is still going on?

Since deciding which of the two outcomes will take place is generally impossible, we will not try to distinguish between them. We will not concern ourselves with what is going on when an attempt is made to compute an undefined function value. We will be satisfied with a new definition, which, in essence, says that

> A *partial* function is "computable" if there is an algorithm
> that can compute its value *whenever the function is defined.*

When it is known that such a function is *total*, we will drop the adjective partial. We can now give a formalization of the concept of "computable" partial function.

Formalization. (cont'd from p. 98)   *The intuitive concept of "computable" partial function* $\varphi : \mathcal{A} \to \mathcal{B}$ *is formalized as follows:*

φ is **"computable"**   *if there exists a TM that can compute the value*
$$\varphi(x) \text{ for any } x \in \mathrm{dom}(\varphi)$$
and $\mathrm{dom}(\varphi) = \mathcal{A};$

φ is **partial "computable"**   *if there exists a TM that can compute the value*
$$\varphi(x) \text{ for any } x \in \mathrm{dom}(\varphi);$$

φ is **"incomputable"**   *if there is no TM that can compute the value*
$$\varphi(x) \text{ for any } x \in \mathrm{dom}(\varphi).$$

Note that Fig. 5.12 (p. 98) is valid for the new definition of "computable" functions.

NB *Since the concept of a "computable" function is now formalized, we will no longer use quotation marks to distinguish between its intuitive and formal meanings. From now on, when we will say that a function is* computable, *it will be tacitly understood that it is total (by definition). And when we say that a function is* partial computable *(or p.c. for short), we will be aware of the fact that the function may or may not be total.*

Remarks. 1) In the past, the naming of computable and partial computable functions was not uniform. In 1996, Soare suggested to modernize and unify the terminology. According to this, we use the terms *computable* function (instead of the older term *recursive* function) and *partial computable* function (instead of the older term *partial recursive* function). The reasons for the unification will be described on p. 153. 2) Above, we have defined the concept of an *incomputable* function. But, do such functions really exist? A clue that they do exist was given in Sect. 5.3.3, where it was noticed that numerical functions outnumber the constructions of μ-recursive (i.e., p.c.) functions. We will be able to construct a particular such function later (see Sects. 8.2.3 and 8.3.1).

A Generalization

Later, a slight generalization of the above formalization will ease our expression. Let $\varphi : \mathcal{A} \to \mathcal{B}$ be a partial function and $\mathcal{S} \subseteq \mathcal{A}$ an arbitrary set. Observe that even if φ is incomputable, there may still exist a Turing machine that is capable of computing the value $\varphi(x)$ for arbitrary $x \in \mathcal{S} \cap \mathrm{dom}(\varphi)$. (See Fig. 5.16.) In this case, we will say that φ is *partial computable on the set* $\mathcal{S}$.[18] If, in addition, $\mathcal{S} \subseteq \mathrm{dom}(\varphi)$, then we will say that φ is *computable on the set* $\mathcal{S}$.[19] Here is the official definition.

Fig. 5.16 $\varphi : \mathcal{A} \to \mathcal{B}$ is partial computable (p.c.) on $\mathcal{S} \subseteq \mathcal{A}$ if it can be computed everywhere on $\mathcal{S} \cap \mathrm{dom}(\varphi)$

Definition 5.3. (Computability on a Set) Let $\varphi : \mathcal{A} \to \mathcal{B}$ be a partial function and $\mathcal{S} \subseteq \mathcal{A}$. We say that:

φ is **computable on** $\mathcal{S}$ *if* there exists a TM that can compute the value
$$\varphi(x) \text{ for any } x \in \mathcal{S} \cap \mathrm{dom}(\varphi)$$
$$\text{and } \mathcal{S} \subseteq \mathrm{dom}(\varphi);$$

φ is **partial computable on** $\mathcal{S}$ *if* there exists a TM that can compute the value
$$\varphi(x) \text{ for any } x \in \mathcal{S} \cap \mathrm{dom}(\varphi);$$

φ is **incomputable on** $\mathcal{S}$ *if* there is *no* TM that can compute the value
$$\varphi(x) \text{ for any } x \in \mathcal{S} \cap \mathrm{dom}(\varphi).$$

If we take $\mathcal{S} = \mathcal{A}$, the definition transforms into the above formalization.

---

[18] Equivalently, $\varphi : \mathcal{A} \to \mathcal{B}$ is *p.c. on the set* $\mathcal{S} \subseteq \mathcal{A}$ if the restriction $\varphi|_{\mathcal{S}}$ is a p.c. function.

[19] Equivalently, $\varphi : \mathcal{A} \to \mathcal{B}$ is *computable on the set* $\mathcal{S} \subseteq \mathrm{dom}(\varphi)$ if $\varphi|_{\mathcal{S}}$ is a computable function.

5.3.5 Applications of the Thesis

The *Computability Thesis* (CT) is useful in proving the existence or non-existence of certain "algorithms". Such proofs are called *proofs by CT*. There are two cases:

1. Suppose we want to prove that a given function φ is p.c. We might try to construct a TM and prove that it is capable of computing the values of φ. However, this approach is cumbersome and prone to mistakes. Instead, we can do the following:

   a. informally describe an "algorithm" that "computes" the values of φ;
   b. refer to the CT (saying: By CT the "algorithm" *can* be replaced by *some* Turing program; hence φ is p.c.).

2. Suppose we want to prove that a function φ is "incomputable". To do this, we

   a. prove that φ is not Turing-computable (i.e., there exists no TM for computing the values of φ);
   b. refer to the CT (saying: Then, by CT, φ is "incomputable").

5.4 Chapter Summary

Hilbert's Program left open the *Entscheidungsproblem*, the problem that was calling for an algorithm that would, for any mathematical formula, decide whether the formula can be derived.

Soon it became clear that the problem could not be solved unless the intuitive, loose definition of the concept of the algorithm was replaced by a rigorous, formal definition. Such a definition, called the model of computation, should characterize the notions of "algorithm", "computation", and "computable".

In 1930, the search for an appropriate model of computation started. Different ideas arose and resulted in several totally different models of computation: the μ-recursive functions, (general) recursive functions, λ-definable functions, the Turing machine, the Post machine, and Markov algorithms. Although diverse, the models shared two important properties, namely that they were *reasonable* and they fulfilled the *Effectiveness Requirement*.

The Turing machine was accepted by many as the most appropriate model of computation. Surprisingly, it turned out that all the models are equivalent in the sense that what can be computed by one can also be computed by the others.

Finally, the *Computability Thesis* equated the informally defined concepts of "algorithm", "computation", and "computable" with the counterparts that were formally defined by the models of computation. In effect, the thesis declared that all these models of computation also fulfill the *Completeness Requirement*. It was also found that the definition of a "computable" function must be founded on partial functions instead of only on total functions.

All in all, the *Computability Thesis* made it possible to mathematically treat the intuitive concepts of computation.

Problems

5.1. Prove that these functions are primitive recursive:

(a) $\mathrm{const}_j^k(n_1,\ldots,n_k) \overset{\text{def}}{=} j$, for $j \geqslant 0$ and $k \geqslant 1$;

(b) $\mathrm{add}(m,n) \overset{\text{def}}{=} m + n$;

(c) $\mathrm{mult}(m,n) \overset{\text{def}}{=} mn$;

(d) $\mathrm{power}(m,n) \overset{\text{def}}{=} m^n$;

(e) $\mathrm{fact}(n) \overset{\text{def}}{=} n!$;

(f) $\mathrm{tower}(m,n) \overset{\text{def}}{=} m^{m^{\cdot^{\cdot^{m}}}} \Big\} n$ levels;

(g) $\mathrm{minus}(m,n) \overset{\text{def}}{=} m \dot- n = \begin{cases} m-n & \text{if } m \geqslant n; \\ 0 & \text{otherwise.} \end{cases}$

(h) $\mathrm{div}(m,n) \overset{\text{def}}{=} m \div n = \lfloor \frac{m}{n} \rfloor$;

(i) $\mathrm{floorlog}(n) \overset{\text{def}}{=} \lfloor \log_2 n \rfloor$;

(j) $\log^*(n) \overset{\text{def}}{=}$ the smallest k such that $\overbrace{\log(\log(\cdots(\log(n))\cdots))}^{k \text{ times}} \leqslant 1$;

(k) $\gcd(m,n) \overset{\text{def}}{=}$ greatest common divisor of m and n;

(l) $\mathrm{lcm}(m,n) \overset{\text{def}}{=}$ least common multiple of m and n;

(m) $\mathrm{prime}(n) \overset{\text{def}}{=}$ the nth prime number;

(n) $\pi(x) \overset{\text{def}}{=}$ the number of primes not exceeding x;

(o) $\phi(n) \overset{\text{def}}{=}$ the number of positive integers that are $\leqslant n$ and relatively prime to n (*Euler funct.*);

(p) $\max^k(n_1,\ldots,n_k) \overset{\text{def}}{=} \max\{n_1,\ldots,n_k\}$, for $k \geqslant 1$;

(q) $\min^k(n_1,\ldots,n_k) \overset{\text{def}}{=} \min\{n_1,\ldots,n_k\}$, for $k \geqslant 1$;

(r) $\mathrm{neg}(x) \overset{\text{def}}{=} \begin{cases} 0 & \text{if } x \geqslant 1; \\ 1 & \text{if } x = 0. \end{cases}$

(s) $\mathrm{and}(x,y) \overset{\text{def}}{=} \begin{cases} 1 & \text{if } x \geqslant 1 \wedge y \geqslant 1; \\ 0 & \text{otherwise.} \end{cases}$

(t) $\mathrm{or}(x,y) \overset{\text{def}}{=} \begin{cases} 1 & \text{if } x \geqslant 1 \vee y \geqslant 1; \\ 0 & \text{otherwise.} \end{cases}$

(u) $\text{if-then-else}(x,y,z) \overset{\text{def}}{=} \begin{cases} y & \text{if } x \geqslant 1; \\ z & \text{otherwise.} \end{cases}$

(v) $\mathrm{eq}(x,y) \overset{\text{def}}{=} \begin{cases} 1 & \text{if } x = y; \\ 0 & \text{otherwise.} \end{cases}$

(w) $\mathrm{gr}(x,y) \overset{\text{def}}{=} \begin{cases} 1 & \text{if } x > y; \\ 0 & \text{if } x \leqslant y. \end{cases}$

(x) $\mathrm{geq}(x,y) \overset{\text{def}}{=} \begin{cases} 1 & \text{if } x \geqslant y; \\ 0 & \text{if } x < y. \end{cases}$

(y) $\mathrm{ls}(x,y) \overset{\text{def}}{=} \begin{cases} 1 & \text{if } x < y; \\ 0 & \text{if } x \geqslant y. \end{cases}$

(z) $\mathrm{leq}(x,y) \overset{\text{def}}{=} \begin{cases} 1 & \text{if } x \leqslant y; \\ 0 & \text{if } x > y. \end{cases}$

5.2. Prove: If $f : \mathbb{N} \to \mathbb{N}$ is primitive recursive, then so is $g(n,m) \overset{\text{def}}{=} f^{(n)}(m) = \overbrace{f(f(\cdots(f(m))\cdots))}^{n \text{ times}}$.

5.3. Prove: Every primitive recursive function is total.

5.4. Prove: Every μ-recursive function can be obtained from the initial functions ζ, σ, π_i^k by a finite number of compositions and primitive recursions and *at most one* μ-operation.

Definition 5.4. (Ackermann Function) A version $A : \mathbb{N}^2 \to \mathbb{N}$ of the **Ackermann function** is defined as follows:

$$A(0,n) = \ n+1 \tag{1}$$
$$A(m+1,0) = \ A(m,1) \tag{2}$$
$$A(m+1,n+1) = A(m,A(m+1,n)) \tag{3}$$

Remark. What is the intuition behind the Ackermann function? Imagine a sequence of binary operations $o_1, o_2, o_3, \ldots$ on $\mathbb{N}^2$, where each operation is defined by the preceding one as follows: *Given arbitrary $x, y \in \mathbb{N}$, the value $o_k(x,y)$ is obtained by applying x to itself y times using o_{k-1}; in particular, o_1 applies x to itself y times using the successor function, i.e., $o_1(x,y) = x + y$.*

The first operations are $o_1(x,y) = x + \overbrace{1 + \ldots + 1}^{y \text{ times}} = x + y$, $o_2(x,y) = \overbrace{x + x + \ldots + x}^{y \text{ times}} = x \cdot y$, $o_3(x,y) = \underbrace{x \cdot x \cdot \ldots \cdot x}_{y \text{ times}} = x^y$, and $o_4(x,y) = \underbrace{x \hat{\ }(x \hat{\ }(\ldots \hat{\ } x))}_{y \text{ times}} = x^{x^{\cdot^{\cdot^{x}}}}$.

As k increases, the values $o_k(x,y)$ grow extremely fast. We can view k as the third variable and define a new function $\mathrm{ack} : \mathbb{N}^3 \to \mathbb{N}$ by $\mathrm{ack}(k,x,y) \overset{\text{def}}{=} o_k(x,y)$. This is the so-called *Ackermann generalized exponential*. The function A in Definition 5.4 can be obtained from ack.

5.5. Let A be the Ackermann function as defined above.

(a) Prove: A is intuitively computable.

[*Hint.* $A(0,n)$ is the successor function, hence intuitively computable. Suppose that, for all n, $A(m,n)$ is intuitively computable. To see that $A(m+1,n)$ is intuitively computable we repetitively apply (3) (to obtain $A(m+1,0)$) and then (2) (to decrease $m+1$).]

(b) Prove: A is a μ-recursive function.

(c) Try to compute $A(k,k)$ for $k = 0, 1, 2, 3$.

(d) The function A grows faster than *any* primitive recursive function in the following sense: For every primitive recursive $f(n)$, there is an $n_0 \in \mathbb{N}$ such that $f(n) < A(n,n)$ for all $n > n_0$. Can you prove that?
(*Remark.* This is why A is *not primitive* recursive.)

Bibliographic Notes

- For a treatise on *physical aspects* of information and computation, see Feynman [71]. The use of quantum phenomena in computation is described in Aaronson [1].
- *Primitive* recursive functions were first described in Gödel [81], and (*general*) recursive functions in Herbrand [98] and Gödel [82]. Based on these, μ-*recursive functions* were defined in Kleene [121].
- Ackermann functions were defined in Ackermann [2] and were extensively studied in the 1930s by Péter [176]. See also Kleene [124, §55] and Hermes [100].
- The λ-*calculus* was introduced in Church [35]. Hindley and Seldin [105] is a nice introduction to the λ-*calculus* and Haskell Curry's related model of computation called *combinatory logic* (CL). For an extensive treatise on the two models, see Barendregt [13].
- Turing introduced his *a-machine* in [262]. In [87], Gödel gave the reason why he thought that Turing machine established beyond any doubt the formal definition of the intuitive notion of computability. See also Sieg [224] for a discussion on this. The equivalence of the λ-calculus and the Turing machine was proved in Turing [263].
- The *Post machine* was introduced in Post [181]. Post suspected (but did not prove) that it is equivalent to the Turing machine and the λ calculus. See Pettorossi [177] and Manna [146] for a description of its workings and for the proof of its equivalence with the Turing machine. The importance of having a queue (instead of a stack) in the Post machine is explained in Rich [198, Sect. 18.2.3].
- *Markov algorithms* were described in Markov [147, 148].
- *Cellular automata* were first investigated by von Neumann [271, 272] in the beginning 1950s. When trying to conceive a system capable of producing exact copies of itself he focused on *discrete two*-dimensional systems. Although the resulting cellular automaton used 29 different states and a rather complicated dynamics directed by transition rules, it was capable of self-reproduction. This was also the first discrete parallel computational model in history that was formally shown to be a universal computer. In the 1970, Conway introduced a cellular automaton called *Life* (or, *Game of Life*) that is one of the simplest computational models ever proved to be universal computer. The game first appeared in Gardner [77]. An initial configuration of *Life* can encode an input string. One can let *Life* run, read at some point the current configuration as the result of the computation performed so far, and decode the configuration to output string. Conway et al. [18, Chap. 25] proved that *Life* can compute everything that UTM can. In the 1980s, Wolfram explored *one*-dimensional cellular automata and conjectured that an elementary one-dimensional cellular automaton known as *Rule 110* is capable of universal computation. In 2002, Wolfram [279, pp.675–691] sketched the proof of the conjecture, and a full proof was given by Cook [40] in 2004.

- DNA computing, which uses chemical reactions on biological molecules to perform computation, was discovered in 1993 by Adleman [4, 5].
- A lot of information about the reasons for searching for models of computation is given in Kleene [124]. A comprehensive historical view of the development of models of computation, their properties, and applications can be found in Odifreddi [173] and Adams [3]. A concise introduction to models of computation, from the standard Turing machine and μ-recursive functions to the modern models inspired by quantum physics is Fernández [70]. For the proofs that all these models of computation are equivalent see, for example, Machtey and Young [144] (for the Turing machine, μ-recursive functions, RAM, and Markov algorithms) and Enderton [67] (for the register machine).
- Church published his thesis in [37] and Turing proposed his thesis in [262]. The genesis of the *Computability Thesis*, Gödel's reservations about Church's Thesis, and flaws that were recently (1988) discovered in Church's development of his thesis, are detailed in Gandy [76], Sieg [221], and Soare [242, 244].
- Based on his paper [180], Post later introduced *canonical* and *normal systems* in [182] and [183, 184]. Here he introduced his thesis about set generation.
- Several cited papers are in Cooper and van Leeuwen [44], van Heijenoort [268], and Davis [57].
- An interesting approach to models if computation is given in Floyd and Beigel [73]. The authors first systematically describe and formalize the general properties of the most important components (devices, as they call them, such as finite memories, stacks, queues, counters, tapes, control units) that are used to define and construct reasonable abstract computing machines. Then the authors give a single definition that encompasses all the familiar abstract computing machines, such as TMs and RAMs, and their possible uses (as recognizers, acceptors, transducers).
- Many historical notes about different models of computation are scattered in a number of excellent general monographs on *Computability Theory*. For such monographs, see the Bibliographic Notes to Chapter 6 on p. 152.

Chapter 6
The Turing Machine

A machine is a mechanically, electrically, or electronically operated device for performing a task. A program is a sequence of coded instructions that can be inserted into a machine to control its operation.

Abstract The Turing machine convincingly formalized the concepts of "algorithm", "computation", and "computable". It convinced researchers by its simplicity, generality, mechanical operation, and resemblance to human activity when solving computational problems, and by Turing's reasoning and analysis of "computable" functions and his argumentation that partial "computable" functions are exactly Turing-computable functions. Turing considered several variants that are generalizations of the basic model of his machine. But he also proved that they add nothing to the computational power of the basic model. This strengthened belief in the Turing machine as an appropriate model of computation. Turing machines can be encoded and consequently enumerated. This enabled the construction of the universal Turing machine, which is capable of computing anything that can be computed by any other Turing machine. This seminal discovery laid the theoretical grounds for several all-important practical consequences, the general-purpose computer and the operating system being the most notable. The Turing machine is a versatile model of computation: It can be used to compute values of a function, or to generate elements of a set, or to decide about the membership of an object in a set. The last led to the notions of decidable and semi-decidable sets, which would later prove to be very important in solving general computational problems.

6.1 Turing Machine

Most researchers accepted the Turing machine (TM) as the most appropriate model of computation. We described the Turing machine briefly in Sect. 5.2.2. In this section we will go into detail. First, we will describe the basic model of the TM. We will then introduce several other variants that are generalizations of the basic model. Finally, we will prove that, from the viewpoint of general computability, they add nothing to the computational power of the basic model. This will strengthen our belief in the basic Turing machine as a simple yet highly powerful model of computation.

© The Author(s), under exclusive license to Springer-Verlag GmbH, DE,   111
part of Springer Nature 2020
R. Robič, *The Foundations of Computability Theory*, https://doi.org/10.1007/978-3-662-62421-0_6

Fig. 6.1 Alan Turing
(Courtesy: See Preface)

6.1.1 Basic Model

Definition 6.1. (Turing Machine) The basic variant of the **Turing machine**
has the following components (Fig. 6.2): a *control unit* containing a *Turing
program*; a *tape* consisting of *cells*; and a movable *window* over the tape,
which is connected to the control unit. (See also the details below.)

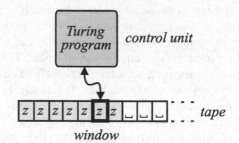

Fig. 6.2 Turing machine
(basic model)

The details are:

1. The *tape* is used for writing and reading the input data, intermediate data, and
   output data (results). It is divided into equally sized cells, and is potentially infinite in one direction (i.e., whenever needed, it can be extended in that direction
   with a finite number of cells).

   In each cell there is a *tape symbol* belonging to a finite *tape alphabet* $\Gamma = \{z_1, \ldots, z_t\}$, $t \geqslant 3$. The symbol z_t is special, for it indicates that a cell is empty; for
   this reason it is denoted by $\sqcup$ and called the *empty space*. In addition to $\sqcup$ there
   are at least two[1] additional symbols: 0 and 1. We will take $z_1 = 0$ and $z_2 = 1$.

---

[1] The reasons for at least two additional symbols are mostly practical (leaving out of consideration
the non-polynomial relation between the lengths of unary and binary representation of data, which
is important in *Computational Complexity Theory*). Only once, in Sect. 8.3.1, we will come across
Turing machines that need just *one* additional tape symbol (which will be the symbol 1). There,
we will simply ignore the other additional symbol.

The input data is contained in the *input word*. This is a word over some finite *input alphabet* Σ such that $\{0,1\} \subseteq \Sigma \subseteq \Gamma - \{\sqcup\}$. Initially, all the cells are empty (i.e., each contains $\sqcup$) except for the leftmost cells, which contain the input word.

The *control unit* is always in some *state* belonging to a finite *set of states* $Q = \{q_1, \ldots, q_s\}$, where $s \geqslant 1$. We call q_1 the *initial state*. Some states are said to be *final*; they are gathered in the set $F \subseteq Q$. All the other states are non-final. When the index of a state will be of no importance, we will use q_{yes} and q_{no} to refer to *any* final and non-final state, respectively.

There is a program called the *Turing program (TP)* in the control unit. The program directs the components of the machine. It is characteristic of the particular Turing machine, that is, different TMs have different TPs. Formally, a Turing program is a partial function $\delta : Q \times \Gamma \to Q \times \Gamma \times \{\text{Left}, \text{Right}, \text{Stay}\}$. It is also called the *transition function*. We can view δ as a table $\Delta = Q \times \Gamma$, where the component $\Delta[q_i, z_r] = (q_j, z_w, D)$ if $\delta(q_i, z_r) = (q_j, z_w, D)$ is an instruction of δ, and $\Delta[q_i, z_r] = 0$ if $\delta(q_i, z_r)\uparrow$ (see Fig. 6.3). Without loss of generality we assume that $\delta(q_{no}, z)\downarrow$ for some $z \in \Gamma$, and $\delta(q_{yes}, z)\uparrow$ for all $z \in \Gamma$. That is, there is always a transition from a non-final state, and none from a final state.

The *window* can move over any single cell, thus making the cell accessible to the control unit. The control unit can then *read a symbol* from the cell under the window, and *write a symbol* to the cell, replacing the previous symbol. In one step, the window can only move to a neighboring cell.

Δ	z_1	z_2	$\cdots$	z_r	$\cdots$	z_t
q_1	•	•	$\cdots$	•	$\cdots$	•
q_2	•	•	$\cdots$	•	$\cdots$	•
$\vdots$	$\vdots$	$\vdots$	$\ddots$	$\vdots$		$\vdots$
q_i	•	•	$\cdots$	(q_j, z_w, D)	$\cdots$	
$\vdots$	$\vdots$	$\vdots$		$\vdots$	$\ddots$	$\vdots$
q_s	•	•	$\cdots$	•	$\cdots$	•

Fig. 6.3 Turing program δ represented as a table Δ. Instruction $\delta(q_i, z_r) = (q_j, z_w, D)$ is described by the component $\Delta[q_i, z_r]$

2. Before the Turing machine is started, the following must take place:

   a. an input word is written to the beginning of the tape;
   b. the window is shifted to the beginning of the tape;
   c. the control unit is set to the initial state.

3. From now on the Turing machine operates independently, in a mechanical step wise fashion as instructed by its Turing program δ. Specifically, if the TM is in a state $q_i \in Q$ and it reads a symbol $z_r \in \Gamma$, then:

- *if* q_i is a final state, *then* the TM halts;
- *else, if* $\delta(q_i, z_r)\uparrow$ (i.e., TP has no next instruction), *then* the TM halts;
- *else, if* $\delta(q_i, z_r)\downarrow = (q_j, z_w, D)$, *then* the TM does the following:
  a. changes the state to q_j;
  b. writes z_w through the window;
  c. moves the window to the next cell in direction $D \in \{\text{Left}, \text{Right}\}$, or leaves the window where it is ($D = \text{Stay}$).

Formally, a Turing machine is a seven-tuple $T = (Q, \Sigma, \Gamma, \delta, q_1, \sqcup, F)$. To fix a particular Turing machine, we must fix $Q, \Sigma, \Gamma, \delta$, and F.   □

Remarks. 1) Because δ is a partial function, it may be undefined for certain arguments q_i, z_r, that is, $\delta(q_i, z_r)\uparrow$. In other words, a partial Turing program has no instruction $\delta(q_i, z_r) = (q_j, z_w, D)$ that would tell the machine what to do when in the state q_i it reads the symbol z_r. Thus, for such pairs, the machine halts. This always happens in final states (q_{yes}) and, for some Turing programs, also in some non-final states (q_{no}). 2) The interpretation of these two different ways of halting (i.e., what they tell us about the input word or the result) will depend on what purpose the machine will be used for. (We will see later that the Turing machine can be used for computing function values, generating sets, or recognizing t sets.)

Computation.   What does a *computation* on a Turing machine look like? Recall (Sect. 5.2.4) that the internal configuration of an abstract computing machine describes all the relevant information the machine possesses at a particular step of the computation. We now define the internal configuration of a Turing machine.

Definition 6.2. (Internal Configuration) Let T be a basic TM and w an arbitrary input word. Start T on w. The **internal configuration** of T after a finite number of computational steps is the word uq_iv, where

- q_i is the current state of T;
- $uv \in \Gamma^*$ are the contents of T's tape up to (a) the rightmost non-blank symbol or (b) the symbol to the left of the window, whichever is rightmost. We assume that $v \neq \varepsilon$ in the case (a), and $v = \varepsilon$ in the case (b).
- T is scanning the leftmost symbol of v in the case (a), and the symbol $\sqcup$ in the case (b).

Not every sequence can be an internal configuration of T (given input w). Clearly, the configuration prior to the first step of the computation is q_1w; we call it the *initial* configuration. After that, only sequences uq_iv that can be *reached* from the initial configuration by executing the program δ are internal configurations of T. So, if uq_iv is an internal configuration, then the *next* internal configuration can easily be constructed using the instruction $\delta(q_i, z_r)$, where z_r is the scanned symbol.

The computation of T on w is represented by a sequence of internal configurations starting with the initial configuration. Just as the computation may not halt, the sequence may also be infinite. (We will use internal configurations in Theorem 9.2.)

Example 6.1. (Addition on TM) Let us construct a Turing machine that transforms an input word $1^{n_1}01^{n_2}$ into $1^{n_1+n_2}$, where n_1, n_2 are natural numbers. For example, 111011 is to be transformed into 11111. Note that the input word can be interpreted as consisting of two unary-encoded natural numbers n_1, n_2 and the resulting word interpreted as containing their sum. Thus, the Turing machine is to compute the function $\mathrm{sum}(n_1 + n_2) = n_1 + n_2$.

First, we give an intuitive description of the Turing program. If the first symbol of the input word is 1, then the machine deletes it (instruction 1), and then moves the window to the right over all the symbols 1 (instruction 2) until the symbol 0 is read. The machine then substitutes this symbol with 1 and halts (instruction 3). However, if the first symbol of the input word is 0, then the machine deletes it and halts (instruction 4).

Formally, the Turing machine is $T = (Q, \Sigma, \Gamma, \delta, q_1, \sqcup, \{q_3\})$, where:

- $Q = \{q_1, q_2, q_3\}$;   //$s = 3$; q_1 is the initial state, q_3 is the final state (hence $F = \{q_3\}$);
- $\Sigma = \{0, 1\}$;
- $\Gamma = \{0, 1, \sqcup\}$;   // $t = 3$;
- the Turing program consists of the following instructions:

    1. $\delta(q_1, 1) = (q_2, \sqcup, \mathrm{Right})$;
    2. $\delta(q_2, 1) = (q_2, 1, \mathrm{Right})$;
    3. $\delta(q_2, 0) = (q_3, 1, \mathrm{Stay})$;
    4. $\delta(q_1, 0) = (q_3, \sqcup, \mathrm{Stay})$.

The state q_3 is final, because $\delta(q_3, z)\uparrow$ for all $z \in \Gamma$ (there are no instructions $\delta(q_3, z) = \ldots$).
For the input word 111011, the computation is illustrated in Fig. 6.4.

Fig. 6.4 Computing $3+2$ on a TM. Over the arrows are written the numbers of applied instructions

The corresponding sequence of internal configurations is:

$$q_1 111011 \rightarrow q_2 11011 \rightarrow 1q_2 1011 \rightarrow 11q_2 011 \rightarrow 11q_3 111.$$

If the input was 011, instruction 4 would execute, leaving the result 11. For the input 1110, the computation would proceed as in the figure above; only the result would be 111. $\square$

Example 6.2. (Addition on Another TM) Let us solve the same problem with another Turing machine that has a different Turing program. First, the window is moved to the right until $\sqcup$ is reached. Then the window is moved to the left (i.e., to the last symbol of the input word) and the symbol is deleted. If the deleted symbol is 0, the machine halts. Otherwise, the window keeps moving to the left and upon reading 0 the symbol 1 is written and the machine halts.

The Turing machine is $T' = (Q, \Sigma, \Gamma, \delta, q_1, \sqcup, \{q_5\})$, where:

- $Q = \{q_1, q_2, q_3, q_4, q_5\}$;   //$s = 5$; q_1 is the initial and q_5 the final state ($F = \{q_5\}$);
- $\Sigma = \{0, 1\}$;
- $\Gamma = \{0, 1, \sqcup\}$;   // $t = 3$;
- Turing program:

  1. $\delta(q_1, 1) = (q_2, 1, \text{Right})$ 6. $\delta(q_3, 0) = (q_5, \sqcup, \text{Stay})$
  2. $\delta(q_1, 0) = (q_2, 0, \text{Right})$ 7. $\delta(q_3, 1) = (q_4, \sqcup, \text{Left})$
  3. $\delta(q_2, 1) = (q_2, 1, \text{Right})$ 8. $\delta(q_4, 1) = (q_4, 1, \text{Left})$
  4. $\delta(q_2, 0) = (q_2, 0, \text{Right})$ 9. $\delta(q_4, 0) = (q_5, 1, \text{Stay})$
  5. $\delta(q_2, \sqcup) = (q_3, \sqcup, \text{Left})$

  The are no instructions of the form $\delta(q_5, z) = \ldots$, so the state q_5 is final.

For the input word 111011, the computation is illustrated in Fig. 6.5.

Fig. 6.5 Computing $3 + 2$ on a different TM. The numbers of applied instructions are on the arrows

The corresponding sequence of internal configurations is:

$q_1 111011 \to 1q_2 11011 \to 11q_2 1011 \to 111q_2 011 \to$
$\to 1110q_2 11 \to 11101q_2 1 \to 111011q_2 \to 11101q_3 1 \to$
$\to 1110q_4 1 \to 111q_4 01 \to 111q_5 11$ □

6.1.2 Generalized Models

Turing considered different variants of the basic model; each is a generalization of the basic model in some respect. The variants differ from the basic model in their *external configurations*. For example, finite memory can be added to the control unit; the tape can be divided into parallel tracks, or it can become unbounded in both directions, or it can even be multi-dimensional; additional tapes can be introduced; and nondeterministic instructions can be allowed in Turing programs. The variants V of the basic model are:

- **Finite-Storage TM.** This variant V has in its control unit a finite storage capable of memorizing $k \geqslant 1$ tape symbols and using them during the computation. The Turing program is formally $\delta_V : Q \times \Gamma \times \Gamma^k \to Q \times \Gamma \times \{\text{Left}, \text{Right}, \text{Stay}\} \times \Gamma^k$.

Fig. 6.6 Finite-storage TM can store k tape symbols

- **Multi-track TM.** This variant V has the tape divided into $tk \geqslant 2$ *tracks*. On each track there are symbols from the alphabet Γ. The window displays tk-tuples of symbols, one symbol for each track. Formally the Turing program is the function $\delta_V : Q \times \Gamma^{tk} \to Q \times \Gamma^{tk} \times \{\text{Left}, \text{Right}, \text{Stay}\}$.

Fig. 6.7 Multi-track TM has tk tracks on its tape

- **Two-Way Unbounded TM.** This variant V has the tape unbounded in both directions. Formally, the Turing program is $\delta_V : Q \times \Gamma \to Q \times \Gamma \times \{\text{Left}, \text{Right}, \text{Stay}\}$.

Fig. 6.8 Two-way TM has unbounded tape in both directions

- **Multi-tape TM.** This variant V has $tp \geqslant 2$ unbounded tapes. Each tape has its own window that is independent of other windows. Formally, the Turing program is $\delta_V : Q \times \Gamma^{tp} \to Q \times (\Gamma \times \{\text{Left}, \text{Right}, \text{Stay}\})^{tp}$.

$tp = 2$

Fig. 6.9 Multi-tape TM has tp tapes with separate windows

- **Multidimensional TM.** This variant V has a d-dimensional tape, $d \geqslant 2$. The window can move in d dimensions, i.e., $2d$ directions $L_1, R_1, L_2, R_2, \ldots, L_d, R_d$. The Turing program is $\delta_V : Q \times \Gamma \to Q \times \Gamma \times \{L_1, R_1, L_2, R_2, \ldots, L_d, R_d, \text{Stay}\}$.

$d = 2$

Fig. 6.10 Multidimensional TM has a d-dimensional tape

- **Nondeterministic TM.** This variant V has a Turing program δ_V that assigns to each (q_i, z_r) a finite *set* of alternative transitions $\{(q_{j_1}, z_{w_1}, D_1), (q_{j_2}, z_{w_2}, D_2), \ldots\}$. In each (q_i, z_r), the machine V can *miraculously pick out* from the set $\delta_V(q_i, z_r)$ a transformation—if such exists—that can lead the remaining computation to a *final* state q_{yes}. Accordingly, we define that the machine V *accepts* a given input word x if there *exists* a computation of V on x that terminates in a final state q_{yes}; otherwise, the machine V immediately *rejects* x and halts.

Obviously, the nondeterministic TM is not a reasonable model of computation because it can foretell the future computation from each of the alternative transitions. Nevertheless, it is a very useful tool that makes it possible to define the minimum number of steps needed to compute the solution (when there is one). Again, this is important when we investigate the *computational complexity* of problem solving. For *computability* on unrestricted models of computation, this is irrelevant. So we will not be using nondeterministic Turing machines.

6.1.3 Equivalence of Generalized and Basic Models

Although each generalized model V of the TM seems to be more powerful than the basic model T, it is not so; T can compute anything that V can compute. We will prove this by describing how T can simulate V. (The other way round is obvious as T is a special case of V.) In what follows we describe the main ideas of the proofs and leave the details to the reader as an exercise.

- **Simulation of a Finite-Storage TM.** Let V be a finite-storage TM with the storage capable of memorizing $k \geqslant 1$ tape symbols. The idea of the simulation is that T can memorize a finite number of tape symbols by *encoding the symbols in the indexes of its states*. Of course, this may considerably enlarge the number of T's states, but recall that in the definition of the Turing machine there is no limitation on the number of states, as long as this number is finite.

  To implement the idea, we do the following: 1) We redefine the indexes of T's states by enforcing an internal structure on each index (i.e., by assuming that certain data is encoded in each index). 2) Using the redefined indexes, we are able to describe how T's instructions make sure that the tape symbols are memorized in the indexes of T's states. 3) Finally, we show that the redefined indexes can be represented by natural numbers. This shows that T is still the basic model of the Turing machine.

  The details are as follows:

  1. Let us denote by $[i, m_1, \ldots, m_k]$ any index that encodes k symbols $z_{m_1}, \ldots, z_{m_k}$ that were read from T's tape (not necessarily the last-read symbols). Thus, the state $q_{[i, m_1, \ldots, m_k]}$ represents some usual T state q_i as well as the contents $z_{m_1}, \ldots, z_{m_k}$ of V's storage. Since at the beginning of the computation there are no memorized symbols, i.e., $z_{m_1} = \ldots = z_{m_k} = \sqcup$, and since $\sqcup = z_t$, where $t = |\Gamma|$, we see that the initial state is $q_{[1, t, \ldots, t]}$.

  2. The Turing program of T consists of two kinds of instructions. Instructions of the first kind *do not* memorize the symbol z_r that has been read from the current cell. Such instructions are of the form $\delta(q_{[i, m_1, \ldots, m_k]}, z_r) = (q_{[j, m_1, \ldots, m_k]}, z_w, D)$. Note that they leave the memorized symbols $z_{m_1}, \ldots, z_{m_k}$ unchanged. In contrast, the instructions of the second kind *do* memorize the symbol z_r. This symbol can replace any of the memorized symbols (as if z_r had been written into any of the k locations of V's storage). Let ℓ, $1 \leqslant \ell \leqslant k$, denote which of the memorized symbols is to be substituted by z_r. The general form of the instruction is now $\delta(q_{[i, m_1, \ldots, m_k]}, z_r) = (q_{[j, m_1, \ldots, m_{\ell-1}, r, m_{\ell+1}, \ldots, m_k]}, z_w, D)$. After the execution of such an instruction, T will be in a new state that represents some usual state q_j as well as the new memorized symbols $z_{m_1}, \ldots, z_{m_{\ell-1}}, z_r, z_{m_{\ell+1}}, \ldots, z_{m_k}$.

  3. There are st^k indexes $[i, m_1, \ldots, m_k]$, where $1 \leqslant i \leqslant s = |Q|$ and $1 \leqslant m_\ell \leqslant t = |\Gamma|$ for $\ell = 1, \ldots, k$. We can construct a bijective function f from the set of all the indexes $[i, m_1, \ldots, m_k]$ onto the set $\{1, 2, \ldots, st^k\}$. (We leave the construction

of f as an exercise.) Using f, the states $q_{[i,m_1,\ldots,m_k]}$ can be renamed into the usual form $q_{f([i,m_1,\ldots,m_k])}$, where the indexes are natural numbers.

In summary, a finite-storage TM V can be simulated by a basic Turing machine T if the indexes of T's states are appropriately interpreted.

- **Simulation of a Multi-track TM.** Let V be a tk-track TM, $tk \geqslant 2$. The idea of the simulation is that T considers tk-tuples of V's symbols as single symbols (see Fig. 6.11). Let V have on the ith track symbols from Γ_i, $i = 1, 2, \ldots, tk$. Then let T's tape alphabet be $\Gamma_1 \times \Gamma_2 \times \ldots \times \Gamma_{tk}$. If V has in its program δ_V an instruction $\delta_V(q_i, z_{r_1}, \ldots, z_{r_{tk}}) = (q_j, z_{w_1}, \ldots, z_{w_{tk}}, D)$, then let T have in its program δ_T the instruction $\delta_T(q_i, (z_{r_1}, \ldots, z_{r_{tk}})) = (q_j, (z_{w_1}, \ldots, z_{w_{tk}}), D)$. Then T simulates V.

Fig. 6.11 Each pair of symbols in V's window is considered as a single symbol by T

- **Simulation of a Two-Way Unbounded TM.** Let V be a two-way unbounded TM. Assume that, initially, V has its window positioned over the leftmost symbol of the input word. Denote the cell with this symbol by c_0. Fold the left part of the V's tape (the part to the left of c_0) so that the cell c_{-1} moves under c_0, the cell c_{-2} under c_1, and so on (see Fig. 6.12). The result can be viewed as a one-way two-track tape of a new TM V', having the input word written on its upper track.

The machine V' can simulate V as follows: 1) Initially, V' writes a delimiter • to c_{-1}; 2) whatever V does on the *right* part $c_0, c_1 \ldots$ of its tape, V' does on the *upper* track of its tape; and whatever V does on the *left* part $\ldots c_{-2}, c_{-1}$ of its tape, V' does on the part $c_{-1}, c_{-2} \ldots$ on the *lower* track of its tape, while moving its window in the direction opposite to that of V; and 3) whenever V's window moves from c_0 to c_{-1} or from c_{-1} to c_0, the machine V' passes to the opposite track and moves its window one cell to the right or left, respectively. Finally, we know that V' can be simulated by a basic TM T. Hence V can be simulated by T.

Fig. 6.12 The left part of V's tape is folded under the right one to obtain the tracks of V''s tape

- **Simulation of a Multi-tape TM.** Let V be a tp-tape TM, $tp \geqslant 2$. Note that after $k \geqslant 1$ steps the leftmost and rightmost window can be at most $2k + 1$ cells apart. (The distance is maximized when, in each step, the two windows move apart by two cells.) Other windows are between (or as far as) the two outermost windows.

Let us now imagine a two-way unbounded, $2tp$-track Turing machine V' (see Fig. 6.13). The machine V' can simulate V as follows. Each two successive tracks of V' describe the situation on one tape of V. That is, the situation on the ith tape of V is described by the tracks $2i-1$ and $2i$ of V'. The contents of the track $2i-1$ are the same as would be the contents of the ith tape of V. Track $2i$ is empty, except for one cell containing the symbol X. The symbol X is used to mark what V would see on its ith tape (i.e., where the window on the ith tape would be).

Fig. 6.13 Each tape of V is represented by two successive tracks of V''s tape

In one step, V would read tp symbols through its windows, write tp new symbols, move tp windows, and change the state. How can V' simulate all of this? The answer is by moving its single window to and fro and changing the contents of its tracks until they reflect the situation after V's step. Actually, V' can do this in *two sweeps*, first by moving its window from the leftmost X to the rightmost one, and then back to the leftmost X. When, during the *first* sweep, V' reads an X, it records the symbol above the X. In this way, V' records all the tp symbols that V would read through its windows. During the *second* sweep, V' uses information about V's Turing program: If V' detects an X, it substitutes the symbol above the X with another symbol (the same symbol that V would write on the corresponding tape) and, if necessary, moves the X to the neighboring cell (in the direction in which V would move the corresponding window). After the second sweep is complete, V' changes to the new state (corresponding to V's new state).

Some questions are still open. How does V' know that an outermost X has been reached so that the sweep is completed? The machine V' must have a counter (on additional track) that tells how many Xs are still to be detected in the current sweep. Before a sweep starts, the counter is set to tp, and during the sweep the counter is decremented upon each detection of an X. When the counter reaches 0, the window is over an outermost X. This happens in finite time because the outermost Xs are at most $2k+1$ cells apart. Since each single move of V can be simulated by V' in finite time, every computation of V can also be simulated by V'.

The machine V' is a two-way unbounded, multi-track TM. However, from the previous simulations it follows that V' can be simulated by the basic model of the Turing machine T. Hence, V can be simulated by T.

- **Simulation of a Multidimensional TM.** Let V be a d-dimensional TM, $d \geqslant 2$. Let us call the minimal rectangular d-polytope containing every nonempty cell of V's tape the *d-box*. For example, the 2-box is a rectangle (see Fig. 6.14) and the 3-box is a rectangular hexahedron. For simplicity we continue with $d = 2$. A 2-box contains rows (of cells) of equal length (the number of cells in the row).

Fig. 6.14 The rows of V's current 2-box are delimited by # on V''s tape

The machine V can be simulated by a two-way unbounded TM V'. The tape of V' contains finitely many rows of V's current 2-box, delimited by the symbol #. If V were to move its window *within the current* 2-*box*, then V' moves its window either to the neighboring cell in the same row (when V would move its window in the same row), or to the corresponding cell in the left/right neighboring row (when V would move its window to the upper/lower row). If, however, V would move its window *across the border of the current* 2-*box*, then V' either adds a new row before/after the existing rows (when V would cross the upper/lower border), or adds one empty cell to the beginning/end of each existing row and moves # as necessary (when V would cross the left/right border). We see that in order to extend a row by one cell, V' has to shift a part of the contents of its tape. V' can achieve this by moving the contents in a stepwise fashion, cell after cell. (Alternatively, V' can be supplied with a finite storage in its control unit. Using this, V' can shift the contents in a single sweep in a caterpillar-type movement.) The generalization to higher dimensions d is left to the reader as an exercise.

The machine V' is a two-way unbounded TM. As we know, V' can be simulated by the basic model of the TM.

- **Simulation of a Nondeterministic TM.**   Let V be a nondeterministic TM with a Turing program δ_V. The instructions are of the form

$$\delta_V(q_i, z_r) = \{(q_{j_1}, z_{w_1}, D_1), (q_{j_2}, z_{w_2}, D_2), \ldots, (q_{j_k}, z_{w_k}, D_k)\},$$

where $(q_{j_1}, z_{w_1}, D_1), (q_{j_2}, z_{w_2}, D_2), \ldots$ are alternative transitions and k depends on i and r. Call $|\delta_V(q_i, z_r)|$ the *indeterminacy of the instruction* $\delta_V(q_i, z_r)$, and $u \stackrel{\text{def}}{=} \max_{q_i, z_r} |\delta_V(q_i, z_r)|$ the *indeterminacy of the program* δ_V. We call a sequence of numbers $i_1, i_2, \ldots, i_\ell$, $1 \leqslant i_j \leqslant u$, the *scenario* of the execution of δ_V. We say that δ_V *executes alongside the scenario* $i_1, i_2, \ldots, i_\ell$ if the first instruction makes the i_1th transition, the second instruction makes the i_2th transition, and so on.

Let us now describe a three-tape (deterministic) TM V' that will simulate the machine V. The first tape contains the input word x (the same as V). The second tape is used for systematic generation of the scenarios of the execution of δ_V (in shortlex order). The third tape is used to simulate the execution of δ_V on x as if δ_V executed alongside the current scenario.

The machine V' operates as follows: 1) V' generates the next scenario on the second tape, clears the contents of the third tape, and copies x from the first tape to the third one. 2) On the third tape, V' simulates the execution of δ_V alongside the current scenario. If the simulation terminates in q_{yes} (i.e., V would halt on x in q_{yes}), then V' accepts x and halt (as V would accept x and halt). If, during the simulation, the scenario requires a nonexisting transition, or the simulation terminates in a q_{no} state, then V' returns to the first step. 3) If none of the generated scenarios terminates in q_{yes}, V' rejects x and halts (as V would reject x and halt).

If V accepts x, i.e., halts on x in q_{yes}, then it does so after executing a certain *miraculously guessed* scenario. But V' eventually generates this scenario and simulates V according to it. So, V' too halts on x in q_{yes} and accepts x.

It remains to simulate V' with T. But we have already proved that this is possible.

Remark. There is another useful view of the simulation. With each Turing program δ_V and input word x we can associate a directed tree, called the *decision tree of* δ_V for the input x. Vertices of the tree are the pairs $(q_i, z_r) \in Q \times \Gamma$ and there is an arc $(q_i, z_r) \rightsquigarrow (q_j, z_w)$ *iff* there is a transition from (q_i, z_r) to (q_j, z_w), i.e., $(q_j, z_w, D) \in \delta_V(q_i, z_r)$ for some D. The root of the tree is the vertex (q_1, a), where a is the first symbol of x. A scenario is a path from the root to a vertex of the tree. During the execution of δ_V, the nondeterministic TM V starts at the root and then *miraculously picks out*, at each vertex, an arc that is on some path to some final vertex (q_{yes}, z). If there is no such arc, the machine miraculously detects that and immediately halts. In contrast, the *simulator* V' has no miraculous capabilities, so it must systematically generate and check scenarios until one ending in a vertex (q_{yes}, z) is found, or no such scenario exists. If there are finitely many scenarios, the simulation terminates .

6.1.4 Reduced Model

In Definition 6.1 (p. 112) of the basic TM, certain parameters are fixed; e.g., q_1 denotes the initial state; z_1, z_2, z_t denote the symbols $0, 1, \sqcup$, respectively; and the tape is a one-way unbounded single-track tape. We could also fix other parameters, e.g. Σ, Γ, and F (with the exception of δ and Q, because fixing either of these would result in a finite number of different Turing programs). We say that, by fixing these parameters, the basic TMs are *reduced*.

But why do that? The answer is that reduction simplifies many things, because reduced TMs differ only in their δs and Qs, i.e., in their Turing programs. So let us fix Γ and Σ, while fulfilling the condition $\{0, 1\} \subseteq \Sigma \subseteq \Gamma - \{\sqcup\}$ from Definition 6.1. We choose the simplest option, $\Sigma = \{0, 1\}$ and $\Gamma = \{0, 1, \sqcup\}$. No generality is lost by doing so, because any other Σ and Γ can be encoded by 0s and 1s. In addition, by merging the final states into one, say q_2, we can fix the set of final states to a singleton $F = \{q_2\}$. Then, a *reduced Turing machine* is a seven-tuple

$$T = (Q, \{0, 1\}, \{0, 1, \sqcup\}, \delta, q_1, \sqcup, \{q_2\})$$

To obtain a particular reduced Turing machine, we must choose only Q and δ.

6.1.5 Equivalence of Reduced and Basic Models

Is the reduced model of the TM less powerful than the basic one? The answer is no. The two models are equivalent, as they can simulate each other. Let us describe this.

Given an arbitrary reduced TM $R = (Q_R, \{0,1\}, \{0,1,\sqcup\}, \delta_R, q_{1R}, \sqcup, \{q_{2R}\})$, there is a basic TM $T = (Q_T, \Sigma_T, \Gamma_T, \delta_T, q_{1T}, \sqcup, F_T)$ capable of simulating R. Just take $Q_T := Q_R$, $\Sigma_T := \{0,1\}$, $\Gamma_T := \{0,1,\sqcup\}$, $\delta_T := \delta_R$, $q_{1T} := q_{1R}$, and $F_T := \{q_{2R}\}$.

Conversely, let $T = (Q_T, \Sigma_T, \Gamma_T, \delta_T, q_{1T}, \sqcup, F_T)$ be an arbitrary basic TM. We can describe a *finite-storage* TM $S = (Q_S, \{0,1\}, \{0,1,\sqcup\}, \delta_S, q_{1S}, \sqcup, \{q_{2S}\})$ that simulates T on an arbitrary input $w \in \Sigma_T^*$. Since S does not recognize T's tape symbols, we assume that w is binary-encoded, i.e., each symbol of w is replaced by its binary code. Let $n = \lceil \log_2 |\Gamma_T| \rceil$ be the length of this code. Thus, the binary input to S is of length $n|w|$. The machine S has storage in its control unit to record the code of T's current symbol (i.e., the symbol that would be scanned by T at that time). In addition, S's control unit has storage to record T's current state (i.e., the state in which T would be at that time). This storage is initiated to q_{1T}, T's initial state. Now S can start simulating T. It repeatedly does the following: (1) It reads and memorizes n consecutive symbols from its tape. (2) Suppose that T's current state is q_{iT} and the memorized symbols encode the symbol $z_r \in \Gamma_T$. If $\delta_T(q_{iT}, z_r) = (q_{jT}, z_w, D_T)$ is an instruction of δ_T, then S writes the code of z_w into n cells of its tape (thus replacing the code of z_r with the code of z_w), memorizes q_{jT}, and moves the window to the beginning of the neighboring group of n cells (if D_T is Left or Right) or to the beginning of the current group of n cells (if $D_T = $ Stay). As we have seen in Sect. 6.1.3, we can replace S with an equivalent TM R with no storage in its control unit. This is the sought-for reduced TM that is equivalent to T.

NB *The reduced model of the Turing machine enables us to identify Turing machines with their Turing programs. This can simplify the discussion.*

6.1.6 Use of Different Models

Computations on different models of the Turing machine can differ considerably in terms of time (i.e., the number of steps) and space (i.e., the number of visited cells). But this becomes important only when we are interested in the *computational complexity* of problem solving. As concerns *general computability*, i.e., computability on unrestricted models of computation, this is irrelevant.

So the question is: Are different models of the Turing machine of any use in *Computability Theory*? The answer is yes. The generalized models are useful when we try to prove the *existence* of a TM for solving a given problem. Usually, the construction of such a TM is easier if we choose a more versatile model of the Turing machine. In contrast, if we must prove the *nonexistence* of a TM for solving a given problem, then it is usually better to choose a more primitive model (e.g., the basic or reduced model).

Fortunately, by referring to *Computability Thesis*, we can avoid the cumbersome and error-prone constructing of TMs (see Sect. 5.3.5). We will do this only whenever the existence of a TM will be important (and not a particular TM).

6.2 Universal Turing Machine

Recall that Gödel enumerated formulas of *Formal Arithmetic* **A** and, in this way, enabled formulas to express facts about other formulas and, eventually, about themselves. As we have seen (Sect. 4.2.3), such a self-reference of formulas revealed the most important facts about formal axiomatic systems and their theories. Can a similar approach reveal important facts about Turing machines as well? The idea is this: If Turing machines were somehow enumerated (i.e., each TM described by a characteristic natural number, called the *index*), then each Turing machine T could compute with other Turing machines simply by including their indexes in T's input word. Of course, certain questions would immediately arise: How does one enumerate Turing machines with natural numbers? What kind of computing with indexes makes sense? How does one use computing with indexes to discover new facts about Turing machines? In this section we will explain how, in 1936, Turing answered these questions.

6.2.1 Coding and Enumeration of Turing Machines

In order to enumerate Turing machines, we must define how Turing machines will be encoded, that is, represented by words over some coding alphabet. The idea is that we only encode Turing *programs* δ, but in such way that the other components Q, Σ, Γ, F, which determine the particular Turing machine, can be restored from the program's code. An appropriate coding alphabet is $\{0, 1\}$, because it is included in the input alphabet Σ of every TM. In this way, every TM would also be able to read codes of other Turing machines and compute with them, as we suggested above. So, let us see how a Turing machine is encoded in the alphabet $\{0, 1\}$.

Let $T = (Q, \Sigma, \Gamma, \delta, q_1, \sqcup, F)$ be an arbitrary basic Turing machine. If

$$\delta(q_i, z_j) = (q_k, z_\ell, D_m)$$

is an instruction of its Turing program, we encode the instruction by the word

$$K = 0^i 10^j 10^k 10^\ell 10^m,$$

where $D_1 =$ Left, $D_2 =$ Right, and $D_3 =$ Stay.

In this way, we encode each instruction of the program δ. Then, from the obtained codes $K_1, K_2, \ldots, K_r$ we construct the *code* $\langle T \rangle$ of the TM T as follows:

$$\langle T \rangle = 111\, K_1\, 11\, K_2\, 11 \ldots 11\, K_r\, 111. \tag{$*$}$$

We can interpret $\langle T \rangle$ to be the binary code of some natural number. Let us call this number the *index* of the Turing machine T (i.e., its program). Note, however, that some natural numbers are not indexes, because their binary codes are not structured as $(*)$. To avoid this, we make the following **convention**: Any natural number whose binary code is not of the form $(*)$ is an index of a special Turing machine called the *empty* Turing machine. The program of this machine is everywhere undefined, i.e., for each possible pair of state and tape symbol. Thus, for every input, the empty Turing machine immediately halts, in zero steps.

Consequently, the following proposition holds.

Proposition 6.1. *Every natural number is the index of exactly one Turing machine.*

Given an arbitrary index $\langle T \rangle$, we can easily restore from it the components Σ, Γ, Q, and F of the corresponding basic TM $T = (Q, \Sigma, \Gamma, \delta, q_1, \sqcup, F)$; see Sect. 7.2. So, the above construction of $\langle T \rangle$ implicitly defines a *total* mapping $g : \mathbb{N} \to \mathcal{T}$, where $\mathcal{T}$ is the set of all basic Turing machines. Given an arbitrary $n \in \mathbb{N}$, $g(n)$ can be viewed as the nth basic TM and be denoted by T_n. By letting n run through $0, 1, 2, \ldots$ we obtain the sequence $T_0, T_1, T_2, \ldots$, i.e., an *enumeration* of all basic TMs. Later, on p. 139, g will be called the *enumeration function* of the set $\mathcal{T}$. Of course, corresponding to $T_0, T_1, T_2, \ldots$ is the enumeration $\delta_0, \delta_1, \delta_2, \ldots$ of Turing programs.

Remark. We could base the coding and enumeration of TMs on the *reduced* model. In this case, instructions $\delta(q_i, z_j) = (q_k, z_\ell, D_m)$ would be encoded by somewhat shorter words $K = 0^i 10^j 10^k 10^\ell 10^m$, since $j, \ell \in \{0, 1, 2\}$, and no restoration of Σ and Γ from $\langle T \rangle$ would be needed. However, due to the simulation, the programs of the reduced TMs would contain more instructions than the programs of the equivalent basic models. Consequently, their codes would be even longer.

Example 6.3. (TM Code) The code of the TM T in Example 6.1 (p. 115) is

$$\langle T \rangle = 111\, \underbrace{01001001000100}_{K_1}\, 11\, \underbrace{00100100100100}_{K_2}\, 11\, \underbrace{001010001001000}_{K_3}\, 11\, \underbrace{010100010001000}_{K_4}\, 111$$

and the corresponding index is 1075142408958020240455, a large number.

The code $\langle T' \rangle$ of the TM T' in Example 6.2 (p. 116) is

$$\langle T' \rangle = 111\, \underbrace{010010010010 0}_{K_1}\, 11\, \underbrace{010100101 00}_{K_2}\, 11\, \underbrace{001001001001 00}_{K_3}\, 11\, \underbrace{001010010100}_{K_4}\, 11$$

$$\underbrace{001000100010001 0}_{K_5}\, 11\, \underbrace{0001010000010001000}_{K_6}\, 11\, \underbrace{000100100001000 10}_{K_7}\, 11\, \underbrace{0000100100001001 0}_{K_8}\, 11$$

$$\underbrace{0000101000001001000}_{K_9}\, 111.$$

This code is different from $\langle T \rangle$ because T' has a different Turing program. The corresponding index is 133101620081782477988519940592328780481970375943 1. $\square$

Obviously, the indexes of Turing machines are huge numbers and hence not very practical. Luckily, we will almost never need their actual values, and they will not be operands of any arithmetic operation.

6.2.2 The Existence of a Universal Turing Machine

In 1936, using the enumeration of his machines, Turing discovered a seminal fact about Turing machines. We state the discovery in the following proposition.

Proposition 6.2. *There is a Turing machine that can compute whatever is computable by any other Turing machine.*

Proof. The idea is to construct a Turing machine U that is capable of simulating *any* other TM T. To achieve this, we use the method of proving by CT (see Sect. 5.3.5): (a) we describe the concept of the machine U and informally describe the algorithm executed by its Turing program, and (b) we refer to CT to prove that U exists.

(a) The concept of the machine U:

Fig. 6.15 Universal TM: tapes and their contents

- Tapes of the machine U (see Fig. 6.15):

1. The first is the *input tape*. This tape contains an input word consisting of two parts: the code $\langle T \rangle$ of an arbitrary TM $T = (Q, \Sigma, \Gamma, \delta, q_1, \sqcup, F)$, and an arbitrary word w.
2. The second is the *work tape*. Initially, it is empty. The machine U will use it in exactly the same way as T would use its own tape when given the input w.
3. The third is the *auxiliary tape*. Initially, it is empty. The machine U will use it to record the current state in which the simulated T would be at that time, and for comparing this state with the final states of T.

- The Turing program of U should execute the following informal algorithm:

  1. Check whether the input word is $\langle T, w \rangle$, where $\langle T \rangle$ is a code of some TM. If it is not, halt.
  2. From $\langle T \rangle$ restore the set F and write the code $\langle q_1, F \rangle$ to the auxiliary tape.
  3. Copy w to the work tape and shift the window to the beginning of w.
  4. *// Let the aux. tape have $\langle q_i, F \rangle$ and the work tape window scan a symbol z_r.*
     If $q_i \in F$, halt.   *// T would halt in a final state.*
  5. On the input tape, search in $\langle T \rangle$ for the instruction beginning with "$\delta(q_i, z_r) =$"
  6. If not found, halt.   *// T would halt in a non-final state.*
  7. *// Suppose that the found instruction is $\delta(q_i, z_r) = (q_j, z_w, D)$.*
     On the work tape, write the symbol z_w and move the window in direction D.
  8. On the auxiliary tape, replace $\langle q_i, F \rangle$ by $\langle q_j, F \rangle$.
  9. Continue with step 4.

(b) The above algorithm can be executed by a human. So, according to the *Computability Thesis*, there is a Turing machine $U = (Q_U, \Sigma_U, \Gamma_U, \delta_U, q_{1U}, \sqcup, F_U)$ whose program δ_U executes this algorithm. We call U the *Universal Turing Machine*.   □

Small Universal Turing Machines

The universal Turing machine U was actually described in detail. It was to be expected that $\langle U \rangle$ would be a huge sequence of 0s and 1s. Indeed, for example, the code of U constructed by Penrose[2] and Deutsch[3] in 1989 had about 5,500 bits.

Shannon[4] was aware of this when in 1956 he posed the problem of the construction of the *simplest* universal Turing machine U. He was interested in the simplest two-way unbounded model of such a machine. Thus, U was to be deterministic with no storage in its control unit, and have a single two-way infinite tape with one track. To measure the complexity of U Shannon proposed the product $|Q_U| \cdot |\Gamma_U|$. The product is an upper bound on the number of instructions in the program δ_U (see Fig. 6.3 on p. 113). Alternatively, the complexity of U could be measured more realistically by the number of *actual* instructions in δ_U.

Soon it became clear that there is a trade-off between $|Q_U|$ and $|\Gamma_U|$: the number of states can be decreased if the number of tape symbols is increased, and vice versa. So the researchers focused on different *classes* of universal Turing machines. Such a class is denoted by UTM(s,t), for some $s, t \geqslant 2$, and by definition contains all the universal Turing machines with s states and t tape symbols (of the above model).

In 1996, Rogozhin[5] found universal Turing machines in the classes UTM(2,18), UTM(3,10), UTM(4,6), UTM(5,5), UTM(7,4), UTM(10,3), and UTM(24,2). Of these, the machine $U \in$ UTM$(4,6)$ has the smallest number of instructions: 22.

---

[2] Roger Penrose, b. 1931, British physicist, mathematician and philosopher.

[3] David Elieser Deutsch, b. 1953, British physicist.

[4] Claude Elwood Shannon, 1916–2001, American mathematician and electronics engineer.

[5] Yurii Rogozhin, b. 1949, Moldavian mathematician and computer scientist.

6.2.3 The Importance of the Universal Turing Machine

We can now upgrade Fig. 5.13 (p. 99) to Fig. 6.16. The existence of the universal TM indicated that it might be possible to design a general-purpose computing machine—something that is today called the general-purpose computer.

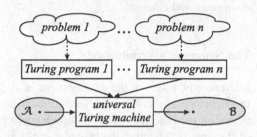

Fig. 6.16 The universal Turing machine can execute any Turing program and thus compute the solution of any problem

6.2.4 Practical Consequences: Data vs. Instructions

Notice that the machine U uses both the *program* of T and the input *data* to T as its own input *data*, i.e., as two input words $\langle T \rangle$ and w written in the *same* alphabet Σ_U. U interprets the word $\langle T \rangle$ as a sequence of instructions to be simulated (i.e., executed), and the word w as input to the simulated T. This consequence is one of the important discoveries of Turing.

Consequence 6.1. (Data vs. Instructions) *There is no a priori difference between data and instructions; the distinction between the two is established by their interpretation.*

6.2.5 Practical Consequences: General-Purpose Computer

Turing's proof of the existence of a universal Turing machine was a theoretical proof that *a general-purpose computing machine* is possible. This answers the question raised by Babbage a century earlier (see p. 6). Thus, the following practical consequence of Turing's discovery was evident.

Consequence 6.2. (General-Purpose Computer) *It is possible to construct a physical computing machine that can compute whatever is computable by any other physical computing machine.*

The construction of a general-purpose computing machine started at the beginning of the 1940s. After initial unsuccessful trials, which were mainly due to the teething troubles of electronics, researchers developed the first, increasingly efficient general-purpose computing machines, now called *computers*. These included ENIAC, EDVAC, and IAS, which were developed by research teams led by Mauchly,[6] Eckert,[7] von Neumann, and others. By the mid-1950s, a dozen other computers had emerged.

Von Neumann's Architecture

Interestingly, much of the development of early computers did not closely follow the structure of the universal Turing machine. The reasons for this were both the desire for the efficiency of the computing machine and the technological conditions of the time. Abstracting the essential differences between these computers and the universal TM, and describing the differences in terms of Turing machines, we find the following:

- Cells are now *enumerated*.
- The control unit does not access cells by a time-consuming movement of the window. Indeed, there is no window. Instead, the *control unit directly accesses an arbitrary cell in constant time* by using an additional component.
- The program is no longer in the control unit. Instead, the *program is written in cells* of the tape (as is the input data to the program).
- The control unit still executes the program in a stepwise fashion, but the *control unit has different duties*. Specifically, in each step, it typically does the following:

  1. reads an instruction from a cell;
  2. reads operands from cells;
  3. executes the operation on the operands;
  4. writes the result to a cell.

To do this, the control unit uses additional components: the *program counter*, which describes from which cell the next instruction of the program will be read, *registers*, which store operands, and a special register, called the *accumulator*, where the result of the operation is left.

Of course, due to these differences, terminological differences also arose. For example, *main memory* ($\approx$ tape), *program* ($\approx$ Turing program), *processor* ($\approx$ control unit), *memory location* ($\approx$ cell), and *memory address* ($\approx$ cell number). The general structure of these computers was called the *von Neumann architecture* (after an influential report on the logical design of the EDVAC computer, in which von Neumann described the key findings of its design team).

---

[6] John William Mauchly, 1907–1980, American physicist.
[7] John Adam Presper Eckert, Jr., 1919–1995, American engineer and computer scientist.

6.2.6 Practical Consequences: Operating System

What takes care of loading a program P, which is to be executed by a computer, into the memory? This is the responsibility of the *operating system* (*OS*). The operating system is a special program that is *resident* in memory. When it executes, it takes care of everything needed to execute any other program P. In particular:

1. It *reads* the program P and its input data from the computer's environment.
2. It *loads* P into the memory.
3. It *sets apart* additional memory space, which P will be using during its execution. This space contains the *data region* with P's input data and other global variables; the *runtime stack* for local variables of procedures and procedure-linkage information; and the *heap* for dynamic allocation of space when explicitly demanded by P.
4. It *initiates* P's execution by transferring control to it, i.e., by writing to the program counter the address of P's first instruction.
5. When P halts, it takes over and gives a chance to the next waiting program.

In time, additional goals and tasks were imposed on operating systems, mainly because of the desire to improve the efficiency of the computer and its user-friendliness. Such tasks include multiprogramming (i.e., supporting the concurrent execution of several programs), memory management (i.e., simulating a larger memory than available in reality), file system (i.e., supporting permanent data storage), input/output management (i.e., supporting communication with the environment), protection (i.e., protecting programs from other programs), security (i.e., protecting programs from the environment), and networking (i.e., supporting communication with other computers).

But there remains the question of what loads the very OS into the memory and starts it. For this, *hardware* is responsible. There is a small program, called the *bootstrap loader*, embedded in the hardware. This program

1. starts automatically when the computer is turned on,
2. loads the OS to the memory, and
3. transfers the control to the OS.

In terms of Turing machines, the bootstrap loader is the Turing program in the control unit of the universal Turing machine U. Its role, however, is to read the *simulator* (i.e., OS) into the control unit, thus turning U into a true universal TM. (The reading of finitely many data and hence the whole simulator into the control unit of a Turing machine is possible, as we described in Sect. 6.1.3.)

We conclude that a modern general-purpose computer can be viewed as a universal Turing machine.

6.2.7 Practical Consequences: RAM Model of Computation

After the first general-purpose computers emerged, researchers tried to abstract the
essential properties of these machines in a suitable model of computation. They
suggested several models of computation, of which we mention the *register machine* and its variants *RAM* and *RASP*.[8] All of these were proved to be equivalent
to the Turing machine: What can be computed on the Turing machine can also be
computed on any of the new models, and vice versa. In addition, the new models displayed a property that was becoming increasingly important as solutions to more and
more problems were attempted using the computers. Namely, these models proved
to be more suitable for and realistic in estimating the computational resources (e.g.,
time and space) needed for computations on computers. This is particularly true of
the RAM model. Analysis of the *computational complexity* of problems, i.e., an estimation of the amount of computing resources needed to solve a problem, initiated
Computational Complexity Theory, a new area of *Computability Theory*. So, let us
describe the RAM model of computation.

While the Turing machine has *sequential access* to data on its tape, RAM accesses data *randomly*. There are several variants of RAM but they differ only in the
level of detail in which they reflect the von Neumann architecture. We present a
more detailed one in Fig. 6.17.

Fig. 6.17 RAM model of
computation

Definition 6.3. (RAM) The **random access machine** (**RAM**) model of computation has several components: the *processor* with *registers*, two of which
are the *program counter* and the *accumulator*; the *main memory*; the *input* and
output memory; and a *program* with *input data*. Also the following holds:

---

[8] The *register machine* and its variants RAM (random access machine) and RASP (random access
stored program) were gradually defined by Wang (1954), Melzak and Minsky (1961), Shepherdson
and Sturgis (1963), Elgot and Robinson (1964), Hartmanis (1971), and Cook and Rechow (1973).

1. *a. Input* and *Output Memory:* During the computation, input data is read sequentially from the input memory, and the results are written to the output memory. Each memory is a sequence of equally sized *locations*. The location size is arbitrary, but finite. A location is empty or contains an integer.

   b. Main Memory: This is a potentially infinite sequence of equally sized locations $m_0, m_1, \ldots$. The index i is called the *address* of m_i. Each location is *directly accessible* by the processor: Given an arbitrary i, reading from m_i or writing to m_i is accomplished in constant time.

   c. Registers: This is a sequence of locations $r_1, r_2, \ldots, r_n$, $n \geqslant 2$, in the processor. Registers are directly accessible. Two of them have special roles. *Program counter* $pc(= r_1)$ contains the address of the location in the main memory that contains the instruction to be executed next. *Accumulator* $a(= r_2)$ is involved in the execution of each instruction. Other r_i are given roles as needed.

   d. Program: The program is a finite sequence of *instructions*. The details of the instructions are not very important as long as the RAM is of limited capability (see Sect. 5.2.4). So, it is assumed that the instructions are similar to the instructions of real computers. Thus, there are arithmetical, logic, input/output, and (un)conditional jump instructions. If $n = 2$, each instruction contains the information op about the operation and, depending on op, the information i about the operand. There may be additional modes of addressing, e.g., indirect (denoted by $*i$) and immediate (denoted by $=i$). Examples of instructions are:

read	(read data from input memory to accumulator a)		
load i	($a := m_i$)	load $*i$	($a := m_{m_i}$)
add i	($a := a + m_i$)	add $=i$	($a := a + i$)
jmp i	($pc := i$)		
jz i	(if $a = 0$ then $pc := i$)	jgz i	(if $a > 0$ then $pc := i$)
store i	($m_i := a$)		
write	(write data from accumulator a to output memory)		
halt	(halt)		

2. Before the RAM is started, the following is done: (a) a program is loaded into the main memory (into successive locations starting with m_0); (b) input data are written to the input memory; (c) the output memory and registers are cleared.

3. From this point on, the RAM operates independently in a mechanical step-wise fashion as instructed by its program. Let $pc = k$ at the beginning of a step. (Initially, $k = 0$.) From the location m_k, the instruction I is read and started. At the same time pc is incremented. So, when I is completed, the next instruction to be executed is in m_{k+1}, *unless* one of the following holds: a) I was jmp i; b) I was jz i or jgz i and pc was assigned i; c) I was halt; d) I changed pc so that it contains an address outside the program. In c) and d) the program halts.

We now state the following important proposition.

Proposition 6.3. (RAM vs. TM) *The RAM and the Turing machine are equivalent: What can be computed on one of them can be computed on the other.*

Box 6.1 (Proof of Proposition 6.3). We show how TM and RAM simulate each other.

Simulation of a TM with a RAM. Let $T = (Q, \Sigma, \Gamma, \delta, q_1, \sqcup, F)$ be an arbitrary TM. The RAM will have its main memory divided into three parts. The first part, consisting of the first p locations $m_0, \ldots, m_{p-1}$, will contain the RAM's program. The second part will contain the Turing program δ. The third part will be the rest of the main memory; during the simulation, it will contain the same data as T would have on its tape. Two of the RAM's registers will play special roles: r_3 will reflect the current state of T, and r_4 will contain the address of the location in the RAM's main memory corresponding to the cell under T's window.

The RAM is initialized as follows. Let us view δ as a table $\Delta = Q \times \Gamma$, where the component $\Delta[q, z] = (q', z', D)$ if $\delta(q, z) = (q', z', D)$ is an instruction of δ, and $\Delta[q, z] = 0$ if $\delta(q, z)\!\uparrow$. Since there are $d = |Q| \cdot |\Gamma|$ components in Δ, we can bijectively map them to the d locations $m_p, \ldots, m_{p+d-1}$. So, we choose a bijection $\ell : \Delta \to \{p, \ldots, p+d-1\}$ and write each $\Delta[q, z]$ into the location $m_{\ell(q,z)}$. (A possible ℓ would map row after row of Δ into the memory.) The third part of the RAM's memory is cleared and T's input word is written to the beginning of it, i.e., into the locations $m_{p+d}, m_{p+d+1}, \ldots$. The registers r_3 and r_4 are set to q_1 and $p + d$, respectively.

Now, the simulation of T starts. Each step of T is simulated as follows. Based on the values $q = (r_3)$ and $z = m_{(r_4)}$, the RAM reads the value of $\delta(q, z)$ from $m_{\ell(q,z)}$. If this value is 0, the RAM halts because $\delta(q, z)\!\uparrow$. Otherwise, the value read is (q', z', D), so the RAM must simulate the instruction $\delta(q, z) = (q', z', D)$. This it can do in three steps: 1) $r_3 := q'$; 2) $m_{(r_4)} := z'$; 3) depending on D, it decrements r_4 ($D = \text{Left}$), increments r_4 ($D = \text{Right}$), or leaves r_4 unchanged ($D = \text{Stay}$).

Remark. Note that the RAM could simulate any other TM. To do this, it would only change δ in the second part of the main memory and, of course, adapt the value of d.

Simulation of a RAM with a TM. Let R be an arbitrary RAM. R will be simulated by the following multi-tape TM T. For each register r_i of R there is a tape, called the r_i-tape, whose contents will be the same as the contents of r_i. Initially, it contains 0. There is also a tape, called the m-tape, whose contents will be the same as the contents of R's main memory. If, for $i = 0, 1, 2, \ldots$, the location m_i would contain c_i, then the m-tape will have written $|\, 0\!:\!c_0 \mid 1\!:\!c_1 \mid 2\!:\!c_2 \mid \ldots \mid i\!:\!c_i \mid \ldots$ (up to the last nonempty word).

T operates as follows. It reads from the pc-tape the value of the program counter, say k, and increments this value on the pc-tape. Then it searches the m-tape for the subsequence $\mid k\!:\,$. If the subsequence is found, T reads c_k, i.e., R's instruction I, and extracts from I both the information op about the operation and the information i about the operand. What follows depends on the addressing mode: (a) If I is op $= i$ (immediate addressing), then the operand is i. (b) If I is op i (direct addressing), then T searches the m-tape for the subsequence $\mid i\!:\,$. If the subsequence is found, then the operand is c_i. (c) If I is op $*i$ (indirect addressing), then, after T has found c_i, it searches the m-tape for the subsequence $\mid c_i\!:\,$. If the subsequence is found then the operand is c_{c_i}. When the operand if known, T executes op using the operand and the contents of the a-tape. $\quad\square$

NB *General-purpose computers are capable of computing exactly what Turing machines can—assuming there are no limitations on the time and space consumed by the computations. Because of this, we will continue to investigate computability by using the Turing machine as the model of computation. The conclusions that we will come to will also hold for modern general-purpose computers.*

6.3 Use of a Turing Machine

In this section we will describe three elementary tasks for which we can use a Turing machine: 1) to compute the values of a function; 2) to generate elements of a set; and 3) to find out whether objects are members of a set.

6.3.1 Function Computation

A Turing machine is implicitly associated, for each natural $k \geqslant 1$, with a k-ary function mapping k words into one word. Since the function is *induced* by the Turing machine, we will call it the k-ary *proper function* of the Turing machine. Here are the details.

Definition 6.4. (Proper Function) Let $T = (Q, \Sigma, \Gamma, \delta, q_1, \sqcup, F)$ be an arbitrary Turing machine and $k \geqslant 1$ a natural number. The k-**ary proper function** of T is a partial function $\varphi_T^{(k)} : (\Sigma^*)^k \to \Sigma^*$, defined as follows:

If the input word to T consists of words $u_1, \ldots, u_k \in \Sigma^*$, then

$$\varphi_T^{(k)}(u_1, \ldots, u_k) \overset{\text{def}}{=} \begin{cases} v, & \text{if} \quad T \text{ halts in any state } \textit{and} \text{ the tape contains} \\ & \qquad \text{only the word } v \in \Sigma^*; \\ \uparrow, & \text{else, i.e., } T \text{ doesn't halt } \textit{or} \text{ } T \text{ halts but the tape} \\ & \qquad \text{doesn't contain a word in } \Sigma^*. \end{cases}$$

If e is the index of T, we also denote the k-ary proper function of T by $\varphi_e^{(k)}$. When k is known from the context or it is not important, we write φ_T or φ_e. The domain of φ_e is also denoted by $\mathcal{W}_e$, i.e., $\mathcal{W}_e = \text{dom}(\varphi_e) = \{x \mid \varphi_e(x)\downarrow\}$.

So, given k words $u_1, \ldots, u_k$ written in the input alphabet of the Turing machine, we write the words to the tape of the machine, start it and wait until the machine halts and leaves a single word on the tape written in the same alphabet. If this does happen, and the resulting word is denoted by v, then we say that the machine has computed the value v of its k-ary proper function for the arguments $u_1, \ldots, u_k$.

The interpretation of the words $u_1, \ldots, u_k$ and v is left to us. For example, we can view the words as the encodings of natural numbers. In particular, we may use the alphabet $\Sigma = \{0, 1\}$ to encode $n \in \mathbb{N}$ by 1^n, and use 0 as a delimiter between the different encodings on the tape. For instance, the word 11101011001111 represents the numbers $3, 1, 2, 0, 4$. (The number 0 was represented by the empty word $\varepsilon = 1^0$.) In this case, $\varphi_T^{(k)}$ is a k-ary numerical function with values represented by the words $1 \ldots 1$. When the function value is 0, the tape is empty (i.e., contains $1^0 = \varepsilon$). Another encoding will be given in Sect. 6.3.6.

TM as a Computer of a Function

Usually we face the opposite task: *Given a function* $\varphi : (\Sigma^*)^k \to \Sigma^*$, *find a TM capable of computing* φ's *values*. Thus, we must find a TM T such that $\varphi_T^{(k)} \simeq \varphi$. Depending on how powerful, if at all, such a T can be, i.e., depending on the extent to which φ can possibly be computed, we distinguish between three kinds of functions φ (in accordance with the formalization on p. 104).

Definition 6.5. Let $\varphi : (\Sigma^*)^k \to \Sigma^*$ be a function. We say that

φ is **computable**   *if*   there is a TM that can compute φ
anywhere on $\mathrm{dom}(\varphi)$
and $\mathrm{dom}(\varphi) = (\Sigma^*)^k$;

φ is **partial computable**   *if*   there is a TM that can compute φ
anywhere on $\mathrm{dom}(\varphi)$;

φ is **incomputable**   *if*   there is *no* TM that can compute φ
anywhere on $\mathrm{dom}(\varphi)$.

A TM that can compute φ anywhere on $\mathrm{dom}(\varphi)$ is also called the **computer** of φ.

Example 6.4. (Addition in Different Ways) The two Turing machines in Examples 6.1 (p. 115) and 6.2 (p. 116) are computers of the function $\mathrm{sum}(n_1, n_2) = n_1 + n_2$, where $n_1, n_2 \geqslant 0$.   □

A slight generalization defines (in accordance with Definition 5.3 on p. 105) what it means when we say that such a function is computable *on a set* $S \subseteq (\Sigma^*)^k$.

Definition 6.6. Let $\varphi : (\Sigma^*)^k \to \Sigma^*$ be a function and $S \subseteq (\Sigma^*)^k$. We say that

φ is **computable on** S *if*   there is a TM that can compute φ
anywhere on S;

φ is **partial computable on** S *if*   there is a TM that can compute φ
anywhere on $S \cap \mathrm{dom}(\varphi)$;

φ is **incomputable on** S *if*   there is *no* TM that can compute φ
anywhere on $S \cap \mathrm{dom}(\varphi)$.

6.3.2 Set Generation

When can elements of a set be enumerated? That is, when can elements of a set be listed in a finite or infinite sequence so that each and every element of the set sooner or later appears in the sequence? Moreover, when can the sequence be generated by an algorithm? These questions started the quest for the formalization of set generation. The answers were provided by Post, Church, and Turing.

Post's Discoveries

Soon after Hilbert's program appeared in the 1920s, Post started investigating the decidability of formal theories. In 1921, he proved that the *Propositional Calculus* **P** is a decidable theory by showing that there is a decision procedure (i.e., algorithm) that, for an arbitrary proposition of **P**, decides whether or not the proposition is provable in **P**. The algorithm uses truth-tables. Post then started investigating the decidability of the formal theory developed in *Principia Mathematica* (see Sect. 2.2.3). Confronted with *sets* of propositions, he realized that there is a strong similarity between the *process of proving* propositions in a formal theory and the *process of "mechanical, algorithmic generating"* of the elements of a set. Indeed, proving a proposition is, in effect, the same as "generating" a new element of the set of all theorems. This led Post to the questions,

> *What does it mean to "algorithmically generate" elements of a set?*
> *Can every countable set be algorithmically generated?*

Consequently, he started searching for a *model* that would formally define the intuitive concept of the "algorithmic generation of a set" (i.e., an algorithmic listing of all the elements of a set). Such a model is called the **generator** of a set.

In 1920–1921, Post developed *canonical systems* and *normal systems* and proposed these as generators. Informally, a canonical system consists of a symbol S, an alphabet Σ, and a finite set $\mathcal{P}$ of transformation rules, called productions. Starting with the symbol S, the system gradually transforms S through a sequence of intermediate words into a word over Σ. We say that the word has been generated by the canonical system. In each step of the generation, the last intermediate word is transformed by a production applicable to it. Since there may be several applicable productions, one must be selected. As different selections generally lead to different generated words, the canonical system can generate a *set* of words over Σ. Post also showed that productions can be simplified while retaining the generating power of the system. He called canonical systems with simplified productions *normal systems*. The set generated by a normal system he called a *normal set.*[9]

---

[9] Post described his ideas in a less abstract and more readable way than was the usual practice at the time. The reader is advised to read his influential and informative paper from 1944 (see References). Very soon, Post's user-friendly style was adopted by other researchers. This speeded up the exchange of ideas and results and, hence, the development of *Computability Theory*.

Box 6.2 (Normal Systems).

A canonical system is a quintuple $(\mathcal{V}, \Sigma, \mathcal{X}, \mathcal{P}, S)$, where $\mathcal{V}$ is a finite set of symbols and $\Sigma \subset \mathcal{V}$. The symbols in Σ are said to be *final*, and the symbols in $\mathcal{V} - \Sigma$ are *non-final*. There is a distinguished non-final symbol, S, called the *start* symbol. $\mathcal{X}$ is a finite set of *variables*. A variable can be assigned a value that is an element of $\mathcal{V}^*$, i.e., any finite sequence of final and non-final symbols.

The remaining component of the quintuple is $\mathcal{P}$. This is a finite set of transformation rules called *productions*. A production describes the conditions under which, and the manner in which, subwords of a word can be used to build a new word. The general form of a production is

$$\alpha_0 x_1 \alpha_1 x_2 \ldots \alpha_{n-1} x_n \alpha_n \rightarrow \beta_0 x_{i_1} \beta_1 x_{i_2} \ldots \beta_{m-1} x_{i_m} \beta_m,$$

where $\alpha_i, \beta_j \in \mathcal{V}^*$ and $x_k \in \mathcal{X}$. When and how can a production p be applied to a word $v \in \mathcal{V}^*$? If each variable x_k in the left-hand side of p can be assigned a subword of v so that the left-hand side of p becomes equal to v, then v can be transformed into a word v', which is obtained from the right-hand side of p by substituting variables x_k with their assigned values. Note that p may radically change the word v: Some subwords of v may disappear; new subwords may appear in v'; and all these constituents may be arbitrarily permuted in v'.

We write $S \xrightarrow{*}_{\mathcal{P}} v$ to denote that the start symbol S can be transformed into v by a finite number of applications of productions of $\mathcal{P}$. We say that S *generates* the word w if $S \xrightarrow{*}_{\mathcal{P}} w$ and $w \in \Sigma^*$. The set of words generated by $\mathcal{P}$ is denoted by $G(\mathcal{P})$, that is, $G(\mathcal{P}) = \{w \in \Sigma^* \mid S \xrightarrow{*}_{\mathcal{P}} w\}$.

Post proved that the productions of any canonical system can be substituted by productions of the form $\alpha_i x_k \rightarrow x_k \beta_j$. Canonical systems with such productions are said to be *normal*.

Post proved that the formal theory developed in *Principia Mathematica* can be represented as a normal system. Consequently, the set of theorems of *Principia Mathematica* is a normal set. Encouraged by this result, in 1921 he proposed the following formalization of the intuitive notion of set "generation":

Post Thesis. *A set S can be "generated"* $\longleftrightarrow$ *S is normal*

Post did not prove this proposition. He was not aware that it cannot be proved. The reasons are the same as those that, 15 years later, prevented the proving of the *Computability Thesis* (see Sect. 5.3). Namely, the proposition is a variant of the *Computability Thesis*.

In order to move on, he used the proposition as a working hypothesis (now called the *Post Thesis*), i.e., something that he will eventually prove. The thesis enabled him to progress in his research. Indeed, he made several important findings that were, 15 years later, independently, and in different ways discovered by Gödel, Church and Turing. (In essence, these are the existence of undecidable sets, which we will come to in the following sections, and the existence of the universal Turing machine.) Unfortunately, Post did not publish his results because he felt sure that he should prove his thesis first. As a byproduct of his attempts to do that, in 1936 he proposed a model of computation that is now called the Post machine (see Sect. 5.2.3).

Church's Approach

In 1936, Church also became interested in the questions of set "generation". While investigating sets whose elements are values of computable functions, he noticed: If a function g is computable on $\mathbb{N}$, then one can successively compute the values $g(0), g(1), g(2), \ldots$ and hence generate the set $\{g(i) \mid i \in \mathbb{N}\}$.

But, the opposite task is more interesting: Given a set S, find a computable function $g : \mathbb{N} \rightarrow S$ so that $\{g(i) \mid i \in \mathbb{N}\} = S$. If g exists, then S can be listed, i.e., all the elements of S are $g(0), g(1), g(2), \ldots$, and enumerated, i.e., an element $x \in S$ is said to be nth in order if n is the smallest $i \in \mathbb{N}$ for which $g(i) = x$. Such a g is said to be an *enumeration function* of the set S. (For example, the mapping g on p. 126 is an enumeration function of the set of all Turing programs.)

These ideas were also applied by Kleene. He imagined a function g that maps natural numbers into systems of equations $\mathscr{E}$, as defined by Herbrand and Gödel (see p. 84). If g is computable, then by computing $g(0), g(1), g(2), \ldots$ one generates systems $\mathscr{E}_0, \mathscr{E}_1, \mathscr{E}_2, \ldots$, each of which defines a computable function. Kleene then proved that there is *no* (total) computable function g on $\mathbb{N}$ such that g would generate *all* systems of equations that define computable functions. The reader may (correctly) suspect that Kleene's proof is connected with the diagonalizaton in Sect. 5.3.3.

Naturally, Post was interested in seeing how his normal systems compared with Church's generator (i.e., (total) computable functions). He readily proved the following theorem. (We omit the proof.)

Theorem 6.1 (Normal Set). *A set S is normal $\Longleftrightarrow$ $S = \emptyset \lor S$ is the range of a computable function on $\mathbb{N}$.*

TM as a Generator

In addition to Post's normal systems and Church's computable functions, also Turing machines can be generators. A Turing machine that generates a set S will be denoted by G_S (see Fig. 6.18). The machine G_S writes to its tape, in succession, the elements of S and nothing else. The elements are delimited by the appropriate tape symbol in $\Gamma - \Sigma$ (e.g., #).

Fig. 6.18 A set S is generated by the Turing machine G_S

It turned out that the three generators are equivalent in their generating power. That is, if a set can be generated by one of them, it can be generated by any other. Because the Turing machine most convincingly formalized the basic notions of computation, we restate the *Post Thesis* in terms of this model of computation.

Post Thesis. (TM) *A set S can be "generated" $\longleftrightarrow$ S can be generated by a TM*

Due to the power of the Turing machine, the intuitive concept of set "generation" was finally formalized. Therefore, we will no longer use quotation marks.

Sets that can be algorithmically generated were called normal by Post. Today, we call them *computably enumerable* sets. Here is the official definition.

Definition 6.7. (c.e. Set) A set S is **computably enumerable** (for short **c.e.**)[10] if S can be generated by a TM.

From Theorem 6.1 and the above discussion we immediately deduce the following important corollary.

Corollary 6.1. *A set S is c.e. $\Longleftrightarrow$ $S = \emptyset \vee S$ is the range of a computable function on $\mathbb{N}$.*

Remarks. 1) Note that the order in which the elements of a c.e. set are generated is not prescribed. So, any order will do. 2) An element may be generated more than once; what matters is that it be generated at least once. 3) Each element of a c.e. set is generated in finite time (i.e., a finite number of steps of a TM). This, however, does not imply that the *whole* set can be generated in finite time, as the set may be infinite.

6.3.3 Set Recognition

Let T be a Turing machine and w an arbitrary word. Let us write w as the input word to T's tape and start T. There are three possible outcomes. If T after reading w eventually halts in a final state q_{yes}, then we say that T *accepts* w. If T after reading w eventually halts in a non-final state q_{no}, we say that T *rejects* w. If, however, T after reading w never halts, then we say that T *does not recognize* w. Thus, a Turing machine T is implicitly associated with the set of all the words that T accepts. Since the set is induced by T, we will call it the *proper set* of T. Here is the definition.

---

[10] The older name is *recursively enumerable (r.e.)* set. See more about the renaming on p. 153.

Definition 6.8. (Proper Set) Let $T = (Q, \Sigma, \Gamma, \delta, q_1, \sqcup, F)$ be a Turing machine. The **proper set**[11] of T is the set $L(T) \overset{\text{def}}{=} \{w \in \Sigma^* \mid T \text{ accepts } w\}$.

Usually we are confronted with the opposite task: *Given a set S, find a Turing machine T such that $L(T) = S$.* Put another way, we must find a Turing machine that accepts exactly the given set. Such a T is called an **acceptor** of the set S. However, we will see that the existence of an acceptor of S is closely connected with S's amenability to *set recognition*. So let us focus on this notion.

TM as a Recognizer of a Set

Informally, to *completely recognize a set* in a given environment, also called the *universe*, is to determine which elements of the universe are members of the set and which are not. Finding this out separately for each and every element of the universe is impractical because the universe may be too large. Instead, it suffices to exhibit an algorithm capable of deciding this for an arbitrary element of the universe.

What can be said about such an algorithm? Let us be given the universe $\mathcal{U}$, an arbitrary set $S \subseteq \mathcal{U}$, and an arbitrary element $x \in \mathcal{U}$. We ask: "Is x a member of the set S?", or, for short, $x \in ?S$. The answer is either YES or NO, as there is no third possibility besides $x \in S$ and $x \notin S$.

However, the answer may not be obvious (Fig. 6.19).

Fig. 6.19 Is x in S or in $\overline{S} = \mathcal{U} - S$? The border between S and $\overline{S}$ is not clear

So, let us focus on the construction of an algorithm A that will be capable of answering the question $x \in ?S$. First, recall the definition of the characteristic function of a set.

Definition 6.9. The **characteristic function** of a set S, where $S \subseteq \mathcal{U}$, is a function $\chi_S : \mathcal{U} \to \{0, 1\}$ defined by

$$\chi_S(x) \overset{\text{def}}{=} \begin{cases} 1 (\equiv \text{YES}), & \text{if } x \in S; \\ 0 (\equiv \text{NO}), & \text{if } x \notin S. \end{cases}$$

---

[11] Also called the *language* of the Turing machine T.

By definition, χ_S is *total* on $\mathcal{U}$, that is, $\chi_S(x)$ is defined for every $x \in \mathcal{U}$. If the sought-for algorithm A could compute χ_S's values, then it would answer the question $x \in ?S$ simply by computing the value $\chi_S(x)$. In this way the task of set recognition would be reduced to the task of function computation.

But how would A compute the value $\chi_S(x)$? The general definition of the characteristic function χ_S reveals nothing about how to compute χ_S and, consequently, how to construct A. What is more, the definition reveals nothing about the computability of the function χ_S. So, until $\mathcal{U}$ and S are defined in greater detail, nothing particular can be said about the design of A and the computability of χ_S.

Nevertheless, we *can* distinguish between three kinds of sets S, based on the extent to which the values of χ_S can possibly be computed on $\mathcal{U}$ (and, consequently, how S can possibly be recognized in $\mathcal{U}$).

Definition 6.10. Let $\mathcal{U}$ be the universe and $S \subseteq \mathcal{U}$ be an arbitrary set. We say that the set

S is **decidable** (or **computable**[12]) in $\mathcal{U}$ *if* χ_S is a computable function on $\mathcal{U}$;
S is **semi-decidable** in $\mathcal{U}$ *if* χ_S is a computable function on S;
S is **undecidable** (or **incomputable**) in $\mathcal{U}$ *if* χ_S is an incomputable function on $\mathcal{U}$.

(Remember that $\chi_S : \mathcal{U} \to \{0,1\}$ is total.)

This, in turn, tells us how powerful the algorithm A can be:

- When a set S is decidable in $\mathcal{U}$, there exists an algorithm (Turing program) A capable of deciding, for an *arbitrary* element $x \in \mathcal{U}$, whether or not $x \in S$. We call such an algorithm a **decider** of the set S in $\mathcal{U}$ and denote it by D_S. A decider D_S makes it possible to completely recognize S in $\mathcal{U}$, that is, to determine what is in S and what is in $\overline{S} = \mathcal{U} - S$.
- When a set S is semi-decidable in $\mathcal{U}$, there is an algorithm A capable of determining, for an *arbitrary* $x \in S$, that x *is* a member of S. If, however, in truth $x \in \overline{S}$, then A may or may not find this out (because it may not halt). We call such an algorithm the **recognizer** of the set S in $\mathcal{U}$ and denote it by R_S. The recognizer R_S makes it possible to completely determine what is in S, *but* it may or may not be able to completely determine what is in $\overline{S}$.
- When a set S is undecidable in $\mathcal{U}$, there is no algorithm A that is capable of deciding, for *arbitrary* element $x \in \mathcal{U}$, whether or not $x \in S$. In other words, *any* candidate algorithm A fails to decide, for *at least one* element $x \in \mathcal{U}$, whether or not $x \in S$ (because on such an input A gives an incorrect answer or never halts). This can happen for an x that is in truth a member of S or $\overline{S}$. In addition, there can be several such elements. Hence, we cannot completely determine what is in S, or in $\overline{S}$, or in both.

---

[12] The older name is *recursive set*. We describe the reasons for a new name on p. 153.

A word of caution. Observe that a decidable set is by definition also semi-decidable. Hence, a decider of a set is also a recognizer of the set. The inverse does not hold in general; we will prove later that there are semi-decidable sets that are not decidable.

Remarks. The adjectives *decidable* and *computable* will be used as synonyms and depend on the context. For example, we will say "decidable set" in order to stress that the set is completely recognizable in its universe. But we will say "computable set" in order to stress that the characteristic function of the set is computable on the universe. The reader should always bear in mind the alternative adjective and its connotation. We will use the adjectives *undecidable* and *incomputable* similarly. Interestingly, the adjective *semi-decidable* will find a synonym in the adjective *computably enumerable (c.e.)*. In the following we explain why this is so.

6.3.4 Generation vs. Recognition

In this section we show that set generation and set recognition are closely connected tasks. We do this in two steps: First, we prove that every c.e. set is also semi-decidable; then we prove that every semi-decidable set is also c.e. So, let $\mathcal{U}$ be the universe and $S \subseteq \mathcal{U}$ be an arbitrary set.

Fig. 6.20 R_S checks whether G_S generated x

Suppose that the set S is c.e. We can use the generator G_S to answer the question "Is x a member of S?" for arbitrary $x \in \mathcal{U}$. To do this, we enhance G_S so that the resulting algorithm A checks each generated word to see whether or not it is x (see Fig. 6.20). If and when this happens, the algorithm A outputs the answer YES (i.e., $x \in S$) and halts; otherwise, the algorithm A continues generating and checking the members of S. (Clearly, when in truth $x \notin S$ and S is infinite, A never halts.) It is obvious that A is R_S, a recognizer of the set S. Thus, S is semi-decidable.

We have proven the following theorem.

Theorem 6.2. *A set S is c.e.* $\Longrightarrow$ *S is semi-decidable in $\mathcal{U}$.*

What about the other way around? Is a semi-decidable set also c.e.? In other words, given a recognizer R_S of S, can we construct a generator G_S of S? The answer is yes, but the proof is trickier.

Fig. 6.21 a) A naive construction of G_S that works only for decidable sets S. b) An improved construction of G_S that works for any semi-decidable set S

Suppose that the set S is semi-decidable. The *naive* construction of G_S is as follows (see Fig. 6.21a). Assuming that $\mathcal{U}$ is c.e., G_S uses 1) a generator $G_{\mathcal{U}}$ to generate elements of $\mathcal{U}$ and 2) a recognizer R_S to find out, for each generated $x \in \mathcal{U}$, whether or not $x \in S$. If R_S answers YES, then G_S outputs (generates) x, and if the answer is NO, then G_S outputs nothing. *However*, there is a pitfall in this approach: If in truth $x \notin S$, then the answer NO is not guaranteed (because S is semi-decidable). So G_S may wait indefinitely long for it. In that case, the operation of G_S drags on for an infinitely long time, although there are still elements of $\mathcal{U}$ that should be generated and checked for membership in S.

This trap can be avoided by a technique called *dovetailing* (see Fig. 6.21b). The idea is to ensure that G_S waits for R_S's answer for only a finitely long time. To achieve this, G_S must allot to R_S a finite number of steps, say j, to answer the question $x \in ?S$. If R_S answers YES (i.e., $x \in S$) in exactly j steps, then G_S generates (outputs) x. Otherwise, G_S asks R_S to recognize, in the same controlled fashion, some other candidate element of $\mathcal{U}$. Of course, later G_S must start again the recognition of x, but this time allotting to R_S a larger number of steps.

To implement this idea, G_S must systematically label and keep track of each candidate element of $\mathcal{U}$ (e.g., with the natural number i) as well as of the currently allotted number of steps (e.g., with the natural number j). For this reason, G_S uses a generator $G_{\mathbb{N}^2}$, which systematically generates pairs $(i, j) \in \mathbb{N}^2$ in such a way that each pair is generated exactly once. The details are given in the proof of the next theorem (see Box 6.3).

Theorem 6.3. *Let the universe $\mathcal{U}$ be c.e. Then:*
A set S is semi-decidable in $\mathcal{U} \Longrightarrow S$ is c.e.

Box 6.3 (Proof).

Recall that a c.e. set is the range of a computable function on $\mathbb{N}$ (Corollary 6.1, p. 140). Hence, there is a computable function $f : \mathbb{N} \to \mathcal{U}$ such that, for every $x \in \mathcal{U}$, there is an $i \in \mathbb{N}$ for which $x = f(i)$. Thus, the set $\mathcal{U}$ can be generated by successive computing of the values $f(0), f(1), f(2), \ldots$ After $f(i)$ is computed, it is fed into the recognizer R_S for a *controlled recognition*, i.e., a finite number j of steps are allotted to R_S in which it tries to decide $x \in ?S$. (This prevents R_S from getting trapped in an endless computation of $\chi_S(x)$.) If the decision is not made in the jth step, it might be made in the following step. So R_S tries again later to recognize x, this time with $j+1$ allotted steps.

Obviously, we need a generator capable of generating all the *pairs* (i, j) of natural numbers in such a way that each pair is generated exactly once. This can be done by a generator that generates pairs in the order of visiting dots (representing pairs) described in Fig 6.22.

Fig. 6.22 The order of generating pairs $(i, j) \in \mathbb{N}^2$. Each pair is generated exactly once

The initial generated pairs are $(0,0)$, $(0,1), (1,0)$, $(0,2), (1,1), (2,0)$, $(0,3), (1,2), (2,1), (3,0), \ldots$

The generator must output pairs $(i, j) \in \mathbb{N}^2$ so that, for each $k = 0, 1, 2, 3, \ldots$, it systematically generates all the pairs having $i + j = k$. These pairs correspond to the dots on the same diagonal in the table. It is not hard to conceive such a generator. The generator successively increments the variable k and, for each k, first outputs $(0, k)$ and then all the remaining pairs up to and including $(k, 0)$, where each pair is constructed from the previous one by incrementing its first component and decrementing its second component by 1.

This was but an intuitive description of the generator's algorithm (program). However, according to the *Computability Thesis*, there exists an actual Turing machine performing all the described tasks. We denote this Turing machine by $G_{\mathbb{N}^2}$. It is easy to see that, for arbitrary $i, j \in \mathbb{N}$, the machine $G_{\mathbb{N}^2}$ generates every (i, j) exactly once.

How will $G_{\mathbb{N}^2}$ be used? With the pair generator $G_{\mathbb{N}^2}$, the function f, and the recognizer R_S, we can now construct an improved generator G_S of the set S. (See Fig. 6.21b.) This generator repeats the following four steps: 1) it demands from $G_{\mathbb{N}^2}$ the next pair (i, j); 2) it generates an element $x \in \mathcal{U}$ by computing $x := f(i)$; 3) it demands from R_S the answer to the question $x \in ?S$ in exactly j steps; 4) if the answer is YES (i.e., $x \in S$), then G_S outputs (generates) x; otherwise it generates nothing and returns to 1). $\square$

6.3.5 The Standard Universes Σ^* and $\mathbb{N}$

In the rest of the book, the universe $\mathcal{U}$ will be either Σ^*, the set of all the words over the alphabet Σ, or $\mathbb{N}$, the set of all natural numbers. In what follows we show that both Σ^* and $\mathbb{N}$ are c.e. sets. This will closely link the tasks of set generation and set recognition.

Theorem 6.4. Σ^* *and* $\mathbb{N}$ *are c.e. sets.*

Proof. We intuitively describe the generators of the two sets.

a) The generator G_{Σ^*} will output words in *shortlex* order (i.e., in order of increasing length, and, in the case of equal length, in lexicographical order; see Appendix A, p. 369). For example, for $\Sigma = \{a,b,c\}$, the first generated words are

$$\varepsilon,$$
$$a,b,c,$$
$$aa,ab,ac,ba,bb,bc,ca,cb,cc,$$
$$aaa,aab,aac,aba,abb,abc,aca,acb,acc,baa,bab,bac,bba,bbb,bbc,bca,bcb,bcc,caa,cab,cac,cba,cbb,cbc,cca,ccb,ccc,$$
$$\vdots$$

To generate the words of length $\ell + 1$, G_{Σ^*} does the following: For each previously generated word w of length ℓ, it outputs the words ws for each symbol $s \in \Sigma$.

b) The generator $G_{\mathbb{N}}$ will generate binary representations of natural numbers $n = 0, 1, \ldots$. To achieve this, it operates as follows. First, it generates the two words of length $\ell = 1$, that is, 0 and 1. Then, to generate all the words of length $\ell + 1$, it outputs, for each previously generated word w of length ℓ *that starts with* 1, the words $w0$ and $w1$. (In this way the words with leading 0s are not generated.) For example, the first generated binary representations and the corresponding natural numbers are:

$$0,1, \qquad\qquad\qquad 0,1,$$
$$10,11, \qquad\qquad\qquad 2,3,$$
$$100,101,110,111, \qquad\qquad\qquad 4,5,6,7,$$
$$1000,1001,1010,1011,1100,1101,1110,1111 \qquad 8,9,10,11,12,13,14,15,$$
$$\vdots \qquad\qquad\qquad\qquad \vdots$$

The existence of the Turing machines G_{Σ^*} and $G_{\mathbb{N}}$ is assured by the *Computability Thesis*. $\qquad\square$

Combining Theorems 6.2, 6.3, and 6.4 we find that when the universe is Σ^* or $\mathbb{N}$, set generation and set recognition are closely linked tasks.

Corollary 6.2. *Let the universe* $\mathcal{U}$ *be* Σ^* *or* $\mathbb{N}$*. Then:*
A set S is semi-decidable in $\mathcal{U}$ $\Longleftrightarrow$ S is c.e.

Remark. Since in the following the universe will be either Σ^* or $\mathbb{N}$, the corollary allows us to use the adjectives *semi-decidable* and *computably enumerable* as synonyms. Similarly to the pairs of adjectives *decidable–computable* and *undecidable–incomputable*, we will use them in accordance with our wish to stress the amenability of a set of interest to recognition or generation.

6.3.6 Formal Languages vs. Sets of Natural Numbers

In this section we show that it is not just by chance that both Σ^* and $\mathbb{N}$ are c.e. sets. This is because there is a bijective function from Σ^* to $\mathbb{N}$. We prove this in Box 6.4.

Theorem 6.5. *There is a bijection* $f_{|\Sigma|} : \Sigma^* \to \mathbb{N}$.

Box 6.4 (Proof).

We prove the theorem for the alphabet $\Sigma = \{a_0, a_1\}$ and then for a general $\Sigma = \{a_0, a_1 \ldots, a_{p-1}\}$.

a) Let $\Sigma = \{a_0, a_1\}$. Imagine that a_0 and a_1 represent the numbers 0 and 1, respectively. In general, we will say that $u \in \{a_0, a_1\}^*$ *represents* a natural number if u has no leading symbols a_0 (with the exception of $u = a_0$). For example, $a_0 a_1$ represents no natural number. We can now define the function $\#_2 : \{a_0, a_1\}^* \to \mathbb{N}$ as follows: *If $u \in \{a_0, a_1\}^*$ represents a natural number, then let $\#_2(u)$ be that number.* For example, $\#_2(a_1 a_0 a_1) = 5$ while $\#_2(a_0 a_1)$ is undefined. Now let $w \in \{a_0, a_1\}^*$ be an *arbitrary* word. We associate w with a natural number $f_2(w)$, where

$$f_2(w) \stackrel{\text{def}}{=} \#_2(a_1 w) - 1.$$

For example, $f_2(\varepsilon) = 0; f_2(a_0) = 1; f_2(a_1) = 2; f_2(a_0 a_0) = 3; f_2(a_0 a_1) = 4; f_2(a_1 a_0) = 5; f_2(a_1 a_1) = 6$. Next, we prove that $f_2 : \{a_0, a_1\}^* \to \mathbb{N}$ is bijective. The function f_2 is injective: If $w_1 \neq w_2$, then $a_1 w_1 \neq a_1 w_2$, so $\#_2(a_1 w_1) - 1 \neq \#_2(a_1 w_2) - 1$, and $f_2(w_1) \neq f(w_2)$. The function f_2 is surjective: If n is an arbitrary natural number, then $n = f_2(w)$, where w is obtained from the binary code of $n + 1$ by canceling the leading symbol (which is a_1). For example, for $n = 4$, the binary code of $n + 1$ is $a_1 a_0 a_1$, so $w = 01$. Thus, $f_2(01) = 4$.

b) Let $\Sigma = \{a_0, a_1 \ldots, a_{p-1}\}$ be an arbitrary alphabet. If we interpret each a_i as the number i, then we can define that a word $u \in \{a_0, a_1, \ldots, a_{p-1}\}^*$ *represents* a natural number if u has no leading symbols a_0 (with the exception of $u = a_0$). In other words, u represents a natural number if u is the code of a natural number in the positional number system with base p. Now define the function $\#_p : \{a_0, a_1, \ldots, a_{p-1}\}^* \to \mathbb{N}$ as follows: *If $u \in \{a_0, a_1, \ldots, a_{p-1}\}^*$ represents a natural number, then let $\#_p(u)$ be that number.* For example, if $p = 3$ and $u = a_2 a_1$, then $\#_3(a_2 a_1) = 2 \cdot 3^1 + 1 \cdot 3^0 = 7$. Consider the word $u = a_0 a_0 \ldots a_0 \in \{a_0, a_1, \ldots, a_{p-1}\}^*$. Then $a_1 u$ represents the number $\#_p(a_1 u) = \#_p(a_1 a_0 a_0 \ldots a_0) = p^{|u|}$. In canonical (i.e., shortlex) order there are $p^0 + p^1 + \ldots + p^{|u|-1}$ words preceding the word u. Assuming that the searched-for function f_p maps these words in the same order to consecutive numbers $0, 1, \ldots$, then u is mapped to the number $\sum_{i=0}^{|u|-1} p^i = \frac{p^{|u|}-1}{p-1}$. Finally, let $w \in \{a_0, a_1, \ldots, a_{p-1}\}^*$ be an *arbitrary* word. The function f_p must map w to the number $\frac{p^{|w|}-1}{p-1} + \#_p(a_1 w) - \#_p(a_1 \underbrace{a_0 \ldots a_0}_{|w|})$. After rearranging we obtain

$$f_p(w) = \#_p(a_1 w) - p^{|w|} + \frac{p^{|w|}-1}{p-1}.$$

Take, for example, $\Sigma = \{a_0, a_1, a_2\}$. Then $w = a_2 a_1$ is mapped to $f_3(a_2 a_1) = \#_3(a_1 a_2 a_1) - 3^2 + \frac{8}{2} = 16 - 9 + 4 = 11$. The function $f_p : \Sigma^* \to \mathbb{N}$ is bijective. We leave the proof of this as an exercise. $\square$

A subset of the set Σ^* is said to be a *formal language* over the alphabet Σ. The above theorem states that every formal language $\mathcal{S} \subseteq \Sigma^*$ is associated with exactly one set $f_{|\Sigma|}(\mathcal{S}) \subseteq \mathbb{N}$ of natural numbers—and vice versa. How can we use this? The answer is given in the following remark.

NB *When a property of sets is independent of the nature of their elements, we are allowed to choose whether to study the property using formal languages or sets of natural numbers. The results will apply to the alternative, too. For Computability Theory, three properties of this kind are especially interesting: the decidability, semi-decidability, and undecidability of sets. We will use the two alternatives based on the context and the ease and clarity of the presentation.*

6.4 Chapter Summary

In addition to the basic model of the Turing machine there are several variants. Each is a generalization of the basic model in some respect. Nevertheless, the basic model is capable of computing anything that can be computed by any other variant.

Turing machines can be encoded by words consisting of 0s and 1s. This enables the construction of the universal Turing machine, a machine that is capable of simulating any other Turing machine. Thus, the universal Turing machine can compute anything that can be computed by any other Turing machine. Practical consequences of this are the existence of the general-purpose computer and the operating system.

RAM is a model of computation that is equivalent to the Turing machine but is more appropriate in *Computational Complexity Theory*, where time and space are bounded computational resources.

Turing machines can be enumerated and generated. This allows us to talk about the nth Turing machine, for any natural n. The Turing machine can be used as a computer (to compute values of a function), or as a generator (to generate elements of a set), or as a recognizer (to find out which objects are members of a set and which are not).

A function $\varphi : \mathcal{A} \to \mathcal{B}$ is said to be partial computable on $\mathcal{A}$ if there exists a Turing machine capable of computing the function's values wherever the function is defined. A partial computable (p.c.) function $\varphi : \mathcal{A} \to \mathcal{B}$ is said to be computable on $\mathcal{A}$ if it is defined for every member of $\mathcal{A}$. A partial function is incomputable if there exists no Turing machine capable of computing the function's values wherever the function is defined.

A set is decidable in a universe if the characteristic function of the set is computable on the universe. A set is undecidable in a universe if the characteristic function of the set is incomputable on the universe. A set is semi-decidable if the characteristic function of the set is computable on this set. A set is computably enumerable (c.e.) if there exists a Turing machine capable of generating exactly the members of the set.

There is a bijective function between the sets Σ^* and $\mathbb{N}$; so instead of studying the decidability of formal languages, we can study the decidability of sets of natural numbers.

Problems

6.1. Given Q and δ, how many reduced TMs $T = (Q, \{0,1\}, \{0,1,\sqcup\}, \delta, q_1, \sqcup, \{q_2\})$ are there?

6.2. Informally design an algorithm that restores Σ, Γ, Q, and F from the code $\langle T \rangle$ of a basic TM.

6.3. Informally design the following Turing machines:

(a) a basic TM that accepts the language $\{0^n 1^n \mid n \geq 1\}$;
   [*Hint.* Repeatedly replace the leftmost 0 by X and the rightmost 1 by Y.]

(b) a TM that accepts the language $\{0^n 1^n 0^n \mid n \geq 1\}$;

(c) a basic TM that accepts an input if its first symbol does not appear elsewhere on the input;
   [*Hint.* Use a finite-storage TM.]

(d) a TM that recognizes the set of words with an equal number of 0s and 1s;

(e) a TM that decides whether or not a binary input greater than 2 is a prime;
   [*Hint.* Use a multi-track TM. Let the input be on the first track. Starting with the number 1, repeatedly generate on the second track the next larger number less than the input; for each generated number copy the input to the third track and then subtract the second-track number from the third-track number as many times as possible.]

(f) a TM that recognizes the language $\{wcw \mid w \in \{0,1\}^*\}$;
   [*Hint.* Check off the symbols of the input by using a multi-track TM.]

(g) a TM that recognizes the language $\{ww \mid w \in \{0,1\}^*\}$;
   [*Hint.* Locate the middle of the input.]

(h) a TM that recognizes the language $\{ww^R \mid w \in \{0,1\}^*\}$ of palindromes;
   [*Hint.* Use a multi-tape TM.]

(i) a TM that shifts its tape contents by three cells to the right;
   [*Hint.* Use a finite-storage TM to perform caterpillar-type movement.]

(j) Adapt the TM from (i) to move its tape contents by three cells to the left.

6.4. Prove: A nondeterministic d-dimensional tp-tape tk-track TM can be simulated by a basic TM.

6.5. Prove:

(a) A set is c.e. *iff* it is generated by a TM.

(b) If a set is c.e. then there is a generator that generates each element in the set exactly once.

(c) A set is computable *iff* it is generated by a TM in shortlex order.

(d) Every c.e. set is accepted by a TM with two nonaccepting states and one accepting state.

6.6. Informally design TMs that compute the following functions (use any version of TM):

(a) $\mathrm{const}_j^k(n_1, \ldots, n_k) \stackrel{\text{def}}{=} j$, for $j \geq 0$ and $k \geq 1$;

(b) $\mathrm{add}(m,n) \stackrel{\text{def}}{=} m+n$;

(c) $\mathrm{mult}(m,n) \stackrel{\text{def}}{=} mn$;
   [*Hint:* start with $0^m 1 0^n$, put 1 after it, and copy 0^n onto the right end m times.]

(d) $\mathrm{power}(m,n) \stackrel{\mathrm{def}}{=} m^n$;

(e) $\mathrm{fact}(n) \stackrel{\mathrm{def}}{=} n!$;

(f) $\mathrm{tower}(m,n) \stackrel{\mathrm{def}}{=} \left. m^{m^{\cdot^{\cdot^{\cdot^{m}}}}} \right\} n$ levels;

(g) $\mathrm{minus}(m,n) \stackrel{\mathrm{def}}{=} m \dot{-} n = \begin{cases} m-n & \text{if } m \geqslant n; \\ 0 & \text{otherwise.} \end{cases}$

(h) $\mathrm{div}(m,n) \stackrel{\mathrm{def}}{=} m \div n = \lfloor \frac{m}{n} \rfloor$;

(i) $\mathrm{floorlog}(n) \stackrel{\mathrm{def}}{=} \lfloor \log_2 n \rfloor$;

(j) $\log^*(n) \stackrel{\mathrm{def}}{=}$ the smallest k such that $\overbrace{\log(\log(\cdots(\log(n))\cdots))}^{k \text{ times}} \leqslant 1$;

(k) $\mathrm{gcd}(m,n) \stackrel{\mathrm{def}}{=}$ greatest common divisor of m and n;

(l) $\mathrm{lcm}(m,n) \stackrel{\mathrm{def}}{=}$ least common multiple of m and n;

(m) $\mathrm{prime}(n) \stackrel{\mathrm{def}}{=}$ the nth prime number;

(n) $\pi(x) \stackrel{\mathrm{def}}{=}$ the number of primes not exceeding x;

(o) $\phi(n) \stackrel{\mathrm{def}}{=}$ the number of positive integers that are $\leqslant n$ and relatively prime to n (*Euler funct.*);

(p) $\max^k(n_1,\ldots,n_k) \stackrel{\mathrm{def}}{=} \max\{n_1,\ldots,n_k\}$, for $k \geqslant 1$;

(r) $\mathrm{neg}(x) \stackrel{\mathrm{def}}{=} \begin{cases} 0 & \text{if } x \geqslant 1; \\ 1 & \text{if } x = 0. \end{cases}$

(s) $\mathrm{and}(x,y) \stackrel{\mathrm{def}}{=} \begin{cases} 1 & \text{if } x \geqslant 1 \wedge y \geqslant 1; \\ 0 & \text{otherwise.} \end{cases}$

(t) $\mathrm{or}(x,y) \stackrel{\mathrm{def}}{=} \begin{cases} 1 & \text{if } x \geqslant 1 \vee y \geqslant 1; \\ 0 & \text{otherwise.} \end{cases}$

(u) $\text{if-then-else}(x,y,z) \stackrel{\mathrm{def}}{=} \begin{cases} y & \text{if } x \geqslant 1; \\ z & \text{otherwise.} \end{cases}$

(v) $\mathrm{eq}(x,y) \stackrel{\mathrm{def}}{=} \begin{cases} 1 & \text{if } x = y; \\ 0 & \text{otherwise.} \end{cases}$

(w) $\mathrm{gr}(x,y) \stackrel{\mathrm{def}}{=} \begin{cases} 1 & \text{if } x > y; \\ 0 & \text{if } x \leqslant y. \end{cases}$

(x) $\mathrm{geq}(x,y) \stackrel{\mathrm{def}}{=} \begin{cases} 1 & \text{if } x \geqslant y; \\ 0 & \text{if } x < y. \end{cases}$

(y) $\mathrm{ls}(x,y) \stackrel{\mathrm{def}}{=} \begin{cases} 1 & \text{if } x < y; \\ 0 & \text{if } x \geqslant y. \end{cases}$

(z) $\text{leq}(x,y) \overset{\text{def}}{=} \begin{cases} 1 & \text{if } x \leqslant y; \\ 0 & \text{if } x > y. \end{cases}$

6.7. Prove the following *Basic Theorem*, which provides an alternative characterization of c.e. sets.

Theorem 6.6. (Basic Theorem on C.E. Sets) *A set S is c.e.* $\Longleftrightarrow$ *S is the domain of a p.c. function.*

[*Hint.*($\Rightarrow$) Let S be c.e. We will use Corollary 6.1, p. 140. If $S = \emptyset$, then $S = \text{dom}(\varphi)$, where φ is the everywhere undefined p.c. function. Otherwise, $S = \text{rng}(f)$ of a computable function f. Define a p.c. function φ as follows: $\varphi(x) := x$ if x eventually appears in $f(0), f(1), f(2), \ldots$; else $\varphi(x) := \uparrow$. Then $S = \text{dom}(\varphi)$.
($\Leftarrow$) Let $S = \text{dom}(\varphi)$ where φ is a p.c. function. The set $\text{dom}(\varphi)$ is semi-decidable. (Given an x, the recognizer $R_{\text{dom}(\varphi)}$ starts computing $\varphi(x)$ and, if the computation terminates, answers YES.) Now apply Theorem 6.3, p. 145.]

Remark. There is a similar statement about the range of a p.c. function (see Problem 7.2c, p. 171).

Definition 6.11. (Pairing Function) A **pairing function** is any computable bijection $f : \mathbb{N}^2 \to \mathbb{N}$ whose inverse functions f_1^{-1}, f_2^{-1}, defined by $f(f_1^{-1}(n), f_2^{-1}(n)) = n$, are computable. Therefore, $f(i,j) = n$ iff $f_1^{-1}(n) = i$ and $f_2^{-1}(n) = j$. The **standard** (or Cantor) pairing function $p : \mathbb{N}^2 \to \mathbb{N}$ is defined by

$$p(i,j) \overset{\text{def}}{=} \frac{1}{2}(i+j)(i+j+1) + i.$$

6.8. Prove: The function $p(i,j)$ is a pairing function with inverse functions

$$p_1^{-1}(n) = i = n - \frac{1}{2}w(w+1) \quad \text{and}$$

$$p_2^{-1}(n) = j = w - i$$

where

$$w = \left\lfloor \frac{\sqrt[2]{8n+1} - 1}{2} \right\rfloor.$$

6.9. What is the connection between p and the generation of pairs in Fig. 6.22 (p.145)?

6.10. We can use pairing functions to define bijective functions that map $\mathbb{N}^k$ onto $\mathbb{N}$.

(a) Describe how we can use p from Prob. 6.8 to define and compute a bijection $p^{(3)} : \mathbb{N}^3 \to \mathbb{N}$?

(b) Can you find the corresponding inverse functions $p_1^{(3)-1}, p_2^{(3)-1}, p_3^{(3)-1}$?

(c) How would you generalize to bijections from $\mathbb{N}^k$ onto $\mathbb{N}$, where $k \geqslant 3$?
[*Hint.* Define $p^{(k)} : \mathbb{N}^k \to \mathbb{N}$ by $p^{(k)}(i_1, \ldots, i_k) = p(p^{(k-1)}(i_1, \ldots, i_{k-1}), i_k)$ for $k \geqslant 3$, and $p^{(2)} = p$.]

6.11. Informally design the following generators (use any version of TM):

(a) $G_{\mathbb{N}^3}$, the generator of 3-tuples $(i_1, i_2, i_3) \in \mathbb{N}^3$;

(b) $G_{\mathbb{N}^k}$, the generator of k-tuples $(i_1, \ldots, i_k) \in \mathbb{N}^k$ (for $k \geqslant 3$);

(c) $G_{\mathbb{Q}}$, the generator of rational numbers $\frac{i}{j}$.

Bibliographic Notes

- Turing introduced his machine in [262]. Here, he also described in detail the construction of a universal Turing machine.
- The search for the smallest universal Turing machine was initiated in Shannon [214]. The existence and implementation of several small universal Turing machines is described in Rogozhin [203].
- The RAM model was formally considered in Cook and Reckhow [41]. In describing the RAM model we leaned on Aho et al. [8]. Von Neumann's influential report on the design of the EDVAC computer is [270].
- The concept of the *computably enumerable* (*c.e.*) set was formally defined in Kleene [121] (where he used the term recursively enumerable). Fifteen years before Kleene, Post discovered the equivalent concept of the *generated set*, but it remained unpublished until Post [182].
- Actually, many of Post's investigations and results were ahead of their time. Unfortunately, this contributed to his fear of being not understood properly and the delay of publication of his results. For the significance of Post's work in mathematical logic and computability theory see Murawski [167].
- The most significant works of Turing, along with key commentaries from leading scholars in the field, are collected in Cooper and van Leeuwen [44]. A comprehensive biography of Alan Turing is Hodges [107].

The following excellent books were consulted and are sources for the subjects covered in Part II of this book:

- General monographs on *Computability Theory* from the 1950s are Kleene [124] and Davis [54]. The first one contains a lot of information about the reasons for searching for models of computation.
- In the mid-1960s, a highly influential monograph on *Computability Theory* was Rogers [202]. The treatise contains all the results up to the mid-1960s, presented in a less formal style. Due to this, the book remained among the main sources in *Computability Theory* for the next twenty years.
- Some of the influential monographs from the 1970s and 1980s are Machtey and Young [144], Hopcroft and Ullman [113], Cutland [51], and Lewis and Papadimitriou [142].
- In the mid-1980s, the classic monograph of Rogers was supplemented and substituted by Soare [241]. This is a collection of the results in *Computability Theory* from 1931 to the mid-1980s. Another influential monograph of that decade was Boolos and Jeffrey [22].
- In the mid-1990s, computability theorists defined a uniform terminology (see below). Some of the influential general monographs are Davis et al. [59] and Kozen [128]. Shoenfield [217] gives a condensed overview of as many as possible topics, without going deeply in any of them, which were of research interest in the 1990s.
- The 2000s were especially fruitful and bore Boolos et al. [21], Shen and Vereshchagin [215], Cooper [43], Harel and Feldman [96], Kozen [129], Sipser [234], Rich [198], Epstein and Carnielli [68], and Rosenberg [204].
- The first half of the 2010s brought the monographs of Enderton [67], Weber [275], and Soare [246].
- The conceptual foundations of *Computability Theory* are discussed in light of our modern understanding in Copeland et al. [46]. An overview of *Computability Theory* until the end of the millennium is Griffor [91].
- Many of the recent general books on *Computational Complexity Theory* offer the basics of *Computability Theory*. These include Goldreich [89, 90] and Homer and Selman [112]. Two recent monographs that investigate the impact of randomness on computation and complexity are Nies [170] and Downey and Hirschfeldt [63]. For an intuitive introduction to this area, see Weber [275, Sect. 9.2].

Remark. (On the new terminology in *Computability Theory*.) The terminology in the papers of the 1930s is far from uniform. This is not surprising, though, as the theory was just born. Initially, *partial recursive function* was used for any function constructed using the rules of Gödel and Kleene (see p. 81). However, after the *Computability Thesis* was accepted, the adjective "partial recursive" expanded to eventually designate any computable function, *regardless of the model of computation on which the function was defined.* At the same time, the Turing machine became widely accepted as the most convincing model of computation. But the functions computable by Turing machines do not exhibit recursiveness (i.e., self-reference) as explicitly as the functions computable in Gödel-Kleene's model of computation. (We will explain this in the *Recursion Theorem*, Sect. 7.4.) Consequently, the adjective "partial recursive" lost its sense and was groundless when the subjects under discussion were Turing-computable functions. A more appropriate adjective would be "computable" (which was used by Turing himself). Nevertheless, with time the whole research field took the name *Recursion Theory*, in spite of the fact that its prime interest has always been computability (and not only recursion).

In 1996, Soare[13] proposed corrections to the terminology so that notions and concepts would regain their original meanings, as intended by the first researchers (see Soare [242]). In summary:

* the term *computable* and its variants should be used in connection with notions: computation, algorithm; Turing machine, register machine; function (defined on one of these models), set (generated or recognized on one of these models); relative computability;
* the term *recursive* and its variants should be used in connection with notions: recursive (inductive) definition; (general) recursive function (Herbrand-Gödel), primitive recursive and μ-recursive function (Gödel-Kleene) and some other notions from the theory of recursive functions.

---

[13] Robert Irving Soare, b. 1940, American mathematician.

Chapter 7
The First Basic Results

> *Recursion is a method of defining objects in which the object*
> *being defined is applied within its own definition.*

Abstract In the previous chapters we have defined the basic notions and concepts of a theory that we are interested in, *Computability Theory*. In particular, we have rigorously defined its basic notions, i.e., the notions of algorithm, computation, and computable function. We have also defined some new notions, such as the decidability and semi-decidability of a set, that will play key roles in the next chapter (where we will further develop *Computability Theory*). As a side product of the previous chapters we have also discovered some surprising facts, such as the existence of the universal Turing machine. It is now time to start using this apparatus and deduce the first theorems of *Computability Theory*. In this chapter we will first prove several simple but useful theorems about decidable and semi-decidable sets and their relationship. Then we will deduce the so-called *Padding Lemma* and, based on it, introduce the extremely important concept of the index set. This will enable us to deduce two influential theorems, the *Parameter Theorem* and the *Recursion Theorem*. We will not be excessively formal in our deductions; instead, we will equip them with meaning and motivation wherever appropriate.

7.1 Some Basic Properties of Semi-decidable (C.E.) Sets

For starters, we prove in this section some basic properties of semi-decidable sets. In what follows, $\mathcal{A}$, $\mathcal{B}$, and $\mathcal{S}$ are sets.

Theorem 7.1. $\mathcal{S}$ *is decidable* $\Longrightarrow$ $\mathcal{S}$ *is semi-decidable*

Proof. This is a direct consequence of Definition 6.10 (p. 142) of (semi-)decidable sets. $\square$

R. Robič, *The Foundations of Computability Theory*, https://doi.org/10.1007/978-3-662-62421-0_7

Theorem 7.2. S *is decidable* $\implies \overline{S}$ *is decidable*

Proof. The decider $D_{\overline{S}}$ starts D_S and reverses its answers. $\square$

The next theorem is due to Post and is often used.

Theorem 7.3. (Post's Theorem) S *and* $\overline{S}$ *are semi-decidable* $\iff S$ *is decidable*

Proof. Let S and $\overline{S}$ be semi-decidable sets. Then, there are recognizers R_S and $R_{\overline{S}}$. Every $x \in \mathcal{U}$ is a member of either S or $\overline{S}$; the former situation can be detected by R_S and the latter by $R_{\overline{S}}$. So let us combine the two recognizers into the following algorithm: 1) Given $x \in \mathcal{U}$, simultaneously start R_S and $R_{\overline{S}}$ on x and wait until one of them answers YES. 2) If the YES came from R_S, output YES (i.e., $x \in S$) and halt; otherwise (i.e., the YES came from $R_{\overline{S}}$) output NO (i.e., $x \notin S$) and halt. This algorithm decides, for arbitrary $x \in \mathcal{U}$, whether or not x is in S. Thus, the algorithm is D_S, a decider for S. $\square$

Theorem 7.4. S *is semi-decidable* $\iff S$ *is the domain of a computable function*

Proof. If S is semi-decidable, then it is (by Definition 6.10 on p. 142) the domain of the characteristic function χ_S, which is computable on S. Inversely, if S is the domain of a computable function φ, then the characteristic function, defined by $\chi_S(x) = 1 \iff \varphi(x)\downarrow$, is computable on S (see Definition 6.6 on p. 136). Hence, S is semi-decidable. $\square$

Theorem 7.5.

$\mathcal{A}$ *and* $\mathcal{B}$ *are semi-decidable* $\implies \mathcal{A} \cup \mathcal{B}$ *and* $\mathcal{A} \cap \mathcal{B}$ *are semi-decidable*
$\mathcal{A}$ *and* $\mathcal{B}$ *are decidable* $\implies \mathcal{A} \cup \mathcal{B}$ *and* $\mathcal{A} \cap \mathcal{B}$ *are decidable*

Proof. a) Let $\mathcal{A}$ and $\mathcal{B}$ be semi-decidable sets. Their characteristic functions $\chi_\mathcal{A}$ and $\chi_\mathcal{B}$ are computable on $\mathcal{A}$ and $\mathcal{B}$, respectively. The function $\chi_{\mathcal{A} \cup \mathcal{B}}$, defined by

$$\chi_{\mathcal{A} \cup \mathcal{B}}(x) \stackrel{\text{def}}{=} \begin{cases} 1, & \text{if } \chi_\mathcal{A}(x) = 1 \vee \chi_\mathcal{B}(x) = 1, \\ 0, & \text{otherwise} \end{cases}$$

is the characteristic function of the set $\mathcal{A} \cup \mathcal{B}$ and is computable on $\mathcal{A} \cup \mathcal{B}$. Hence, $\mathcal{A} \cup \mathcal{B}$ is semi-decidable. Similarly, the function $\chi_{\mathcal{A} \cap \mathcal{B}}$, defined by

$$\chi_{\mathcal{A} \cap \mathcal{B}}(x) \stackrel{\text{def}}{=} \begin{cases} 1, & \text{if } \chi_\mathcal{A}(x) = 1 \wedge \chi_\mathcal{B}(x) = 1, \\ 0, & \text{otherwise} \end{cases}$$

is the characteristic function of the set $\mathcal{A} \cap \mathcal{B}$ and is computable on $\mathcal{A} \cap \mathcal{B}$. So, $\mathcal{A} \cap \mathcal{B}$ is semi-decidable.

b) Let $\mathcal{A}$ and $\mathcal{B}$ be decidable sets. Then, $\chi_\mathcal{A}$ and $\chi_\mathcal{B}$ are computable functions on $\mathcal{U}$. Also, the functions $\chi_{\mathcal{A} \cup \mathcal{B}}$ and $\chi_{\mathcal{A} \cap \mathcal{B}}$ are computable on $\mathcal{U}$. Hence, $\mathcal{A} \cup \mathcal{B}$ and $\mathcal{A} \cap \mathcal{B}$ are decidable sets. $\square$

7.2 Padding Lemma and Index Sets

We have seen in Sect. 6.2.1 that each natural number can be viewed as the index of exactly one Turing machine (see Fig. 7.1). Specifically, given an arbitrary $e \in \mathbb{N}$, we can find the corresponding Turing machine $T_e = (Q, \Sigma, \Gamma, \delta, q_1, \sqcup, F)$ by the following algorithm:

> **if** the binary code of e is of the form $111 K_1 11 K_2 11 \ldots 11 K_r 111$, where each
> K is of the form $0^i 10^j 10^k 10^\ell 10^m$ for some i, j, k, ℓ, m,
> **then** determine $\delta, Q, \Sigma, \Gamma, F$ by inspecting $K_1, K_2, \ldots, K_r$ and taking into account
> that $K = 0^i 10^j 10^k 10^\ell 10^m$ encodes the instruction $\delta(q_i, z_j) = (q_k, z_\ell, D_m)$;
> **else** T_e is the empty Turing machine.

But remember that the Turing machine T_e is implicitly associated, for each $k \geqslant 1$, with a k-ary partial computable function $\varphi_e^{(k)}$, the k-ary *proper function of* T_e (see Sect. 6.3.1). Consequently, each $e \in \mathbb{N}$ is the index of exactly one k-ary partial computable function for any fixed $k \geqslant 1$. See Fig. 7.1.

Fig. 7.1 Each natural number e is the index of exactly one partial computable function (denoted by $\varphi_e^{(k)}$)

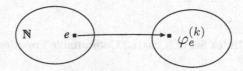

Padding Lemma

What about the other way round? Is each Turing machine represented by exactly one index? The answer is no; a Turing machine has several indexes. To see this, let T be an arbitrary Turing machine and

$$\langle T \rangle = 111 K_1 11 K_2 11 \ldots 11 K_r 111$$

its code. Here, each subword K encodes an instruction of T's program δ. Let us now permute the subwords $K_1, K_2, \ldots, K_r$ of $\langle T \rangle$. Of course, we get a different code, but notice that the new code still represents the same Turing program δ, i.e., the same *set* of instructions, and hence the same Turing machine, T. Thus, T has several different indexes (at least $r!$, where r is the number of instructions in δ).

Still more: We can insert into $\langle T \rangle$ new subwords $K_{r+1}, K_{r+2}, \ldots$, where each of them represents a *redundant* instruction, i.e., an instruction that will *never* be executed. By such *padding* we can construct an unlimited number of new codes each of which describes a different Turing program. But notice that, when started, each of these programs *behaves* (i.e., executes) in the same way as T's program δ. In other words, each of the constructed indexes defines the same partial computable function, T's proper function. Formally: If e is the index of T and x an arbitrary index constructed from e as described, then $\varphi_x \simeq \varphi_e$. (See Fig. 7.2.)

We have just proved the so-called *Padding Lemma*.

Lemma 7.1. (Padding Lemma) *A partial computable function has countably infinitely many indexes. Given one of them, countably infinitely many others can be generated.*

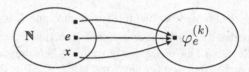

Fig. 7.2 Each p.c. function (denoted by $\varphi_e^{(k)}$) has countably infinitely many indexes

Index Set of a Partial Computable Function

There may exist Turing machines whose indexes cannot be transformed one to another simply by permuting their instructions or by instruction padding, and yet the machines compute the same partial computable function (i.e., solve the same problem). For instance, we described two such machines in Examples 6.1 and 6.2 (see pp. 115,116). The machines compute the sum of two integers in two different ways. In other words, the machines have different programs but they compute the same partial computable function $\varphi^{(2)} : \mathbb{N}^2 \to \mathbb{N}$, that is, the function $\varphi^{(2)}(m,n) = m+n$. We say that these machines are equal in their *global behavior* (because they compute the same function) but they differ in their *local behavior* (as they do this in different ways).

All of this leads in a natural way to the concept of the *index set of a partial computable function*. Informally, the index set of a p.c. function consists of the indexes of all the Turing machines that compute this function. (See Fig. 7.3.) So let φ be an arbitrary p.c. function. Then there exists at least one Turing machine T capable of computing the values of φ. Let e be the index of T. Then $\varphi_e \simeq \varphi$, where φ_e is the proper function of T (having the same number of arguments as φ, i.e., the same k-arity). Now let us collect *all* the indexes of *all* the Turing machines that compute the function φ. We call this set the *index set* of φ and denote it by $\mathrm{ind}(\varphi)$.

Definition 7.1. (Index Set) The **index set** of a p.c. function φ is the set $\text{ind}(\varphi) \stackrel{\text{def}}{=} \{x \in \mathbb{N} \mid \varphi_x \simeq \varphi\}$.

Informally, $\text{ind}(\varphi)$ contains all the (encoded) Turing programs that compute φ. Taking into account the *Computability Thesis*, we can say that $\text{ind}(\varphi)$ contains all the (encoded) algorithms that compute (values of) the function φ. There are countably infinitely many algorithms in $\text{ind}(\varphi)$. Although they may differ in their local behavior, they all exhibit the same global behavior.

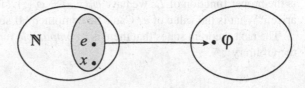

Fig. 7.3 Index set $\text{ind}(\varphi)$ of a p.c. function φ

Index Set of a Semi-decidable Set

We can now in a natural way define one more useful notion. Let S be an arbitrary semi-decidable set. Of course, its characteristic function χ_S is partial computable. But now we know that the index set $\text{ind}(\chi_S)$ of the function χ_S contains all the (encoded) algorithms that are capable of computing χ_S. In other words, $\text{ind}(\chi_S)$ consists of all the (encoded) *recognizers* of the set S. For this reason we also call $\text{ind}(\chi_S)$ the *index set of the set S* and denote it by $\text{ind}(S)$.

7.3 Parameter (s-m-n) Theorem

Consider an arbitrary multi-variable partial computable function φ. Select some of its variables and assign arbitrary values to them. This changes the role of these variables, because the fixed variables become *parameters*. In this way, we obtain a new function ψ of the rest of the variables. The *Parameter Theorem*, which is the subject of this section, tells us that the index of the new function ψ depends only on the index of the original function φ and its parameters; what is more, the index of ψ can be computed, for any φ and parameters, with a computable function s.

There are important practical consequences of this theorem. First, since the index of a function represents the Turing program that computes the function's values, the *Parameter Theorem* tells us that the parameters of φ can always be *incorporated into* the program for φ to obtain the program for ψ. Second, since parameters are natural numbers, they can be viewed as indexes of Turing programs, so the incorporated parameters can be interpreted as *subprograms* of the program for ψ.

7.3.1 Deduction of the Theorem

First, we develop the basic form of the theorem. Let $\varphi_x(y,z)$ be an arbitrary partial computable function of two variables, y and z. Its values can be computed by a Turing machine T_x. Let us pick an arbitrary natural number—call it $\bar{y}$—and substitute each occurrence of y in the expression of $\varphi_x(y,z)$ by $\bar{y}$. We say that the variable y has been changed to the *parameter* $\bar{y}$. The resulting expression represents a new function $\varphi_x(\bar{y},z)$ of one variable z. Note that $\varphi_x(\bar{y},z)$ is a partial computable function (otherwise $\varphi_x(y,z)$ would not be partial computable). Therefore, there is a Turing machine—call it T_e—that computes $\varphi_x(\bar{y},z)$. (At this point e is not known.) Since φ_e is the proper function of T_e, we have $\varphi_x(\bar{y},z) = \varphi_e(z)$. Now, the following questions arise: "What is the value of e? Can e be computed? If so, how can e be computed?"

The next theorem states that there is a *computable* function, whose value is e, for the arbitrary $x,\bar{y}$.

Theorem 7.6. (Parameter Theorem) *There is injective computable function* $s : \mathbb{N}^2 \to \mathbb{N}$ *such that, for every* $x,\bar{y} \in \mathbb{N}$,

$$\varphi_x(\bar{y},z) = \varphi_{s(x,\bar{y})}(z).$$

Proof idea. Let $x,\bar{y} \in \mathbb{N}$ be given. First, we conceive a Turing machine T that operates as follows: Given an arbitrary z as the input, T inserts $\bar{y}$ before z and then starts simulating Turing machine T_x on the inputs $\bar{y},z$. Obviously, T computes the value $\varphi_x(\bar{y},z)$. We then show that such a T can be *constructed* for *any* pair $x,\bar{y}$. Therefore, the index of the constructed T can be denoted by $s(x,\bar{y})$, where s is a computable function mapping $\mathbb{N}^2$ into $\mathbb{N}$. The details of the proof are given in Box 7.1.

Box 7.1 (Proof of the *Parameter Theorem*).

Let us describe the actions of the machine T_e that will entail the equality $\varphi_x(\bar{y},z) = \varphi_e(z)$ for every $x,\bar{y} \in \mathbb{N}$. Initially, there is only z written on T_e's tape. Since $\bar{y}$ will also be needed, T_e prepares it on its tape. To do this, it shifts z to the right by $\bar{y}+1$ cells, and writes unary-encoded $\bar{y}$ (and a separator symbol) to the emptied space. After this, T_e starts simulating T_x's program on inputs $\bar{y}$ and z and thus computing the value $\varphi_x(\bar{y},z)$. Consequently, $\varphi_x(\bar{y},z) = \varphi_e(z)$.

But where is the function s? It follows from the above description that the program P_e of the machine T_e is a sequence $P_1 ; P_2$ of two programs P_1 and P_2 such that P_1 inserts $\bar{y}$ on the tape and then leaves the control to P_2, which is responsible for computing the values of the function $\varphi_x(y,z)$. Hence, $\langle P_e \rangle = \langle P_1 ; P_2 \rangle$. Now we see the role of s: The function s must compute $\langle P_1 ; P_2 \rangle$ for arbitrary x and $\bar{y}$. To do this, s must perform the following tasks (informally):

1. Construct the code $\langle P_1 \rangle$ from the parameter $\bar{y}$.
   (Note that P_1 is simple: It must shift the contents of the tape $\bar{y}+1$ times to the right and write the symbol 1 to the first $\bar{y}$ of the emptied cells.)
2. Construct the code $\langle P_2 \rangle$ from the index x.

3. Construct the code $\langle P_1; P_2 \rangle$ from $\langle P_1 \rangle$ and $\langle P_2 \rangle$.
   (To do this, take the word $\langle P_1 \rangle \langle P_2 \rangle$ and change it so that, instead of halting, P_1 starts the first instruction of P_2.)
4. Compute the index e from $\langle P_1; P_2 \rangle$ and return $s(x, \bar{y}) := e$.

All the tasks are computable. So, s is a computable function. It is injective, too.   □

Next, we generalize the theorem. Let the function φ_x have $m \geqslant 1$ parameters and $n \geqslant 1$ variables. That is, the function is $\varphi_x(\bar{y}_1, \ldots, \bar{y}_m, z_1, \ldots, z_n)$. The proof of the following generalization of the *Parameter Theorem* uses the same ideas as the previous one, so we omit it.

Theorem 7.7. (s-m-n Theorem) *For arbitrary $m, n \geqslant 1$ there is an injective computable function $s_n^m : \mathbb{N}^{m+1} \to \mathbb{N}$ such that, for every $x, \bar{y}_1, \ldots, \bar{y}_m \in \mathbb{N}$,*

$$\varphi_x(\bar{y}_1, \ldots, \bar{y}_m, z_1, \ldots, z_n) = \varphi_{s_n^m(x, \bar{y}_1, \ldots, \bar{y}_m)}(z_1, \ldots, z_n).$$

Summary. If the variables $y_1, \ldots, y_m$ of a function φ_x are assigned fixed values $\bar{y}_1, \ldots, \bar{y}_m$, respectively, then we can build the values in T_x's Turing program and obtain a Turing program for computing the function φ of the rest of the variables. The new program is represented by the index $s_n^m(x, \bar{y}_1, \ldots, \bar{y}_m)$, where s_n^m is an injective and computable function. It can be proved that s_n^m is a primitive recursive function.

7.4 Recursion (Fixed-Point) Theorem

The construction of the universal Turing machine brought, as a byproduct, the cognizance that a Turing machine can compute with other Turing machines simply by manipulating their indexes and simulating programs represented by the indexes. But, can a Turing machine manipulate its own index and consequently compute with itself? Does the question make sense? The answer to both questions is yes, as will be explained in this section. Namely, we have seen that there is another model of computation, the μ-recursive functions of Gödel and Kleene (see Sect. 5.2.1), that allows functions to be defined and hence computed by referring to themselves, i.e., recursively (see Box 5.1). Since the Turing machine and μ-recursive functions are equivalent models of computation, it is reasonable to ask whether Turing machines also allow for, and can make sense of, recursiveness (i.e., self-reference). That this is actually so follows from the *Recursion Theorem*, which was proved by Kleene in 1938. The theorem is also called the *Fixed-Point Theorem*.

7.4.1 Deduction of the Theorem

1. Let $i \in \mathbb{N}$ be an arbitrary number and $f : \mathbb{N} \to \mathbb{N}$ an arbitrary computable function.

2. **Definition.** Given i and f, let $T^{(i)}$ be a TM that performs the following steps:

   a. $T^{(i)}$ has two input data: i (as picked above) and an arbitrary x;
   b. $T^{(i)}$ interprets input i as an index and constructs the TP of T_i;
   c. $T^{(i)}$ simulates T_i on input i;   // i.e., $T^{(i)}$ tries to compute $\varphi_i(i)$
   d. if the simulation halts, then $T^{(i)}$ performs steps (e)–(g):   // i.e., if $\varphi_i(i)\downarrow$
   e.   it applies the function f to the result $\varphi_i(i)$;   // i.e., $T^{(i)}$ computes $f(\varphi_i(i))$
   f.   it interprets $f(\varphi_i(i))$ as an index and constructs the TP of $T_{f(\varphi_i(i))}$;
   g.   it simulates $T_{f(\varphi_i(i))}$ on input x.   // In summary: $T^{(i)}$ computes $\varphi_{f(\varphi_i(i))}(x)$

   We have seen that $T^{(i)}$ on input i, x computes the value $\varphi_{f(\varphi_i(i))}(x)$. Thus, the proper function of $T^{(i)}$ depends on input i.

3. But, the steps a–g are only a definition of the machine $T^{(i)}$. (We can define whatever we want.) So the question arises: "For which $i \in \mathbb{N}$ does $T^{(i)}$ actually *exist*?" And what is more: "For which $i \in \mathbb{N}$ can $T^{(i)}$ be *constructed*?"

   Proposition. $T^{(i)}$ *exists for every* $i \in \mathbb{N}$. *There is an algorithm that constructs* $T^{(i)}$ *for arbitrary i.*

   Proof. We have already proved: *i)* every natural number is an index of a TM; *ii)* every index can be transformed into a Turing program; and *iii)* every TM can be simulated by another TM. Therefore, each of the steps (a)–(g) is *computable*. Consequently, there is, for every $i \in \mathbb{N}$, a Turing machine that executes steps (a)–(g). We call this machine $T^{(i)}$.
   Can $T^{(i)}$ be constructed by an *algorithm*? To answer in the affirmative, it suffices to describe the algorithm with which we construct the program of $T^{(i)}$. Informally, we do this as follows: We write the program of $T^{(i)}$ as a sequence of five calls of subprograms, a call for each of the tasks (b), (c), (e), (f), (g). We need only three different subprograms: (1) to convert an index to the corresponding Turing program (for steps b, f); (2) to simulate a TM (for steps c, g); and (3) to compute the value of the function f (for step e). We already know that subprograms (1) and (2) can be constructed (see Sect. 6.2). The subprogram (3) is also at our disposal because, by assumption, f is computable. Thus, we have informally described an algorithm for constructing the program of $T^{(i)}$.   □

4. According to the *Computability Thesis*, there is a Turing machine T_j that does the same thing as the informal algorithm in the above proof, i.e., T_j constructs $T^{(i)}$ for arbitrary $i \in \mathbb{N}$. But this means that T_j computes a computable function that maps $\mathbb{N}$ into the set of all Turing machines (or rather, their indexes). In other words, T_j's proper function φ_j is total.

   Consequence. *There is a computable function* $\varphi_j : \mathbb{N} \to \mathbb{N}$ *such that* $T^{(i)} = T_{\varphi_j(i)}$.

Note that φ_j does not execute steps a–g; it only returns the index of $T^{(i)}$, which does execute these steps.

5. The unary proper function of $T_{\varphi_j(i)}$ is denoted by $\varphi_{\varphi_j(i)}$. Applied to x, it computes the same value as $T_{\varphi_{j(i)}} = T^{(i)}$ in steps a–g, that is, the value $\varphi_{f(\varphi_i(i))}(x)$. Consequently,

$$\varphi_{\varphi_j(i)}(x) = \varphi_{f(\varphi_i(i))}(x).$$

6. Recall that i is arbitrary. So, let us take $i := j$. After substituting i by j, we obtain

$$\varphi_{\varphi_j(j)}(x) = \varphi_{f(\varphi_j(j))}(x)$$

and, introducing $n := \varphi_j(j)$, we finally obtain

$$\varphi_n(x) = \varphi_{f(n)}(x).$$

This equation was our goal. Recall that in the equation f is an *arbitrary* computable function. We have proven the *Recursion Theorem*.

Theorem 7.8. (Recursion Theorem) *For every computable function f there is a natural number n such that $\varphi_n \simeq \varphi_{f(n)}$. The number n can be computed from the index of the function f.*

7.4.2 Interpretation of the Theorem

What does the *Recursion Theorem* tell us? First, observe that from $\varphi_n \simeq \varphi_{f(n)}$ it does not necessarily follow that $n = f(n)$. For example, f defined by $f(k) = k+1$ is a computable function, yet there is no natural number n such that $n = f(n)$. Rather, the equality $\varphi_n \simeq \varphi_{f(n)}$ states that two partial computable functions, whose indexes are n and $f(n)$, are equal (i.e., φ_n and $\varphi_{f(n)}$ have equal domains and return equal values).

Second, recall that the indexes represent Turing programs (and hence Turing machines). So we can interpret f as a transformation of a Turing program represented by some $i \in \mathbb{N}$ into a Turing program represented by $f(i)$. Since f is supposed to be total, it represents a transformation that modifies *every* Turing program in some way. Now, in general, the original program (represented by i) and the modified program (represented by $f(i)$) are not equivalent, i.e., they compute different proper functions φ_i and $\varphi_{f(i)}$. This is where the *Recursion Theorem* makes an entry; it states that there exists a program (represented by some n) which is an exception to this rule. Specifically:

If a transformation f modifies every Turing program, then
some Turing program n is transformed into an equivalent Turing program $f(n)$.

In other words, if a transformation f modifies every Turing machine, there is always some Turing machine T_n for which the modified machine $T_{f(n)}$ computes the same function as T_n. Although the Turing programs represented by n and $f(n)$ may completely differ in their instructions, they nevertheless return equal results, i.e., compute the same function $\varphi\,(\simeq \varphi_n \simeq \varphi_{f(n)})$. We have said that such machines differ in their *local behavior* but are equal in their *global behavior* (see p. 158).

Can we find out the value of n, i.e., the (index of the) Turing program that is modified by f into the equivalent program? According to step 6 of the deduction we have $n = \varphi_j(j)$, while according to step 3 j can be computed and depends on the function f only. So n depends only on f. Since f is computable, n can be computed.

7.4.3 Fixed Points of Functions

The *Recursion Theorem* is also called the *Fixed-Point Theorem*, because the number n for which $\varphi_n = \varphi_{f(n)}$ has become known as a *fixed point of the function* f.[1] Using this terminology we can restate the theorem as follows.

Theorem 7.9. (Fixed-Point Theorem) *Every computable function has a fixed point.*

(We will use this version in Sect. 9.3.) Now the following question naturally arises: Is the number n that we constructed from the function f the *only* fixed point of f? The answer is *no*. Actually, there are countably infinitely many others.

Theorem 7.10. *A computable function has countably infinitely many fixed points.*

Proof. Assume that there exists a computable function f with only *finitely many* fixed points. Denote the finite set of f's fixed points by $\mathcal{F}$. Choose any partial computable function φ_e with the property that none of its indexes is in $\mathcal{F}$, i.e., $\varphi_e \not\simeq \varphi_x$ for every $x \in \mathcal{F}$. (In short, $\mathrm{ind}(\varphi_e) \cap \mathcal{F} = \emptyset$.) Now comes the tricky part: Let $g : \mathbb{N} \to \mathbb{N}$ be a function that is implicitly defined by

$$\varphi_{g(x)} \simeq \begin{cases} \varphi_e, & \text{if } x \in \mathcal{F}; \\ \varphi_{f(x)}, & \text{otherwise.} \end{cases}$$

Although g is not explicitly defined, we can determine the following two relevant properties of g:

1. The function g is *computable*. The algorithm to compute $g(x)$ for an arbitrary x has two steps: (a) Decide $x \in ? \mathcal{F}$. (This can always be done because $\mathcal{F}$ is finite.) (b) If $x \in \mathcal{F}$, then $g(x) := e$; otherwise, compute $f(x)$ and assign it to $g(x)$. (This can always be done as f is computable.)

2. The function g has *no fixed point*. If $x \in \mathcal{F}$, then $\varphi_{g(x)} \simeq \varphi_e$ (by definition) and $\varphi_e \not\simeq \varphi_x$ (property of φ_e), so $\varphi_{g(x)} \not\simeq \varphi_x$. This means that none of the elements of $\mathcal{F}$ is a fixed point of g. On the other hand, if $x \notin \mathcal{F}$, then $\varphi_{g(x)} \simeq \varphi_{f(x)}$ (by def.) and $\varphi_{f(x)} \not\simeq \varphi_x$ (as x is not a fixed point of f), which implies that $\varphi_{g(x)} \not\simeq \varphi_x$. This means that none of the elements of $\overline{\mathcal{F}}$ is a fixed point of g.

So, g is a computable function with no fixed points; this contradicts the *Fixed-Point Theorem*. □

---

[1] In fact, what remains fixed under f is the *function* φ_n (or global behavior of T_n), not the index n.

Consequently, we can upgrade the interpretation of the *Fixed-Point Theorem*:

If a transformation f modifies every Turing program, then countably infinitely
many Turing programs are transformed into equivalent Turing programs.

7.4.4 Practical Consequences: Recursive Program Definition

The *Recursion Theorem* allows a partial computable function to be defined using its
own index, as described in the following schematic definition:

$$\varphi_i(x) \stackrel{\text{def}}{=} \underbrace{[\dots i \dots x \dots]}_{P}.$$

On the left-hand side there is a function φ_i, which is being defined, and on the right-
hand side there is a Turing program P, i.e., a sequence $[\dots i \dots x \dots]$ of instructions,
describing how to compute the value of φ_i at the argument x. Consider the variable
i in P. Obviously, i is also the index of the function being defined, φ_i. But, an index of a function is just a description of the program computing that function (see
Sect. 6.2.1). Thus, it looks as if the variable i in P describes the very program P,
and hence the index i of the function being defined. But this is a circular definition,
because the object being defined (i.e., i) is a part of the definition!

On the other hand, we know that μ-recursive functions, i.e., the Gödel-Kleene
model of computation, allow constructions, and hence definitions, of functions in a
self-referencing manner by using the rule of primitive recursion (see Box 5.1, p. 82).
As a Turing machine is an equivalent model of computation, we expect that it, too,
allows self-reference. To prove that this is indeed so, we need both the *Recursion
Theorem* and the *Parameter Theorem*. Let us see how.

Notice first that the variables i and x have different roles in P. While x represents an *arbitrary* natural number, i.e., the *argument* of the function being defined,
the variable i represents an unknown but *fixed* natural number, i.e., the *index* of this
function. Therefore, i can be treated as a (yet unknown) *parameter* in P. Now, viewing the program P as a function of one variable x and one parameter i, we can apply
the *Parameter Theorem* and move the parameter i to the index of some new function,
$\varphi_{s(i)}(x)$. Observe that this function is still defined by P:

$$\varphi_{s(i)}(x) \stackrel{\text{def}}{=} \underbrace{[\dots i \dots x \dots]}_{P}. \tag{$*$}$$

The index $s(i)$ describes a program computing $\varphi_{s(i)}(x)$, but it does not appear in the
program P. Due to the *Parameter Theorem*, the function s is computable. But then
the *Recursion Theorem* tells us that there is a natural number n such that $\varphi_{s(n)} \simeq \varphi_n$.
So, let us take $i := n$ in the definition ($*$). Taking into account that $\psi_{s(n)} \simeq \varphi_n$, we
obtain the following definition:

$$\varphi_n(x) \stackrel{\text{def}}{=} [\ldots n \ldots x \ldots]. \tag{**}$$

Recall that n is a particular natural number that is dependent on the function s. Consequently, φ_n is a particular partial computable function. This function has been defined in $(**)$ with its own index.

We have proved that there are partial functions computable according to Turing that can be defined recursively. This finding is most welcome, because the Turing machine as the model of computation does not exhibit the capability of supporting self-reference and recursiveness as explicitly as the Gödel-Kleene model of computation (i.e., the μ-recursive functions).

7.4.5 Practical Consequences: Recursive Program Execution

Suppose that a partial computable function is defined recursively. Can the function be computed by a Turing machine? If so, what does the computation look like?

Let us be given a function defined by $\varphi(x) \stackrel{\text{def}}{=} x!$, where $x \in \mathbb{N}$. How can $\varphi(x)$ be defined and computed recursively by a Turing program δ? For starters, recall that $\varphi(x)$ can be recursively (inductively) defined by the following system of equations:

$$\begin{aligned} \varphi(x) &= x \cdot \varphi(x-1), && \text{if } x > 0 \\ \varphi(0) &= 1. \end{aligned} \tag{*}$$

For example, $\varphi(4) = 4 \cdot \varphi(3) = 4 \cdot 3 \cdot \varphi(2) = 4 \cdot 3 \cdot 2 \cdot \varphi(1) = 4 \cdot 3 \cdot 2 \cdot 1 \cdot \varphi(0) = 4 \cdot 3 \cdot 2 \cdot 1 \cdot 1 = 24$.

The definition $(*)$ uncovers the actions that should be taken by the Turing program δ in order to compute the value $\varphi(x)$. Let $\delta(a)$ denote an instance of the execution of δ on input a. We call $\delta(a)$ the *activation* of δ on a; in this case, a is called the *actual parameter* of δ. To compute the value $\varphi(a)$, the activation $\delta(a)$ should do, in succession, the following:

1. *activate* $\delta(a-1)$, i.e., call δ on $a-1$ to compute $\varphi(a-1)$;
2. multiply its actual parameter $(= a)$ and the result r $(= \varphi(a-1))$ of $\delta(a-1)$;
3. return the product $(=$ the result r of $\delta(a))$ to the caller.

In this fashion the Turing machine should execute every $\delta(a)$, $a = x, x-1, \ldots, 2, 1$. The exception is the activation $\delta(0)$, whose execution is trivial and immediately returns the result $r = 1$. Returns to whom? To the activation $\delta(1)$, which activated $\delta(0)$ and has, since then, been waiting for $\delta(0)$'s result. This will enable $\delta(1)$ to resume execution at step 2 and then, in step 3, return its own result. Similarly, the following should hold in general: Every *callee* $\delta(a-1)$ should return its result r $(= \varphi(a-1))$ to the awaiting *caller* $\delta(a)$, thus enabling $\delta(a)$ to continue its execution.

We now describe how, in principle, activations are represented in a Turing machine, what actions the machine must take to start a new activation of its program, and how it collects the awaited result. For each activation $\delta(a), a = x, \ldots, 2, 1$, the following *runtime actions* must be taken:

a. *Before control is transferred to $\delta(a)$, the activation record of $\delta(a)$ is created.*

The activation record (AR) of $\delta(a)$ is a finite sequence of tape cells set aside to hold the information that will be used or computed by $\delta(a)$. In the case of $\varphi(x) = x!$, the AR of $\delta(a)$ contains the actual parameter a and an empty field r for the result $(= \varphi(a))$ of $\delta(a)$. We denote the AR of $\delta(a)$ by $[a, r]$. (See Fig. 7.4 (a).)

b. *$\delta(a)$ activates $\delta(a-1)$ and waits for its result.*

To do this, $\delta(a)$ creates, next to its AR, a new AR $[a-1, r]$ to be used by $\delta(a-1)$. Note that $a - 1$ is a value and r is empty. Then, $\delta(a)$ moves the tape window to $a-1$ of $[a-1, r]$ and changes the state of the control unit to the initial state q_1. In this way, $\delta(a)$ stops its own execution and invokes a new instance of δ. Since the input to the invoked δ will be $a-1$ from $[a-1, r]$, the activation $\delta(a-1)$ will start. (See Fig. 7.4 (b).)

c. *$\delta(a-1)$ computes its result, stores it in its AR, and returns the control to $\delta(a)$.*

The details are as follows. After $\delta(a-1)$ has computed its result $\varphi(a-1)$, it writes it into r of $[a-1, r]$. Then, $\delta(a-1)$ changes the state of the control unit to some previously designated state, say q_2, called the *return state*. In this way, $\delta(a-1)$ informs its caller that it has terminated. (See Fig. 7.4 (c).)

d. *$\delta(a)$ resumes its execution. It reads $\delta(a-1)$'s result from $\delta(a-1)$'s AR, uses it in its own computation, and stores the result in its AR.*

Specifically, when the control unit has entered the return state, $\delta(a)$ moves the tape window to $[a-1, r]$ of $\delta(a-1)$, copies r $(= \varphi(a-1))$ into its own $[a, r]$, and deletes $[a-1, r]$. Then $\delta(a)$ continues executing the rest of its program (steps 2 and 3) and finally writes the result $(= \varphi(a))$ into r of its AR. (See Fig. 7.4 (d).)

activation record

(a)   (b)   (c)   (d)

Fig. 7.4 (a) An activation $\delta(a)$ and its activation record. (b) $\delta(a-1)$ starts computing $\varphi(a-1)$. (c) $\delta(a)$ reads $\delta(a-1)$'s result. (d) $\delta(a)$ computes $\varphi(a)$ and stores it in its activation record

Example 7.1. (Computation of $\varphi(x) = x!$ on TM) Figure 7.5 shows the steps in the computation of the value $\varphi(4) = 4! = 24$. □

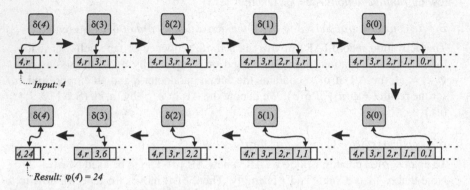

Fig. 7.5 Recursive computation of the value $\varphi(4) = 4!$ on a Turing machine

Clearly, δ must take care of all runtime actions described in *a, b, c, d*. Consequently, δ will consist of two sets of instructions:

- the instructions implementing the actions explicitly dictated by the algorithm (∗);
- the instructions implementing the *runtime actions a, b, c, d*, implicit in (∗).

To prove that there exists a TM that operates as described above, we refer to the *Computability Thesis* and leave the construction of the corresponding program δ as an exercise to the reader.

Looking at Fig. 7.5, we notice that adding and deleting of ARs on the tape of a Turing machine follow the rules characteristic of a data structure called the *stack*. The first AR that is created on the tape is at the *bottom* of the stack, and the rightmost existing AR is at the *top* of the stack. A new AR can be created next to the top AR; we say that it is *pushed* onto the stack. An AR can be deleted from the stack only if it is at the top of the stack; we say that it is *popped* off the stack. We will use these mechanisms shortly.

Later, we will often say that a TM T *calls* another TM T' to do a task for T. The call-and-return mechanism described above can readily be used to make T's program δ call the program δ' of T', pass actual parameters to δ', and collect δ''s results.

7.4.6 Practical Consequences: Procedure Calls in General-Purpose Computers

The mechanisms described in the previous section are used in general-purpose computers to handle procedure calls during program execution.

The Concept of the Procedure

In general-purpose computers we often use *procedures* to decompose large programs into components. The idea is that procedures have other procedures do subtasks. A procedure (*callee*) executes its task when it is *called* (invoked) by another procedure (*caller*). The caller can also be the operating system. A callee may return one or several values to its caller. Each procedure has its *private storage*, where it can access its private (*local*) variables that are needed to perform its task.

High-level procedural programming languages support procedures. They make *linkage conventions* in order to define and implement in standard ways the key mechanisms for procedure management. These mechanisms are needed to:

- *invoke* a procedure and map its actual parameters to the callee's private space, so that the callee can use them as its input;
- *return* control to the caller after the callee's termination, so that the caller can continue its execution immediately after the point of the call.

Most languages allow a procedure to return one or more values to the caller.

After a program (source code) is written, it must be compiled. One of the tasks of the compiler is to *embed* into the generated code all the *runtime* algorithms and data structures that are necessary to implement the call-and-return behavior implicitly dictated by the source code.

The code is then linked into the executable code, which is ready to be loaded into the computer's main memory for execution.

When the program is started, its call-and-return behavior can be modeled with a *stack*. In the simplest case, the caller pushes the return address onto the stack and transfers the control to the callee; when the callee returns, it pops the return address off the stack and transfers the control back to the caller at the return address. In reality, besides the return address there is more information passed between the two procedures.

The Role of the Compiler

We now see what the compiler must do. First, it must embed in the generated code runtime algorithms that will correctly push onto and pop off the stack the information that must be transferred between a caller and callee via the stack. Second, it must establish a data structure that will contain this information. The structure is called the *activation record* (AR). This must be implemented as a private block of memory associated with a specific invocation of a specific procedure. Consequently, the AR has a number of fields, each for one of the following data:

- *return address*, where the caller's execution will resume after the callee's termination;
- *actual parameters*, which are input data to the callee;
- *register contents*, which the caller preserves to resume its execution after return;
- *return values*, which the callee returns to the caller;
- *local variables*, which are declared and used by the callee;
- *addressability information*, which allows the callee to access non-local variables.

The Role of the Operating System

As described in Sect. 6.2.5, modern general-purpose computers and their operating systems allocate to each executable program, prior to its execution, a memory space that will be used by the program during its execution. Part of this memory is the *runtime stack*. During the execution of the program, an AR is pushed onto (popped off) the runtime stack each time a procedure is called (terminated). The current top AR contains the information associated with the currently executing procedure. In case the size of the runtime stack exceeds the limits set by the operating system, the operating system suspends the execution of the program and takes over. This can happen, for example, in a recursive program when a recursive procedure keeps invoking itself (because it does not meet its recursion-termination criterion).

Execution of Recursive Programs

So how would the program for computing the function $\varphi(x) = x!$ execute on a general-purpose computer?

First, we write the program in a high-level procedural programming language, such as C or Java. In this we follow the recursive definition (∗) on p. 166 of the function $\varphi(x)$. We obtain a source program similar to P below. Of course, P is recursive (i.e., self-referencing) because the activation $P(x)$ starts a new activation $P(x-1)$, i.e., a new instance of P with a decreased actual parameter. The invocations of P come to an end when the recursion-termination criterion ($x = 0$) is satisfied.

```
program P(x : integer) return integer;
  {
    if x > 0 return x * P(x − 1) else
    if x = 0 return 1
    else Error(illegal_input)
  }
```

7.5 Chapter Summary

A decidable set is also semi-decidable. If a set is decidable, so is its complement. If a set and its complement are semi-decidable, they are both decidable. A set is semi-decidable if and only if it is the domain of a computable function. If two sets are decidable, their union and intersection are decidable. If two sets are semi-decidable, their union and intersection are semi-decidable.

The *Padding Lemma* states that a partial computable function has countably infinitely many indexes; given one of them, countably many other indexes of the function can be generated. The *index set* of a partial computable function contains all of its indexes, that is, all the encoded Turing programs that compute the function.

The *Parameter (s-m-n) Theorem* states that the parameters of a function can be built into the function's Turing program to obtain a new Turing program computing the function without parameters. The new Turing program can be algorithmically constructed from the old program and the parameters only. The generalization of the theorem is called the *s-m-n Theorem*.

The *Recursion Theorem* states that if a transformation transforms every Turing program, then some Turing program is transformed into an equivalent Turing program. The *Recursion Theorem* allows a Turing-computable function to be defined recursively, i.e., with its own Turing program. The Turing machine model of computation supports the recursive execution of Turing programs. This lays the theoretical grounds for the recursive execution of programs in general-purpose computers and, more generally, for the call-and-return mechanism that allows the use of procedures in programs.

Problems

7.1. Show that:

(a) All tasks in the proof of the *Parameter Theorem* are computable;
[*Hint.* See Box 7.1.]

(b) The function s in the *Parameter Theorem* is injective.

7.2. Prove the following consequences of the *Parameter Theorem*:

(a) There is a computable function f such that $\mathrm{rng}(\varphi_{f(e)}) = \mathrm{dom}(\varphi_e)$;

[*Hint.* Suppose that there existed a p.c. function defined by $\varphi(e,x) \stackrel{\text{def}}{=} \begin{cases} x & \text{if } x \in \mathrm{dom}(\varphi_e); \\ \uparrow & \text{otherwise.} \end{cases}$

Viewing $\varphi(e,x)$ as a function of x, we would have $\mathrm{rng}(\varphi(e,x)) = \mathrm{dom}(\varphi_e)$. But $\varphi(e,x)$ does exist; it is computed by a TM that simulates universal Turing machine U on e,x and outputs x if U halts.]

(b) There is a computable function g such that $\mathrm{dom}(\varphi_{g(e)}) = \mathrm{rng}(\varphi_e)$;

(c) A set $\mathcal{S}$ is c.e. $\Longleftrightarrow$ $\mathcal{S}$ is the range of a p.c. function.

Remark. Similar is true of the domain of a p.c. function (see Theorem 6.6, p. 151).

7.3. Prove the following consequences of the *Recursion Theorem*:

(a) There is a TM that on any input outputs its own index and nothing else.

(In other words: There is an $n \in \mathbb{N}$ such that φ_n is the constant function with value n, i.e., $\mathrm{rng}(\varphi_n) = \{n\}$.)

[*Hint.* Denote by κ_c the constant function $\kappa_c : \mathbb{N} \to \{c\}$, where $c \in \mathbb{N}$ is an arbitrary number. Since κ_c is a computable function, it is also a p.c. and hence $\kappa_c \simeq \varphi_k$, for some $k \in \mathbb{N}$. Clearly, k depends on c. Prove that there is a computable function f such that $k = f(c)$. Then $\kappa_c \simeq \varphi_{f(c)}$, for arbitrary c. Now, since f is computable, the *Recursion Theorem* guarantees that there exists an n for which $\varphi_n \simeq \varphi_{f(n)}$. But $\varphi_{f(n)} \simeq \kappa_n$ and hence $\varphi_n \simeq \kappa_n$. It follows that $\mathrm{rng}(\varphi_n) = \{n\}$.]

Remark. We can view such a TM as a Turing program that prints out its own code. Generally, any program that prints out its own copy is called a *quine*.[2] Quines exist in any programming language that has the ability to output any computable string.

(b) There are two different Turing machines that output each other's indexes and nothing else.

(That is, there are $m, n \in \mathbb{N}$, $m \neq n$, such that φ_m is the constant function with value n, and φ_n is the constant function with value m, i.e., $\mathrm{rng}(\varphi_m) = \{n\} \wedge \mathrm{rng}(\varphi_n) = \{m\}$.)

7.4. Prove the following consequences of the *Recursion Theorem*:

(a) For every computable function f, there exists an $n \in \mathbb{N}$ such that $\mathcal{W}_n = \mathcal{W}_{f(n)}$.

(Recall from Definition 6.4, p. 135, that $\mathcal{W}_i \stackrel{\text{def}}{=} \mathrm{dom}(\varphi_i)$.)

(b) There is an $n \in \mathbb{N}$ such that $\mathcal{W}_n = \{n\}$.

(c) There is an $n \in \mathbb{N}$ such that $\mathcal{W}_n = \{n^2\}$.

(d) There is an $n \in \mathbb{N}$ such that TMs T_n and T_{n+1} both compute the same function, that is, $\varphi_n \simeq \varphi_{n+1}$.

(e) For every p.c. function $\varphi(e, x)$, there is an $n \in \mathbb{N}$ such that $\varphi(n, x) = \varphi_n(x)$.

7.5. Theorem 7.10, p. 164, states that a computable function f has countably infinitely many fixed points. Recall that we proved the theorem in a non-constructive way; that is, we didn't describe a generator that would—given one fixed point of f—list (i.e. enumerate) countably infinitely many other fixed points of f.

(a) Can you describe an algorithm to generate an infinite set of fixed points of f?

[*Hint.* Using f, construct a strictly increasing computable function $s : \mathbb{N} \to \mathbb{N}$ such that if i is a fixed point of f then $s(i)$ is a fixed point of f, i.e., $\varphi_i = \varphi_{f(i)} \Rightarrow \varphi_{s(i)} = \varphi_{f(s(i))}$.]

(b) Problem (a) has asked you to construct, given a computable function f, a Turing machine that will generate an infinite set $\{i, s(i), s(s(i)), \dots\}$ of *some* of the fixed points of f. But can we generate the set of *all* fixed points of f? The answer is that this cannot be done for every f. Try to prove the following:

There exists a computable function f whose set of fixed points is not a c.e. set.

7.6. We say that a TM T_1 *calls* another TM T_2 to perform a subtask for T_1. We view this as the program δ_1 calling the program (procedure) δ_2. Explain in detail how such a procedure call can be implemented on Turing machines.

[*Hint.* Merge programs δ_1 and δ_2 into a program δ_1', which will use activation records to activate δ_2 (see Sect. 7.4.5).]

---

[2] In honour of Willard Van Orman Quine, 1908–2000, American philosopher and logician.

Bibliographic Notes

- Theorem 7.3 was proved in Post [183].
- The *Parameter Theorem* and *Recursion Theorem* were first proved in Kleene [122]. An extensive description of both theorems and their variants as well as their applications is in Odifreddi [173]. Problem 7.5 a is from Hopcroft and Ullman [113] and Problem 7.5 b is from Rogers [202].
- For runtime environments, activation records, and procedure activation in general-purpose computers, see Aho et al. [9, Chap. 7] and Cooper and Torczon [42, Chap. 6].
- Exercise 7.3 a applies the *Recursion Theorem* to prove the existence of quines, programs that replicate themselves by outputting their own codes. The first self-reproducing computer programs were devised in 1972 by Bratley and Millo [25] and in 1980 by Burger et al. [28].
- But self-replication is not restricted to programs only. The research on *self-replication* was initiated already in the late 1940s by von Neumann, who had become interested in the questions of whether a machine can produce copies of itself and what is the logic necessary for such self-replication of machines. See von Neumann [272] for some of his results. Rogers [202, Sect. 11.4] describes the application of the *Recursion Theorem* to prove a general theorem that yields various other results on self-replicating machines. Sipper [233] gives an overview of research carried out in the domain of self-replication over the next 50 years, starting from von Neumann's work.

Chapter 8
Incomputable Problems

A problem is unsolvable if there is no single procedure that can construct a solution for an arbitrary instance of the problem.

Abstract After the basic notions of computation had been formalized, a close link between computational problems—in particular, decision problems—and sets was discovered. This was important because the notion of the set was finally settled, and sets made it possible to apply diagonalization, a proof method already discovered by Cantor. Diagonalization, combined with self-reference, made it possible to discover the first incomputable problem, i.e., a decision problem called the *Halting Problem*, for which there is no single algorithm capable of solving every instance of the problem. This was simultaneously and independently discovered by Church and Turing. After this, *Computability Theory* blossomed, so that in the second half of the 1930s one of the main questions became, "Which computational problems are computable and which ones are not?" Indeed, using various proof methods, many incomputable problems were discovered in different fields of science. This showed that incomputability is a constituent part of reality.

8.1 Problem Solving

In previous chapters we have discussed how the values of functions can be computed, how sets can be generated, and how sets can be recognized. All of these are *elementary* computational tasks in the sense that they are all closely connected with the computational model, i.e., the Turing machine. However, in practice we are also confronted with other kinds of problems that require certain computations to yield their solutions.[1] All such problems we call *computational problems*. Now the following question immediately arises: "Can we use the accumulated knowledge about how to solve the three elementary computational tasks to solve other kinds of computational problems?" In this section we will explain how this can be done.

---

[1] Obviously, we are not interested in psychological, social, economic, political, philosophical, and other related problems whose solutions require reasoning beyond (explicit) computation.

© The Author(s), under exclusive license to Springer-Verlag GmbH, DE,   175
part of Springer Nature 2020
R. Robič, *The Foundations of Computability Theory*, https://doi.org/10.1007/978-3-662-62421-0_8

8.1.1 Decision Problems and Other Kinds of Problems

Before we start searching for the answer to the above question, we must define precisely what we mean by other "kinds" of computational problems. It is a well-known fact, based upon our everyday experience, that there is a myriad of different computational problems. By grouping all the "similar" problems into classes, we can try to put this jungle of computational problems in order (a rather simple one, actually). The "similarity" between problems can be defined in several ways. For example, we could define a class of all the problems asking for numerical solutions, and another class of all the other problems. However, we will proceed differently. We will define two problems to be *similar* if their solutions are "equally simple." Now, "equally simple" is a rather fuzzy notion. Fortunately, we will not need to formalize it. Instead, it will suffice to define the notion informally. So let us define the following four *kinds (i.e., classes) of computational problems*:

- **Decision** problems (also called **yes/no** problems). The solution of a decision problem is the answer YES or NO. The solution can be represented by a *single bit* (e.g., $1 =$ YES, $0 =$ NO).
  Examples: Is there a prime number in a given set of natural numbers? Is there a Hamiltonian cycle in a given graph?

- **Search** problems. The solution of a search problem is an element of a given set S such that the element has a given property P. The solution is an *element of a set*.
  Examples: Find the largest prime number in a given set of natural numbers. Find the shortest Hamiltonian cycle in a given weighted graph.

- **Counting** problems. The solution of a counting problem is the number of elements of a given set S that have a given property P. The solution is a *natural number*.
  Examples: How many prime numbers are in a given set of natural numbers? How many Hamiltonian cycles are in a given graph?

- **Generation** problems (also called **enumeration** problems). The solution of a generation problem is a list of elements of a given set S that have a given property P. The solution is a *sequence of elements of a set*.
  Examples: List all the prime numbers in a given set of natural numbers. List all the Hamiltonian cycles of a given graph.

Which of these kinds of problems should we focus on? Here we make a pragmatic choice and focus on the *decision* problems, because these problems ask for the simplest possible solutions, i.e., solutions representable by a single bit. Our choice does not imply that other kinds of computational problems are not interesting—we only want to postpone their treatment until the decision problems are better understood.

8.1.2 Language of a Decision Problem

In this subsection we will show that there is a close link between *decision problems* and *sets*. This will enable us to *reduce* questions about decision problems to questions about sets. We will uncover the link in four steps.

1.   Let $\mathcal{D}$ be a decision problem.

2. In practice we are usually confronted with a particular *instance d* of the problem $\mathcal{D}$. The instance d can be obtained from $\mathcal{D}$ by replacing the variables in the definition of $\mathcal{D}$ with actual data. Thus, the problem $\mathcal{D}$ can be viewed as the set of all the possible instances of this problem. We say that an instance $d \in \mathcal{D}$ is *positive* or *negative* if the answer to d is YES or NO, respectively.

Example 8.1. (Problem Instance) Let $\mathcal{D}_{Prime} \equiv$ "Is n a prime number?" be a decision problem. If we replace the variable n by a particular natural number, say 4, we obtain the instance $d \equiv$ "Is 4 a prime number?" of $\mathcal{D}_{Prime}$. This instance is negative because its solution is the answer NO. In contrast, since 2009 we have known that the solution to the instance $d \equiv$ "Is $2^{43,112,609} - 1$ a prime number?" is YES, so the instance is positive.   □

So:

                                Let d be an instance of $\mathcal{D}$.

3. Any instance of a decision problem is either positive or negative (due to the *Law of Excluded Middle*; see Sect. 2.1.2). So the answer to the instance d is either YES or NO. But how can we *compute* the answer, say on the Turing machine? In the natural-language description of d there can be various data, e.g., numbers, matrices, graphs. But in order to compute the answer on a machine—be it a modern computer or an abstract model such as the Turing machine—we must rewrite these data in a form that is understandable to the machine. Since any machine uses its own alphabet Σ (e.g., $\Sigma = \{0, 1\}$), we must choose a function that will transform every instance of $\mathcal{D}$ into a word over Σ. Such a function is called the *coding function* and will be denoted by *code*. Therefore, code : $\mathcal{D} \to \Sigma^*$, so code($\mathcal{D}$) is the set of codes of all instances of $\mathcal{D}$. We will usually write $\langle d \rangle$ instead of the longer code(d). There are many different coding functions. From the point of view of *Computability Theory*, we can choose any of them as long as the function is *computable* on $\mathcal{D}$ and *injective*. These requirements[2] are natural because we want the coding process to halt and we do not want a word to encode different instances of $\mathcal{D}$.

Example 8.2. (Instance Code) The instances of the problem $\mathcal{D}_{Prime} \equiv$ "Is n a prime number?" can be encoded by the function code : $\mathcal{D}_{Prime} \to \{0, 1\}^*$ that rewrites n in its binary representation; e.g., code("Is 4 a prime number?") = 100 and code("Is 5 a prime number?") = 101. □

So:

                        Let code : $\mathcal{D} \to \Sigma^*$ be a coding function.

---

[2] In *Computational Complexity Theory* we also require that the coding function does not produce unnecessarily long codes.

4. After we have encoded d into $\langle d \rangle$, we could start searching for a Turing machine capable of computing the answer to d when given the input $\langle d \rangle \in \text{code}(\mathcal{D})$. But we proceed differently: We gather the *codes of all the positive instances* of $\mathcal{D}$ in a *set* and denote it by $L(\mathcal{D})$. This is a subset of Σ^*, so it is a formal language (see Sect. 6.3.6). Here is the definition.

> **Definition 8.1.** (Language of a Decision Problem) The **language of a decision problem** $\mathcal{D}$ is the set $L(\mathcal{D}) \overset{\text{def}}{=} \{ \langle d \rangle \in \Sigma^* \mid d \text{ is a positive instance of } \mathcal{D} \}$.

Example 8.3. (Language of a Decision Problem)   The language of the decision problem $\mathcal{D}_{Prime}$ is the set $L(\mathcal{D}_{Prime}) = \{10,11,101,111,1011,1101,10001,10011,10111,11101,\ldots\}$.   □

5. Now the following is obvious:

$$\text{An instance } d \text{ of } \mathcal{D} \text{ is positive} \iff \langle d \rangle \in L(\mathcal{D}). \qquad (*)$$

What did we gain by this? The equivalence $(*)$ tells us that computing the answer to d can be substituted with deciding whether or not $\langle d \rangle$ is in $L(\mathcal{D})$. (See Fig. 8.1.) Thus we have found a connection between decision problems and sets:

Solving a decision problem $\mathcal{D}$ can be reduced to deciding membership of the set $L(\mathcal{D})$ in Σ^*.

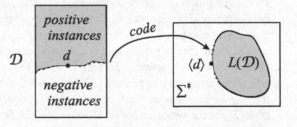

Fig. 8.1 The answer to the instance d of a decision problem $\mathcal{D}$ can be obtained by determining where the word $\langle d \rangle$ is relative to the set $L(\mathcal{D})$

The connection $(*)$ is important because it enables us to apply (when solving decision problems) all the theory developed to decide sets. Recall that deciding a set means determining what is in the set and what is in its complement (Sect. 6.3.3). Now let us see what the decidability of $L(\mathcal{D})$ in Σ^* tells us about the solvability of the decision problem $\mathcal{D}$:

- *Let $L(\mathcal{D})$ be decidable.* Then there exists a decider $D_{L(\mathcal{D})}$, which, for arbitrary $\langle d \rangle \in \Sigma^*$, answers the question $\langle d \rangle \in ? \ L(\mathcal{D})$. Because of $(*)$, the answer tells us whether d is a positive or a negative instance of $\mathcal{D}$. Consequently, there is an algorithm that, for the arbitrary instance $d \in \mathcal{D}$, decides whether d is positive or negative. The algorithm is $D_{L(\mathcal{D})}(\text{code}(\cdot))$, the composition of the coding function and the decider of $L(\mathcal{D})$.

- *Let $L(\mathcal{D})$ be semi-decidable.* Then, for arbitrary $\langle d \rangle \in L(\mathcal{D})$, the recognizer $R_{L(\mathcal{D})}$ answers the question $\langle d \rangle \in? L(\mathcal{D})$ with YES. However, if $\langle d \rangle \notin L(\mathcal{D})$, $R_{L(\mathcal{D})}$ may or may not return NO in finite time. So: There is an algorithm that for an arbitrary *positive* instance $d \in \mathcal{D}$ finds that d *is* positive. The algorithm is $R_{L(\mathcal{D})}(\mathrm{code}(\cdot))$.

- *Let $L(\mathcal{D})$ be undecidable.* Then there is no algorithm capable of answering, for arbitrary $\langle d \rangle \in \Sigma^*$, the question $\langle d \rangle \in? L(\mathcal{D})$. Because of $(*)$, there is no algorithm capable of deciding, for an arbitrary instance $d \in \mathcal{D}$, whether d is positive or negative.

So we can extend our terminology about sets (Definition 6.10) to decision problems.

Definition 8.2. Let $\mathcal{D}$ be a decision problem. We say that the problem

$\mathcal{D}$ is **decidable** (or **computable**)   *if* $L(\mathcal{D})$ is a decidable set;

$\mathcal{D}$ is **semi-decidable**   *if* $L(\mathcal{D})$ is a semi-decidable set;

$\mathcal{D}$ is **undecidable** (or **incomputable**)   *if* $L(\mathcal{D})$ is an undecidable set.

NB *Instead of a(n) (un)decidable problem we can say (in)computable problem. But bear in mind that the latter notion is more general: It can be used with all kinds of computational problems, not only with decision problems (Sects. 9.3 and 9.5). So, an (un)decidable problem is the same as an (in)computable decision problem. The term* (un)solvable *from this section's motto is even more general: It addresses all kinds of computational and non-computational problems. (See Fig. 8.2.)*

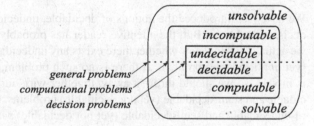

Fig. 8.2 The relationship between different kinds of problems

8.1.3 Subproblems of a Decision Problem

Often we encounter a decision problem that is a special version of another, more general decision problem. Is there any connection between the languages of the two problems? Is there any connection between the decidabilities of the two problems? Let us start with a definition.

Definition 8.3. (Subproblem) A decision problem $\mathcal{D}_{Sub}$ is a **subproblem** of a decision problem $\mathcal{D}_{Prob}$ if $\mathcal{D}_{Sub}$ is obtained from $\mathcal{D}_{Prob}$ by imposing additional restrictions on (some of) the variables of $\mathcal{D}_{Prob}$.

A restriction can be put in various ways: A variable can be assigned a particular value; or it can be restricted to take only the values from a given set of values; or a relation between a variable and another variable can be imposed. For instance, if $\mathcal{D}_{Prob}$ has the variables x and y, then we might want to impose the equality $x = y$. (We will do this in the next section.) It should be clear that the following hold:

$$\mathcal{D}_{Sub} \text{ is a subproblem of } \mathcal{D}_{Prob} \implies \text{code}(\mathcal{D}_{Sub}) \subseteq \text{code}(\mathcal{D}_{Prob})$$
$$\mathcal{D}_{Sub} \text{ is a subproblem of } \mathcal{D}_{Prob} \implies L(\mathcal{D}_{Sub}) \subseteq L(\mathcal{D}_{Prob})$$

Of course, we assume that we can decide whether or not $d \in \mathcal{D}_{Sub}$, for any $d \in \mathcal{D}_{Prob}$. It is easy to prove the next theorem.

Theorem 8.1. *Let $\mathcal{D}_{Sub}$ be a subproblem of a decision problem $\mathcal{D}_{Prob}$. Then:*
$$\mathcal{D}_{Sub} \text{ is undecidable} \implies \mathcal{D}_{Prob} \text{ is undecidable}$$

Proof. Let $\mathcal{D}_{Sub}$ be undecidable and suppose that $\mathcal{D}_{Prob}$ is decidable. We could use the algorithm for solving $\mathcal{D}_{Prob}$ to solve (any instance of) $\mathcal{D}_{Sub}$, so $\mathcal{D}_{Sub}$ would be decidable. Contradiction. $\square$

8.2 There Is an Incomputable Problem — Halting Problem

We have just introduced the notions of decidable, undecidable, and semi-decidable decision problems. But, the attentive reader has probably noticed that at this point we actually *do not know* whether there exists any undecidable or any semi-decidable (yet *not* decidable) problem. If there is no such problem, then the above definition is in vain, so we should throw it away, together with our attempt to explore undecidable and semi-decidable (yet not decidable) problems. (In this case the definition of undecidable and semi-decidable (yet not decidable) *sets* would also be superfluous.) In other words, the recently developed part of *Computability Theory* should be abandoned. Therefore, the important question at this point is this: "Is there any decision problem that is undecidable or semi-decidable (yet not decidable)?"

To prove the existence of such a problem, we should find a decision problem $\mathcal{D}$ such that the set $L(\mathcal{D})$ is undecidable or semi-decidable (but not decidable). But how can we find such a $\mathcal{D}$ (if there is one at all)? In 1936, Turing succeeded in this.[3] How did he do that?

Turing was well aware of the fact that difficulties in obtaining computational results are caused by those Turing machines that may not halt on some input data.

---

[3] Independently and at the same time, Church also found such a problem.

It would therefore be beneficial, he reckoned, if we could *check, for any Turing machine T and any input word w, whether or not T halts on w*. If such a check were possible, then, given an arbitrary pair (T, w), we would first check the pair (T, w) and then, depending on the outcome of the check, we would either start T on w, or try to improve T so that it would halt on w, too. This led Turing to define the following decision problem.

Definition 8.4. (Halting Problem) The **Halting Problem** $\mathcal{D}_{Halt}$ is defined by
$\mathcal{D}_{Halt} \equiv$ "Given a Turing machine T and a word $w \in \Sigma^*$, does T halt on w?"

Turing then proved the following theorem.

Theorem 8.2. *The Halting Problem $\mathcal{D}_{Halt}$ is undecidable.*

Before we go into the proof, we introduce two sets that will play an important role in the proof and, indeed, in the rest of the book. The sets are called the *universal* and *diagonal languages*, respectively.

Definition 8.5. (Universal Language $\mathcal{K}_0$) The **universal language**, denoted[4] $\mathcal{K}_0$, is the language of the *Halting Problem*, that is,

$$\mathcal{K}_0 \stackrel{\text{def}}{=} L(\mathcal{D}_{Halt}) = \{\langle T, w \rangle \mid T \text{ halts on } w\}$$

The second language, $\mathcal{K}$, is obtained from $\mathcal{K}_0$ by imposing the restriction $w = \langle T \rangle$.

Definition 8.6. (Diagonal Language $\mathcal{K}$) The **diagonal language**, denoted $\mathcal{K}$, is defined by
$$\mathcal{K} \stackrel{\text{def}}{=} \{\langle T, T \rangle \mid T \text{ halts on } \langle T \rangle\}$$

Observe that $\mathcal{K}$ is the language $L(\mathcal{D}_H)$ of the decision problem

$$\mathcal{D}_H \equiv \text{"Given a Turing machine } T, \text{ does } T \text{ halt on } \langle T \rangle?"$$

The problem $\mathcal{D}_H$ is a *subproblem* of $\mathcal{D}_{Halt}$, since it is obtained from $\mathcal{D}_{Halt}$ by restricting the variable w to $w = \langle T \rangle$.

We can now proceed to the proof of Theorem 8.2.

---

[4] The reasons for the notation $\mathcal{K}_0$ are historical. Namely, Post proved that this language is *Turing-complete*. We will explain the notion of Turing-completeness in Part III of the book; see Sect. 14.1.

Proof. (of the theorem)   The plan is to prove (in a lemma) that $\mathcal{K}$ is an undecidable set; this will then imply that $\mathcal{K}_0$ is also an undecidable set and, consequently, that $\mathcal{D}_{Halt}$ is an undecidable problem. The lemma is instructive because it applies a cunningly defined Turing machine to its own code.

Lemma 8.1. (Undecidability of $\mathcal{K}$) *The set $\mathcal{K}$ is undecidable.*

Proof. (of the lemma) *Suppose that $\mathcal{K}$ is a decidable set.* Then there must exist a decider $D_{\mathcal{K}}$, which, for arbitrary T, answers the question $\langle T, T \rangle \in ? \mathcal{K}$ with

$$D_{\mathcal{K}}(\langle T,T \rangle) = \begin{cases} \text{YES, if } T \text{ halts on } \langle T \rangle; \\ \text{NO, if } T \text{ does not halt on } \langle T \rangle. \end{cases}$$

Now we construct a new Turing machine S. The intention is to construct S in such a way that, when given as input its *own* code $\langle S \rangle$, S will expose the incapability of $D_{\mathcal{K}}$ to predict whether or not S will halt on $\langle S \rangle$. The machine S is depicted in Fig. 8.3.

The shrewd Turing machine S uncovers the incapability of $D_{\mathcal{K}}$ to answer whether S halts on its own code <S>.

The supposed machine $D_{\mathcal{K}}$ answers, for arbitrary T, whether T halts on its own code <T>.

Fig. 8.3 S exposes the incapability of the decider $D_{\mathcal{K}}$ to predict the halting of S on $\langle S \rangle$

The machine S operates as follows. The input to S is the code $\langle T \rangle$ of an arbitrary Turing machine T. The machine S doubles $\langle T \rangle$ into $\langle T, T \rangle$, sends this to the decider $D_{\mathcal{K}}$, and starts it. The decider $D_{\mathcal{K}}$ eventually halts on $\langle T, T \rangle$ and answers either YES or NO to the question $\langle T, T \rangle \in ?\mathcal{K}$. If $D_{\mathcal{K}}$ has answered YES, then S asks $D_{\mathcal{K}}$ again the same question. If, however, $D_{\mathcal{K}}$ has answered NO, then S outputs its own answer YES and *halts*.

But S is *shrewd*: if given as input its *own* code $\langle S \rangle$, it puts the supposed decider $D_{\mathcal{K}}$ in insurmountable trouble. Let us see why. Given the input $\langle S \rangle$, S doubles it into $\langle S, S \rangle$ and hands it over to $D_{\mathcal{K}}$, which in finite time answers the question $\langle S, S \rangle \in ?\mathcal{K}$ with either YES or NO. Now let us analyze the consequences of each of the answers:

a. *Suppose that $D_{\mathcal{K}}$ has answered with $D_{\mathcal{K}}(\langle S,S \rangle)$ = YES.* Then S repeats the question $\langle S, S \rangle \in ?\mathcal{K}$ to $D_{\mathcal{K}}$, which in turn stubbornly repeats its answer $D_{\mathcal{K}}(\langle S,S \rangle)$ = YES. It is obvious that S cycles and will *never* halt. But note that, at the same time, $D_{\mathcal{K}}$ repeatedly predicts just the opposite, i.e., that S *will* halt on $\langle S \rangle$. We must conclude that in the case (a) the supposed decider $D_{\mathcal{K}}$ fails to compute the correct answer.

b. *Suppose that $D_{\mathcal{K}}$ has answered with $D_{\mathcal{K}}(\langle S, S \rangle)$ = NO. Then S returns to the* environment its own answer and *halts*. But observe that just as before $D_{\mathcal{K}}$ has computed the answer $D_{\mathcal{K}}(\langle S, S \rangle)$ = NO and thus predicted that S will *not* halt on $\langle S \rangle$. We must conclude that in the case (b) the supposed decider $D_{\mathcal{K}}$ fails to compute the correct answer.

Thus, the supposed decider $D_{\mathcal{K}}$ is unable to correctly decide the question $\langle S, S \rangle \in ?\mathcal{K}$. This contradicts our supposition that $\mathcal{K}$ is a decidable set and $D_{\mathcal{K}}$ its decider. We must conclude that $\mathcal{K}$ is an undecidable set. *The lemma is proved.*

Because $\mathcal{K}$ is undecidable, so is the problem $\mathcal{D}_H$. Now $\mathcal{D}_H$ is a subproblem of the *Halting Problem* $\mathcal{D}_{Halt}$, and code$(\mathcal{D}_H)$ is a subset of code$(\mathcal{D}_{Halt})$. Then, by Theorem 8.1, the *Halting Problem* is undecidable too. This completes the proof of Theorem 8.2. $\square$

As noted, the sets $\mathcal{K}_0$ and $\mathcal{K}$ are called the universal and diagonal languages, respectively. The reasons for such a naming are explained in the following subsection.

8.2.1 Consequences: The Basic Kinds of Decision Problems

Now we know that besides decidable problems there also exist undecidable problems. What about semi-decidable problems? Do they exist? The answer is clearly *yes*, because every decidable set is by definition semi-decidable. So the real question is: Are there *undecidable* problems that are *semi-decidable*? That is, are there decision problems such that only their positive instances are guaranteed to be solvable? The answer is *yes*. Let us see why.

Theorem 8.3. *The set $\mathcal{K}_0$ is semi-decidable.*

Proof. We must prove that there is a recognizer for $\mathcal{K}_0$ (see Sect. 6.3.3). Let us conceive the following machine: Given an arbitrary input $\langle T, w \rangle \in \Sigma^*$, the machine must simulate T on w, and if the simulation halts, the machine must return YES and halt. So, if such a machine exists, it will answer YES *iff* $\langle T, w \rangle \in \mathcal{K}_0$. But we already know that such a machine exists: It is the *Universal Turing machine U* (see Sect. 6.2.2). Hence, $\mathcal{K}_0$ is semi-decidable. $\square$

Theorems 8.2 and 8.3 imply that $\mathcal{K}_0$ is an undecidable semi-decidable set.

The proof of Theorem 8.3 has also revealed that $\mathcal{K}_0$ is the proper set of the Universal Turing machine. This is why $\mathcal{K}_0$ is called the *universal language*.

What about the set $\overline{\mathcal{K}_0}$, the complement of $\mathcal{K}_0$?

Theorem 8.4. *The set $\overline{\mathcal{K}}_0$ is not semi-decidable.*

Proof. If $\overline{\mathcal{K}}_0$ were semi-decidable, then both $\overline{\mathcal{K}}_0$ and $\mathcal{K}_0$ would be semi-decidable. Then $\mathcal{K}_0$ would be decidable because of Theorem 7.3 (p. 156). But this would contradict Theorem 8.2. So, the set $\overline{\mathcal{K}}_0$ is not semi-decidable. $\square$

Similarly, we prove that the undecidable set $\mathcal{K}$ is semi-decidable, and that $\overline{\mathcal{K}}$ is not semi-decidable.

We have proved the existence of undecidable semi-decidable sets (e.g., $\mathcal{K}_0$ and $\mathcal{K}$) and the existence of undecidable sets that are not even semi-decidable (e.g., $\overline{\mathcal{K}}_0$ and $\overline{\mathcal{K}}$). We now know that the class of all sets partitions into three nonempty subclasses, as depicted in Fig. 8.4. (Of course, we continue to talk about sets that are subsets of Σ^* or $\mathbb{N}$, as explained in Sect. 6.3.5.)

Fig. 8.4 The three main classes of sets according to their decidability

The Class of All Decision Problems

We can view sets as languages of decision problems. Then, $\mathcal{K}_0$ and $\mathcal{K}$ (Fig. 8.4) are the languages of decision problems $\mathcal{D}_{Halt}$ and $\mathcal{D}_H$, respectively. What about $\overline{\mathcal{K}}_0$ and $\overline{\mathcal{K}}$? These are the languages of decision problems $\mathcal{D}_{\overline{Halt}}$ and $\mathcal{D}_{\overline{H}}$, respectively, where $\mathcal{D}_{\overline{Halt}} \equiv$ "Given a Turing machine T and a word $w \in \Sigma^*$, does T *never* halt on w?" and $\mathcal{D}_{\overline{H}} \equiv$ "Given a Turing machine T, does T *never* halt on $\langle T \rangle$?" Now we see (Fig. 8.5) that *the class of all decision problems* partitions into two nonempty subclasses: the class of decidable problems and the class of undecidable problems. There is also a third class, the class of semi-decidable problems, which contains all the decidable problems and some, but not all, of the undecidable problems.

In other words, a decision problem $\mathcal{D}$ can be of one of three kinds:

- $\mathcal{D}$ is *decidable*. This means that there is an algorithm D that can solve an *arbitrary* instance $d \in \mathcal{D}$. Such an algorithm D is called a *decider of the problem $\mathcal{D}$*.
- $\mathcal{D}$ is *semi-decidable and undecidable*. This means that no algorithm can solve an arbitrary instance $d \in \mathcal{D}$, but there is an algorithm R that can solve an arbitrary *positive $d \in \mathcal{D}$*. The algorithm R is called a *recognizer of the problem $\mathcal{D}$*.
- $\mathcal{D}$ is *not semi-decidable*. Now, for any algorithm there exist a positive instance and a negative instance of $\mathcal{D}$ such that the algorithm cannot solve either of them.

Fig. 8.5 The three main classes of decision problems according to their decidability

8.2.2 Consequences: Complementary Sets and Decision Problems

From the previous theorems it follows that there are only three possibilities for the decidability of a set S and its complement $\overline{S}$:

1. S and $\overline{S}$ are decidable (e.g., ▲, △ in Fig. 8.6);
2. S and $\overline{S}$ are undecidable; one is semi-decidable and the other is not (e.g., ○, ●);
3. S and $\overline{S}$ are undecidable and neither is semi-decidable (e.g., □, ■).

Fig. 8.6 Decidability of complementary sets

Formally, two complementary sets S and $\overline{S}$ are associated with two complementary decision problems $\mathcal{D}_S$ and $\overline{\mathcal{D}}_S$, where $\overline{\mathcal{D}}_S \overset{\text{def}}{=} \mathcal{D}_{\overline{S}}$. Of course, for the decidability of these two problems there are also only three possibilities (Fig. 8.7).

The class of all decision problems:

●	■ □ } undecidable
○	
▲ △	} decidable

semi-decidable {

Fig. 8.7 Decidability of complementary decision problems

Remark. Because of the situation ●,○, two undecidable problems can differ in their undecidability. Does this mean that ● is more difficult than ○? Are there different *degrees* of undecidability? The answer is *yes*. Indeed, there is much to be said about this phenomenon, but we postpone the treatment of it until Part III.

8.2.3 Consequences: There Is an Incomputable Function

We are now able to easily prove that there exists an incomputable function. Actually, we foresaw this much earlier, in Sect. 5.3.3, by using the following simple argument. There are c different functions $f : \Sigma^* \to \Sigma^*$, where c is the cardinality of the set of real numbers. Among these functions there can be at most $\aleph_0$ *computable* functions (as there are at most $\aleph_0$ different Turing machines, which compute functions' values). Since $\aleph_0 < c$, there must be functions $f : \Sigma^* \to \Sigma^*$ that are not computable. However, this proof is not constructive: It neither constructs a particular incomputable function nor shows how such a function could be constructed, at least in principle. In the intuitionistic view, this proof is not acceptable (see Sect. 2.2.2).

But now, when we know that $\mathcal{K}_0$ is undecidable, it is an easy matter to exhibit a particular incomputable function. This is the characteristic function $\chi_{\mathcal{K}_0} : \Sigma^* \to \{0, 1\}$, where

$$\chi_{\mathcal{K}_0}(x) = \begin{cases} 1 \, (\equiv \text{YES}), & \text{if } x \in \mathcal{K}_0; \\ 0 \, (\equiv \text{NO}), & \text{if } x \notin \mathcal{K}_0. \end{cases}$$

Corollary 8.1. (Incomputable Function) *The characteristic function $\chi_{\mathcal{K}_0}$ is incomputable.*

Proof. If $\chi_{\mathcal{K}_0}$ were computable, then $\mathcal{K}_0$ would be decidable, contradicting Theorem 8.2. □

By the same argument we also see that $\chi_{\mathcal{K}}$ is incomputable.

Now the following question arises: Are there incomputable functions that are defined less generally than the characteristic functions of undecidable sets? The answer is yes. We will define such a function, the so-called *Busy Beaver function*, in the next section.

8.3 Some Other Incomputable Problems

We have seen that the formalization of the basic notions of computability led to the surprising discovery of a computational problem, the *Halting Problem*, an arbitrary instance of which no single algorithm is capable of solving. Informally, this means that this problem is in general unsolvable; see Fig. 8.2 on p. 179. Of course, the following question was immediately raised: Are there any other incomputable problems? The answer is *yes*. Indeed, since the 1940s many other computational problems have been proved to be incomputable. The first of these problems were somewhat unnatural, in the sense that they referred to the properties and the operations of models of computation. After 1944, however, more realistic incomputable problems were (and are still being) discovered in different fields of science and in other nonscientific fields. In this section we will list some of the known incomputable problems, grouped by the fields in which they occur. (We will postpone the discussion of the methods for proving their incomputability to the next section.)

8.3.1 Problems About Turing Machines

Halting of Turing Machines. We already know two incomputable problems about Turing machines: the *Halting Problem* $\mathcal{D}_{Halt}$ and its subproblem $\mathcal{D}_H$. But there are other incomputable problems about the halting of Turing machines.

▲ HALTING OF TURING MACHINES

Let T and T' be arbitrary Turing machines and U the universal Turing machine. Let $w \in \Sigma^*$ be an arbitrary word and $w_0 \in \Sigma^*$ an arbitrary fixed word. Questions:

$$\mathcal{D}_{Halt} \equiv \text{``Does } T \text{ halt on } w?\text{''}$$
$$\mathcal{D}_H \equiv \text{``Does } T \text{ halt on } \langle T \rangle?\text{''}$$
$$\text{``Does } T \text{ halt on empty input?''}$$
$$\text{``Does } T \text{ halt on every input?''}$$
$$\text{``Does } T \text{ halt on } w_0?\text{''}$$
$$\text{``Does } U \text{ halt on } w?\text{''}$$
$$\text{``Do } T \text{ and } T' \text{ halt on the same inputs?''}$$

These problems are all undecidable; no algorithm can solve any of them in general.

Properties of Turing Machine Languages. Recall from Sect. 6.3.3 that $L(T)$, the proper set of a Turing machine T, contains exactly those words in Σ^* on which T halts in the accepting state. So, which Turing machines have empty and which ones nonempty proper sets? Next, by Definition 6.10 (p. 142), $L(T)$ is decidable if we can algorithmically decide, for an arbitrary word, whether or not the word is in $L(T)$. Which Turing machines have decidable languages and which ones do not?

▲ PROPERTIES OF TM LANGUAGES

Let T be an arbitrary Turing machine. Question:

$$\mathcal{D}_{Emp} \equiv \text{``Is the language } L(T) \text{ empty?''}$$
$$\text{``Is the language } L(T) \text{ decidable?''}$$

Both problems are undecidable; no algorithm can solve either of them for arbitrary T.

Busy Beaver Problems. Informally, a busy beaver is the most productive Turing machine of its kind. Let us define the kind of Turing machines we are interested in. Let $n \geqslant 1$ be a natural number. Define $\mathcal{T}_n$ to be the set of all two-way unbounded Turing machines $T = (Q, \Sigma, \Gamma, \delta, q_1, \sqcup, F)$, where $Q = \{q_1, \ldots, q_{n+1}\}$, $\Sigma = \{0, 1\}$, $\Gamma = \{0, 1, \sqcup\}$, $\delta : Q \times \Gamma \to Q \times \{1\} \times \{\text{Left}, \text{Right}\}$ and $F = \{q_{n+1}\}$. Informally, $\mathcal{T}_n$ contains all Turing machines that have 1) the tape unbounded in both directions; 2) $n \geqslant 1$ non-final states (including the initial state q_1) and one final state q_{n+1}; 3) tape symbols $0, 1, \sqcup$ and input symbols $0, 1$; 4) instructions that write to a cell only the symbol 1 and move the window either to the left or right. Such TMs have unbounded space on their tape for writing the symbol 1, and they do not waste time either writing symbols other than 1 or leaving the window as it is.

It can be shown that there are finitely many Turing machines in $\mathcal{T}_n$. (See Problem 8.3, p. 204.) Now, let us pick an arbitrary $T \in \mathcal{T}_n$ and start it on an empty input (i.e., with the tape containing $\sqcup$ in each cell.) If T halts, we define its *score* $\sigma(T)$ to be the number of symbols 1 that remain on the tape after halting. If, however, T does not halt, we let $\sigma(T)$ be undefined. In this way we have defined a partial function $\sigma : \mathcal{T}_n \to \mathbb{N}$, where

$$\sigma(T) \stackrel{\text{def}}{=} \begin{cases} k & \text{if } T \text{ halts on empty input and leaves on its tape } k \text{ symbols 1;} \\ \uparrow & \text{otherwise.} \end{cases}$$

We will say that a Turing machine $T \in \mathcal{T}_n$ is a *stopper* if it halts on an empty input. Intuitively, we expect that, in general, different stoppers in $\mathcal{T}_n$—*if they exist*—attain different scores. Since there are only finitely many stoppers in $\mathcal{T}_n$, at least one of them must attain the highest possible score in $\mathcal{T}_n$. But, what if there were *no* stoppers in $\mathcal{T}_n$? Then, the highest score would not be defined for $\mathcal{T}_n$ (i.e., for that n). Can this happen at all? The answer is no; in 1962 Radó[5] proved that for *every* $n \geqslant 1$ there *exists* a stopper in $\mathcal{T}_n$ that attains, among all the stoppers in $\mathcal{T}_n$, the maximum value of σ. Such a stopper is called an *n-state Busy Beaver* and will be denoted by *n-BB*. Consequently, the function $s : \mathbb{N} \to \mathbb{N}$, defined by

$$s(n) \stackrel{\text{def}}{=} \sigma(n\text{-}BB)$$

is well defined (i.e., $s(n){\downarrow}$ for every $n \in \mathbb{N}$). We call it the *Busy Beaver function*.

At last we can raise the following question: Given an arbitrary Turing machine of the above kind (i.e., a machine in $\bigcup_{i \geqslant 1} \mathcal{T}_i$), can we algorithmically decide whether or not the machine is an *n*-state Busy Beaver for some *n*? Interestingly, we cannot.

▲ BUSY BEAVER PROBLEM

Let $T \in \bigcup_{i \geqslant 1} \mathcal{T}_i$ be an arbitrary Turing machine. Question:

$$\mathcal{D}_{BB} \equiv \text{``Is } T \text{ a Busy Beaver?''}$$

The problem is undecidable. There is no algorithm that can solve it for arbitrary T.

What about the Busy Beaver function s? Is it computable? The answer is *no*.

▲ BUSY BEAVER FUNCTION

The Busy Beaver function is incomputable.

No algorithm can compute, for arbitrary $n \geqslant 1$, the score of the *n*-state Busy Beaver.

---

[5] Tibor Radó, 1895–1965, Hungarian-American mathematician.

8.3.2 Post's Correspondence Problem

This problem was defined and proved to be undecidable by Post in 1946 as a result of his research into normal systems (see Sect. 6.3.2). It is important because it was one of the first realistic undecidable problems to be discovered and because it was used to prove the undecidability of several other decision problems. The problem can be defined as follows (Fig. 8.8).

Fig. 8.8 A positive instance of *Post's Correspondence Problem*

Let C be a finite set of elements, called *card templates*. Each card template is divided into an upper and lower half. Each half contains a word over some alphabet Σ. We call the two words the upper and the lower word of the card template. For each card template there is a stack with a potentially infinite number of exact copies of the template, called the *cards*. Cards from any stack can be put one after another in an arbitrary order to form a finite sequence of cards, called the *C-sequence*. Each C-sequence defines two compound words, called *sentences*. The upper sentence U is the concatenation of all the upper words in the C-sequence. Similarly, the lower sentence L is the concatenation of all the lower words in the C-sequence. Now the question is: Is there a C-sequence such that $U = L$?

▲ POST'S CORRESPONDENCE PROBLEM

Let C be a finite set of card templates, each with an unlimited number of copies. Question:

$$\mathcal{D}_{PCP} \equiv \text{"Is there a finite } C\text{-sequence such that } U = L?\text{"}$$

This problem is undecidable; no algorithm can solve it for arbitrary C.

8.3.3 Problems About Algorithms and Computer Programs

Termination of Algorithms and Programs. Algorithm designers and computer programmers are usually interested in whether their algorithms and programs always terminate (i.e., do not get trapped into endless cycling). Since the Turing machine formalizes the notion of algorithm, we can easily restate two of the above halting problems as the following undecidable problems.

▲ TERMINATION OF ALGORITHMS (PROGRAMS)

Let A be an arbitrary algorithm and d be arbitrary input data. Questions:

$$\mathcal{D}_{Term} \equiv \text{``Does an algorithm } A \text{ terminate on every input data?''}$$
$$\text{``Does an algorithm } A \text{ terminate on input data } d?''$$

Both problems are undecidable. Consequently, there is no computer program capable of checking, for an arbitrary computer program P, whether or not P eventually terminates (even if the input data is fixed). This holds irrespectively of the programming language used to encode the algorithms.

Correctness of Algorithms and Programs. Algorithm designers and computer programmers are most interested in their algorithms and programs correctly solving their problems of interest. It would therefore be highly advantageous to construct a computer program V that would verify, for an arbitrary problem $\mathcal{P}$ and an arbitrary algorithm A, whether or not A correctly solves $\mathcal{P}$. In order to be read by such a would-be verifier, the problem $\mathcal{P}$ and the algorithm A must be appropriately encoded. So let $\mathrm{code}(\mathcal{P})$ denote the word encoding the problem $\mathcal{P}$, and let $\mathrm{code}(A)$ denote the computer program describing the algorithm A.

▲ CORRECTNESS OF ALGORITHMS (PROGRAMS)

Let $\mathcal{P}$ be an arbitrary computational problem and A an arbitrary algorithm. Question:

$$\mathcal{D}_{Corr} \equiv \text{``Does the algorithm } \mathrm{code}(A) \text{ correctly solve the problem } \mathrm{code}(\mathcal{P})?''$$

The problem is undecidable; there is no algorithm (verifier) capable of solving it for arbitrary $\mathcal{P}$ and A. This is true irrespective of the programming language used to write the algorithms. Indeed, the problem remains undecidable even if $\mathrm{code}(A)$ is allowed to use only the most elementary data types (e.g., integers, character strings) and perform only the most elementary operations on them. This is bad news for those who work in the field of program verification.

Shorter Equivalent Programs. Sometimes, programmers try to shorten their programs in some way. They may want to reduce the number of program statements, or the total number of symbols, or the number of variables in their programs. In any case, they use some reasonable measure of program length. While striving for a shorter program they want to retain its functionality; that is, a shortened program should be functionally equivalent to the original one. Here, we define two programs as being *equivalent* if, for every input, they return equal results. So, the question is: Is there a shorter equivalent program? In general, this is an undecidable problem.

▲ EXISTENCE OF SHORTER EQUIVALENT PROGRAMS

Let $\mathrm{code}(A)$ be a program describing an algorithm A. Question:

"Given a program code(A), is there a shorter equivalent program?"

This is an undecidable problem; no algorithm can solve it in general. This holds for any reasonable definition of program length and for any programming language used to code algorithms.

8.3.4 Problems About Programming Languages and Grammars

The syntax of a language—be it natural, programming, or purely formal—is described by means of a *grammar*. This can be viewed as a Post canonical system (see Box 6.2 on p. 138) whose productions have been restricted (i.e., simplified) in any of several possible ways. For example, natural languages, such as English, developed complex syntaxes that demand so-called *context-sensitive grammars* to deal with. In contrast, programming languages, which are artificial, have their syntaxes defined by simpler grammars, the so-called *context-free grammars* (CFGs). This is because such grammars simplify the recognition (parsing) of computer programs (i.e., checking whether the programs are syntactically correct). This in turn allows us to construct simpler and faster parsers (and compilers). Finally, the basic building blocks of computer programs, the so-called *tokens*, are words with even simpler syntax. Tokens can be described and recognized with so-called *regular grammars*. In summary, a grammar G can be used in several ways: to *define* a language, which we will denote by $L(G)$; or to *generate* (some) elements of $L(G)$; or to *recognize* the language $L(G)$. For example, given a string w of symbols (be it a natural-language sentence or a computer program), we can answer the question $w \in ?L(G)$ by trying to generate w using G. In particular, the parser tries to generate the given computer program using G. When G is a CFG, we say that $L(G)$ is a *context-free language* (CFL); and when G is regular, $L(G)$ is said to be a *regular language*. Many problems about grammars and their languages have been defined in the fields of programming language design and translation. Some of these are incomputable.

Ambiguity of CFG Grammars. Programming language designers are only interested in grammars that do not allow a computer program to be parsed in two different ways. If this happened, it would mean that the structure and meaning of the program could be viewed in two different ways. Which of the two was intended by the programmer? Such a grammar would be *ambiguous*. So the detection of ambiguous grammars is an important practical problem.

▲ AMBIGUITY OF CFG GRAMMARS
   Let G be a context-free grammar. Question:

"Is there a word that can be generated by G in two different ways?"

The problem is undecidable; there is no algorithm capable of solving it for arbitrary G. As a consequence, programming language designers must invent and apply, for different Gs, different approaches to prove that G is unambiguous. To ease this, they additionally restrict feasible CFGs and deal with, for example, the so-called LL(k) grammars. Since CFGs are a subclass of the class of context-sensitive grammars, the above decision problem is a subproblem of the problem "Is there a word that can be generated by a context-sensitive grammar G in two different ways?" Hence, the latter is undecidable as well.

Equivalence of CFG Grammars. Sometimes a programming language designer wants to improve a grammar at hand while retaining the old generated language. So, after improving a grammar G_1 to a grammar G_2, the designer asks whether or not $L(G_2) = L(G_1)$. In other words, he or she asks whether or not G_1 and G_2 are *equivalent* grammars.

▲ EQUIVALENCE OF CFG GRAMMARS
  Let G_1 and G_2 be CFGs. Question:

$$\textit{"Do } G_1 \textit{ and } G_2 \textit{ generate the same language?"}$$

This problem is undecidable; no algorithm can solve it for arbitrary G_1 and G_2. Language designers must invent, for different pairs G_1, G_2, different approaches to prove their equivalence. As above, the equivalence of context-sensitive grammars is an undecidable problem too.

Some Other Properties of CFGs and CFLs. There are other practical problems about context-free grammars and languages that turned out to be undecidable. Some are listed below. We leave it to the reader to interpret each of them as a problem of programming language design.

▲ OTHER PROPERTIES OF CFGs AND CFLs
  Let G and G' be arbitrary CFGs, and let $\mathcal{C}$ and $\mathcal{R}$ be an arbitrary CFL and a regular language, respectively. As usual, Σ is the alphabet. Questions:

$$\textit{"Is } L(G) = \Sigma^* \textit{?"}$$
$$\textit{"Is } L(G) \textit{ regular?"}$$
$$\textit{"Is } \mathcal{R} \subseteq L(G) \textit{?"}$$
$$\textit{"Is } L(G) = \mathcal{R} \textit{?"}$$
$$\textit{"Is } \overline{L(G)} \textit{ a CFL?"}$$
$$\textit{"Is } L(G) \subseteq L(G') \textit{?"}$$
$$\textit{"Is } L(G) \cap L(G') = \emptyset \textit{?"}$$
$$\textit{"Is } L(G) \cap L(G') \textit{ a CFL?"}$$
$$\textit{"Is } \mathcal{C} \textit{ an ambiguous CFL?"}$$
$$\textit{"Is there a palindrome in } L(G) \textit{?"}$$

Each of these problems is undecidable. No algorithm can solve it in general. Of course, these problems are undecidable for context-sensitive G, G', and C.

Remark. As we saw, all the above undecidability results extend to *context-sensitive* grammars and languages. This was bad news for researchers in linguistics, such as Chomsky,[6] who did not expect a priori limitations on the mechanical, algorithmic processing of natural languages.

8.3.5 Problems About Computable Functions

For many properties about computable functions the following holds: The property depends neither on *how* the function's values are computed (i.e., with what program or algorithm) nor on *where* the computation is performed (i.e., on what computer or model of computation). Such a property is intrinsic to the function itself, that is, it belongs to the function as a correspondence $\varphi : A \to B$ between two sets. For instance, totality is such a property, because whether or not $\varphi : A \to B$ is total depends *only* on the definition of φ, A, and B, and it has nothing to do with the actual computation of φ's values. Deciding whether or not an arbitrary computable function has such an intrinsic property is usually an undecidable problem.

▲ INTRINSIC PROPERTIES OF COMPUTABLE FUNCTIONS
Let $\varphi : A \to B$ and $\psi : A \to B$ be arbitrary computable functions. Questions:

$$\mathcal{D}_{\mathcal{K}_1} \equiv \text{``Is } \mathrm{dom}(\varphi) \text{ empty?''}$$
$$\mathcal{D}_{\mathcal{F}in} \equiv \text{``Is } \mathrm{dom}(\varphi) \text{ finite?''}$$
$$\mathcal{D}_{\mathcal{I}nf} \equiv \text{``Is } \mathrm{dom}(\varphi) \text{ infinite?''}$$
$$\mathcal{D}_{\mathcal{C}of} \equiv \text{``Is } A - \mathrm{dom}(\varphi) \text{ finite?''}$$
$$\mathcal{D}_{\mathcal{T}ot} \equiv \text{``Is } \varphi \text{ total?''}$$
$$\mathcal{D}_{\mathcal{E}xt} \equiv \text{``Can } \varphi \text{ be extended to a total computable function?''}$$
$$\mathcal{D}_{\mathcal{S}ur} \equiv \text{``Is } \varphi \text{ surjective?''}$$
$$\text{``Is } \varphi \text{ defined at } x?''$$
$$\text{``Is } \varphi(x) = y \text{ for at least one } x?''$$
$$\text{``Is } \mathrm{dom}(\varphi) = \mathrm{dom}(\psi)?''$$
$$\text{``Is } \varphi \simeq \psi?''$$

All of the above problems are undecidable; no algorithm can solve any of them.

Remark. We can now understand the difficulties, described in Sect. 5.3.3, that would have arisen if we had used only *total* functions to formalize the basic notions of computation: The basic notions of computation would have been defined in terms of an undecidable mathematical notion (i.e., the property of being a total function).

---

[6] Avram Noam Chomsky, b. 1928, American linguist and philosopher.

8.3.6 Problems from Number Theory

Solvability of Diophantine Equations. Let $p(x_1, x_2, \ldots, x_n)$ be a polynomial with unknowns x_i and integer coefficients. A *Diophantine equation* is the equation $p(x_1, x_2, \ldots, x_n) = 0$, for some $n \geqslant 1$, where only integer solutions $x_1, x_2, \ldots, x_n$ are sought. For example, the Diophantine equation $2x^4 - 4xy^2 + 5yz + 3z^2 - 6 = 0$ asks for all the triples $(x, y, z) \in \mathbb{Z}^3$ that satisfy it. Depending on the polynomial $p(x_1, x_2, \ldots, x_n)$, the Diophantine equation may or may not have a solution. For example, $6x + 8y - 24 = 0$ and $x^4 z^2 + y^4 z^2 - 2z^4 = 0$ both have a solution, while $6x + 8y - 25 = 0$ and $x^2 + y^2 - 3 = 0$ do not. Since Diophantine equations find many applications in practice, it would be beneficial to devise an algorithm that would decide, for arbitrary $p(x_1, x_2, \ldots, x_n)$, whether or not the Diophantine equation $p(x_1, x_2, \ldots, x_n) = 0$ has a solution. This is known as *Hilbert's tenth problem*, the tenth in his list of 23 major problems that were unresolved in 1900.

▲ SOLVABILITY OF DIOPHANTINE EQUATIONS
Let $p(x_1, x_2, \ldots, x_n)$ be an arbitrary polynomial with unknowns $x_1, x_2, \ldots, x_n$ and rational integer coefficients. Question:

"Does a Diophantine equation $p(x_1, x_2, \ldots, x_n) = 0$ have a solution?"

This problem is undecidable. There is no algorithm capable of solving it for an arbitrary polynomial p. This was proved in 1970 by Matiyasevič,[7] who built on the results of Davis,[8] Putnam,[9] and Robinson.[10]

8.3.7 Problems from Algebra

Mortal Matrix Problem. Let $m, n \geqslant 1$ and let $\mathcal{M} = \{M_1, \ldots, M_m\}$ be a set of $n \times n$ matrices with integer entries. Choose an arbitrary finite sequence $i_1, i_2, \ldots, i_k$ of numbers i_j, where $1 \leqslant i_j \leqslant m$, and multiply the matrices in this order to obtain the product $M_{i_1} \times M_{i_2} \times \ldots \times M_{i_k}$. Can it happen that the product is equal to O, the zero matrix of order $n \times n$? The question is known as the *Mortal Matrix Problem*.

▲ MORTAL MATRIX PROBLEM
Let $\mathcal{M}$ be a finite set of $n \times n$ matrices with integer entries. Question:

"Can the matrices of $\mathcal{M}$ be multiplied in some order, possibly with repetition, so that the product is the zero matrix O?"

The problem is undecidable. No algorithm can solve it for arbitrary $\mathcal{M}$.

---

[7] Juri Vladimirovič Matiyasevič, b. 1947, Russian mathematician and computer scientist.
[8] Martin David Davis, b. 1928, American mathematician.
[9] Hilary Whitehall Putnam, 1926–2016, American philosopher, mathematician, computer scientist.
[10] Julia Hall Bowman Robinson, 1919–1985, American mathematician.

Word Problems. Let Σ be an arbitrary alphabet. A *word on* Σ is any finite sequence of the symbols of Σ. As usual, Σ^* is the set of all the words on Σ, including the empty word ε. If $u, v \in \Sigma^*$, then uv denotes the word that is the concatenation (i.e., juxtaposition) of the words u and v. A *rule over* Σ is an expression $x \to y$, where $x, y \in \Sigma^*$. Let $\mathcal{R}$ be an arbitrary finite set of rules over Σ. The pair $(\Sigma, \mathcal{R})$ is called the *semi-Thue system*.[11] In a semi-Thue system $(\Sigma, \mathcal{R})$ we can investigate whether and how a word of Σ^* can be transformed, using only the rules of $\mathcal{R}$, into another word of Σ^*. Given two words $u, v \in \Sigma^*$, a *transformation* of u into v in the semi-Thue system $(\Sigma, \mathcal{R})$ is a finite sequence of words $w_1, \ldots, w_n \in \Sigma^*$ such that 1) $u = w_1$ and $w_n = v$ and 2) for each $i = 1, \ldots n - 1$ there exists a rule in $\mathcal{R}$, say $x_i \to y_i$, such that $w_i = px_i s$ and $w_{i+1} = py_i s$, where the prefix p and suffix s are words in Σ^*, possibly empty. We write $u \overset{*}{\to} v$ to assert that there exists a transformation of u into v in the semi-Thue system $(\Sigma, \mathcal{R})$. Then the *word problem for semi-Thue system* $(\Sigma, \mathcal{R})$ is the problem of determining, for arbitrary words $u, v \in \Sigma^*$, whether or not $u \overset{*}{\to} v$. We can impose additional requirements on semi-Thue systems and obtain Thue systems. Specifically, a *Thue system* is a semi-Thue system $(\Sigma, \mathcal{R})$ in which $x \to y \in \mathcal{R} \Longleftrightarrow y \to x \in \mathcal{R}$. Thus, in a Thue system we have $u \overset{*}{\to} v \Longleftrightarrow v \overset{*}{\to} u$.

▲ WORD PROBLEM FOR SEMI-GROUPS

Let $(\Sigma, \mathcal{R})$ be an arbitrary Thue system and $u, v \in \Sigma^*$ be arbitrary words. Question:

Can u be transformed into v in the Thue system $(\Sigma, \mathcal{R})$*?*

This problem is undecidable. No algorithm can solve it for arbitrary u, v and $\Sigma, \mathcal{R}$.

Let $(\Sigma, \mathcal{R})$ be a Thue system in which the following holds: For every $a \in \Sigma$ there is a $b \in \Sigma$ such that both $\varepsilon \to ba$ and $ba \to \varepsilon$ are rules of $\mathcal{R}$. Informally, every symbol a has a symbol b that annihilates it. (It follows that every word has an annihilating "inverse" word.) For such Thue systems the following problem arises.

▲ WORD PROBLEM FOR GROUPS

Let $(\Sigma, \mathcal{R})$ be an arbitrary Thue system where $\forall a \in \Sigma \; \exists b \in \Sigma (\varepsilon \to ba \in \mathcal{R} \; \wedge \; ba \to \varepsilon \in \mathcal{R})$, and let $u, v \in \Sigma^*$ be arbitrary words. Question:

"Is $u \overset{*}{\to} v$ in $(\Sigma, \mathcal{R})$*?"*

The problem is undecidable; no algorithm can solve it for arbitrary $u, v, \Sigma, \mathcal{R}$.

Remark. This is fine, but where are the semi-groups and groups? In a Thue system $(\Sigma, \mathcal{R})$ the relation $\overset{*}{\to}$ is an equivalence relation, so Σ^* is partitioned into equivalence classes. The set $\Sigma^*/\overset{*}{\to}$ of all the classes, together with the operation of concatenation of classes, is a *semi-group* $(\Sigma^*/\overset{*}{\to}, \cdot)$. If, in addition, the rules $\mathcal{R}$ fulfill the above "annihilation" requirement, then $(\Sigma^*/\overset{*}{\to}, \cdot)$ is a *group*.

---

[11] Axel Thue, 1863–1922, Norwegian mathematician.

Let $\mathcal{E}$ be a finite set of equations between words. Sometimes, a new equation can be deduced from $\mathcal{E}$ by a finite number of substitutions and concatenations and by using the transitivity of the relation "=". Here is an example. Let $\mathcal{E}$ contain the equations

$$1.\ bc = cba$$
$$2.\ ba = abc$$
$$3.\ ca = ac$$

We can deduce the equation $abcc = cacacbaa$. How? By concatenating the symbol a to both sides of equation 1 we get the equation $bca = cbaa$. Its left-hand side can be transformed by a series of substitutions as follows: $b\underline{ca} \overset{3.}{=} \underline{ba}c \overset{2.}{=} abcc$. Similarly, we transform its right-hand side: $c\underline{ba}a \overset{1.}{=} cab\underline{ca} \overset{2.}{=} cac\underline{ba}a \overset{2.}{=} caca\underline{bc}a \overset{1.}{=} cacacbaa$. Since "=" is transitive, we finally obtain the equation $abcc = cacacbaa$. Can we algorithmically check whether or not an equation follows from a set of equations?

▲ EQUALITY OF WORDS
   Let $\mathcal{E}$ be an arbitrary finite set of equations between words and let u, v be two arbitrary words. Question:

$$\textit{"Does } u = v \textit{ follow from } \mathcal{E}?"$$

The problem is undecidable. No algorithm can solve it for arbitrary $u, v, \mathcal{E}$.

8.3.8 Problems from Analysis

Existence of Zeros of Functions. We say that a function $f(x)$ is *elementary* if it can be constructed from a finite number of exponentials $e^{(\cdot)}$, logarithms $\log(\cdot)$, roots $\sqrt[(\cdot)]{(\cdot)}$, real constants, and the variable x by using function composition and the four basic operations $+$, $-$, $\times$, and $\div$. For example, $\sqrt[2]{\frac{a}{\pi}}e^{-ax^2}$ is an elementary function. If we allow these functions and the constants to be complex, then also trigonometric functions and their inverses become elementary. Now, often we are interested in zeros of functions. Given a function, before we attempt to compute its zeros, it is a good idea to check whether or not they exist. Can we do the checking algorithmically?

▲ EXISTENCE OF ZEROS OF FUNCTIONS
   Let $f : \mathbb{R} \to \mathbb{R}$ be an arbitrary elementary function. Question:

$$\textit{"Is there a real solution to the equation } f(x) = 0?"$$

The problem is undecidable. No algorithm can answer this question for arbitrary f. Consequently, the problem is undecidable for general functions f.

8.3.9 Problems from Topology

Classification of Manifolds. Topological *manifolds* play an important role in the fields of mathematics and physics (e.g., topology, general relativity). Intuitively, the simplest topological manifolds are like curves and surfaces, but they can also be of higher dimension and impossible to be pictured in $\mathbb{R}^3$. Here the dimension is the number of independent numbers needed to specify an arbitrary point in the manifold. An n-dimensional topological manifold is called an *n-manifold* for short.

The crucial property of n-manifolds is, roughly speaking, that they *locally* "look like" Euclidian space $\mathbb{R}^n$. Let us make this more precise. First, two sets $U \subseteq \mathbb{R}^k$ and $V \subseteq \mathbb{R}^n$ are said to be *homeomorphic* if there is a bijection $h : U \to V$ such that both h and h^{-1} are continuous. Such a function h is called a *homeomorphism*. Intuitively, if U and V are homeomorphic, then they can be continuously deformed one into the other without tearing or collapsing. Second, a set $\mathcal{M} \subseteq \mathbb{R}^k$ is said to be *locally Euclidean of dimension n* if every point of $\mathcal{M}$ has a neighborhood in $\mathcal{M}$ that is homeomorphic to a ball in $\mathbb{R}^n$. At last we can define: An n-manifold is a subset of some $\mathbb{R}^k$ that is locally Euclidean of dimension n.

The Euclidean space $\mathbb{R}^3$ is a 3-manifold. But there are many more: 1-manifolds are curves (e.g., line, circle, parabola, other curves), while 2-manifolds are surfaces (e.g., plane, sphere, cylinder, ellipsoid, paraboloid, hyperboloid, torus). For $n \geqslant 3$, n-manifolds cannot be visualized (with the exception of $\mathbb{R}^3$ or parts of it).

There is one more important notion that we will need. A *topological invariant* is a property that is preserved by homeomorphisms. For example, the dimension of a manifold is a topological invariant, but there are others too. If two manifolds have different invariants, they are not homeomorphic.

One of the most important problems of topology is to *classify* manifolds up to topological equivalence. This means that we want to produce, for each dimension n, a *list* of n-manifolds, called *n-representatives*, such that every n-manifold is homeomorphic to exactly one n-representative. Ideally, we would also like to devise an algorithm that would compute, for any given n-manifold, the corresponding n-representative (or, rather, its location in the list). Unfortunately, the problem is incomputable for dimensions $n \geqslant 4$. Specifically, in 1958 Markov proved that for $n \geqslant 4$ the question of whether or not two arbitrary n-manifolds are homeomorphic is undecidable.

▲ CLASSIFICATION OF MANIFOLDS

Let $n \geqslant 4$ and let $\mathcal{M}_1, \mathcal{M}_2$ be arbitrary topological n-manifolds. Question:

"Are topological n-manifolds $\mathcal{M}_1$ and $\mathcal{M}_2$ homeomorphic?"

The problem is undecidable. There is no algorithm capable of distinguishing two arbitrary manifolds with four or more dimensions.

8.3.10 Problems from Mathematical Logic

Decidability of First-Order Theories. In mathematical logic there were also many problems that turned out to be undecidable. After Hilbert's program was set, one of the main problems of mathematical logic was to solve the *Decidability Problem* for **M**, where **M** denotes the theory belonging to the sought-for formal axiomatic system for all mathematics. Then, the *Decidability Problem* for **M** is denoted by $\mathcal{D}_{Dec(\mathbf{M})}$ and defined as $\mathcal{D}_{Dec(\mathbf{M})} \equiv$ "Is F a theorem of **M**?", where F can be an arbitrary formula of **M**. If this problem were decidable, then the corresponding decider D_{Entsch} could today be programmed and run on modern computers. Such programs, called *automatic theorem provers*, would ease research in mathematics: One would only need to construct a relevant mathematical proposition, submit it to the automatic prover, and wait for the positive or negative answer. However, in 1936, Hilbert's challenge ended with an unexpected result that was simultaneously found by Church and Turing. In short, the problem $\mathcal{D}_{Dec(\mathbf{M})}$ is undecidable. What is more, this holds for any first-order formal axiomatic system **F**.

▲ DECIDABILITY OF FIRST-ORDER THEORIES

Let **F** be an arbitrary first-order theory and F an arbitrary formula of **F**. Question:

$$\mathcal{D}_{Dec(\mathbf{F})} \equiv \text{``Is F a theorem of } \mathbf{F}?\text{''}$$

This problem is undecidable; no algorithm can solve it for arbitrary $F \in \mathbf{F}$. Specifically, there is no algorithm D_{Entsch} that can decide whether or not an arbitrary mathematical formula is provable.

Remark. So yet another goal of Hilbert's program, the *Entscheidungsproblem* (goal D, Sect. 4.1.2) has received a negative answer. We will describe the proof on p. 217. Let us add that certain particular theories *can be* decidable; for instance, in 1921, Post proved that the *Propositional Calculus* **P** is a decidable theory. He devised an algorithm that, for an arbitrary proposition of **P**, decides whether or not the proposition is provable in **P**. The algorithm uses truth-tables. An obvious practical consequence of the undecidability of $\mathcal{D}_{Dec(\mathbf{F})}$ is that designers of automatic theorem provers are aware of the hazard that their programs, no matter how improved, may not halt on some inputs.

Satisfiability and Validity of First-Order Formulas. Recall that to interpret a theory one has to choose a particular set $\mathcal{S}$ and particular functions and relations defined on $\mathcal{S}$, and define, for every formula F of **F**, how it is to be understood as a statement about the members, functions, and relations of $\mathcal{S}$ (see Sect. 3.1.3). So let us be given a theory and an interpretation ι of it. If F has no free variable symbols, then its interpretation $\iota(F)$ is either a true or a false statement about the state of affairs in $\mathcal{S}$. If, however, F contains free variable symbols, then, as soon as all the free variable symbols are assigned values, the formula $\iota(F)$ becomes either a true or a false statement about $\mathcal{S}$. Obviously, the assignment of values to free variable symbols can *affect* the truth-value of $\iota(F)$. In general, there are many possible assignments of values to free variable symbols. A formula F is said to be *satisfiable* under the interpretation ι if $\iota(F)$ becomes true for *at least one* assignment of values to its free variable symbols. And a formula F is said to be *valid* under the interpretation ι if $\iota(F)$ becomes true for *every* assignment of values to its free variable symbols. Finally, if a formula is valid under *every* interpretation, it is said to be *logically valid*. For instance, in a first-order theory the formula $\neg\forall x\, P(x) \Rightarrow \exists x \neg P(x)$ is logically valid. Satisfiability and validity are nice properties; it would be beneficial to be able to algorithmically recognize formulas with such properties. Can we do this? Unfortunately, we cannot.

▲ SATISFIABILITY OF FIRST-ORDER FORMULAS

Let **F** be a first-order theory, ι an interpretation of **F**, and F an arbitrary formula of **F**. Question:

$$\mathcal{D}_{Sat(\mathbf{F},\iota)} \equiv \text{``Is F satisfiable under } \iota\text{?''}$$

This problem is undecidable. There is no algorithm capable of solving it for an arbitrary F. However, if **F** = **P**, the *Propositional Calculus*, the problem *is* decidable. In that case, the problem is stated as follows:

$$\mathcal{D}_{Sat(\mathbf{P})} \equiv \text{``Can variable symbols of a Boolean expression F be assigned}$$
$$\textit{truth-values in such a way that F attains the truth-value 'true'?''}$$

This is the so-called *Satisfiability Problem for Boolean Expressions*, which plays an important role in *Computational Complexity Theory*.

▲ VALIDITY OF FIRST-ORDER FORMULAS

Let **F** be a first-order theory, ι an interpretation of **F**, and F an arbitrary formula of **F**. Question:

$$\mathcal{D}_{Val(\mathbf{F},\iota)} \equiv \text{``Is F valid under } \iota\text{?''}$$

The problem is undecidable; no algorithm can solve it for an arbitrary F.

8.3.11 Problems About Games

Tiling Problems. Let $\mathcal{T}$ be a finite set of elements, called *tile templates*. Each tile template is a 1×1 square object with each of its four sides colored in one of finitely many colors (see Fig. 8.9). For each tile template there is a stack with a potentially infinite number of exact copies of the template, called the *tiles*.

Next, let the plane be partitioned into 1×1 squares, that is, view the plane as the set $\mathbb{Z}^2$. Any finite subset of $\mathbb{Z}^2$ can be viewed as a *polygon* with a border consisting only of horizontal and vertical sides—and vice versa. Such a polygon can be $\mathcal{T}$-*tiled*, which means that every 1×1 square of the polygon is covered with a tile from an arbitrary stack associated with $\mathcal{T}$. Without any loss of generality, we assume that the tiles cannot be rotated or reflected.

Now, a $\mathcal{T}$-tiling of a polygon is said to be *regular* if every two neighboring tiles of the tiling match in the color of their common sides. The question is: Does the set $\mathcal{T}$ suffice to regularly $\mathcal{T}$-tile an arbitrary polygon? Can we decide this algorithmically for an arbitrary $\mathcal{T}$? The answer is *no*.

▲ DOMINO TILING PROBLEM

Let $\mathcal{T}$ be a finite set of tile templates, each with an unlimited number of copies (tiles). Question:

"Can every finite polygon be regularly $\mathcal{T}$-tiled?"

This problem is undecidable. No algorithm can solve it for an arbitrary set $\mathcal{T}$.

a regular $\mathcal{T}$-tiling of the polygon

Fig. 8.9 A positive instance of the domino tiling problem

A path $p(A,B,X)$ is any finite sequence of 1×1 squares in $\mathbb{Z}^2$ that connects the square A to the square B and does not cross the square X (see Fig. 8.10).

▲ DOMINO SNAKE PROBLEM

Let $\mathcal{T}$ be a finite set of tile templates and let A, B, and X be arbitrary 1×1 squares in $\mathbb{Z}^2$. Question:

"Is there a path $p(A,B,X)$ that can be regularly $\mathcal{T}$-tiled?"

The problem is undecidable; there is no algorithm capable of solving it for an arbitrary $\mathcal{T}, A, B, X$.

a regular $\mathcal{T}$-tiling of a path $p(A,B,X)$

Fig. 8.10 A positive instance of the domino snake problem

8.4 Can We Outwit Incomputable Problems?

We said that for an incomputable (undecidable) problem $\mathcal{P}$ there exists no *single* algorithm A capable of solving an *arbitrary* instance p of the problem. However, in practice it may happen that we are only confronted with a *particular* instance p_i of $\mathcal{P}$. In this case there exists an algorithm A_i that can solve the particular instance p_i. The algorithm A_i must initially check whether or not the input data actually describe the instance p_i and, if so, start solving p_i. If we succeed in constructing the algorithm A_i, it will necessarily be designed specially for the particular instance p_i. In general, A_i will not be useful for solving any other instance of the problem $\mathcal{P}$.

In the following, we discuss two attempts to get around the harsh statement that A does not exist:

1. It seems that we could construct the sought-for general algorithm A simply by combining all the particular algorithms A_i for all the instances $p_i \in \mathcal{P}$ into one unit. That is,

Algorithm $A(p)$:
  begin
    if $p = p_1$ **then** A_1 **else**
    if $p = p_2$ **then** A_2 **else**
      $\vdots$
    if $p = p_i$ **then** A_i **else**
      $\vdots$
  end.

However, there is a pitfall in this approach. In general, the particular algorithms A_i differ one from another—we say that they are *not uniform*. This is because we lack a single method for constructing the particular algorithms. (If such a method existed, then the construction of A would go along with this method, and the problem $\mathcal{P}$ would be computable.) Consequently, we must be sufficiently ingenious, for each instance $p_i \in \mathcal{P}$ separately, to discover and construct the corresponding particular algorithm A_i.

But there are infinitely many instances of $\mathcal{P}$ (otherwise, $\mathcal{P}$ would be computable), so the construction of all of the particular algorithms would never end. Even if the construction somehow finished, the encoding of such A would not be finite, and this would violate the fundamental assumption that algorithms are representable by finite programs. (Which Turing program would correspond to A?) Such an A would not be an algorithm in the sense of the *Computability Thesis*.

2. The second attempt avoids the construction of infinitely many particular algorithms A_i. To do this, we first construct a *generator* that is capable of generating, in succession, all of the programs. Specifically, we can easily construct a generator G_{Σ^*} that generates the programs in *shortlex* order (that is, in order of increasing length, and, in the case of equal length, in lexicographical order; see Sect. 6.3.5 and Appendix A, p. 369). Then we can use G_{Σ^*} as an element in the construction of some other algorithm. In particular, it seems that we can construct a single, finite, and general algorithm B that is capable of solving the problem $\mathcal{P}$:

> **Algorithm** $B(p)$:
>   **begin**
>   **repeat**
> $P := $ call G_{Σ^*} and generate the next program;
>   **until** P can solve p ;
>   Start P on input p ;
>   **end.**

Unfortunately, there are pitfalls in this attempt too. First, we have tacitly assumed that, for each instance $p \in \mathcal{P}$, there exists a particular program P that solves p. Is this assumption valid? Well, such a program might use a table containing the expected input data (i.e., those that define the instance p), and the answer to p. Upon receiving the actual input the program would check whether or not these define the instance p and, if they do, it would return the associated answer. But now another question arises: Who will (pre)compute the answers to various instances and fill them in the tables? Secondly, who will decide (i.e., verify) whether or not the next generated program P can solve the instance $p \in \mathcal{P}$? The tables would enable such a verification, but the assumption that such tables are known is unrealistic. So, the verification should be done by a single algorithm. Such a verifier should be capable of deciding, for an *arbitrary* pair P and $p \in \mathcal{P}$, whether or not P correctly solves p. But the existence of such a verifier would imply that B correctly solves $\mathcal{P}$, and hence $\mathcal{P}$ is computable. This would be a contradiction. Moreover, if the verifier could decide, for an *arbitrary* problem $\mathcal{P}$, the question "Does P solve $p \in \mathcal{P}$?", then also the incomputable problem CORRECTNESS OF ALGORITHMS (PROGRAMS) would be computable (see Sect. 8.3.3). Again this would be a contradiction.

NB *Should anyone construct an algorithm and claim that the algorithm can solve an incomputable problem, then we will be able—by just appealing to the* Computability Thesis—*to rebut the claim right away. Indeed, without any analysis of the algorithm, we will be able to predict that the algorithm fails for at least one instance of the problem; when solving such an instance, the algorithm either never halts or returns a wrong result. Moreover, we will know that this holds for any present or future programming language (to code the algorithm), and for any present or future computer (to run the program).*

8.5 Chapter Summary

There are several kinds of computational problems: *decision, search, counting,* and *generation* problems. The solution of a decision problem is one of the answers YES or NO. An instance of a decision problem is *positive* or *negative* if the solution of the instance is YES or NO, respectively.

There is a close link between decision problems and *sets*. Because of this we can *reduce* questions about decision problems to questions about sets. The *language of a decision problem* is the set of codes of all the positive instances of the problem. The question of whether an instance of a decision problem is positive reduces to the question of whether the code of the instance is in the corresponding language.

A decision problem is *decidable, undecidable,* or *semi-decidable* if its language is decidable, undecidable, or semi-decidable, respectively.

Given a Turing machine T and a word w, the *Halting Problem* asks whether or not T eventually halts on w. The *Halting Problem* is undecidable. There are undecidable problems that are semi-decidable; such is the *Halting Problem*. There are also undecidable problems that are not even semi-decidable; such is the *non-Halting Problem*, which asks whether or not T *never* halts on w.

An n-state *stopper* is a TM that has $n \geqslant 1$ non-final states and one final state, and a program that always writes the symbol 1 and moves the window and which, if started on an empty input, eventually halts and leaves a number of 1s on its tape. An n-state *Busy Beaver* is an n-state stopper that outputs, from among all the n-state stoppers, the maximum number of 1s. The problem of deciding whether or not a TM is an n-state Busy Beaver for any n is undecidable. The number of 1s output by an n-state Busy Beaver is a function of n that is *incomputable*.

There are many other and more practical undecidable decision problems (and incomputable computational problems) in various fields of science.

Problems

8.1. Given a *search problem* (see Sect. 8.1.1), we can derive from it the corresponding decision problem. For example, from the search problem $\mathcal{P}(\mathcal{S}) \equiv$ "Find the largest prime in $\mathcal{S}$." we derive a decision problem $\mathcal{D}(\mathcal{S}, n) \equiv$ "Does $\mathcal{S}$ contain a prime larger than n?" Discuss the following questions:

(a) Can we construct an algorithm for $\mathcal{P}(\mathcal{S})$, given a decider for $\mathcal{D}(\mathcal{S}, n)$? How?

(b) Can a search problem be computable if its derived decision problem is undecidable?

8.2. Similarly to search problems, *counting* and *generation problems* can also be associated in a natural way with the corresponding decision problems.

(a) Give examples of counting and generation problems and the derived decision problems.

(b) Can a counting or generation problem be computable if its decision counterpart is undecidable?

8.3. Let $n \geqslant 1$. Define $\mathcal{T}_n$ to be the set of all the two-way unbounded TMs $T = (Q, \Sigma, \Gamma, \delta, q_1, \sqcup, F)$, where $Q = \{q_1, \ldots, q_{n+1}\}$, $\Sigma = \{0, 1\}$, $\Gamma = \{0, 1, \sqcup\}$, $\delta : Q \times \Gamma \to Q \times \{1\} \times \{\text{Left}, \text{Right}\}$ and $F = \{q_{n+1}\}$.

(a) Prove: $\mathcal{T}_n$ contains finitely many TMs, for any $n \geqslant 1$.

(b) Compute $|\mathcal{T}_n|$.

Bibliographic Notes

- The undecidability of the *Halting Problem* was proved in Turing [262]. Simultaneously, an explicit undecidable problem was given in Church [37].
- The *Busy Beaver Problem* was introduced in Radó [195]; here, also the incomputability of the Busy Beaver functions was proved.
- Post proved the undecidability of a certain problem about normal systems in [184]. Then, in [185], he reduced this problem to the *Post Correspondence Problem* and thus proved the undecidability of the latter.
- The reasons for the undecidability of the *Program Verification Problem* are explained in Harel and Feldman [96].
- The missing step in the (negative) solution of *Hilbert's tenth problem* appeared in Matiyasevič [151, 152]. A complete account of Matiyasevič's contribution to the final solution of Hilbert's tenth problem is given in Davis [53] and Davis [54, Appendix 2].
- The undecidability of the *Mortal Matrix Problem* as well as the undecidability of several other problems about matrices are discussed in Halava et al. [93].
- The undecidability of the problem of the *Existence of Zeros* of real elementary functions was proved in Wang [274]. Some other undecidable problems involving elementary functions of a real variable appeared in Richardson [199]. For some of the first discovered undecidable problems about systems of polynomials, differential equations, partial differential equations, and functional equations, see Adler [6].
- In describing topological manifolds we leaned on Lee [138] and Boothby [24]. See also Tu [261] for an excellent introduction to manifolds. The undecidability of the problem of *Classifying Manifolds* was proved in Markov [149]. The concepts and ideas used in Markov's proof are described in Stillwell [251] and the sketch of the proof is given in Massey [150, p.144].
- The *Domino Tiling Problem* that asks whether the whole plane can be regularly $\mathcal{T}$-tiled was introduced and proved to be undecidable in Wang [273]. The undecidability of a general version of this problem was proved in Berger [17]. The undecidability results of various variants of the *Domino Snake Problem* can be found in Etzion-Petruschka et al. [69]. For a nice discussion of domino problems, see Harel and Feldman [96].
- The undecidability of the *Word Problem for Semigroups* was proved in Post [186]. See also Kleene [124] for the proof. That the much more difficult *Word Problem for Groups* is also undecidable was proved by Novikov [171] and independently by Boone [23].

Chapter 9
Methods of Proving Incomputability

A method is a particular way of doing something.

Abstract How can we prove the undecidability of the problems listed in the previous chapter? Are there any general methods of proving the undecidability of decision problems? The answer is yes: Today we have at our disposal several such methods. These are: 1) proving by *diagonalization*, 2) proving by *reduction*, 3) proving by the *Recursion Theorem*, and 4) proving by *Rice's Theorem*. In addition, we can apply the results discovered by the *relativization* of the notion of computability. In this chapter we will describe the first four methods and postpone the discussion of *relativized computability*—which needs a separate, more extensive treatment—to the following chapters.

9.1 Proving by Diagonalization

We have already encountered diagonalization twice: first, when we proved the existence of an intuitively computable numerical function that is not both μ-recursive and total (Sect. 5.3.3), and second, when we proved the undecidability of the *Halting Problem* (Sect. 8.2). In these two cases diagonalization was used in two different ways, directly and indirectly. In this section we will give general descriptions of both.

9.1.1 Direct Diagonalization

This method was first used by Cantor in 1874 in his proof that $2^{\mathbb{N}}$, the power set of $\mathbb{N}$, is uncountable. The generalized idea of the method is as follows.

Let P be a property and $\mathcal{S} = \{x \mid P(x)\}$ be the class of *all* the elements with property P. Suppose that we are given a countable set $\mathcal{T} \subseteq \mathcal{S}$ such that $\mathcal{T} = \{e_0, e_1, e_2, \ldots\}$ and each e_i can be uniquely represented by an ordered set, that is, $e_i = (c_{i,0}, c_{i,1}, c_{i,2}, \ldots)$, where $c_{i,j}$ are members of a set $\mathcal{C}$ (say, natural numbers).

We may ask whether the elements of S can *all* be exhibited simply by listing the elements of T. In other words, we may ask

$$\text{"Is } T = S\text{?"}$$

If we *doubt* this, we can embark on a proof that $T \subsetneq S$. To prove this, we can try the following method, called *diagonalization*.

Imagine a table T (see Fig. 9.1) with the elements $e_0, e_1, e_2, \ldots$ on its vertical axis, the numbers $0, 1, 2, \ldots$ on the horizontal axis, and the entries $T(i, j) = c_{i,j}$.

Fig. 9.1 By switching each component of the diagonal $d = (c_{0,0}, c_{1,1}, c_{2,2}, \ldots)$ we obtain $sw(d)$, the switched diagonal. But $sw(d)$ differs from every e_i, because $sw(c_{i,i}) \neq c_{i,i}$ for every $i \in \mathbb{N}$

The diagonal components $c_{i,i}$ define an object d, called the *diagonal*:

$$d = (c_{0,0}, c_{1,1}, c_{2,2}, \ldots).$$

Let $sw : C \to C$ be a function such that $sw(c) \neq c$ for every $c \in C$. We will call sw the *switching function*. Define the *switched diagonal* to be the object

$$sw(d) = (sw(c_{0,0}), sw(c_{1,1}), sw(c_{2,2}), \ldots).$$

Now observe that, for every i, the switched diagonal $sw(d)$ differs from the element e_i because $sw(c_{i,i}) \neq c_{i,i}$ (see Fig. 9.1). Consequently, $sw(d) \notin T = \{e_0, e_1, e_2, \ldots\}$. This, however, does not yet prove $T \subsetneq S$ *unless* $sw(d) \in S$. Therefore, if $sw(d)$ *has* the property P, then $sw(d) \in S$, and hence $T \subsetneq S$.

In sum, if we can find a switching function sw such that the switched diagonal $sw(d)$ has the property P, then $T \neq S$. The steps of the method are then as follows.

Method. (Proof by Diagonalization) Given a property P and a countable set $\mathcal{T} = \{e_0, e_1, e_2, \ldots\}$ of elements with property P, then to prove that $\mathcal{T} \subsetneq \mathcal{S}$, where $\mathcal{S}$ is the class of *all* the elements with property P, proceed as follows:

1. Uniquely represent each element e_i of the set $\mathcal{T}$ by an ordered set, that is, $e_i = (c_{i,0}, c_{i,1}, c_{i,2}, \ldots)$, where $c_{i,j} \in \mathcal{C}$ and $\mathcal{C}$ is a set;
2. Let T be a table with the elements $e_0, e_1, e_2, \ldots$ of $\mathcal{T}$ on its vertical axis, the numbers $0, 1, 2, \ldots$ on its horizontal axis, and entries $T(i, j) = c_{i,j}$;
3. Try to find a function $sw : \mathcal{C} \to \mathcal{C}$ with the following properties:
   - $sw(c) \neq c$ for every $c \in \mathcal{C}$;
   - The object $(sw(c_{0,0}), sw(c_{1,1}), sw(c_{2,2}), \ldots)$ has property P;
4. If such a function sw is found, then $\mathcal{T} \subsetneq \mathcal{S}$.

Example 9.1. ($\mathbb{N}$ vs. $\mathbb{R}$) Let us prove that there are more real numbers than natural numbers. We focus on real numbers in the open interval $(0,1) \subset \mathbb{R}$ and show that the interval is uncountable.

Define the property $P(x) \equiv$ "x is a real number in $(0,1)$." So $\mathcal{S} = (0,1)$. Let $\mathcal{T} = \{e_0, e_1, e_2, \ldots\}$ be an *arbitrary* countable set, where $e_i \in \mathcal{S}$ for all i, and each e_i is uniquely represented as $e_i = 0.c_{i,0} c_{i,1} c_{i,2} \ldots$, where $c_{i,j} \in \mathcal{C} = \{0,1,2,3,4,5,6,7,8,9\}$. The uniqueness of the representation is achieved if those e_i that would have finitely many non-zero digits (e.g., 0.25) are represented by using an infinite sequence of 9s (e.g., $0.24999\ldots$). Clearly, $\mathcal{T} \subseteq \mathcal{S}$. Using diagonalization we then prove that $\mathcal{T} \neq \mathcal{S}$. How? Imagine a table T (see Fig. 9.1) with the elements $e_0, e_1, e_2, \ldots$ on its vertical axis, the numbers $0, 1, 2, \ldots$ on the horizontal axis, and the entries $T(i, j) = c_{i,j}$. Define the switching function by $sw : c \mapsto c + 1 \pmod{10}$. The switched diagonal $sw(d)$ represents a number $0.c'_{0,0} c'_{1,1} c'_{2,2} \ldots$, where $c'_{i,i} = sw(c_{i,i})$, which is in $(0,1) = \mathcal{S}$ but differs from every e_i (as $c'_{i,i} \neq c_{i,i}$). Hence, $sw(d) \in \mathcal{S} - \mathcal{T}$ and $\mathcal{T} \neq \mathcal{S}$. Since $\mathcal{T}$ was an arbitrary countable subset of $\mathcal{S} = (0,1)$, it follows that $(0,1)$—and therefore $\mathbb{R}$—cannot be countable. □

Example 9.2. (Intuitively Computable but Not Total μ-Recursive Function) Let us try to prove the existence of numerical functions that are intuitively computable but not total μ-recursive.

Define the property $P(f) \equiv$ "f is an intuitively computable numerical function" and let $\mathcal{S}$ be the class of all such functions. Recall (Box 5.1, p. 82) that every (total or non-total) μ-recursive function is uniquely defined by its construction. Since a construction is a finite word over a finite alphabet, say $\{0,1\}$, it can be interpreted as a natural number. Thus, given arbitrary $n \in \mathbb{N}$, the nth *construction* (and the corresponding total or non-total μ-*recursive function*) is precisely defined. Suppose that we are given a set $\mathcal{T} = \{f_0, f_1, f_2, \ldots\}$, where f_i is the ith *total* μ-recursive function in the sequence of all (total and non-total) μ-recursive functions. Each f_i is perfectly defined by its values on $\mathbb{N}$, so we can represent it by $(c_{i,0}, c_{i,1}, c_{i,2}, \ldots)$, where $c_{i,j} = f_i(j) \in \mathbb{N} = \mathcal{C}$ and $j \in \mathbb{N}$. Let T be a table with $f_0, f_1, f_2, \ldots$ on its vertical axis, $0, 1, 2, \ldots$ on the horizontal axis, and the entries $T(i, j) = f_i(j)$. Let the switching function be $sw : c \mapsto c + 1$. Clearly, $sw(d)$ represents a total function. To compute its value at some $n \in \mathbb{N}$ one locates n-th row of T and adds 1 to $T(n,n)$. So $sw(d) \in \mathcal{S}$. But $sw(d)$ differs from every f_i as $sw(d)$ and f_i attain different values at i. Thus, $sw(d) \in \mathcal{S} - \mathcal{T}$, and the function $sw(d)$ is intuitively computable but not total μ-recursive. □

Remark. We began with a proof that the set $\mathcal{T}$ of all total μ-recursive functions is *enumerable*, i.e., $\mathcal{T} = \{f_0, f_1, f_2, \ldots\}$. In the Platonic view, the set $\{f_0, f_1, f_2, \ldots\}$ *exists* (and so does exist T), even if the constructions of the functions f_i are not known to us. Thus, only the supposition that $\mathcal{T}$ can be given as $\{f_0, f_1, f_2, \ldots\}$ enabled us to prove the existence[1] of $sw(d) \in \mathcal{S} - \mathcal{T}$.

---

[1] But to really compute the *value* of $sw(d)$ at $n \in \mathbb{N}$, we would need *actual* contents of $T(n,n)$, so we should *know* how f_n is constructed. Can we algorithmically find the construction of f_n, given n? Can we decide whether or not a given construction defines a *total* μ-recursive function?

9.1.2 Indirect Diagonalization

First, recall that we encoded Turing machines T by encoding their programs by words $\langle T \rangle \in \{0,1\}^*$ and, consequently, by natural numbers that we called indexes (see Sect. 6.2.1). We then proved that every natural number is the index of exactly one Turing program. Hence, we can speak of the first, second, third, and so on Turing program, or, due to the *Computability Thesis*, of the first, second, third, and so on algorithm. So there is a sequence $A_0, A_1, A_2, \ldots$ such that *every* algorithm that we can conceive of is an A_i for some $i \in \mathbb{N}$. Remember that A_i is encoded (described) by its index i.

Fig. 9.2 Table T contains the results $A_i(j)$ of applying every algorithm A_i on every input j. The diagonal values are $A_i(i)$, $i = 0,1,2,\ldots$. The shrewd algorithm is S. If applied on its own code, S uncovers the inability of the alleged decider D_P to decide whether or not S has the property P

Second, imagine an infinite table T (see Fig. 9.2) with $A_0, A_1, A_2, \ldots$ on the vertical axis, and $0, 1, 2, \ldots$ on the horizontal axis. Define the components of T as follows: For each pair $i, j \in \mathbb{N}$, let the component $T(i,j) = A_i(j)$, the result of algorithm A_i when given the input j.

Now let P be a property that is sensible for algorithms. Consider the question

> "Is there an algorithm D_P capable of deciding, for an arbitrary algorithm A, whether or not A has property P?"

If it exists, the algorithm D_P is the *decider* of the class of algorithms that have property P. If we doubt that D_P exists, we can try to prove this with the following method, which is a generalization of the method used to prove Lemma 8.1 (p. 182).

Method. (Proof by Diagonalization) To prove that there is no algorithm D_P capable of deciding whether or not an arbitrary algorithm has a given property P, proceed as follows:

1. Suppose that the decider D_P exists.

2. Try to construct a *shrewd* algorithm S with the following properties:

   - S uses D_P;
   - if S is given as an input its own code $\langle S \rangle$, it uncovers the inability of D_P to decide whether or not S has the property P.

3. If such an algorithm S is constructed, then D_P does not exist, and the property P is undecidable.

The second step requires S to expose (probably in some shrewd way) the inability of the alleged D_P to output the correct answer on the input $\langle S \rangle$. But how do we construct such an S? Initially, S must call D_P and hand over to it the input $\langle S \rangle$. By supposition, D_P will answer either YES (i.e., S has property P) or NO (i.e., S does not have property P). Then—and this is the hard part of the proof—we must construct the rest of S in such a way that the resulting S will have the property P *if and only if* D_P has decided the contrary. *If* we succeed in constructing such an S, then the alleged D_P errs on at least one input, namely $\langle S \rangle$, contradicting our supposition in the first step. Consequently, the decider D_P does not exist and the decision problem "Does A have the property P?" is undecidable.

Of course, the existence of S and its construction both depend on property P. If S exists, it may take considerable ingenuity to construct it.

Example 9.3. (Shrewd Algorithm for the *Halting Problem*) Is there a decider D_P capable of deciding, for an arbitrary TM, whether TM halts on its own code?

We construct the shrewd TM S as follows. When S reads $\langle S \rangle$, it sends it to D_P to check whether S halts on $\langle S \rangle$. If D_P has answered YES (i.e., saying that S halts on $\langle S \rangle$), we must make sure that the following execution of S will *never* halt. This is why we loop the execution back to D_P. Otherwise, if D_P has answered NO (i.e., indicated that S does not halt on $\langle S \rangle$), we force S to do just the contrary: So S must halt immediately. The algorithm S is depicted in Fig. 8.3 on p. 182. □

Remark. We can apply this method even if we use some other equivalent model of computation. In that case, A_i still denotes an algorithm, but the notion of the algorithm is now defined according to the chosen model of computation. For example, if the model of computation is the λ-calculus, then A_i is a λ-term.

9.2 Proving by Reduction

Until the undecidability of the first decision problem was proved, diagonalization was the only method capable of producing such a proof. Indeed, we have seen that the undecidability of the *Halting Problem* was proven by diagonalization. However, after the undecidability of the first problem was proven, one more method of proving undecidability became applicable. The method is called *reduction*. In this section we will first describe reduction in general, and then focus on two special kinds of this method, the so-called *m*-reduction and 1-reduction.

9.2.1 Reductions in General

Suppose we are given a problem $\mathcal{P}$. Instead of solving $\mathcal{P}$ directly, we might try to solve it *indirectly* by executing the following scenario (Fig. 9.3):

1. *express* $\mathcal{P}$ in terms of some other problem, say $\mathcal{Q}$;
2. solve $\mathcal{Q}$;
3. *construct* the solution to $\mathcal{P}$ by using the solution to $\mathcal{Q}$ only.

If 1 and 3 have actually been implemented, we say that $\mathcal{P}$ has been *reduced* to $\mathcal{Q}$.

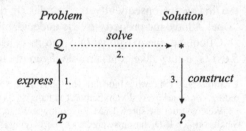

Fig. 9.3 Solving the problem $\mathcal{P}$ is reduced to (or substituted by) solving the problem $\mathcal{Q}$

How can we implement steps 1 and 3? To express $\mathcal{P}$ in terms of $\mathcal{Q}$ we need a function—call it r—that maps every instance $p \in \mathcal{P}$ into an instance $q \in \mathcal{Q}$. Such an r must meet two *basic conditions*:

a. Since problem instances are encoded by words in Σ^* (see Sect. 8.1.2), r must be a function $r : \Sigma^* \to \Sigma^*$ that assigns to every code $\langle p \rangle$ a code $\langle q \rangle$, where $p \in \mathcal{P}$ and $q \in \mathcal{Q}$.
b. Since we want to be able to compute $\langle q \rangle = r(\langle p \rangle)$ for arbitrary $\langle p \rangle$, the function r must be computable on Σ^*.

To be able to construct the solution to $\mathcal{P}$ from the solution to $\mathcal{Q}$ we add one more basic condition:

c. The function r must be such that, for arbitrary $p \in \mathcal{P}$, the solution to p can be computed from the solution to $q \in \mathcal{Q}$, where $\langle q \rangle = r(\langle p \rangle)$.

If such a function r is found, it is called the *reduction* of problem $\mathcal{P}$ to problem $\mathcal{Q}$. In this case we say that $\mathcal{P}$ is *reducible* to $\mathcal{Q}$, and denote the fact by

$$\mathcal{P} \leq \mathcal{Q}.$$

The informal interpretation of the above relation is as follows.

If $\mathcal{P} \leq \mathcal{Q}$, then the problem $\mathcal{P}$ is *not harder to solve* than the problem $\mathcal{Q}$.

The reduction r may still be too general to serve a particular, intended purpose. In this case we define a set $\mathcal{C}$ of additional conditions to be fulfilled by r. If such a function r is found, we call it the *$\mathcal{C}$-reduction* of problem $\mathcal{P}$ to problem $\mathcal{Q}$, and say that $\mathcal{P}$ is *$\mathcal{C}$-reducible* to $\mathcal{Q}$. We denote this by

$$\mathcal{P} \leq_{\mathcal{C}} \mathcal{Q}.$$

What sort of conditions should be included in $\mathcal{C}$? The answer depends on the *kind* of problems $\mathcal{P}, \mathcal{Q}$ and on our *intentions*. For example, $\mathcal{P}$ and $\mathcal{Q}$ can be decision, search, counting, or generation problems. Our intention may be to employ reduction in order to see whether some of the problems $\mathcal{P}$ and $\mathcal{Q}$ have a certain property. For instance, some properties of interest are (i) the decidability of a problem, (ii) the computability of a problem, (iii) the computability of a problem in polynomial time, and (iv) the approximability of a problem. The $\mathcal{C}$-reductions that correspond to these properties are the *m-reduction* ($\leq_m$), *Turing* reduction ($\leq_T$), *polynomial-time bounded* Turing reduction ($\leq_T^P$), and *L*-reduction ($\leq_L$), respectively.

In the following we will focus on the *m*-reduction ($\leq_m$) and postpone the discussion of the Turing reduction ($\leq_T$) and some of its *stronger* cases to Chaps. 11 and 14. The *resource-bounded* reductions, such as $\leq_m^P$, $\leq_T^P$, and $\leq_L$, are used in *Computational Complexity Theory*, so we will not discuss them.

9.2.2 The m-Reduction

If $\mathcal{P}$ and $\mathcal{Q}$ are decision problems, then they are represented by the languages $L(\mathcal{P}) \subseteq \Sigma^*$ and $L(\mathcal{Q}) \subseteq \Sigma^*$, respectively. We use this in the next definition.

Definition 9.1. (*m*-Reduction) Let $\mathcal{P}$ and $\mathcal{Q}$ be decision problems. A reduction $r : \Sigma^* \to \Sigma^*$ is said to be an *m-**reduction** of $\mathcal{P}$ to $\mathcal{Q}$ if the following additional condition is met:

$$\mathcal{C} : \ \langle p \rangle \in L(\mathcal{P}) \Longleftrightarrow r(\langle p \rangle) \in L(\mathcal{Q}), \ \text{for every } p \in \mathcal{P}.$$

In this case we say that $\mathcal{P}$ is *m-**reducible** to $\mathcal{Q}$ and denote this by $\mathcal{P} \leq_m \mathcal{Q}$.

The condition $\mathcal{C}$ enforces that r transforms the codes of the *positive* instances $p \in \mathcal{P}$ into the codes of the *positive* instances $q \in \mathcal{Q}$, and the codes of the *negative* instances $p \in \mathcal{P}$ into the codes of the *negative* instances $q \in \mathcal{Q}$. This is depicted in Fig. 9.4.

Fig. 9.4 *m*-reduction of a decision problem $\mathcal{P}$ to a decision problem $\mathcal{Q}$. Instead of solving $\mathcal{P}$ directly, we show how to express an arbitrary instance p of $\mathcal{P}$ with an instance q of $\mathcal{Q}$, so that the answer to q will also be the answer to p. The mapping $r : \langle p \rangle \mapsto \langle q \rangle$ must be computable, but not necessarily injective

Obviously, *m*-reduction r maps the set $L(\mathcal{P})$ into the set $L(\mathcal{Q})$, i.e., $r(L(\mathcal{P})) \subseteq L(\mathcal{Q})$. Here, we distinguish between two possible situations:

1. $r(L(\mathcal{P})) \subsetneq L(\mathcal{Q})$. In this case r transforms $\mathcal{P}$ into a (proper) *subproblem* of $\mathcal{Q}$. We say that problem $\mathcal{P}$ is properly *contained* in problem $\mathcal{Q}$.

2. $r(L(\mathcal{P})) = L(\mathcal{Q})$. In this case r merely restates $\mathcal{P}$ into $\mathcal{Q}$. We say that problem $\mathcal{P}$ is *equal* to problem Q.

But Fig. 9.4 reveals even more:

a) Suppose that $L(\mathcal{Q})$ is a decidable set. Then also $L(\mathcal{P})$ is a decidable set.

> *Proof.* Given an arbitrary $\langle p \rangle \in \Sigma^*$, we compute the answer to the question $\langle p \rangle \in ? L(\mathcal{P})$ as follows: 1) compute $r(\langle p \rangle)$ and 2) ask $r(\langle p \rangle) \in ? L(\mathcal{Q})$. Since $L(\mathcal{Q})$ is decidable, the answer to the question can be computed. If the answer is YES, then $r(\langle p \rangle) \in L(\mathcal{Q})$ and, because of the condition $\mathcal{C}$ in the definition of the *m*-reduction, also $\langle p \rangle \in L(\mathcal{P})$. If, however, the answer is NO, then $r(\langle p \rangle) \notin L(\mathcal{Q})$ and hence $\langle p \rangle \notin L(\mathcal{P})$. Thus, the answer to the question $\langle p \rangle \in ? L(\mathcal{P})$ can be computed for an arbitrary $\langle p \rangle \in \Sigma^*$. $\qquad\square$

Summary: Let $\mathcal{P} \leq_m \mathcal{Q}$. Then: $L(\mathcal{Q})$ is decidable $\Longrightarrow L(\mathcal{P})$ is decidable.

b) Suppose that $L(Q)$ is a semi-decidable set. Then $L(P)$ is a semi-decidable set.

Proof. Given an arbitrary $\langle p \rangle \in \Sigma^*$, ask $\langle p \rangle \in? L(P)$. If in truth $\langle p \rangle \in L(P)$, then $r(\langle p \rangle) \in L(Q)$ by C. Since $L(Q)$ is semi-decidable, the question $r(\langle p \rangle) \in? L(Q)$ is answered with YES. At the same time this is the answer to the question $\langle p \rangle \in? L(P)$. *If, however*, in truth $\langle p \rangle \notin L(P)$, then $r(\langle p \rangle) \notin L(Q)$. Since $L(Q)$ is semi-decidable, the question $r(\langle p \rangle) \in? L(Q)$ is *either* answered with NO *or* not answered at all. At the same time this is the outcome of the question $\langle p \rangle \in? L(P)$. □

So: Let $P \leq_m Q$. Then: $L(Q)$ is semi-decidable $\Longrightarrow L(P)$ is semi-decidable.

We have proved the following theorem.

Theorem 9.1. *Let P and Q be decision problems. Then:*

a) $P \leq_m Q \land Q \quad\quad decidable\ problem \Longrightarrow P \quad\quad decidable\ problem$
b) $P \leq_m Q \land Q\ semi\text{-}decidable\ problem \Longrightarrow P\ semi\text{-}decidable\ problem$

Recall that a set that is not semi-decidable must be undecidable (see Fig. 8.4 on p. 184). Using this in the contraposition of the case b) of Theorem 9.1, substituting U for P, and assuming that $U \leq_m Q$, we obtain the following important corollary.

Corollary 9.1. *U undecidable problem $\land U \leq_m Q \Longrightarrow Q$ undecidable problem*

9.2.3 Undecidability and m-Reduction

The above corollary is the backbone of the following method of proving that a decision problem Q is undecidable. Informally, the method proves that a known undecidable problem U is a subproblem of, or equal to, the problem Q.

Method. The undecidability of a decision problem Q can be proved as follows:

1. Select: an undecidable problem U;
2. Prove: $U \leq_m Q$;
3. Conclude: Q is an undecidable problem.

Example 9.4. (*Halting Problem* Revisited) Using this method we proved the undecidability of the *Halting Problem* $\mathcal{D}_{Halt}$ (see Sect. 8.2). To see this, select $U = \mathcal{D}_H$ and observe that U is a subproblem of $\mathcal{D}_{Halt}$. Note also that after we proved the undecidability of the problem $\mathcal{D}_H$ (by diagonalization), $\mathcal{D}_H$ was the only choice for U. □

Another method of proving the undecidability of a decision problem Q combines proof by contradiction and case a) of Theorem 9.1 with U substituted for P.

Method. The undecidability of a decision problem Q can be proved as follows:

1. Suppose: Q is a decidable problem;   // Supposition.
2. Select: an undecidable problem U;
3. Prove: $U \leq_m Q$;
4. Conclude: U is a decidable problem;   // 1 and 3 and Theorem 9.1a.
5. Contradiction between 2 and 4!
6. Conclude: Q is an undecidable problem.

Let us comment on steps 2 and 3 (the other steps are trivial).

- *Step 2:* For U we select an undecidable problem that seems to be the most promising one to accomplish step 3. This demands good knowledge of undecidable problems and of problem Q, and often a mix of inspiration, ingenuity, and luck. Namely, until step 3 is accomplished, it is not clear whether the task of m-reducing U to Q is impossible or possible (but we are not equal to the task of constructing this m-reduction). If we give up, we may want to try with another U.

- *Step 3:* Here, the reasoning is as follows: Since Q is decidable (as supposed in step 1), there is a decider $D_{L(Q)}$ that solves Q (i.e., answers the question $x \in ?\ L(Q)$ for an arbitrary x). We can use $D_{L(Q)}$ as a building block and try to construct a decider $D_{L(U)}$; this would be capable of solving U (by answering the question $w \in ?\ L(U)$ for an arbitrary w). If we succeed in this construction, we will be able to conclude (in step 4) that U is decidable.

We now see how to tackle step 3: Using the supposed decider $D_{L(Q)}$, construct the decider $D_{L(U)}$. A possible construction of $D_{L(U)}$ is depicted in Fig. 9.5.

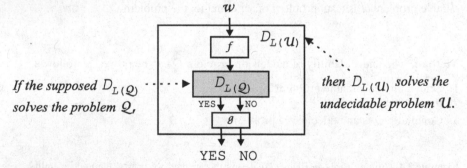

Fig. 9.5 Suppose that $D_{L(Q)}$ solves the decision problem Q. Using $D_{L(Q)}$, construct $D_{L(U)}$, which solves an undecidable problem U. Then $D_{L(Q)}$ cannot exist and Q is undecidable

The machine $D_{L(\mathcal{U})}$ operates as follows: It reads an arbitrary input $w \in \Sigma^*$, transforms w into $f(w)$ and hands this over to the decider $D_{L(\mathcal{Q})}$. The latter understands this as the question $f(w) \in? L(\mathcal{Q})$ and always answers with YES or NO. The answer $D_{L(\mathcal{Q})}(f(w))$ is then transformed by a function g and output to the environment as the answer of the machine $D_{L(\mathcal{U})}$ to the question $w \in? L(\mathcal{U})$.

What do the functions f and g look like? Of course, we expect f and g to depend on the problems $\mathcal{Q}$ and $\mathcal{U}$. However, because we want the constructed $D_{L(\mathcal{U})}$ to *always* halt, we must require the following:

- f and g are *computable* functions;

- $g(D_{L(\mathcal{Q})}(f(w))) = \begin{cases} \text{YES} & \text{if } w \in L(\mathcal{U}), \\ \text{NO} & \text{if } w \notin L(\mathcal{U}), \end{cases}$ for every $w \in \Sigma^*$.

The next (demanding) example highlights the main steps in our reasoning when we use this method. To keep it as clear as possible, we will omit certain details and refer the reader to the Problems later in this chapter to fill in the missing steps.

Example 9.5. (Computable Proper Set)   Let $\mathcal{Q} \equiv$ "Is $L(T)$ computable?" We suspect that $\mathcal{Q}$ is undecidable. Let us prove this by reduction. *Suppose $\mathcal{Q}$ were decidable.* Then there would exist a decider $D_{L(\mathcal{Q})}$ that could tell, for any $\langle T \rangle$, whether or not $L(T)$ is computable. Now, what would be the *undecidable* problem $\mathcal{U}$ for which a *contradictory* TM could be constructed by using $D_{L(\mathcal{Q})}$? This is the tricky part of the proof, and to answer the question, a great deal of inspiration is needed.
   Here it is. First, we show that there is a TM A which on input $\langle T, w \rangle$ constructs (the code of) a TM B such that $L(B) = \mathcal{K}_0$ (if T accepts w) and $L(B) = \emptyset$ (if T does not accept w). (Problem 9.5a.) Then, using A and the alleged $D_{L(\mathcal{Q})}$, we can construct a *recognizer R* for $\overline{\mathcal{K}_0}$. (Problem 9.5b.)
   But this is a contradiction! The existence of R would mean that $\overline{\mathcal{K}_0}$ is c.e. (which it is not). So, our supposition is wrong and $\mathcal{Q}$ is undecidable.   □

9.2.4 The 1-Reduction

Until now we did not discuss whether or not the reduction function r is injective. We do this now. If r is not injective, then several instances, say $m \geqslant 1$, of the problem $\mathcal{P}$ can transform into the same instance $q \in \mathcal{Q}$. If, however, r is injective, then different instances of $\mathcal{P}$ transform into different instances of $\mathcal{Q}$. This is a special case of the previous situation, where $m = 1$. In this case we say that $\mathcal{P}$ is 1-*reducible* to $\mathcal{Q}$ and denote this by

$$\mathcal{P} \leq_1 \mathcal{Q}.$$

Since it holds that

$$\mathcal{P} \leq_1 \mathcal{Q} \Longrightarrow \mathcal{P} \leq_m \mathcal{Q},$$

we can try to prove the undecidability of a problem by using 1-reduction. Notice that the general properties of m-reductions hold for 1-reductions too.

Example 9.6. (*Halting Problem* Revisited)   Let us revisit the proof that the *Halting Problem* $\mathcal{D}_{Halt}(= \mathcal{Q})$ is undecidable (see Sect. 8.2). After we proved that $\mathcal{D}_H$ is undecidable (see Lemma 8.1, p. 182), we took $\mathcal{U} = \mathcal{D}_H$. Then, the reasoning was: *If* $\mathcal{D}_{Halt}$ *were decidable, then the asso-*ciated decider $D_{L(\mathcal{D}_{Halt})}$ *could be used to construct a decider for the problem* $\mathcal{D}_H$. This decider is depicted in Fig. 9.6 a). The function f is total and computable since it only doubles the input $\langle T \rangle$. So is g, because it is the identity function.   □

Fig. 9.6 Undecidability of problems: a) $\mathcal{D}_{Halt}$; b) $\overline{\mathcal{D}}$, if $\mathcal{D}$ undecidable; c) $\mathcal{D}_{Empty}$

Example 9.7. (Complementary Problem)   If a problem $\mathcal{D}$ is undecidable, then the complementary problem $\overline{\mathcal{D}}$ is undecidable. To prove this, take $\mathcal{U} = \mathcal{D}$, $\mathcal{Q} = \overline{\mathcal{D}}$ and suppose that $\overline{\mathcal{D}}$ were decidable. We could then use the supposed decider $D_{L(\overline{\mathcal{D}})}$, which is the same as $D_{\overline{L(\mathcal{D})}}$, to construct the decider $D_{L(\mathcal{D})}$. This is depicted in Fig. 9.6 b). Now f is the identity function and g transforms an arbitrary answer into the opposite one.   □

Example 9.8. (Empty Proper Set)   Recall from Sect. 6.3.3 that the proper set of a Turing machine is the set of all the words accepted by the machine. So, given a Turing machine T, does the proper set $L(T)$ contain any words? This is the decision problem $\mathcal{D}_{Empty} \equiv$ "Is $L(T) = \emptyset$?" Let us prove that it is undecidable. For the undecidable problem $\mathcal{U}$ we pick the *Halting Problem* $\mathcal{D}_{Halt} \equiv$ "Does T halt on w?" and suppose that $\mathcal{D}_{Empty}(= \mathcal{Q})$ is decidable. Then we can use the associated decider $D_{L(\mathcal{D}_{Empty})}$ to construct a decider for the language $L(\mathcal{D}_{Halt})$! The construction is in Fig. 9.6 c). When the constructed machine reads an arbitrary input $\langle T, w \rangle$, it uses the function f to construct the code $\langle T' \rangle$ of a new machine T'. This machine, if started, would operate as follows: Given an arbitrary input x, the machine T' would simulate T on w and, if T halted, then T' would accept x; otherwise, T' would not recognize x. Consequently, we have $L(T') \neq \emptyset \iff T$ halts on w. But the validity of $L(T') \neq \emptyset$ can be decided by the supposed decider $D_{L(\mathcal{D}_{Empty})}$. After swapping the answers (using the function g) we obtain the answer to the question "Does T halt on w?" So the constructed machine is capable of deciding the *Halting Problem*. Contradiction.   □

Example 9.9. (Entscheidungsproblem)   Recall that the *Decidability Problem* for a theory **F** is the question "*Is theory* **F** *decidable?*". In other words, the problem asks whether there exists a decider that, for arbitrary formula F $\in$ **F**, finds out whether or not F is provable in **F**. Hilbert was particularly interested in the *Decidability Problem* for **M**, the sought-for theory that would formalize all mathematics, and, of course, in the associated decider D_{Entsch}. But Church and Turing independently proved the following theorem.

Theorem 9.2. *The Entscheidungsproblem* $\mathcal{D}_{Entsch}$ *is undecidable.*

Proof. (Reduction $\mathcal{D}_H \leq_m \mathcal{D}_{Entsch}$.) Suppose the problem $\mathcal{D}_{Entsch}$ were decidable. Then there would exist a decider D_{Entsch} capable of deciding, for an arbitrary formula of the *Formal Arithmetic* **A**, whether or not the formula is provable in **A** (see Sects. 4.2.1, 4.2.2). Now consider the statement

$$\text{"Turing machine } T \text{ halts on input } \langle T \rangle\text{."} \qquad (*)$$

The breakthrough was made by Turing when he showed that this statement can be represented by a formula in **A**. In the construction of the formula, he used the concept of the *internal configuration* of the Turing machine T (see Definition 6.2 on p. 114). Let us designate this formula by

$$K(T).$$

Then, using the supposed decider D_{Entsch} we could easily construct a decider D capable of deciding, for arbitrary T, whether or not T halts on $\langle T \rangle$. The decider D is depicted in Fig. 9.7. Here, the function f maps the code $\langle T \rangle$ into the formula $K(T)$, while the function g is the identity function and maps the unchanged answer of D_{Entsch} into the answer of D.

If the supposed D_{Entsch} solves the problem "Is K(T) provable in **A**? "

then the algorithm D solves the undecidable problem $\mathcal{D}_H =$ "Does T halt on $\langle T \rangle$? "

Fig. 9.7 Solving the *Halting Problem* with the alleged decider for the *Entscheidungsproblem*

Suppose that the *Formal Arithmetic* **A** were as Hilbert expected, that is, consistent and complete. Then, D's answer would tell us, for arbitrary T, whether or not the statement $(*)$ is true. But this would in turn mean that the halting problem $\mathcal{D}_H$ is decidable—which it is provably not! We must conclude that the problem $\mathcal{D}_{Entsch}$ is undecidable. (*A fortiori*, thanks to Gödel we know that *Formal Arithmetic* **A** is not complete.) This completes the proof. □

Thus, the decider D_{Entsch} does not exist. What about semi-decidability? Is it perhaps that $\mathcal{D}_{Entsch}$ is a semi-decidable problem? The answer is yes.

Theorem 9.3. *The Entscheidungsproblem* $\mathcal{D}_{Entsch}$ *is semi-decidable.*

Proof. Let $F \in \mathbf{M}$ be an arbitrary formula. We have seen in Sect. 4.2.2 that we can systematically generate all the sequences of symbols of **M** and, for each generated sequence, algorithmically check whether or not it is a proof of F in **M**. If the formula F is in truth a theorem of **M**, then a proof of F in **M** exists and will, therefore, be eventually generated and recognized. So the described algorithm is a recognizer of the set of all mathematical theorems. If, however, F is in truth *not* a theorem of **M**, then the recognizer will never halt. Nevertheless, this suffices for the problem $\mathcal{D}_{Entsch}$ to be semi-decidable. This completes the proof.   □

Remark.   After 16 years, yet another goal of *Hilbert's Program* received a negative response: *Formal Arithmetic* **A** *is not a decidable theory.* Thus, the recognition of arithmetical theorems cannot be fully automated, regardless of the kind of algorithms we might use to achieve this goal. Since **A** is a subtheory of **M**, the same holds for mathematics: No matter what kind of algorithms we use, only a proper subclass of all mathematical theorems can be algorithmically recognized. So there will always be mathematical theorems whose theoremship is algorithmically undecidable.

After the researchers learned this, some focused on formal axiomatic systems and theories that are weaker than *Formal Arithmetic* yet algorithmically decidable. An example of such a theory is *Presburger Arithmetic,*[2] introduced in 1929. In this arithmetic the operation of multiplication is omitted, thus allowing on $\mathbb{N}$ only the operation of addition and the equality relation.   □

9.3 Proving by the Recursion Theorem

Recall from Sect. 7.4 that the *Recursion Theorem* can be restated as the *Fixed-Point Theorem*, which tells us that *every computable function has a fixed point.* This reveals the following method for proving the incomputability of functions.

> **Method.** (Incomputability of Functions) Suppose that a function g has no fixed point. Then, g is not computable (i.e., it is not total, or it is incomputable, or both). If we somehow prove that g *is* total, then g must be incomputable.

We further develop this method into a method for proving the undecidability of problems as follows. Let $\mathcal{D}$ be a decision problem that we want to prove is undecidable. *Suppose* that $\mathcal{D}$ is decidable. Then the characteristic function $\chi_{L(\mathcal{D})}$ is computable. We can use $\chi_{L(\mathcal{D})}$ as a (necessary) component and construct a function $g : \mathbb{N} \to \mathbb{N}$ in such a way that g is computable. We then try to prove that g has no fixed point. If we succeed in this, we have a contradiction with the *Fixed-Point Theorem.* Thus, g cannot be computable and, consequently, $\chi_{L(\mathcal{D})}$ cannot be computable either. So $\mathcal{D}$ is undecidable. We summarize this in the following method.

> **Method.** (Undecidability of Problems) Undecidability of a decision problem $\mathcal{D}$ can be proved as follows:
>
> 1. Suppose: $\mathcal{D}$ is a decidable problem;
> 2. Construct: a computable function g using the characteristic function $\chi_{L(\mathcal{D})}$;
> 3. Prove: g has no fixed point;
> 4. Contradiction with the *Fixed-Point Theorem*!
> 5. Conclude: $\mathcal{D}$ is an undecidable problem.

---

[2] Mojżesz Presburger, 1904–1943, Polish mathematician, logician, and philosopher.

Example 9.10. We will use this method to prove *Rice's Theorem* in the next subsection. □

Example 9.11. See the proof of the incomputability of a *search problem* in Sect. 9.5. □

9.4 Proving by Rice's Theorem

All the methods of proving the undecidability of problems that we have described until now may require a considerable degree of inspiration, ingenuity, and even luck. In contrast, the method that we will consider in this section is far less demanding. It is based on a theorem that was discovered and proved in 1951 by Rice.[3] There are three versions of the theorem: for partial computable functions, for index sets, and for computably enumerable sets. They state:

> for p.c. functions: *Every non-trivial property of p.c. functions is undecidable.*
> for index sets: *Every index set different from $\emptyset$ and $\mathbb{N}$ is undecidable.*
> for c.e. sets: *Every non-trivial property of c.e. sets is undecidable.*

The theorem reveals that the undecidability of certain kinds of decision problems is more of a rule than an exception. Let us see the details.

9.4.1 Rice's Theorem for P.C. Functions

Let P be a property that is sensible for functions, and φ an arbitrary partial computable function. We define the following decision problem:

$$\mathcal{D}_P \equiv \text{``Does p.c. function } \varphi \text{ have the property } P?\text{''}$$

We will say that the property P is *decidable* if the problem $\mathcal{D}_P$ is decidable. Thus, if P is a decidable property of p.c. functions, then there exists a Turing machine that decides, for an arbitrary p.c. function φ, whether or not φ has the property P.

What are the properties P that we will be interested in? We will not be interested in a property if it is such that a function has it in some situations and does not have it in other situations. This could happen when the property depends on the way in which function values are computed. For example, whether or not a function φ has the property defined by $P(\varphi) \equiv \text{``}\varphi(n)$ is computed in n^3 steps'' certainly depends on the algorithm used to compute $\varphi(n)$. It may also depend on the program (i.e., algorithm encoding) as well as on the machine where the function values are computed. (The machine can be an actual computer or an abstract computing machine.) For this reason, we will only be interested in properties that are *intrinsic* to functions, i.e., properties of functions where the functions are viewed only as mappings from one set to another. Such properties are because of their basic nature insensitive to

---

[3] Henry Gordon Rice, 1920–2003, American logician and mathematician.

the machine, algorithm, and program that are used to compute function values. For instance, being total is an intrinsic property of functions, because every function φ is either total or not total, irrespective of the way the φ values are computed.

So let P be an arbitrary intrinsic property of p.c. functions. It may happen that *every* p.c. function has this property P; and it also may happen that *no* p.c. function has the property P. In each of these two cases we will say that the property P is *trivial*.

This is where *Rice's Theorem* enters. Rice discovered the following surprising connection between the decidability and the triviality of the intrinsic properties of partial computable functions.

Theorem 9.4. (Rice's Theorem for p.c. Functions) *Let P be an arbitrary intrinsic property of p.c. functions. Then:*

$$P \text{ is decidable} \iff P \text{ is trivial.}$$

Proof. See below.   □

9.4.2 Rice's Theorem for Index Sets

Before we embark on the proof of Theorem 9.4, we restate the theorem in an equivalent form that will be more convenient to prove. The new form will refer to *index sets* instead of p.c. functions.

So let P be an arbitrary intrinsic property of p.c. functions. Define $\mathscr{F}_P$ to be the class of all p.c. functions having the property P,

$$\mathscr{F}_P = \{\psi \mid \psi \text{ is a p.c. function with property } P\}.$$

Then the decision problem $\mathcal{D}_P$ on the previous page can be written as

$$\mathcal{D}_P \equiv \varphi \in? \mathscr{F}_P.$$

Let $\text{ind}(\mathscr{F}_P)$ be the set of indexes of all the Turing machines that compute any of the functions in $\mathscr{F}_P$. Clearly $\text{ind}(\mathscr{F}_P) = \bigcup_{\psi \in \mathscr{F}_P} \text{ind}(\psi)$, where $\text{ind}(\psi)$ is the index set of the function ψ (see Sect. 7.2). Now the critical observation is: Because P is insensitive to the program used to compute φ's values, it must be that $\text{ind}(\mathscr{F}_P)$ contains *either all* of the indexes of the Turing machines computing φ, *or none* of these indexes—depending on whether or not φ has property P. So the question $\varphi \in? \mathscr{F}_P$ is equivalent to the question $x \in? \text{ind}(\mathscr{F}_P)$, where x is the index of an arbitrary Turing machine computing φ. Hence,

$$\mathcal{D}_P \equiv x \in? \text{ind}(\mathscr{F}_P).$$

From the last formulation of $\mathcal{D}_P$ it follows that

$\mathcal{D}_P$ is a decidable problem $\iff$ ind($\mathscr{F}_P$) is a decidable set.

But when is the set ind($\mathscr{F}_P$) decidable? The answer gives the following version of *Rice's Theorem*.

Theorem 9.5. (Rice's Theorem for Index Sets) *Let $\mathscr{F}_P$ be an arbitrary set of p.c. functions. Then:*

$$\text{ind}(\mathscr{F}_P) \text{ is decidable} \iff \text{ind}(\mathscr{F}_P) \text{ is either } \emptyset \text{ or } \mathbb{N}.$$

This form of *Rice's Theorem* can now easily be proved by the *Fixed-Point Theorem* (see Box 9.1).

Box 9.1 (Proof of Theorem 9.5).

Both $\emptyset$ and $\mathbb{N}$ are decidable sets. Consequently, ind($\mathscr{F}_P$) is decidable if it is either of the two. Now, take an arbitrary ind($\mathscr{F}_P$) that is neither $\emptyset$ nor $\mathbb{N}$. *Suppose* that ind($\mathscr{F}_P$) is decidable. Then we can deduce a contradiction as follows. Because ind($\mathscr{F}_P$) is a proper and nonempty subset of $\mathbb{N}$, there must be two different natural numbers a and b such that $a \in \text{ind}(\mathscr{F}_P)$ and $b \in \overline{\text{ind}(\mathscr{F}_P)}$. Define a function $f : \mathbb{N} \to \mathbb{N}$ that maps every $x \in \text{ind}(\mathscr{F}_P)$ into b and every $x \in \overline{\text{ind}(\mathscr{F}_P)}$ into a.

Clearly, f is total. But it is also computable. Why? By supposition ind($\mathscr{F}_P$) is decidable, so in order to compute $f(x)$ for an arbitrary $x \in \mathbb{N}$, we first decide whether $x \in \text{ind}(\mathscr{F}_P)$ or $x \in \overline{\text{ind}(\mathscr{F}_P)}$ and, depending on the answer, assign $f(x) = b$ or $f(x) = a$, respectively.

So, by the *Fixed-Point Theorem*, f should have a fixed point. But it is obvious that there can be *no* fixed point: Namely, for an arbitrary $x \in \mathbb{N}$, the numbers x and $f(x)$ are in different sets ind($\mathscr{F}_P$) and $\overline{\text{ind}(\mathscr{F}_P)}$, so $\psi_x \not\simeq \psi_{f(x)}$. This is a contradiction. Hence, ind($\mathscr{F}_P$) cannot be decidable. $\square$

There is some bad news brought by *Rice's Theorem*. We have just proved that the question $\mathcal{D}_P \equiv$ "Does φ have the property P?" is only decidable for trivial properties P. But trivial properties are not very interesting. A property is more interesting if it is non-trivial, i.e., if it is shared by some functions and not by others. Non-trivial properties are less predictable, so getting the answer to the above question is usually more "dramatic," particularly when the consequences of different answers are very different. In this respect *Rice's Theorem* brings to us disillusion because it tells us that any attempt to algorithmically fully recognize an interesting property of functions is in vain.

But there is good news too. *Rice's Theorem* tells us that determining whether the problem $\mathcal{D}_P$ is decidable or not can be reduced to determining whether or not the property P of functions is trivial. But the latter is usually easy to do. So we can set up the following method of proving the undecidability of decision problems of the kind $\mathcal{D}_P$. In this way we can prove the undecidability of the decision problems listed in Section 8.3.5.

Method. (Undecidability of Problems) Given a property P, the undecidability of the decision problem $\mathcal{D}_P \equiv$ "Does a p.c. function φ have the property P?" can be proved as follows:

1. Try to show that P meets the following conditions:
   a. P is a property sensible for functions;
   b. P is insensitive to the machine, algorithm, or program used to compute φ.
2. If P fulfills the above conditions, then try to show that P is non-trivial. To do this,
   a. find a p.c. function that has property P;
   b. find a p.c. function that does not have property P.

If all the steps are successful, then the problem $\mathcal{D}_P$ is undecidable.

9.4.3 Rice's Theorem for C.E. Sets

Let R be a property that is sensible for sets and let S be an arbitrary c.e. set. We define the following decision problem:

$$\mathcal{D}_R \equiv \text{"Does a c.e. set } S \text{ have the property } R\text{?"}$$

As with functions, we say that the property R of c.e. sets is *decidable* if the problem $\mathcal{D}_R$ is decidable. So, if R is decidable, then there is a Turing machine capable of deciding, for an arbitrary c.e. set S, whether or not R holds for S. For the same reasons as above, we are only interested in the *intrinsic* properties of sets. These are the properties R that are independent of the way of recognizing the set S. Hence, the answer to the question of whether or not S has the property R is insensitive to the *local* behavior of the machine, algorithm, and program that are used to recognize S. For example, being finite is an intrinsic property of sets, because every set S either is or is not finite, irrespective of the way in which S is recognized. Finally, we say that R is *trivial* if it holds for all c.e. sets or for none.

Theorem 9.6. (Rice's Theorem for c.e. Sets) *Let R be an arbitrary intrinsic property of c.e. sets. Then: R is decidable $\Longleftrightarrow R$ is trivial.*

Box 9.2 (Proof of Theorem 9.6).

Let S be an arbitrary c.e. set and define $\mathscr{S}_R = \{\mathcal{X} \mid \mathcal{X} \text{ is a c.e. set with property } R\}$. Then we can restate the above decision problem to

$$\mathcal{D}_R \equiv S \in? \mathscr{S}_R. \tag{1}$$

Since a c.e. set is the domain of a p.c. function (see Theorem 6.6 on p. 151), we have $\mathcal{S} = \text{dom}(\varphi_x)$ for some $x \in \mathbb{N}$, hence $\mathscr{S}_R = \{\text{dom}(\varphi_i) \mid \varphi_i \text{ is a p.c. function} \wedge R(\text{dom}(\varphi_i))\}$. Now let $\mathscr{F}_P$ be the set of all p.c. functions with the property P that their domains have the property R. Note that since R is intrinsic to c.e. sets, P is intrinsic to p.c. functions. Then (1) can be restated to

$$\mathcal{D}_R \equiv \varphi_x \in? \mathscr{F}_P. \tag{2}$$

We now proceed as in Sect. 9.4.2. First, we introduce the set $\text{ind}(\mathscr{F}_P)$ of the indexes of all of the p.c. functions that are in $\mathscr{F}_P$. This allows us to restate (2) to

$$\mathcal{D}_R \equiv x \in? \text{ind}(\mathscr{F}_P). \tag{3}$$

So, $\mathcal{D}_R$ is decidable *iff* $x \in? \text{ind}(\mathscr{F}_P)$ is decidable. But the latter is (by Theorem 9.5) decidable *iff* $\text{ind}(\mathscr{F}_P)$ is either $\emptyset$ or $\mathbb{N}$. Hence, $\mathcal{D}_R$ is decidable *iff* either none or all of the domains of p.c. functions have the property R; that is, *iff* either none or all of the c.e. sets have the property R. In other words, $\mathcal{D}_R$ is decidable *iff* R is trivial. □

Therefore, determining whether the problem $\mathcal{D}_R$ is decidable or not can be reduced to determining whether or not R is a trivial property of c.e. sets. Again the latter is much easier to do.

Based on this we can set up a method for proving undecidability of problems $\mathcal{D}_R$ for different properties R. The method can easily be obtained from the previous one (just substitute R for P and c.e. sets for p.c. functions). In this way we can prove the undecidability of many decision problems, such as: *"Is a in $\mathcal{X}$? Is $\mathcal{X}$ equal to $\mathcal{A}$? Is $\mathcal{X}$ regular? Is $\mathcal{X}$ computable?"*, where a and $\mathcal{A}$ are parameters.

Remarks. (1) Since c.e. sets are semi-decidable and vice versa, *Rice's Theorem* for c.e. sets can be stated in terms of semi-decidable sets: Any nontrivial property of semi-decidable sets is undecidable. (2) What about the properties of computable (i.e. decidable) sets? Is there an analogue of *Rice's Theorem* for these? This does *not* automatically follow from *Rice's Theorem* for c.e. sets, as the class of computable sets is properly contained in the class of c.e. sets. Still, the answer is *yes*. Informally, the analogue states: *If* a property R of computable sets is decidable *then* finitely many integers determine whether a computable set $\mathcal{X}$ has the property R.

9.4.4 Consequences: Behavior of Abstract Computing Machines

Let us return to *Rice's Theorem* for index sets. Since indexes represent Turing programs (and Turing machines), this version of *Rice's Theorem* refers to TPs (and TMs). In essence, it tells us that only trivial properties of TPs (and TMs) are decidable. For example, whether or not an arbitrary TM halts is an undecidable question, because the property of halting is not trivial.

This version of *Rice's Theorem* also discloses an unexpected relation between the local and global behavior of TMs and, consequently, of abstract computing machines in general. Recall from p. 158 that we identified the *global behavior* of a TM T with its proper function (i.e., the function φ_T the machine T computes) or with its proper set (i.e., the set $L(T)$ the machine accepts and hence recognizes). In contrast, the *local behavior* of T refers to the particular way in which T's program δ executes. *Rice's Theorem* tells us that, given an arbitrary T, we know perfectly how T behaves locally, but we are unable to algorithmically determine its global behavior—unless this behavior is trivial (in the above sense). Since the models of computation are equivalent, we conclude: In general, we cannot algorithmically predict the global behavior of an abstract computing machine from the machine's local behavior.

9.5 Incomputability of Other Kinds of Problems

In the previous sections we were mainly interested in *decision* problems. Now that we have developed a good deal of the theory and methods for dealing with decision problems, it is time to take a look at other kinds of computational problems. As mentioned in Sect. 8.1.1, there are also search problems, counting problems, and generation problems. For each of these kinds there exist incomputable problems.

In this section we will prove the incomputability of a certain search problem. Search problems are of particular interest because they frequently arise in practice. For example, the *sorting problem* is a search problem: To sort a sequence $a_1, a_2, \ldots, a_n$ of numbers is the same as to find a permutation $i_1, i_2, \ldots, i_n$ of indexes $1, 2, \ldots, n$ so that $a_{i_1} \leqslant a_{i_2} \leqslant \ldots \leqslant a_{i_n}$. Actually, every *optimization problem* is a search problem because it involves the search for a feasible solution that best fulfills a given set of conditions.

We now focus on the following search problem.

▲ SHORTEST EQUIVALENT PROGRAM
Let code(A) be an arbitrary program describing an algorithm A. The aim is:

"Given a program code(A), *find the shortest equivalent program."*

So the problem asks for the shortest description of an arbitrary given algorithm A.

Before we continue we must clear up several things. First, we consider two programs to be *equivalent* if, for every input, they return the same results (although they compute the results in different ways). Secondly, in order to speak about the shortest programs, we must define the *length* of a program. Since there are several ways to do this, we ask, which of them make sense? Here, the following must hold if a definition of the program length is to be considered reasonable:

If programs get longer, then, eventually,
their shortest equivalent programs get longer too.

To see this, observe that if the lengths of the shortest equivalent programs were bounded above by a constant, then there would be only a finite number of shortest programs. This would imply that all of the programs could be shortened into only a finite number of shortest equivalent programs. But this would mean that only a finite number of computational problems can be algorithmically solved.

Now we are prepared to state the following proposition.

Proposition 9.1. *The problem* SHORTEST EQUIVALENT PROGRAM *is incomputable.*

Box 9.3 (Proof of Proposition 9.1).

(Contradiction with the *Recursion Theorem*) We focus on Turing programs and leave the generalization to arbitrary programming languages to the reader. For the sake of contradiction we make the following supposition.

Supposition ($*$): There exists a Turing program S that transforms every Turing program into the shortest equivalent Turing program.

Three questions immediately arise: (1) What is the transformation of a Turing program? (2) What is the length of a Turing program? (3) When are two Turing programs equivalent? The answers are:

1. The *transformation* of a Turing program: Every Turing program can be represented by a natural number, the index of this Turing program (see Sect. 6.2.1). Considering this, we can view the supposed Turing program S as computing a function $s : \mathbb{N} \to \mathbb{N}$ that *transforms* (i.e., maps) an arbitrary index into another index. The above supposition then states, among other things, that s is a (total) *computable* function.

2. The *length* of a Turing program: Indexes of Turing machines have two important properties. First, *every* Turing program is encoded by an index. Secondly, indexes increase (decrease) simultaneously with the increasing (decreasing) of (a) the number of instructions in Turing programs; (b) the number of different symbols used in Turing programs; and (c) the number of states needed in Turing programs. Therefore, to measure the *length* of a Turing program by its *index* seems to be a natural choice.

3. The *equivalence* of Turing programs: Recall (Sect. 6.3.1) that a Turing program P with index n computes the values of the computable function φ_n (i.e., the proper function of the Turing machine T_n). We say that a Turing program P' is *equivalent* to P if P' computes, for every input, exactly the same values as P. Remember also (Sect. 7.2) that the indexes of all the Turing programs equivalent to P constitute the index set $\mathrm{ind}(\varphi_n)$. Now it is obvious that $s(n)$ must be the *smallest* element in the set $\mathrm{ind}(\varphi_n)$.

Since $s(n)$ is an index (i.e., the code of a Turing program), we see that, by the same argument as above, the function s cannot be bounded above by any constant. In other words, if the lengths of Turing programs increase, so eventually do the lengths of their shortest equivalent programs; that is,

$$\text{For every } n \text{ there exists an } n' > n \text{ such that } s(n) < s(n'). \tag{1}$$

Using the function s we can restate our supposition ($*$) more precisely:

Supposition ($**$): There is a (total) computable function $s : \mathbb{N} \to \mathbb{N}$ such that

$$s(n) \stackrel{\text{def}}{=} \text{the smallest element in } \mathrm{ind}(\varphi_n). \tag{2}$$

We now derive a contradiction from the supposition $(**)$. First, let $f : \mathbb{N} \to \mathbb{N}$ be the function whose values $f(x)$ are computed by the following Turing machine S. Informally, given an arbitrary input $x \in \mathbb{N}$, the machine S computes and generates the values $s(k)$, $k = 0, 1, \ldots$ until an $s(m)$ for which $s(m) > x$ has been generated. At that point S returns the result $f(x) := s(m)$ and halts. Here is a more detailed description. The machine S has one input tape, one output tape, and two work tapes, W_1 and W_2 (see Fig. 9.8). The Turing program of S operates as follows:

1. Let $x \in \mathbb{N}$ be an arbitrary number written on the input tape of S.
2. S generates the numbers $k = 0, 1, 2, \ldots$ on W_1 and after each generated k executes steps 3–6:
3. S computes the value $s(k)$;
4.   if $s(k)$ has not yet been written to W_2,
5.   then S generates $s(k)$ on W_2;
6.   if $s(k) > x$, then S copies $s(k)$ to the output tape (so $f(x) = s(k)$) and halts.

Fig. 9.8 The configuration of the Turing machine S

In step 3, the machine S computes the function s. Because s is supposed to be a computable function (2), and (1) holds, the machine S halts for arbitrary x. Consequently, f is a computable function.

We now show that such an f *cannot* exist. Let x be an arbitrary natural number. We saw that there exists an $m \in \mathbb{N}$ such that $s(m) > x$. But $s(m)$ is by definition the *smallest* index in the set $\mathrm{ind}(\varphi_m)$, so the relation $s(m) > x$ tells us that

$$x \notin \mathrm{ind}(\varphi_m).$$

At the same time, $f(x) = s(m)$, so

$$f(x) \in \mathrm{ind}(\varphi_m).$$

Since x and $f(x)$ are not both in the same index set, the functions φ_x and $\varphi_{f(x)}$ cannot be equal; that is,

$$\varphi_x \neq \varphi_{f(x)}, \text{ for every } x \in \mathbb{N}. \tag{3}$$

The inequality (3) reveals that f has *no* fixed point. But we have seen that f is a computable function, so—according to the *Fixed-Point Theorem*—f *should* have a fixed point. This is a contradiction. Consequently, f is not a computable function. This means that the machine S does not halt for some x. Since (1) holds, it must be that (2) does not hold. Our supposition must be rejected. $\square$

9.6 Chapter Summary

General methods of proving the undecidability of decision problems use diagonalization, reduction, the *Recursion Theorem*, or *Rice's Theorem*.

Diagonalization can be used to prove that a set S of all elements with a given property P is uncountable. It uses a table whose vertical axis is labeled with the members of S and the horizontal axis by all natural numbers. The supposition is that each member of S can be uniquely represented by a sequence of symbols, written in the corresponding row, over an alphabet. If we change each diagonal symbol, and the obtained sequence of diagonal symbols represents an element of S, then the set is uncountable. Another use of diagonalization is to interpret S as the set of all indexes of all algorithms having a given property P. We want to know whether the property is decidable. The labels on the horizontal axis of the table are interpreted as inputs to algorithms. Each entry in the table contains the result of applying the corresponding algorithm on the corresponding input. Of special interest is the diagonal of the table. The idea is to design a shrewd algorithm that will use the alleged decider of the set S, and uncover the inability of the decider to decide whether or not the shrewd algorithm has the property P. Then the property P is undecidable.

Reduction is a method where a given problem is expressed in terms of another problem in such a way that a solution to the second problem would give rise to a solution of the first problem. For decision problems, we can use the m-reduction or 1-reduction. The existence of the m-reduction can be used to prove the undecidability of a decision problem. More powerful is the Turing reduction (see Part III).

The *Recursion Theorem* can be used to prove the undecidability of a decision problem. The method relies on the fact that every computable function has a fixed point. If we prove that a given function has no fixed point, then the function cannot be computable. Since the characteristic function of a set is a function, this method can be used to prove the undecidability of a set.

Rice's Theorem is used to prove that certain properties of partial computable (p.c.) functions or computably enumerable (c.e.) sets are undecidable. The properties must be intrinsic to functions or sets, and thus independent of the way the functions are computed or the sets recognized. *Rice's Theorem* states that every nontrivial intrinsic property is undecidable.

Problems

9.1. Prove:

    (a) $\leq_m$ is a reflexive and transitive relation;

    (b) $\mathcal{A} \leq_m \mathcal{B}$ if $r(\mathcal{A}) \subseteq \mathcal{B}$ and $r(\overline{\mathcal{A}}) \subseteq \overline{\mathcal{B}}$, for some computable function r.

Definition 9.2. (*m*-Complete Set) A set $\mathcal{A}$ is said to be *m*-**complete** if $\mathcal{A}$ is c.e. and $\mathcal{B} \leq_m \mathcal{A}$ for every c.e. set $\mathcal{B}$. A set $\mathcal{A}$ is 1-**complete** if $\mathcal{A}$ is c.e. and $\mathcal{B} \leq_1 \mathcal{A}$ for every c.e. set $\mathcal{B}$.

9.2. Prove:

    (a) $\mathcal{K}$ is *m*-complete;

    (b) $\mathcal{K}_0$ is *m*-complete.

9.3. Prove:

    (a) $\mathcal{K}$ is 1-complete;

    (b) $\mathcal{K}_0$ is 1-complete.

9.4. Prove: $\mathcal{A}$ is 1-complete $\Longrightarrow$ $\mathcal{A}$ is *m*-complete.

9.5. Construct the following Turing machines (which we used in Example 9.5 on p. 215):

    (a) a TM A that takes $\langle T, w \rangle$ as input and outputs a TM B such that

$$L(B) = \begin{cases} \mathcal{K}_0 & \text{if } T \text{ accepts } w; \\ \emptyset & \text{if } T \text{ does not accept } w. \end{cases}$$

    [*Hint.* The actions of B on input x are as follows: B starts T on w; if T accepts w, then B starts U (the recognizer of $\mathcal{K}_0$) on x and outputs U's answer (if it is a YES) as its own; if, however, T doesn't accept w, then U is not started, so B outputs nothing. It remains to be shown that there exists a Turing machine A that is capable of computing the code $\langle B \rangle$.]

    (b) a TM R that recognizes $\overline{\mathcal{K}_0}$ if allowed to call the above A and the alleged decider $D_{L(\mathcal{Q})}$ for the problem $\mathcal{Q} \equiv$ "Is $L(T)$ computable?"

    [*Hint.* The actions of R on input $\langle T, w \rangle$ are: R calls A on the input $\langle T, w \rangle$; hands over the result $\langle B \rangle$ of A to $D_{L(\mathcal{Q})}$; and outputs $D_{L(\mathcal{Q})}$'s answer (if it is a YES) as its own answer.]

9.6. Prove that the following decision problems are undecidable (use any method of this chapter):

    (a) "Does T halt on empty input?"

    (b) "Does T halt on every input?"

    (c) "Does T halt on w_0?" (w_0 is an arbitrary fixed word)

    (d) "Does M halt on w?" (M is a certain fixed TM)

    (e) "Do T and T' halt on the same inputs?"

(f) "Is the language $L(T)$ empty?" ($\equiv \mathcal{D}_{Emp}$)
   [*Hint.* Reduce $\mathcal{D}_{Halt}$ to $\mathcal{D}_{Emp}$.]

(g) "Does an algorithm A terminate on every input data?" ($\equiv \mathcal{D}_{Term}$)

(h) "Does an algorithm A terminate on input data d?"

(i) "Is $dom(\varphi)$ empty?" ($\equiv \mathcal{D}_{\mathcal{K}_1}$)

(j) "Is $dom(\varphi)$ finite?" ($\equiv \mathcal{D}_{\mathcal{F}in}$)

(k) "Is $dom(\varphi)$ infinite?" ($\equiv \mathcal{D}_{\mathcal{I}nf}$)

(l) "Is $\mathcal{A} - dom(\varphi)$ finite?" ($\equiv \mathcal{D}_{Cof}$)

(m) "Is φ total?" ($\equiv \mathcal{D}_{Tot}$)

(n) "Can φ be extended to a (total) computable function?" ($\equiv \mathcal{D}_{\mathcal{E}xt}$)

(o) "Is φ surjective?" ($\equiv \mathcal{D}_{Sur}$)

(p) "Is φ defined at x?"

(q) "Is $\varphi(x) = y$ for at least one x?"

(r) "Is $dom(\varphi) = dom(\psi)$?"

(s) "Is $\varphi \simeq \psi$?"

Bibliographic notes

- The undecidability of the *Entscheidungsproblem* was proved in Church [36] and Turing [262].
- The method of *diagonalization* first appeared in Cantor [30], where he proved that the set of real numbers is uncountable and that there are other uncountable sets. For additional applications of the diagonalization method and its variants, see Odifreddi [173].
- *Reduction* as a method of proving the undecidability of problems was extensively analyzed in Post [183] and Post [184]. Here, in addition to the 1-reduction and *m*-reduction, other so-called *strong* reductions were defined (e.g., the *bounded truth-table* and *truth-table* reductions). The paper also describes the *Turing* reduction as the most general among them.
- For the *resource-bounded* reductions, such as the *polynomial-time bounded* Turing or the *m*-reduction, see Ambos-Spies [10].
- *Rice's Theorem* was proved in Rice [197].
- A number of examples of use of the above methods in incomputability proofs can be found in the general monographs on *Computability Theory* cited in the Bibliographic Notes to Chap. 6.
- *Presburger Arithmetic* was introduced in Presburger [189]. For the proof that this theory is decidable, see Boolos et al. [21].

Part III
RELATIVE COMPUTABILITY

In the previous part we described how the world of incomputable problems was discovered. This resulted in an awareness that there exist computational problems that are unsolvable by any reasonable means of computing, e.g., the Turing machine. In Part III, we will focus on decision problems only. We will raise the questions, "What if an unsolvable decision problem had been somehow made solvable? Would this have turned all the other unsolvable decision problems into solvable problems?" We suspect that this might be possible *if* all the unsolvable decision problems were somehow reducible one to another. However, it will turn out that this is not so; some of them would indeed become solvable, but there would still remain others that are unsolvable. We might speculate even further and suppose that one of the remaining unsolvable decision problems was somehow made solvable. As before, this would turn many unsolvable decision problems into solvable ones; yet, again, there would remain unsolvable decision problems. We could continue in this way, but we would never exhaust the class of unsolvable problems.

Questions of the kind "Had the problem Q been solvable, would this have made the problem P solvable too?" are characteristic of *relativized computability*, a large part of *Computability Theory*. This theory analyzes the solvability of problems *relative to* (or in view of) the solvability of other problems. Although such questions seem to be overly speculative and the answers to be of questionable practical value, they nevertheless reveal a surprising *fact:* Unsolvable decision problems can differ in the degree of their unsolvability. We will show that the class of all decision problems partitions into infinitely many subclasses, called *degrees of unsolvability*, each of which consists of all equally difficult decision problems. It will turn out that, after defining an appropriate relation on the class of all degrees, the degrees of unsolvability are intricately connected in a lattice-like structure. In addition to having many interesting properties *per se*, the structure will reveal many surprising facts about the unsolvability of decision problems.

We will show that there are several approaches to partitioning the class of decision problems into degrees of unsolvability. Each of them will establish a particular *hierarchy* of degrees of unsolvability, such as the *jump* and *arithmetical* hierarchies, and thus offer yet another view of the solvability of computational problems.

Chapter 10
Computation with External Help

*An oracle was an ancient priest who made statements about
future events or about the truth.*

Abstract According to the *Computability Thesis*, *all* models of computation, including those yet to be discovered, are equivalent to the Turing machine and formalize the intuitive notion of computation. In other words, what cannot be solved
on a Turing machine cannot be solved in nature. But what if Turing machines could
get external help from a *supernatural* assistant? In this chapter we will describe the
birth of this idea, its development, and the formalization of it in the concept of the
oracle Turing machine. We will then briefly describe how external help can be added
to other models of computation, in particular to μ-recursive functions. We will conclude with the *Relative Computability Thesis*, which asserts that all such models are
equivalent, one to the other, thus formalizing the intuitive notion of "computation
with external help." Based on this, we will adopt the oracle Turing machine as *the*
model of computation with external help.

10.1 Turing Machines with Oracles

"What if an unsolvable decision problem were *somehow* made solvable?" For instance, what if we somehow managed to construct a procedure that is capable of
mechanically solving an arbitrary instance of the *Halting Problem*? But in making
such a supposition, we must be cautious: Since we have already proved that *no* algorithm can solve the *Halting Problem*, such a supposition would be in contradiction
with this fact *if* we meant by the hypothetical procedure an algorithm as formalized
by the *Computability Thesis*. Making such a supposition would render our theory inconsistent and entail all the consequences described in Box 4.1 (p. 56). So, to avoid
such an inconsistency, we must assume that the hypothetical procedure would *not*
be the ordinary Turing machine.

Well, if the hypothetical procedure is not the ordinary Turing machine, then what
is it? The answer was suggested by Turing himself.

10.1.1 Turing's Idea of Oracular Help

In 1939, Turing published yet another seminal idea, this time of a machine he called the *o-machine*. Here is the citation from his doctoral thesis:

> Let us suppose that we are supplied with some unspecified means of solving number-theoretic problems; a kind of oracle as it were. We shall not go any further into the nature of this oracle apart from saying that it cannot be a machine. With the help of the oracle we could form a new kind of machine (call them *o*-machines), having as one of its fundamental processes that of solving a given number-theoretic problem. More definitely these machines are to behave in this way. The moves of the machine are determined as usual by a table except in the case of moves from a certain internal configuration $\mathfrak{o}$. If the machine is in the internal configuration $\mathfrak{o}$ and if the sequence of symbols marked with l is then the well-formed formula A, then the machine goes into the internal configuration $\mathfrak{p}$ or $\mathfrak{t}$ according as it is or is not true that A is dual. The decision as to which is the case is referred to the oracle.
>
> These machines may be described by tables of the same kind as those used for the description of *a*-machines, there being no entries, however, for the internal configuration $\mathfrak{o}$. We obtain description numbers from these tables in the same way as before. If we make the convention that, in assigning numbers to internal configurations, $\mathfrak{o}$, $\mathfrak{p}$, $\mathfrak{t}$ are always to be q_2, q_3, q_4, then the description numbers determine the behavior of the machines uniquely.

Remarks. Let us add some remarks about the terminology and notation used in the above passage. The "internal configuration" is what we now call the state of the Turing machine. The "sequence of symbols marked with l" is a sequence of symbols on the tape that are currently marked in a certain way. We will see shortly that this marking can be implicit. The "duality" is a property that a formula has or does not have; we will not need this notion in the subsequent text. The "table" is the table Δ representing the Turing program δ as decribed in Fig. 6.3 (see p. 113). To comply with our notation, we will from now on substitute Gothic symbols as follows: $\mathfrak{o} \mapsto q_?$, $\mathfrak{p} \mapsto q_+$ and $\mathfrak{t} \mapsto q_-$.

Now we see that the set Q of the states of the *o*-machine would contain, in addition to the usual states, three distinguished states $q_?, q_+, q_-$. The operation of the *o*-machine would be directed by an ordinary Turing program, i.e., a transition function $\delta : Q \times \Gamma \to Q \times \Gamma \times \{\text{Left}, \text{Right}, \text{Stay}\}$. The function would be undefined in the state $q_?$, for every $z \in \Gamma$, i.e., $\forall z \in \Gamma \, \delta(q_?, z) \uparrow$. However, upon entering the state $q_?$, the *o*-machine would *not* halt; instead, an *oracle* would miraculously supply, in the very next step of the computation, a cue telling the *o*-machine which of the states q_+ and q_- to enter. In this way the oracle would answer the *o*-machine's question asking whether or not the currently marked word on the tape is in a certain, predefined set of words, a set which can be decided by the oracle. The *o*-machine would immediately take up the cue, enter the suggested state, and continue executing its program δ.

Further Development of Turing's Idea

Turing did not further develop his *o*-machine. The idea remained dormant until 1944, when Post awakened it and brought it into use. Let us describe how the idea evolved. We will focus on the interplay between the o-machine and its oracle. In doing so, we will lean heavily on Definition 6.1 of the ordinary TM (see Sect. 6.1.1).

A Turing machine *with an oracle*, i.e., the o-machine, consists of the following components: (1) a control unit with a program; (2) a potentially infinite tape divided into cells, each of which contains a symbol from a tape alphabet $\Gamma = \{z_1, z_2, \dots, z_t\}$, where $z_1 = 0, z_2 = 1$, and $z_t = \sqcup$; (3) a window that can move to the neighboring (and, eventually, to any) cell, thus making the cell accessible to the control unit for reading or writing; and (4) an *oracle*. (See Fig. 10.1.)

Fig. 10.1 The o-machine with the oracle for the set $\mathcal{O}$ can ask the oracle whether or not the word w is in $\mathcal{O}$. The oracle answers in the next step and the o-machine continues its execution appropriately

The control unit is always in some state from a set $Q = \{q_1, \dots, q_s, q_?, q_+, q_-\}$. As usual, the state q_1 is initial and some of $q_1, \dots, q_s$ are final states. But now there are also three additional, *special states*, denoted by $q_?$, q_+, and q_-.

The program is characteristic of the particular o-machine; as usual, it is a partial function $\delta : Q \times \Gamma \to Q \times \Gamma \times \{\text{Left}, \text{Right}, \text{Stay}\}$. But, in addition to the usual instructions of the form $\delta(q_i, z_r) = (q_j, z_w, D)$, for $1 \leqslant i, j \leqslant s$ and $1 \leqslant r, w \leqslant t$, there are *four special instructions for each* $z_r \in \Gamma$:

$$\delta(q_?, z_r) = (q_+, z_r, \text{Stay}) \qquad\qquad\qquad (*)$$
$$\delta(q_?, z_r) = (q_-, z_r, \text{Stay})$$

$$\delta(q_+, z_r) = (q_{j_1}, z_{w_1}, D_1) \qquad\qquad\qquad (**)$$
$$\delta(q_-, z_r) = (q_{j_2}, z_{w_2}, D_2)$$

where $q_{j_1}, q_{j_2} \in Q - \{q_+, q_-\}$, $z_{w_1}, z_{w_2} \in \Gamma$, and $D_1, D_2 \in \{\text{Left}, \text{Right}, \text{Stay}\}$.

Before the o-machine is started the following preparative arrangement is made: (1) an input word belonging to the set Σ^* such that $\{0, 1\} \subseteq \Sigma \subseteq \Gamma - \{\sqcup\}$ is written on the tape; (2) the window is shifted over the leftmost symbol of the input word; (3) the control unit is set to the initial state q_1; and (4) an arbitrary set $\mathcal{O} \subseteq \Sigma^*$ is fixed. We call $\mathcal{O}$ the *oracle set*.

From now on the o-machine operates in a mechanical stepwise fashion as directed *either by the program δ or by the oracle*. The details are as follows. Suppose that the control unit is in a state $q \in Q$ and the cell under the window contains a symbol $z_r \in \Gamma$. (See Fig. 10.1.) Then:

- If $q \neq q_?$, then the o-machine reads z_r into its control unit. If $\delta(q, z_r)\downarrow = (q', z_w, D)$, the control unit executes this instruction as usual (i.e., writes z_w through the window, moves the window in direction D, and enters the state q'); otherwise, the o-machine halts. In short, when the o-machine is not in state $q_?$, it operates as an ordinary Turing machine.

- If $q = q_?$, then the oracle takes this as the question

$$w \in? \mathcal{O}, \qquad\qquad (***)$$

where w is the word starting under the window and ending in the rightmost nonempty cell of the tape. The oracle answers the question $(***)$ in the *next* step of the computation by *advising the control unit which of the special states q_+, q_- to enter* (without changing z_r under the window or moving the window). We denote the oracle's advice by "+" when $w \in \mathcal{O}$, and by "−" when $w \notin \mathcal{O}$. The control unit always takes the advice and enters the state either q_+ or q_-. In short, the oracle helps the program δ to choose the "right" one of the two alternative instructions $(*)$. In this way the oracle miraculously resolves the nondeterminism arising from the state $q_?$.

After that, the o-machine continues executing its program. In particular, if the previous state has been $q_?$, the program will execute the instruction either $\delta(q_+, z_r) = (q_{j_1}, z_{w_1}, D_1)$ or $\delta(q_-, z_r) = (q_{j_2}, z_{w_2}, D_2)$, depending on the oracle's advice. Since the execution continues in one or the other way, this allows the program δ to react differently to different advice. In short, the oracle's advice will be taken into account by the Turing program δ.

The o-machine can ask its oracle, in succession, many questions. Since between two consecutive questions the o-machine can move the window and change the tape's contents, the questions $w \in? \mathcal{O}$, in general, differ in words w. But note that they all refer to the set $\mathcal{O}$, which was fixed before the computation started.

The o-machine halts if the control unit has entered a final state, or the program specifies no next instruction (the machine reads z_r in a state $q(\neq q_?)$ and $\delta(q, z_r)\uparrow$).

Remark. Can the o-machine ask the oracle a question $u \in? \mathcal{O}$, where u is *not* the word currently starting under the window and ending at the rightmost nonempty symbol of the tape? The answer is yes. The machine must do the following: It remembers the position of the window, moves the window past the rightmost nonempty symbol, writes u to the tape, moves the window back to the beginning of u, and enters the state $q_?$. Upon receiving the answer to the question $u \in? \mathcal{O}$, the machine deletes the word u, returns the window to the remembered position, enters the state q_+ or q_- depending on the oracle's answer, and continues the "interrupted" computation.

10.1.2 The Oracle Turing Machine (o-TM)

To obtain a modern definition of the Turing machine with an oracle, we make the following two adaptations to the above o-machine:

1. *We model the oracle with an oracle tape.* The *oracle tape* is a one-way unbounded, read-only tape that contains all the values of the characteristic function $\chi_\mathcal{O}$. (See Fig. 10.2.) We assume that it takes, for arbitrary $w \in \Sigma^*$, only one step to search the tape and return the value $\chi_\mathcal{O}(w)$.

2. *We eliminate the special states $q_?, q_+, q_-$ and redefine the instructions' semantics.* Since the special states behave differently, we will get rid of them. How can we do that? The idea is to 1) adapt the o-machine so that it will interrogate the oracle at *each* instruction of the program, and 2) leave it to the machine to decide whether or not the oracle's advice will be taken into account. To achieve goal 2 we must only use what we learned in the previous subsection: The oracle's advice on the question $\delta(q_?, z_r)$ will have no impact on the computation *iff* $\delta(q_+, z_r) = \delta(q_-, z_r)$. To achieve goal 1, however, we must redefine what actions of the machine will be triggered by the instructions.[1] In particular, we must redefine the instructions in such a way that each instruction will make the machine consult the oracle. Clearly, this will call for a change in the definition of the transition function.

Remark. We do not concern ourselves with how the values $\chi_\mathcal{O}(w)$ emerged on the oracle tape and how it is that any of them can be found and read in just one step. A realistic implementation of such an access to the contents of the oracle tape would be a tremendous challenge when the set $\mathcal{O}$ is infinite. Actually, if we knew how to do this for any set $\mathcal{O}$ then according to the *Computability Thesis* there would exist an ordinary Turing machine capable of providing, in finite time, any data from the oracle tape. Hence, oracle Turing machines could be simulated by ordinary TMs. This would dissolve the supernatural flavor of the oracle Turing machine and bring back the threat of inconsistency (as we explained on p. 233.)

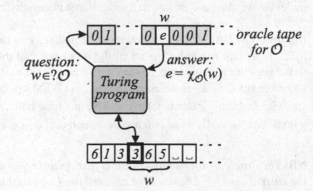

Fig. 10.2 The oracle Turing machine has an oracle tape. Given an oracle set $\mathcal{O} \subseteq \Sigma^*$, the oracle tape contains all the values of the characteristic function $\chi_\mathcal{O}$. Here, $e = \chi_\mathcal{O}(w)$, for any $w \in \Sigma^*$

---

[1] In other words, we must redefine the *operational semantics* of the instructions.

After making these adaptations we obtain a modern definition of the Turing machine that uses external help. We will call it the *oracle Turing machine*, or for short, *o*-TM.

Definition 10.1. (Oracle Turing Machine) The **oracle Turing machine with oracle set** $\mathcal{O}$ (for short, *o*-TM *with oracle set* $\mathcal{O}$, or $\mathcal{O}$-TM) consists of a control unit, an input tape, an oracle Turing program, an oracle tape, and a set $\mathcal{O}$. Formally, $\mathcal{O}$-TM is an eight-tuple $T^{\mathcal{O}} = (Q, \Sigma, \Gamma, \widetilde{\delta}, q_1, \sqcup, F, \mathcal{O})$.

The *control unit* is in a state from the set of states $Q = \{q_1, \ldots, q_s\}$, $s \geqslant 1$. We call q_1 the initial state; some states are final and belong to the set $F \subseteq Q$.

The *input tape* is a usual one-way unbounded tape with a window. The tape alphabet is $\Gamma = \{z_1, \ldots, z_t\}$, where $t \geqslant 3$. For convenience we fix $z_1 = 0$, $z_2 = 1$, and $z_t = \sqcup$. The input alphabet is a set Σ, where $\{0, 1\} \subseteq \Sigma \subseteq \Gamma - \{\sqcup\}$.

The *oracle set* $\mathcal{O}$ is an arbitrary subset of Σ^*.

The *oracle tape* contains, for each $w \in \Sigma^*$, the value $\chi_{\mathcal{O}}(w)$ of the characteristic function $\chi_{\mathcal{O}} : \mathcal{O} \to \{0, 1\}$. It is a read-only tape; although lacking a window, it can immediately find and return the value $\chi_{\mathcal{O}}(w)$, for any $w \in \Sigma^*$.

The *oracle Turing program* (for short, *o*-TP) resides in the control unit; it is a partial function

$$\widetilde{\delta} : Q \times \Gamma \times \{0, 1\} \to Q \times \Gamma \times \{\text{Left}, \text{Right}, \text{Stay}\}.$$

Thus, any instruction is of the form

$$\widetilde{\delta}(q, z, e) = (q', z', D),$$

which is interpreted as follows: If the control unit is in the state q, and reads z and e from the input and oracle tape, respectively, then it changes to the state q', writes z' on the input tape, and moves the window in the direction D. Here, e denotes the value $\chi_{\mathcal{O}}(w)$, where w is the word starting under the window and ending in the rightmost nonempty cell of the input tape.

Before *o*-TM is started, the following take place: 1) an input word belonging to Σ^* is written to the beginning of the input tape; 2) the window is shifted to the beginning of the tape; 3) the control unit is set to the initial state q_1; 4) an oracle set $\mathcal{O} \subseteq \Sigma^*$ is fixed. From now on $\mathcal{O}$-TM operates in a mechanical stepwise fashion, as instructed by $\widetilde{\delta}$. The machine halts when it either enters a final state, or reads z and e in a state q such that $\widetilde{\delta}(q, z, e) \uparrow$.

NB *The oracle is not omnipotent; it is just an unsurpassable expert at recognizing the current set $\mathcal{O} \subseteq \Sigma^*$. Since we adopted the Computability Thesis, we must admit that this capability is supernatural if $\mathcal{O}$ is an undecidable set (in the ordinary sense). In such a case we will not ask where the oracle's expertise comes from.*

10.1.3 Some Basic Properties of o-TMs

Let us list a few basic properties of o-TMs that follow directly from Definition 10.1.

First, looking at Definition 10.1 we find that if we change the oracle set, say from $\mathcal{O}$ to $\mathcal{O}' \subseteq \Sigma^*$, *no* change is needed in the o-TP $\widetilde{\delta}$, i.e., no instruction $\widetilde{\delta}(q,z,e) = (q',z',D)$ must be changed, added, or deleted. The program $\widetilde{\delta}$ remains capable of running with the new oracle set $\mathcal{O}'$. In other words, the o-TP $\widetilde{\delta}$ is *insensitive* to the oracle set $\mathcal{O}$. (By the way, this is why the symbol $\widetilde{\delta}$ has no symbol $\mathcal{O}$ attached.)

Consequence 10.1. *Oracle Turing programs $\widetilde{\delta}$ are not affected by changes in $\mathcal{O}$.*

Second, although changing $\mathcal{O}$ does not affect the o-TP $\widetilde{\delta}$, it does affect, in general, the *execution* of $\widetilde{\delta}$. Namely, if $\mathcal{O}$ changes, $\chi_{\mathcal{O}}$ also changes, so the values e read from the oracle tape change too. But these values are used to select the next instruction to be executed. Nevertheless, we *can* construct o-TPs whose executions are insensitive to changes in the oracle set. Such an o-TP must just strictly ignore the values e read from the oracle tape. To achieve this, the o-TP must be such that

$$o\text{-TP has instr. } \widetilde{\delta}(q,z,0) = (q',z',D) \iff o\text{-TP has instr. } \widetilde{\delta}(q,z,1) = (q',z',D),$$

for every pair $(q,z) \in Q \times \Gamma$. So, if o-TM reads z in state q, it performs (q',z',D) regardless of the value e. If the above only holds for certain pairs (q,z), the o-TP $\widetilde{\delta}$ ignores the oracle for those pairs only. We sum this up in the following statement.

Consequence 10.2. *During the execution, oracle Turing programs $\widetilde{\delta}$ can ignore $\mathcal{O}$.*

Third, we ask: What is the relation between the computations performed by o-TMs and the computations performed by ordinary TMs? Let T be an arbitrary ordinary TM and δ its TP. We can easily construct an o-TM that simulates TM T. Informally, the o-TP $\widetilde{\delta}$ does, while ignoring its oracle set, what the TP δ would do. To achieve this, we must construct the program $\widetilde{\delta}$ in the following way:

> **let** $\widetilde{\delta}$ have no instructions;
> **for** each instruction $\delta(q,z) = (q',z',D)$ in δ **do**
> add instructions $\widetilde{\delta}(q,z,0) = (q',z',D)$ and $\widetilde{\delta}(q,z,1) = (q',z',D)$ to $\widetilde{\delta}$.

So we can state the following conclusion.

Theorem 10.1. *Oracle computation is a generalization of ordinary computation.*

Fourth, we now see that each $\mathcal{O}$-TM, say $T^{\mathcal{O}} = (Q, \Sigma, \Gamma, \widetilde{\delta}, q_1, \sqcup, F, \mathcal{O})$, is characterized by two independent components:

1. *a particular o*-TM, denoted by $T^* = (Q, \Sigma, \Gamma, \widetilde{\delta}, q_1, \sqcup, F, *)$, which is capable of consulting *any* particular oracle set when it is selected and substituted for $*$;
2. *a particular* oracle set $\mathcal{O}$ which is to be "plugged into" T^* in place of $*$.

Formally,

$$T^{\mathcal{O}} = (T^*, \mathcal{O}).$$

10.1.4 Coding and Enumeration of o-TMs

Like ordinary Turing programs, oracle Turing programs can also be encoded and enumerated. The coding proceeds in a similar fashion to that in the ordinary case (see Sect. 6.2.1). But there are differences too: Firstly, we must take into account that oracle Turing programs $\widetilde{\delta}$ are defined differently from ordinary ones δ, and, secondly, oracle sets must be taken into account somehow. Let us see the details.

Coding of o-TMs

To encode an $\mathcal{O}$-TM $T^{\mathcal{O}} = (T^*, \mathcal{O})$ we will only encode the component T^*. The idea is that we only encode *o*-TP $\widetilde{\delta}$ of T^*, but in such a way that the other components Q, Σ, Γ, F, which determine the particular T^*, can be restored from the code. An appropriate coding alphabet is $\{0, 1\}$. This is because $\{0, 1\}$ is contained in the input alphabet of every *o*-TM, so every *o*-TM will be able to read codes of other *o*-TMs as input data. Here are the details.

Let $T^{\mathcal{O}} = (Q, \Sigma, \Gamma, \widetilde{\delta}, q_1, \sqcup, F, \mathcal{O})$ be an arbitrary $\mathcal{O}$-TM. If

$$\widetilde{\delta}(q_i, z_j, e) = (q_k, z_\ell, D_m)$$

is an instruction of the program $\widetilde{\delta}$, we encode the instruction by the word

$$K = 0^i 10^j 10^e 10^k 10^\ell 10^m, \tag{$*$}$$

where $D_1 = $ Left, $D_2 = $ Right, and $D_3 = $ Stay.

In this way, we encode each instruction of the program $\widetilde{\delta}$. From the obtained codes $K_1, K_2, \ldots, K_r$ we construct the *code* of the *o*-TP $\widetilde{\delta}$ as follows:

$$\langle \widetilde{\delta} \rangle = 111 K_1 11 K_2 11 \ldots 11 K_r 111. \tag{$**$}$$

The restoration of Q, Σ, Γ, F would proceed in a similar way to the ordinary case (see Sect. 7.2). We can therefore identify the code $\langle T^* \rangle$ with the code $\langle \widetilde{\delta} \rangle$:

$$\langle T^* \rangle \overset{\text{def}}{=} \langle \widetilde{\delta} \rangle.$$

Enumeration of o-TMs

Since $\langle T^* \rangle$ is a word in $\{0,1\}^*$, we can interpret it as the binary representation of a natural number. We call this number the *index* of the o-TM T^* (and of its o-TP $\widetilde{\delta}$).

Clearly, some natural numbers have binary representations that are not of the form $(**)$. Such numbers are not indexes of any o-TM T^*. This is a weakness of the above coding, because it prevents us from establishing a *surjective* mapping from the set of all o-TMs onto the set $\mathbb{N}$ of natural numbers. However, as in the ordinary case, we can easily patch this by introducing a special o-TM, called the *empty o-TM*, and making the following **convention**: Any natural number whose binary representation is not of the form $(**)$ is an index of the empty o-TM. (The o-TP $\widetilde{\delta}$ of the empty o-TM is everywhere undefined, so it immediately halts on any input word.)

We are now able to state the following proposition.

Proposition 10.1. *Every natural number is the index of exactly one o-TM.*

Given an arbitrary index $\langle T^* \rangle$ we can now restore from it the components Q, Σ, Γ, F of the corresponding o-TM $T^* = (Q, \Sigma, \Gamma, \widetilde{\delta}, q_1, \sqcup, F, *)$. In other words, we have implicitly defined a *total* function $g : \mathbb{N} \to \mathcal{T}^*$ from the set $\mathbb{N}$ of natural numbers to the set $\mathcal{T}^*$ of all o-TMs. Given an arbitrary $n \in \mathbb{N}$, we view $g(n)$ as the nth o-TM and therefore denote this o-TM by T_n^*. By letting $n = 0, 1, 2, \ldots$ we can *generate* the sequence

$$T_0^*, T_1^*, T_2^*, \ldots$$

of all o-TMs. Thus, we have *enumerated* oracle Turing machines. Clearly, the function g enumerates oracle Turing programs too, so we also obtain the sequence

$$\widetilde{\delta}_0, \widetilde{\delta}_1, \widetilde{\delta}_2, \ldots$$

Remarks. (1) Given an arbitrary index n, we can reconstruct from n both the ordinary Turing machine T_n and the oracle Turing machine T_n^*. Of course, the two machines are different objects. Also different are the corresponding sequences $T_1, T_2, T_3, \ldots$ and $T_1^*, T_2^*, T_3^*, \ldots$. The same holds for the sequences $\widetilde{\delta}_0, \widetilde{\delta}_1, \widetilde{\delta}_2, \ldots$ and $\widetilde{\delta}_0, \widetilde{\delta}_1, \widetilde{\delta}_2, \ldots$. (2) Plugging different oracle sets into an o-TM T_n^* affects neither the o-TM nor its index. For this reason we will relax our pedantry and from now on also say that a natural number n is the *index of the $\mathcal{O}$-TM $T_n^{\mathcal{O}}$*. Conversely, given an $\mathcal{O}$-TM $T^{\mathcal{O}}$, we will denote its index by $\langle T^{\mathcal{O}} \rangle$, while keeping in mind that the index is independent of $\mathcal{O}$. Furthermore, we will say that the sequence $T_1^{\mathcal{O}}, T_2^{\mathcal{O}}, T_3^{\mathcal{O}}, \ldots$ is the *enumeration of $\mathcal{O}$-TMs*, and remember that the ordering of the elements of the sequence is independent of the particular $\mathcal{O}$.

10.2 Computation with Oracles

In this section we will continue setting the stage for the theory of oracular computation. Specifically, we will define the basic notions about the *computability* of functions and the *decidability* of sets by oracle Turing machines. Since we will build on Definition 10.1, we will define the new notions using the universe Σ^*. Then we will switch, w.l.o.g., to the equivalent universe $\mathbb{N}$. This will enable us to develop and present the theory in a simpler, unified, and more standard way.

10.2.1 Generalization of Classical Definitions

The new definitions will be straightforward generalizations of definitions of the proper function, computable function, decidable set, and index set for the ordinary TM. We will therefore move at a somewhat faster pace.

Proper Functionals

First, we generalize Definition 6.4 (p. 135) of the proper function. Let T_n^* be an arbitrary o-TM and k a natural number. Then we can define a mapping $\Phi_n^{(k+1)}$ with $k+1$ arguments in the following way:

> Given an arbitrary set $\mathcal{O} \subseteq \Sigma^*$ and arbitrary words $u_1, \ldots, u_k \in \Sigma^*$, write the words to the input tape of the $\mathcal{O}$-TM $T_n^{\mathcal{O}}$ and start it. *If* the machine halts leaving on its tape a single word $v \in \Sigma^*$, *then* let $\Phi_n^{(k+1)}(\mathcal{O}, u_1, \ldots, u_k) \stackrel{\text{def}}{=} v$; otherwise, let $\Phi_n^{(k+1)}(\mathcal{O}, u_1, \ldots, u_k) \stackrel{\text{def}}{=} \uparrow$.

In the above definition, $\mathcal{O}$ is variable and can be instantiated to any subset of Σ^*. Identifying subsets of Σ^* with their characteristic functions we see that $\chi_{\mathcal{O}}$, the characteristic function of the set $\mathcal{O}$, can be any member of $\{0,1\}^{\Sigma^*}$, the set of all the functions from Σ^* to $\{0,1\}$. Consequently, $\Phi_n^{(k+1)}$ can be viewed as a partial function from the set $\{0,1\}^{\Sigma^*} \times (\Sigma^*)^k$ into the set Σ^*. Informally, $\Phi_n^{(k+1)}$ maps a function and k words into a word. Now, functions that map other functions we usually call *functionals*. So, $\Phi_n^{(k+1)}$ is a functional. Since it is associated with a particular o-TM T_n^*, we will call $\Phi_n^{(k+1)}$ the $k+1$-ary *proper functional* of the o-TM T_n^*.

If we fix $\mathcal{O}$ to a particular subset of Σ^*, the argument $\mathcal{O}$ in $\Phi_n^{(k+1)}(\mathcal{O}, x_1, \ldots, x_k)$ turns into a parameter, so the functional $\Phi_n^{(k+1)}(\mathcal{O}, x_1, \ldots, x_k)$ becomes dependent on k arguments $x_1, \ldots, x_k$ only, and hence becomes a function from $(\Sigma^*)^k$ to Σ^*. We will denote this function(al) by $\Phi_n^{\mathcal{O},(k)}(x_1, \ldots, x_k)$, for short $\Phi_n^{\mathcal{O},(k)}$, and call it the *proper functional* of the $\mathcal{O}$-TM $T_n^{\mathcal{O}}$.

In short, each o-TM is associated with a proper functional $\Phi^{(k+1)}$, and each $\mathcal{O}$-TM is associated with a proper functional $\Phi^{\mathcal{O},(k)}$, for arbitrary natural k.

Remarks. (1) It is a common convention to use capital Greek letters in order to distinguish proper functionals of o-TMs from proper functions φ of ordinary TMs. The distinction is needed because the Φs are computed by $\tilde{\delta}$s, while the φs are computed by ordinary δs. (2) When the number k is understood, we will omit the corresponding superscripts in $\Phi_n^{(k+1)}$ and $\Phi_n^{\mathcal{O},(k)}$; for example, we will simply say "Φ_n is a proper functional of T_n^*," or "$\Phi_n^{\mathcal{O}}$ is a proper functional of $T_n^{\mathcal{O}}$."

Corresponding to the enumeration of o-TMs $T_1^*, T_2^*, T_3^*, \ldots$ is, for each natural k, the enumeration of proper functionals $\Phi_1^{(k+1)}, \Phi_2^{(k+1)}, \Phi_3^{(k+1)}, \ldots$. Similarly, corresponding to the enumeration of $\mathcal{O}$-TMs $T_1^{\mathcal{O}}, T_2^{\mathcal{O}}, T_3^{\mathcal{O}}, \ldots$ is, for each natural k, the enumeration of proper functionals $\Phi_1^{\mathcal{O},(k)}, \Phi_2^{\mathcal{O},(k)}, \Phi_3^{\mathcal{O},(k)}, \ldots$.

Function Computation

Next, we generalize Definition 6.5 (p. 136) of the computable function.

Definition 10.2. Let $\mathcal{O} \subseteq \Sigma^*$, $k \geqslant 1$, and $\varphi : (\Sigma^*)^k \to \Sigma^*$ be a function. We say:[2]

φ is $\mathcal{O}$-**computable** *if* there is an $\mathcal{O}$-TM that can compute φ anywhere on $\text{dom}(\varphi)$
and $\text{dom}(\varphi) = (\Sigma^*)^k$;

φ is **partial** $\mathcal{O}$-**computable** (or $\mathcal{O}$-**p.c.**) *if* there is an $\mathcal{O}$-TM that can compute φ anywhere on $\text{dom}(\varphi)$;

φ is $\mathcal{O}$-**incomputable** *if* there is *no* $\mathcal{O}$-TM that can compute φ anywhere on $\text{dom}(\varphi)$.

So, a function $\varphi : (\Sigma^*)^k \to \Sigma^*$ is $\mathcal{O}$-p.c. if it is the k-ary proper functional of some $\mathcal{O}$-TM; that is, if there is a natural n such that $\varphi \simeq \Phi_n^{\mathcal{O},(k)}$.

Let $\mathcal{S} \subseteq (\Sigma^*)^k$ be an arbitrary set. In accordance with Definition 6.6 (p. 136), we will say that

- φ is $\mathcal{O}$-computable *on* $\mathcal{S}$ if φ can be computed by an $\mathcal{O}$-TM for any $x \in \mathcal{S}$;
- φ is $\mathcal{O}$-p.c. *on* $\mathcal{S}$ if φ can be computed by an $\mathcal{O}$-TM for any $x \in \mathcal{S}$ such that $\varphi(x)\downarrow$;
- φ is $\mathcal{O}$-incomputable *on* $\mathcal{S}$ if there is no $\mathcal{O}$-TM capable of computing φ for any $x \in \mathcal{S}$ such that $\varphi(x)\downarrow$.

---

[2] Alternatively, we can say that φ is computable (p.c., incomputable) *relative to* (or, *in*) the set $\mathcal{O}$.

Set Recognition

Finally, we can now generalize Definition 6.10 (p. 142) of decidable set.

Definition 10.3. Let $\mathcal{O} \subseteq \Sigma^*$ be an oracle set. For an arbitrary set $\mathcal{S} \subseteq \Sigma^*$ we say:[3]

$\mathcal{S}$ is $\mathcal{O}$-**decidable** (or $\mathcal{O}$-**computable**) in Σ^* if $\chi_\mathcal{S}$ is $\mathcal{O}$-computable on Σ^*;

$\mathcal{S}$ is $\mathcal{O}$-**semi-decidable** (or $\mathcal{O}$-**c.e.**) in Σ^* if $\chi_\mathcal{S}$ is $\mathcal{O}$-computable on $\mathcal{S}$;

$\mathcal{S}$ is $\mathcal{O}$-**undecidable** (or $\mathcal{O}$-**incomputable**) in Σ^* if $\chi_\mathcal{S}$ is $\mathcal{O}$-incomputable on Σ^*.

(Remember that $\chi_\mathcal{S} : \Sigma^* \to \{0, 1\}$ is total.)

Index Sets

We have seen that each natural number is the index of exactly one o-TM (p. 241). What about the other way round? Is each o-TM represented by exactly one index? The answer is no; an o-TM has countably infinitely many indexes. We prove this in the same fashion as we proved the *Padding Lemma* (see Sect. 7.2): Other indexes of a given o-TM are constructed by padding its code with codes of redundant instructions, and permuting the codes of instructions. So, if e is the index of T^* and x is constructed from e in the described way, then $\Phi_x^{(k+1)} \simeq \Phi_e^{(k+1)}$. This, combined with Definition 10.2, gives us the following generalization of the *Padding Lemma*.

Lemma 10.1. (Generalized Padding Lemma) *An $\mathcal{O}$-p.c. function has countably infinitely many indexes. Given one of them, countably infinitely many others can be generated.*

Similarly to the ordinary case, the *index set of an $\mathcal{O}$-p.c. function* φ contains all the indexes of all $\mathcal{O}$-TMs that compute φ.

Definition 10.4. (Index Set of $\mathcal{O}$-p.c. Function) The **index set** of an $\mathcal{O}$-p.c. function φ is the set $\mathrm{ind}^{\mathcal{O}}(\varphi) \overset{\text{def}}{=} \{x \in \mathbb{N} \mid \Phi_x^{\mathcal{O}} \simeq \varphi\}$.

Let $\mathcal{S}$ be an arbitrary $\mathcal{O}$-c.e. set, and $\chi_\mathcal{S}$ its characteristic function. The index set $\mathrm{ind}^{\mathcal{O}}(\chi_\mathcal{S})$ we will also denote by $\mathrm{ind}^{\mathcal{O}}(\mathcal{S})$ and call the *index set of the $\mathcal{O}$-c.e. set $\mathcal{S}$.*

---

[3] We can also say that $\mathcal{S}$ is decidable (semi-decidable, undecidable) *relative to* the set $\mathcal{O}$.

10.2.2 Convention: The Universe $\mathbb{N}$ and Single-Argument Functions

So far we have been using Σ^* as the universe. But we have shown in Sect. 6.3.6 that there is a bijection from Σ^* onto $\mathbb{N}$, which allows us to arbitrarily choose whether to study the decidability of *subsets of* Σ^* or of *subsets of* $\mathbb{N}$—the findings apply to the alternative too. It is now the right time to make use of this and switch to the universe $\mathbb{N}$. The reasons for this are that, first, $\mathbb{N}$ is often used in the study of relative computability, and second, the presentation will be simpler.

NB *From now on $\mathbb{N}$ will be the universe.*

Some definitions will tacitly (and trivially) adapt to the universe $\mathbb{N}$. For instance:

- an oracle set $\mathcal{O}$ is an arbitrary subset of $\mathbb{N}$ (Definition 10.1);
- a function $\varphi : \mathbb{N}^k \to \mathbb{N}$ is $\mathcal{O}$-computable if there is an $\mathcal{O}$-TM that can compute φ anywhere on $\mathrm{dom}(\varphi) = \mathbb{N}$ (Definition 10.2);
- a set $\mathcal{S} \subseteq \mathbb{N}$ is $\mathcal{O}$-decidable on $\mathbb{N}$ if $\chi_{\mathcal{S}}$ is $\mathcal{O}$-computable on $\mathbb{N}$ (Definition 10.3).

The next two adapted definitions come into play when the impact of different oracle sets on oracular computability is studied:

- a proper functional of the o-TM T_n^* is $\Phi_n^{(k+1)} : \{0,1\}^{\mathbb{N}} \times \mathbb{N}^k \to \mathbb{N}$ (Sect. 10.2.1).
- a proper functional of the $\mathcal{O}$-TM $T_n^{\mathcal{O}}$ is $\Phi_n^{\mathcal{O},(k)} : \mathbb{N}^k \to \mathbb{N}$ (Sect. 10.2.1).

Next, w.l.o.g., we will simplify our discussion by focusing on the case $k = 1$.

NB *From now on we will focus on single-argument functions (the case $k = 1$).*

10.3 Other Ways to Make External Help Available

External help can be made available to other models of computation too.

We can introduce external help to μ-recursive functions. The idea is simple: Given a set $\mathcal{O} \subseteq \mathbb{N}$, add the characteristic function $\chi_{\mathcal{O}}$ to the set $\{\zeta, \sigma, \pi_i^k\}$ of initial functions (see Sect. 5.2.1). Any function that can be constructed from $\{\zeta, \sigma, \pi_i^k, \chi_{\mathcal{O}}\}$ by finitely many applications of composition, primitive recursion, and the μ-operation, uses external help *if* $\chi_{\mathcal{O}}$ appears in the function's construction. Kleene treated (general) recursive functions similarly, where certain auxiliary functions (possibly $\chi_{\mathcal{O}}$) would be made available to systems of equations $\mathcal{E}(f)$.

Post introduced external help into his canonical systems by hypothetically adding primitive assertions expressing (non)membership in $\mathcal{O}$.

Davis proved the equivalence of Kleene's and Post's approaches. The equivalence of each of the two and Turing's approach was proved by Kleene and Post, respectively. Consequently, Turing's, Kleene's, and Post's definitions of functions computable with external help are equivalent, in the sense that if a function is computable according to one definition it is also computable according to the others.

10.4 Relative Computability Thesis

Based on these equivalences and following the example of the *Computability Thesis*, a thesis was proposed stating that the intuitive notion of the "algorithm with external help" is formalized by the concept of the oracle Turing machine.

Relative Computability Thesis.

"algorithm with external help" $\longleftrightarrow$ *oracle Turing program (or equivalent model)*

Also, the intuitive notion of being "computable with external help" is formalized by the notion of being $\mathcal{O}$-computable.

10.5 Practical Consequences: o-TM with a Database or Network

Since oracles are supernatural entities, the following question arises: Can we replace a given oracle for a set $\mathcal{O}$ with a more realistic concept that will simulate the oracle? Clearly, if $\mathcal{O}$ is a decidable set, the replacement is trivial—we only have to replace the oracle with the ordinary TM that is the decider of the set $\mathcal{O}$ (or, alternatively, the computer of the function $\chi_{\mathcal{O}}$).

A different situation occurs when $\mathcal{O}$ is undecidable (and, hence, $\chi_{\mathcal{O}}$ incomputable). Such an $\mathcal{O}$ cannot be finite, because finite sets are decidable. But in practice, we do not really need the *whole* set $\mathcal{O}$, i.e., the values $\chi_{\mathcal{O}}(x)$ for *every* $x \in \mathbb{N}$. This is because the o-TM may only issue the question "$x \in ? \mathcal{O}$" for finitely many different numbers $x \in \mathbb{N}$. (Otherwise, the machine certainly would not halt.) Consequently, there is an $m \in \mathbb{N}$ that is the largest of all xs for which the question "$x \in ? \mathcal{O}$" is issued during the computation. Of course, m depends on the input word that was submitted to the o-TM.

Now we can make use of the ideas discussed in Sect. 8.4. We could compute the values $\chi_{\mathcal{O}}(i)$, $i = 0, 1, \ldots, m$ in advance, where each $\chi_{\mathcal{O}}(i)$ would be computed by an algorithm A_i specially designed to answer only the particular question "$i \in ? \mathcal{O}$". The computed values would then be stored in an external *database* and accessed during the oracular computation. Alternatively, we might compute the values $\chi_{\mathcal{O}}(i)$ on the fly during the oracular computation: Upon issuing a question "$i \in ? \mathcal{O}$", an external *network of computers* would be engaged in the computation of the answer to the question. Of course, the o-TM would idle until the answer arrived.

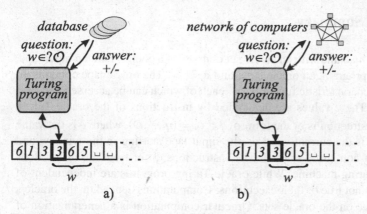

Fig. 10.3 The oracle tape is replaced by a) a database with a finite number of precomputed values $\chi_\mathcal{O}(i), i = 0, 1, \ldots, m$; b) a network of computers that compute each value $\chi_\mathcal{O}(i)$ separately upon the o-TM's request

10.6 Practical Consequences: Online and Offline Computation

In the real world of computer technology and telecommunications, often a device A is associated with some other device B. This association can be dynamic, changing between the states of connectedness and disconnectedness. We say that A is *online* when it is connected to B; that is, either B is under the direct control of A, or B is available for immediate use on A's demand. Otherwise, we say that A is *offline* (disconnected).

Let A be a computer and B any device capable of providing data. We say that A executes an *online algorithm* if the algorithm processes its input data in the order in which the data are fed to the algorithm by B. Thus, an online algorithm cannot know the entire input that will be fed to it, so it may be forced to correct its past actions when new input data arrive. In contrast, we say that A executes an *offline algorithm* if the algorithm is given the whole input data at the beginning of the computation. This allows it to first inspect the whole input and then choose the appropriate computation strategy.

We can now see the parallels between online/offline computing and oracular/ordinary computing. Ordinary Turing machines are offline; they are disconnected from any oracle, or any other device, such as an external database or computer network. Ordinary Turing programs are offline algorithms, because they are given the entire input before they are started. In contrast, oracle Turing machines are online; they are connected to an oracle or a device such as an external database or a computer network. Oracle Turing programs are online algorithms: Although a part of their input is written on the input tape at the beginning of the computation, the other part—the oracle's advice—is processed piece by piece in a serial fashion, without knowing the advice that will be given on future questions.

10.7 Chapter Summary

An oracle Turing machine with an oracle set consists of a control unit, an input tape, an oracle Turing program, an oracle tape, and a set $\mathcal{O}$. The oracle tape contains all the values of the characteristic function $\chi_\mathcal{O}$, each of which can be accessed and read in a single step. These values are demanded by instructions of the oracle Turing program. Each instruction is of the form $\widetilde{\delta}(q,z,e) = (q',z',D)$, where e is the value of $\chi_\mathcal{O}(w)$ and w is the current word on the input tape starting in the window and ending at the last non-space symbol. The instructions differ from the instructions of an ordinary Turing machine. While oracle Turing programs are independent of oracle sets, this is not true of their executions: Computations depend on the oracle's answers and hence on the oracle sets. Oracular computation is a generalization of ordinary computation. Oracle TMs and their programs can be coded and enumerated. With each oracle TM is associated its proper functional. Given an oracle set $\mathcal{O}$, a partial function can be $\mathcal{O}$-computable, $\mathcal{O}$-p.c., or $\mathcal{O}$-incomputable; and a set can be $\mathcal{O}$-decidable, $\mathcal{O}$-c.e., or $\mathcal{O}$-undecidable. The *Relative Computability Thesis* states that the oracle Turing machine formalizes the intuitive notion of the "algorithm with external help." In practice, the contents of the unrealistic oracle tape could be approximated 1) by a finite sequence of precomputed values of $\chi_\mathcal{O}$ (stored in an external database) or 2) by an on-the-fly calculation of each demanded value of $\chi_\mathcal{O}$ separately (computed by a network of computers).

Bibliographic Notes

- The idea of an *oracle* for a possibly incomputable set $\mathcal{O}$ that is attached to a Turing machine was introduced in Turing [264, §4]. The citation on p. 234 is from there.
- Post developed Turing's idea above in the influential paper [184, §11]. See also Post [187].
- Addition of *external help* to other models of computation was considered too. Kleene touched on the addition of external help to (general) recursive functions in [123] and Post to his canonical systems in [187]. See more on this in Soare [246].
- The *equivalence* of Kleene's and Post's approaches to external help was proved in Davis [52]. A proof of the equivalence of Turing's and Kleene's approaches was published in Kleene's book [124, Sects. 68,69]. Later, other models and proofs of their equivalence appeared. For example, the register machine (with external help) is described and proved to be equivalent to partial computable functions (with external help) in Enderton [67].
- The *Relative Computability Thesis* stating that Turing's approach is the formalization of the intuitive notion of "computation with external help" appeared (in terms of Turing and "effective" reducibility) in Post [184, §11].
- The basic notions and concepts of relativized computability obtained their mature form in Kleene and Post [127]. For other important results that appeared in this paper, see the Bibliographic Notes to the following chapters.
- An account of the development of oracle computing is given in Soare [244]. The reader can find out more about Post's research and life in Urquhart [265].
- In defining the modern *oracle Turing machine*, we have leaned on Soare [241, 245].
- The idea of substituting the oracle tape with a *database* or some other kind of *external interaction* was described in Soare [244, 245].

Chapter 11
Degrees of Unsolvability

Degree indicates the extent to which something happens or the amount something is felt.

Abstract In Part II, we proved that besides computable problems there are also incomputable ones. So, given a computational problem, it makes sense to talk about its *degree of unsolvability*. Of course, at this point we only know of *two* such degrees: One is shared by all computable problems, and the other is shared by all incomputable ones. (This will change, however, in the next chapter.) Nevertheless, the main aim of this chapter is to formalize the intuitive notion of degree of unsolvability. Building on the concept of the oracle Turing machine, we will first define the concept of the *Turing reduction*, the most general reduction between computational problems. We will then proceed in a natural way to the definition of *Turing degree*—the formal counterpart of the intuitive notion of degree of unsolvability.

11.1 Turing Reduction

Until now we have been using the generic symbol $\mathcal{O}$ to denote an oracle set. From now on we will be talking about particular oracle sets and denote them by the usual symbols, e.g., $\mathcal{A}, \mathcal{B}$.

What does it mean when we say, for two sets $\mathcal{A}, \mathcal{B} \subseteq \mathbb{N}$, that "$\mathcal{A}$ is $\mathcal{B}$-decidable in $\mathbb{N}$"? By Definition 10.3 (p. 244), it means that the characteristic function $\chi_{\mathcal{A}}$ is $\mathcal{B}$-computable on $\mathbb{N}$. So, by Definition 10.2 (p. 243), there is a $\mathcal{B}$-TM $T^{\mathcal{B}}$ that can compute $\chi_{\mathcal{A}}(x)$ for any $x \in \mathbb{N}$. (Recall that $\chi_{\mathcal{A}}$ is a total function by definition.) Since the oracle can answer the question $n \in ?\mathcal{B}$ for any $n \in \mathbb{N}$, it makes the set $\mathcal{B}$ *appear decidable* in $\mathbb{N}$ in the ordinary sense (see Definition 6.10 on p. 142).

We conclude:

If $\mathcal{A}$ is $\mathcal{B}$-decidable, then the decidability of $\mathcal{B}$ would imply the decidability of $\mathcal{A}$.

This relation between sets deserves a special name. Hence, the following definition.

R. Robič, *The Foundations of Computability Theory*, https://doi.org/10.1007/978-3-662-62421-0_11

251

Definition 11.1. (Turing Reduction) Let $\mathcal{A}, \mathcal{B} \subseteq \mathbb{N}$ be arbitrary sets. We say that $\mathcal{A}$ is **Turing reducible** (for short *T-reducible*) *to* $\mathcal{B}$ if $\mathcal{A}$ is $\mathcal{B}$-decidable. We denote this by

$$\mathcal{A} \leq_T \mathcal{B}. \qquad\qquad (*)$$

Thus, *if $\mathcal{A} \leq_T \mathcal{B}$ then* if $\mathcal{B}$ were decidable also $\mathcal{A}$ would be decidable. The relation $\leq_T$ is called the **Turing reduction** (for short *T-reduction*).

If $\mathcal{B}$ is decidable (in the ordinary sense), then the oracle for $\mathcal{B}$ is no mystery. Such an oracle can be replaced by an ordinary decider $D_\mathcal{B}$ of the set $\mathcal{B}$ (see Sect. 6.3.3). Consequently, the $\mathcal{B}$-TM $T^\mathcal{B}$ is equivalent to an ordinary TM, in the sense that what one can compute the other can also compute. The construction of this TM is straightforward: The TM must simulate the program $\widetilde{\delta}$ of $T^\mathcal{B}$, with the exception that whenever $T^\mathcal{B}$ asks $x \in ?\mathcal{B}$, the TM must call $D_\mathcal{B}$, submit x to it, and wait for its decision.

The situation is quite different when $\mathcal{B}$ is undecidable (in the ordinary sense). In this case, the oracle for $\mathcal{B}$ is *more powerful* than any ordinary TM (because no ordinary TM can decide $\mathcal{B}$). This makes $T^\mathcal{B}$ more powerful than any ordinary TM (because $T^\mathcal{B}$ can decide $\mathcal{B}$, simply by asking whether or not the input is in $\mathcal{B}$). In particular, if we replace a decidable oracle set with an undecidable c.e. (semidecidable) set, this may have "big" implications (see Problem 11.3).

11.1.1 Turing Reduction of a Computational Problem

Why have we named $\leq_T$ a *reduction*? What is reduced here, and what does the reducing? Recall that each subset of $\mathbb{N}$ represents a decision problem (see Sect. 8.1.2). Thus, $\mathcal{A}$ is associated with the decision problem $\mathcal{P} \equiv$ "$x \in ?\mathcal{A}$", and $\mathcal{B}$ with the decision problem $\mathcal{Q} \equiv$ "$x \in ?\mathcal{B}$". The relation $(*)$ can now be interpreted as follows:

If $\mathcal{Q} \equiv$ "$x \in ?\mathcal{B}$" were decidable, then also $\mathcal{P} \equiv$ "$x \in ?\mathcal{A}$" would be decidable.

Because of this we also use the sign $\leq_T$ to relate the associated decision problems:

$$\mathcal{P} \leq_T \mathcal{Q}. \qquad\qquad (**)$$

As regards the oracle for $\mathcal{B}$, we know *what* the oracle does: Since it can answer any question $w \in ?\mathcal{B}$, it *solves the problem* $\mathcal{Q}$. We can view it as a procedure—call it B—for solving the problem $\mathcal{Q}$. But when $\mathcal{B}$ is undecidable, we cannot know *how* the oracle finds the answers. In this case, B is a supernatural "algorithm", whose operation cannot be described and understood by a human; a black box, as it were.

Let us now look at $T^\mathcal{B}$, the machine that computes the values of $\chi_\mathcal{A}$, and let $\widetilde{\delta}$ be its o-TP. For $\widetilde{\delta}$ we know *what* it does and *how* it does it, *when* it asks the oracle and *how* it uses its answers. Hence, we can view $\widetilde{\delta}$ as an ordinary algorithm A for solving $\mathcal{P}$ that can call the mysterious "algorithm" B. There is no limit on the number of calls as long as this number is finite (otherwise, A would not halt).

Considering all this, we can interpret the relation (∗∗) as follows:

> *If there were an algorithm B for solving Q,*
> *then there would be an algorithm A for solving P,*
> *where A could make finitely many calls to B.*

Since $\mathcal{P}$ would in principle be solved *if* $\mathcal{Q}$ were solved, we can focus on the problem $\mathcal{Q}$ and on the design of the algorithm B. We say that we have *reduced $\mathcal{P}$ to $\mathcal{Q}$*. If and when $\mathcal{Q}$ is solved (i.e., B designed), also the problem $\mathcal{P}$ is, in principle, solved (by A). Note that this means that the problem $\mathcal{P}$ is *not more difficult to solve* than the problem $\mathcal{Q}$. This is an alternative interpretation of the relation (∗∗).

The interpretation described above is not bound to decision problems only. Take, for instance, $\mathcal{P} \equiv$ EXISTENCE OF SHORTER EQUIVALENT PROGRAMS (Sect. 8.3.3), $\mathcal{Q} \equiv$ SHORTEST EQUIVALENT PROGRAM (Sect. 9.5), and assume there is an algorithm B for $\mathcal{Q}$. Then, for any program p, the algorithm A (deciding whether or not there is a shorter equivalent program) calls B on p to obtain the shortest program q equivalent to p, compares their lengths, and answers YES or NO. So $\mathcal{P} \leq_T \mathcal{Q}$. Things change if we swap the two problems. Now the existence of B (deciding whether or not there is a shorter equivalent program) does not help us to design A for finding the shortest equivalent program q, because finding q even among the finitely many shorter programs is incomputable—and the oracle used by B cannot offer any help.

NB *From now on we will limit our discussion to set-membership problems. In doing so, we will develop the theory on sets of natural numbers. (Decision problems will be mentioned only to give another view of notions, concepts, and theorems.)*

11.1.2 Some Basic Properties of the Turing Reduction

We now list some of the properties of the T-reduction. First, we check that there indeed exist two sets $\mathcal{A}, \mathcal{B} \subseteq \mathbb{N}$ that are related by $\leq_T$. To see this, consider the situation where $\mathcal{A}$ is an arbitrary decidable set. From Definition 11.1 it immediately follows that $\mathcal{A} \leq_T \mathcal{B}$ for an *arbitrary* set $\mathcal{B}$. Hence the following theorem.

Theorem 11.1. *Let $\mathcal{A}$ be a decidable set. Then $\mathcal{A} \leq_T \mathcal{B}$ for an arbitrary set $\mathcal{B}$.*

But a set $\mathcal{A}$ need not be decidable to be T-reducible to some other set $\mathcal{B}$. This will follow from the next simple theorem.

Theorem 11.2. *For every set S it holds that $\overline{S} \leq_T S$.*

Proof. If S were decidable, then (by Theorem 7.2, p. 156) also $\overline{S} = \mathbb{N} - S$ would be decidable. □

Let S be undecidable in the ordinary sense. Then $\overline{S}$ is undecidable too (Sect. 8.2.2). Now Theorem 11.2 tells us that $\overline{S} \leq_T S$, i.e., making S appear decidable (by using the oracle for S), also makes $\overline{S}$ appear decidable. Thus, a set $\mathcal{A}$ *need not* be decidable to be T-reducible to some other set $\mathcal{B}$.

Is the T-reduction related to the m-reduction, which we defined in Sect. 9.2.2? If so, is one of them more "powerful" than the other? The answer is yes. The next two theorems tell us that the T-reduction is a nontrivial generalization of the m-reduction.

Theorem 11.3. *If two sets are related by $\leq_m$, then they are also related by $\leq_T$.*

Proof. Let $\mathcal{A} \leq_m \mathcal{B}$. Then there is a computable function r such that $x \in \mathcal{A} \iff r(x) \in \mathcal{B}$ (see Sects. 9.2.1 and 9.2.2). Consequently, the characteristic functions $\chi_\mathcal{A}$ and $\chi_\mathcal{B}$ are related by the equation $\chi_\mathcal{A} = \chi_\mathcal{B} \circ r$, where $\circ$ denotes function composition. We now see: If the function $\chi_\mathcal{B}$ were computable, then the composition $\chi_\mathcal{B} \circ r$ would also be computable (because r is computable, and a composition of computable functions is a computable function)—hence, $\mathcal{A} \leq_T \mathcal{B}$. $\square$

However, the converse is not true, as we explain in the next theorem.

Theorem 11.4. *If two sets are related by $\leq_T$, they may not be related by $\leq_m$.*

Proof. Theorem 11.2 states that $\overline{\mathcal{S}} \leq_T \mathcal{S}$ for any set $\mathcal{S}$. Is the same true of the relation $\leq_m$? The answer is no; there exist sets $\mathcal{S}$ such that $\overline{\mathcal{S}} \leq_m \mathcal{S}$ does *not* hold. To see this, let $\mathcal{S}$ be an arbitrary undecidable c.e. set. (Such is the set $\mathcal{K}$; see Sect. 8.2.1.) The set $\overline{\mathcal{S}}$ is not c.e. (otherwise $\mathcal{S}$ would be decidable by Theorem 7.3, p. 156). Now, if $\overline{\mathcal{S}} \leq_m \mathcal{S}$ held, then $\overline{\mathcal{S}}$ would be c.e. (by Theorem 9.1, p. 213), which would be a contradiction.

In particular, $\overline{\mathcal{K}} \leq_T \mathcal{K}$, but $\overline{\mathcal{K}} \not\leq_m \mathcal{K}$. The same is true of $\mathcal{K}_0$: $\overline{\mathcal{K}}_0 \leq_T \mathcal{K}_0$, but $\overline{\mathcal{K}}_0 \not\leq_m \mathcal{K}_0$. $\square$

We can use the T-reduction for proving the undecidability of sets as we use the m-reduction. But the T-reduction must satisfy fewer conditions than the m-reduction (see Definition 9.1, p. 211). This makes T-reductions easier to construct than m-reductions. Indeed, Theorem 11.4 indicates that there are situations where T-reduction is possible while m-reduction is not. So, let us develop a method of proving the undecidability of sets that will use T-reductions. We will closely follow the development of the method for m-reductions (see Sect. 9.2.3). First, from Definition 11.1 we obtain the following theorem.

Theorem 11.5. *For arbitrary sets $\mathcal{A}$ and $\mathcal{B}$ it holds:*

$$\mathcal{A} \leq_T \mathcal{B} \wedge \mathcal{B} \text{ is decidable} \implies \mathcal{A} \text{ is decidable}$$

The contraposition is: $\mathcal{A}$ *is undecidable* $\implies \mathcal{A} \not\leq_T \mathcal{B} \vee \mathcal{B}$ *is undecidable*. Assuming that $\mathcal{A} \leq_T \mathcal{B}$, and using this in the contraposition, we obtain the next corollary.

Corollary 11.1. *For arbitrary sets $\mathcal{A}$ and $\mathcal{B}$ it holds:*

$$\mathcal{A} \text{ is undecidable} \wedge \mathcal{A} \leq_T \mathcal{B} \implies \mathcal{B} \text{ is undecidable}$$

This reveals the following method for proving the undecidability of sets.

Method. The undecidability of a set B can be proved as follows:

1. Suppose: B is decidable;   // Supposition.
2. Select: an undecidable set A;
3. Prove: $A \leq_T B$;
4. Conclude: A is decidable;   // 1 and 3 and Theorem 11.5.
5. Contradiction between 2 and 4!
6. Conclude: B is undecidable.

Remark.   The method can easily be adapted to prove that a decision problem Q is undecidable. First, *suppose* that there is a decider B for Q. Then choose a known undecidable decision problem P and try to construct a decider A for P, where A can make calls to B. If we succeed, P is decidable. Since this is a contradiction, we reject the supposition. So, Q is undecidable.

Turing reduction is a relation that has another two important properties.

Theorem 11.6.   *Turing reduction $\leq_T$ is a reflexive and transitive relation.*

Proof. (Reflexivity) This is trivial. Let S be an arbitrary set. If S were decidable, then, of course, the same S would be decidable. Hence, $S \leq_T S$ by Definition 11.1. (Transitivity) Let A, B, C be arbitrary sets and suppose that $A \leq_T B \wedge B \leq_T C$. So, if C were decidable, B would also be decidable (because $B \leq_T C$), but the latter would then imply the decidability of A (because $A \leq_T B$). Hence, the decidability of C would imply the decidability of A, i.e., $A \leq_T C$.   □

Generally, a reflexive and transitive binary relation is called a *preorder*, and a set equipped with such a relation is said to be preordered by this relation.

Given a preordered set, one of the first things to do is to check whether its preorder qualifies for any of the more interesting orders (see Appendix A). Because these orders have additional properties, they reveal much more about their domains. Two such orders are the equivalence relation and the partial order.

The above theorem tells us that $\leq_T$ is a preorder on $2^{\mathbb{N}}$. Is it, perhaps, even an equivalence relation? To check this we must see whether $\leq_T$ is a symmetric relation, i.e., whether $A \leq_T B$ implies $B \leq_T A$, for arbitrary A, B. But we are already able to point at two sets for which the implication *does not* hold: These are the empty set $\emptyset$ and the diagonal set K (see Definition 8.6, p. 181). Namely, $\emptyset \leq_T K$ (due to Theorem 11.1) while $K \not\leq_T \emptyset$ (because $\emptyset$ is decidable and K undecidable). Thus we conclude:

Turing reduction is not symmetric and, consequently, not an equivalence relation.

Although this result is negative, we will use it in the next subsection.

11.2 Turing Degrees

We are now ready to formalize the intuitive notion of *degree of unsolvability*. Its formal counterpart will be called *Turing degree*. The path to the definition of Turing degree will be a short one: First, we will use the relation $\leq_T$ to define a new relation $\equiv_T$; then we will prove that $\equiv_T$ is an equivalence relation; and finally, we will define Turing degrees to be the equivalence classes of $\equiv_T$.

We have proved in the previous subsection that $\leq_T$ is not symmetric, because there exist sets $\mathcal{A}$ and $\mathcal{B}$ such that $\mathcal{A} \leq_T \mathcal{B}$ and $\mathcal{B} \not\leq_T \mathcal{A}$. However, neither is $\leq_T$ asymmetric, because there do exist sets $\mathcal{A}$ and $\mathcal{B}$ for which $\mathcal{A} \leq_T \mathcal{B}$ and $\mathcal{B} \leq_T \mathcal{A}$. For example, this is the case when both $\mathcal{A}$ and $\mathcal{B}$ are decidable. (This follows from Theorem 11.1.) Thus, for some pairs of sets the relation $\leq_T$ is symmetric, while for others it is not. In this situation, we can define a new binary relation that will tell, for any two sets, whether or not $\leq_T$ is symmetric for them. Here is the definition of the new relation.

Definition 11.2. (Turing Equivalence) Let $\mathcal{A}, \mathcal{B} \subseteq \mathbb{N}$ be arbitrary sets. We say that $\mathcal{A}$ is **Turing-equivalent** (for short *T-equivalent*) to $\mathcal{B}$, if $\mathcal{A} \leq_T \mathcal{B} \wedge \mathcal{B} \leq_T \mathcal{A}$. We denote this by

$$\mathcal{A} \equiv_T \mathcal{B}$$

and read: *If one of $\mathcal{A}, \mathcal{B}$ were decidable, also the other would be decidable.* The relation $\equiv_T$ is called the **Turing equivalence** (for short, *T-equivalence*).

From the above definition it follows that $\mathcal{A} \equiv_T \mathcal{B}$ for any *decidable* sets $\mathcal{A}, \mathcal{B}$. What about undecidable sets? Can two such sets be *T*-equivalent? The answer is yes; it will follow from the next theorem.

Theorem 11.7. *For every set S it holds that $S \equiv_T \overline{S}$.*

Proof. Let S be an arbitrary set. Then $\overline{S} \leq_T S$ (by Theorem 11.2). Now focus on the set $\overline{S}$. The same theorem tells us that $\overline{\overline{S}} \leq_T \overline{S}$, i.e., $S \leq_T \overline{S}$. So $S \equiv_T \overline{S}$. Note that S can be undecidable. *Alternatively:* If one of the characteristic functions $\chi_S, \chi_{\overline{S}}$ were computable, the other would also be computable (because $\chi_{\overline{S}} = 1 - \chi_S$.) □

Calling $\equiv_T$ an "equivalence relation" is justified. The reader should have no trouble in proving that the relation $\equiv_T$ is reflexive, transitive, and symmetric. Therefore, $\equiv_T$ is an equivalence relation on $2^{\mathbb{N}}$, the power set of the set $\mathbb{N}$.

Now, being an equivalence relation, the relation $\equiv_T$ partitions the set $2^{\mathbb{N}}$ into $\equiv_T$-*equivalence classes*. Each $\equiv_T$-equivalence class contains as its elements all the subsets of $\mathbb{N}$ that are *T*-equivalent one to another. It will soon turn out that $\equiv_T$-equivalence classes are one of the central notions of relativized computability. The next definition introduces their naming.

Definition 11.3. (Turing Degree) The **Turing degree** (for short *T-degree*) of a set $\mathcal{S}$, denoted by $\deg(\mathcal{S})$, is the equivalence class $\{\mathcal{X} \in 2^{\mathbb{N}} \mid \mathcal{X} \equiv_T \mathcal{S}\}$.

Given a set $\mathcal{S}$, the T-degree $\deg(\mathcal{S})$ by definition contains as its elements all the sets that are T-equivalent to $\mathcal{S}$. Since $\equiv_T$ is symmetric and transitive, these sets are also T-equivalent to one another. Thus, if one of them were decidable, all would be decidable. In other words, the question $x \in ?\mathcal{X}$ (i.e., the membership problem) is equally (un)decidable for each of the sets $\mathcal{X} \in \deg(\mathcal{S})$. Informally, this means that the information about what is and what is not in one of them is equal to the corresponding information of any other set. In short, the sets $\mathcal{X} \in \deg(\mathcal{S})$ bear the same *information* about their contents.

Remark. We now put $\deg(\mathcal{S})$ in the light of decision problems. Let $\mathcal{X}, \mathcal{Y} \in \deg(\mathcal{S})$ be any sets. Associated with $\mathcal{X}$ and $\mathcal{Y}$ are decision problems $\mathcal{P} \equiv$ "$x \in ?\mathcal{X}$" and $\mathcal{Q} \equiv$ "$y \in ?\mathcal{Y}$", respectively. Since $\mathcal{X} \equiv_T \mathcal{Y}$, we have $\mathcal{P} \leq_T \mathcal{Q}$ and $\mathcal{Q} \leq_T \mathcal{P}$. This means that if either of the problems $\mathcal{P}, \mathcal{Q}$ were decidable, the other one would also be decidable. Thus, $\mathcal{P}$ and $\mathcal{Q}$ are equally (un)decidable. Now we see that the class of all decision problems whose languages are in the T-degree $\deg(\mathcal{S})$ represents a certain *degree of unsolvability*, which we are faced with when we try to solve any of these problems. Problems associated with $\deg(\mathcal{S})$ are equally (un)solvable, i.e., equally difficult.

Based on this we declare the following formalization of the intuitive notion of the degree of unsolvability of decision problems.

Formalization. *The intuitive notion of the degree of unsolvability is formalized by*

$$\text{"degree of unsolvability"} \longleftrightarrow \textit{Turing degree}$$

Remark. Since the concept of "degree of unsolvability" is formalized by the T-degree, we will no longer distinguish between the two. We will no longer use quotation marks to distinguish between its intuitive and formal meaning.

NB *This formalization opens the door to a mathematical treatment of our intuitive, vague awareness that solvable and unsolvable problems differ in something that we intuitively called the degree of unsolvability.*

Intuitively, we expect that the degree of unsolvability of decidable sets differs from the degree of unsolvability of undecidable sets. So let us prove that indeed there are two different corresponding T-degrees.

First, let $\mathcal{S}$ be an arbitrary *decidable* set. Then $\deg(\mathcal{S})$ contains exactly all decidable sets. As the empty set $\emptyset$ is decidable, we have $\emptyset \in \deg(\mathcal{S})$, so $\deg(\mathcal{S}) = \deg(\emptyset)$. This is why we usually denote the class of all decidable sets by $\deg(\emptyset)$.

Second, we have seen (p. 255) that $\emptyset \leq_T \mathcal{K} \wedge \mathcal{K} \not\leq_T \emptyset$. Therefore, $\emptyset \not\equiv_T \mathcal{K}$ and hence $\deg(\emptyset) \neq \deg(\mathcal{K})$.

But $\deg(\emptyset)$ and $\deg(\mathcal{K})$ are $\equiv_T$-equivalence classes, so they share no elements. We can now conclude as we expected to: $\deg(\emptyset)$ and $\deg(\mathcal{K})$ represent two different degrees of unsolvability. We have proved the following theorem.

Theorem 11.8. *There exist at least two T-degrees:*

$$deg(\emptyset) = \{\mathcal{X} \mid \mathcal{X} \equiv_T \emptyset\},$$
$$deg(\mathcal{K}) = \{\mathcal{X} \mid \mathcal{X} \equiv_T \mathcal{K}\}.$$

The Relation $<$

It is natural to say that a decidable decision problem is "less difficult to solve" than an undecidable one. We will now formalize the intuitively understood relation of "being less difficult to solve." To do this, we will introduce a new binary relation, denoted by $<$, which will be capable of expressing formally that $\deg(\emptyset)$ represents a degree of unsolvability that is "lower" than the degree of unsolvability represented by $\deg(\mathcal{K})$. The definition of $<$ will be straightforward. Let us denote the irreflexive reduction of the relation $\leq_T$ as usual by $<_T$, i.e., $\mathcal{A} <_T \mathcal{B} \overset{\text{def}}{\Longleftrightarrow} \mathcal{A} \leq_T \mathcal{B} \wedge \mathcal{A} \not\equiv_T \mathcal{B}$. (Thus $\mathcal{A} <_T \mathcal{B} \Longleftrightarrow \mathcal{A} \leq_T \mathcal{B} \wedge \mathcal{B} \not\leq_T \mathcal{A}$.) Then the sought-for relation $<$ is *induced* by the relation $<_T$, as the following definition describes.

Definition 11.4. (Relation $<$) Let $\deg(\mathcal{A})$ and $\deg(\mathcal{B})$ be arbitrary T-degrees. Then $\deg(\mathcal{A})$ is **lower** than $\deg(\mathcal{B})$, denoted by $\deg(\mathcal{A}) < \deg(\mathcal{B})$, if $\mathcal{A} <_T \mathcal{B}$.

When $\deg(\mathcal{A}) < \deg(\mathcal{B})$, we also say that $\deg(\mathcal{B})$ is *higher* than $\deg(\mathcal{A})$.

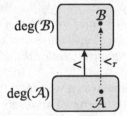

Fig. 11.1 Turing degree $\deg(\mathcal{A})$ is lower than $\deg(\mathcal{B})$, i.e., $\deg(\mathcal{A}) < \deg(\mathcal{B})$. The relation $<$ between the degrees $\deg(\mathcal{A})$ and $\deg(\mathcal{B})$ is induced by the relation $<_T$ between the representatives $\mathcal{A}$ and $\mathcal{B}$

Intuitively, a T-degree should not be lower than itself, because this would not agree with the intended meaning of the relation $<$. So we ask: Is the relation $<$ irreflexive? The answer is yes, as is assured by the next, simple theorem.

Theorem 11.9. *The relation $<$ is irreflexive.*

Proof. Let $\mathcal{S}$ be an arbitrary set. Suppose that $\deg(\mathcal{S}) < \deg(\mathcal{S})$. Then, $\mathcal{S} <_T \mathcal{S}$ (by Definition 11.4), i.e., $\mathcal{S} \leq_T \mathcal{S} \wedge \mathcal{S} \not\equiv_T \mathcal{S}$ (by definition of $<_T$). Hence $\mathcal{S} \not\equiv_T \mathcal{S}$. But this is a contradiction, because $\equiv_T$ is reflexive. □

Since $\emptyset <_T \mathcal{K}$, we can now write the statement

The degree of unsolvability of decidable decision problems is lower than the degree of unsolvability of undecidable decision problems T-equivalent to the Halting Problem

in a formal way as

$$\deg(\emptyset) < \deg(\mathcal{K}).$$

But, we have intuitively anticipated this! What is so special about the above formal statement? Isn't this much ado about nothing? The answer is that the formalization that we have just made will enable us to discover in the next chapter a surprising fact that there are many other degrees of unsolvability. We will show this by the construction of T-degrees that differ from $\deg(\emptyset)$ and $\deg(\mathcal{K})$.

11.3 Chapter Summary

A set $\mathcal{A}$ is Turing reducible to a set $\mathcal{B}$, written $\mathcal{A} \leq_T \mathcal{B}$, if $\mathcal{A}$ is $\mathcal{B}$-decidable. Thus, *if $\mathcal{A} \leq_T \mathcal{B}$ then* if $\mathcal{B}$ were decidable also $\mathcal{A}$ would be decidable.

The T-reduction $\leq_T$ is a generalization of $\leq_m$, the m-reduction; if two sets are related by $\leq_T$, they may not be related by $\leq_m$.

A decidable set is T-reducible to any set. The complement of a set is T-reducible to the set. If a set $\mathcal{A}$ is T-reducible to a decidable set, then $\mathcal{A}$ is decidable. If an undecidable set is T-reducible to a set $\mathcal{B}$, then $\mathcal{B}$ is undecidable.

Correspondingly, a decision problem $\mathcal{P}$ can be T-reducible to a decision problem $\mathcal{Q}$, written $\mathcal{P} \leq_T \mathcal{Q}$. *If $\mathcal{P} \leq_T \mathcal{Q}$ then* if $\mathcal{Q}$ were decidable also $\mathcal{P}$ would be decidable.

If a decision problem $\mathcal{P}$ is T-reducible to a decidable decision problem, then $\mathcal{P}$ is decidable. If an undecidable decision problem is T-reducible to a decision problem $\mathcal{Q}$, then $\mathcal{Q}$ is undecidable.

Generally, a computational problem $\mathcal{P}$ can be T-reducible to a computational problem $\mathcal{Q}$, written $\mathcal{P} \leq_T \mathcal{Q}$. *If $\mathcal{P} \leq_T \mathcal{Q}$ then* the computability of $\mathcal{Q}$ would imply the computability of $\mathcal{P}$.

If a computational problem $\mathcal{P}$ is T-reducible to a computable computational problem, then $\mathcal{P}$ is computable. If an incomputable computational problem is T-reducible to a computational problem $\mathcal{Q}$, then $\mathcal{Q}$ is incomputable.

The T-reduction is reflexive and transitive, but not symmetric.

Two sets are said to be Turing-equivalent if each of them is T-reducible to the other. T-equivalence is an equivalence relation.

A Turing degree of a set $\mathcal{A}$, denoted by $\deg(\mathcal{A})$, is the class of all sets that are T-equivalent to $\mathcal{A}$. T-degree is a formalization of the informal notion of "degree of unsolvability".

There are (at least) two T-degrees: $\deg(\emptyset)$ and $\deg(\mathcal{K})$. The first corresponds to the degree of unsolvability of all decidable sets (i.e., decidable decision problems). The second corresponds to the degree of unsolvability of all undecidable sets that are T-equivalent to the set $\mathcal{K}$, that is, the degree of unsolvability of all undecidable decision problems that are T-equivalent to the *Halting Problem* $\mathcal{D}_H$.

A T-degree $\deg(\mathcal{A})$ is said to be lower than a T-degree $\deg(\mathcal{B})$ if $\mathcal{A}$ is T-reducible to $\mathcal{B}$, but not vice versa.

Problems

11.1. The relations $\leq_T$ and $\subseteq$ are, generally, independent of each other. That is to say, we can have $\mathcal{A} \leq_T \mathcal{B}$ simultaneously with any of the relations $\mathcal{A} \subseteq \mathcal{B}$ and $\mathcal{A} \supseteq \mathcal{B}$, or neither of the two. Can you give examples of such sets $\mathcal{A}$ and $\mathcal{B}$?

11.2. Prove:

(a) $\mathcal{A}$ is a c.e. set $\Longrightarrow \mathcal{A} \leq_T \mathcal{K}$.

(b) $\mathcal{A}, \mathcal{B}$ are disjoint c.e. sets $\Longrightarrow \mathcal{A} \leq_T \mathcal{A} \cup \mathcal{B}$ and $\mathcal{B} \leq_T \mathcal{A} \cup \mathcal{B}$.

11.3. Prove: There exist sets $\mathcal{A}$ and $\mathcal{B}$ such that $\mathcal{A}$ is *not* c.e., $\mathcal{B}$ is (undecidable) c.e., and $\mathcal{A} \leq_T \mathcal{B}$.

[*Hint.* Consider the *Halting Problem* and the *non-Halting Problem*.]

Remark. Therefore, a c.e. oracle set $\mathcal{B}$ can make decidable even a non-c.e. set $\mathcal{A}$.

11.4. Prove: $\equiv_T$ is an equivalence relation.

Bibliographic Notes

- The concept of *Turing reducibility* was first described intuitively in Post [184]. There, he explained how it can be that some decision problem is more difficult than some other undecidable decision problem. He also explored other less general reducibilities, called strong reducibilities, such as 1-reducibility, *m*-reducibility, *btt*-reducibility, and *tt*-reducibility. Although the informal concept of Turing reducibility was used in the following years over and over, it was not formally defined until Kleene [124, §61]. For an exposition on Turing reducibility, see Davis [58].
- The intuitive concept of *degree of unsolvability* was introduced via Turing reducibility in Post [184], and, based on Post [183], used in the abstract of (unpublished paper) Post [187].
- The program to determine the *relative computability* of undecidable decision problems, based on the notions of Turing reducibility, was set forth in Post [184].
- The above concepts were formally defined and became fully understood in Kleene and Post [127].
- Degrees of undecidability are also covered in Ambos-Spies and Fejer [11], Cooper [43], Cutland [51], Enderton [67], Lerman [139], Odifreddi [173], and Soare [241, 246].

Chapter 12
The Turing Hierarchy of Unsolvability

A hierarchy is a system of organizing things into different ranks,
levels, or positions, depending on how important they are.

Abstract At this point we only know of two degrees of unsolvability: the T-degree shared by all the decidable decision problems, and the T-degree shared by all undecidable decision problems that are T-equivalent to the *Halting Problem*. In this chapter we will prove that, surprisingly, for every undecidable decision problem there exists a more difficult decision problem. This will in effect mean that there is an infinite hierarchy of degrees of unsolvability and that there is no most difficult decision problem.

12.1 The Perplexities of Unsolvability

Turing degrees $\deg(\emptyset)$ and $\deg(\mathcal{K})$ are the only T-degrees whose existence we have intuitively anticipated and formally proved. Are $\deg(\emptyset)$ and $\deg(\mathcal{K})$ the only existing T-degrees? Put differently: Is every decision problem *either* decidable *or* exactly as difficult as the *Halting Problem* $\mathcal{D}_H$? If the answer were *no*, then there would be a decision problem that would be undecidable because of some reason essentially different from the reason for which $\mathcal{D}_H$ is undecidable. If so, this would immediately raise the following questions:

1. Are there *undecidable* decision problems that are *more difficult* than $\mathcal{D}_H$?
2. Are there *undecidable* decision problems that are *less difficult* than $\mathcal{D}_H$?

But, would all this make any sense? We could not compare the difficulties of undecidable problems just by comparing the times elapsed to obtain their solutions, as the computations could run indefinitely and return no solutions at all. For the same reason, neither could we use any other measure of the quality of the solutions. The situation we would face is illustrated in Fig. 12.1 (for computational problems).

A way out of this situation is to use the Turing reduction. The idea is that for two undecidable decision problems $\mathcal{P}$ and $\mathcal{Q}$, we consider $\mathcal{Q}$ to be more difficult than $\mathcal{P}$ if $\mathcal{P} <_T \mathcal{Q}$. Equivalently, $\mathcal{Q}$ is considered more difficult than $\mathcal{P}$ if $\deg(\mathcal{P}) < \deg(\mathcal{Q})$.

Fig. 12.1 How much later than never can we obtain a solution to a more difficult incomputable problem? How much less than no solution can we get when solving a more difficult incomputable problem? Is there any sensible definition of the property "to be a more difficult incomputable problem" at all? The answer is yes; the way out of these perplexities is to use the Turing reduction

We will now focus on question 1: Are there *undecidable* decision problems that are *more difficult* than $\mathcal{D}_H$?[1] So, is there a problem $\mathcal{Q}$ such that $\mathcal{D}_H <_T \mathcal{Q}$? If so, can we construct it? The answer is yes; to construct such a $\mathcal{Q}$ we must use a mapping called the Turing jump operator. This is the subject of the next section.

12.2 The Turing Jump

Let $\mathcal{S}$ be an arbitrary set. The *Turing jump* operator is a mapping $' : 2^{\mathbb{N}} \to 2^{\mathbb{N}}$ that assigns (i.e., constructs) to the set $\mathcal{S}$ another set, denoted by $\mathcal{S}'$, whose T-degree is higher than $\deg(\mathcal{S})$. How does the mapping $'$ do this?

First, recall the *Halting Problem* $\mathcal{D}_H$:

$$\text{"Does } T \text{ halt on input } \langle T \rangle ?\text{"}$$

The language of this problem is the set $\mathcal{K}$; it contains the codes of all ordinary Turing machines T that halt on their own codes $\langle T \rangle$. Since $\langle T \rangle$ is just a binary-represented index x, and φ_x is the proper function of the ordinary Turing machine T_x, we can rewrite $\mathcal{K}$ as follows:

$$\mathcal{K} \stackrel{\text{def}}{=} \{\langle T \rangle \,|\, T \text{ halts on input } \langle T \rangle\}$$
$$= \{x \,|\, T_x \text{ halts on input } x\}$$
$$= \{x \,|\, \varphi_x(x)\!\downarrow\}.$$

Secondly, let $\mathcal{S}$ be an arbitrary set. Let us wake up the oracle for $\mathcal{S}$, make it available to each o-TM T^*, and consider the obtained $\mathcal{S}$-TM $T^{\mathcal{S}}$. Recall from Sect. 10.1.4 that we can encode each $T^{\mathcal{S}}$ with $\langle T^{\mathcal{S}} \rangle$ and interpret this as a natural number, the index of $T^{\mathcal{S}}$. Following the above definition of the *Halting Problem*, we now define the halting problem for oracle Turing machines $T^{\mathcal{S}}$:

---

[1] We will answer question 2 in Chapter 13 (Theorem 13.5).

$$\text{"Does } T^S \text{ halt on input } \langle T^S \rangle \text{?"}$$

Let us denote the language of this problem by $\mathcal{K}^S$. The set $\mathcal{K}^S$ contains the codes of all S-TMs T^S that halt on their own codes $\langle T^S \rangle$. Again, each $\langle T^S \rangle$ is a binary-represented index x of T^S. Recalling from Sect. 10.2.1 that Φ_x^S is the proper functional of T_x^S, we rewrite the set $\mathcal{K}^S$ as follows:

$$
\begin{aligned}
\mathcal{K}^S &\stackrel{\text{def}}{=} \{ \langle T^S \rangle \, | \, T^S \text{ halts on input } \langle T^S \rangle \} \\
&= \{ x \, | \, T_x^S \text{ halts on input } x \} \\
&= \{ x \, | \, \Phi_x^S(x) {\downarrow} \}.
\end{aligned}
$$

So, given an arbitrary set $S \subseteq \mathbb{N}$, we have constructed a new set $\mathcal{K}^S \subseteq \mathbb{N}$.

Thirdly, we define $'$ to be the mapping that sends a set S to the set $\mathcal{K}^S$. In plain words, the mapping $'$ operates so that it "elevates" its argument S to an oracle set, i.e., makes S "jump" on the set $\mathcal{K}$. This is why $'$ is called the *Turing jump* operator, and the set $\mathcal{K}^S$ the *Turing jump of the set S*. We also denote $\mathcal{K}^S$ by S'. Here is the official definition.

Definition 12.1. (Turing Jump of a Set) The **Turing jump of a set** S is the set S' defined by

$$S' = \mathcal{K}^S \stackrel{\text{def}}{=} \{ x \, | \, \Phi_x^S(x) {\downarrow} \}.$$

12.2.1 Properties of the Turing Jump of a Set

The main result of this subsection will be Corollary 12.1, which states that S and S' are of different and comparable T-degrees.

First, of course, both S and S' are sets. But the oracle for S' is more powerful than the oracle for S. Indeed, the following lemma tells us that if the oracle for S makes a set appear S-c.e., then the oracle for S' makes the same set appear S'-decidable.

Lemma 12.1. $\mathcal{A}$ *is S-c.e.* $\implies$ $\mathcal{A}$ *is S'-decidable.*

Proof. Let $\mathcal{A}$ be an arbitrary S-c.e. set. Define a binary functional Φ_x^S as follows: $\Phi_x^S(y,z) = 1$ if $y \in \mathcal{A}$, and $\Phi_x^S(y,z){\uparrow}$ if $y \notin \mathcal{A}$. We will not need the actual value of x, but note that x is fixed. The argument z has no impact on Φ_x^S; it is there only because we want it to remain the only argument after the application of the *Parameter Theorem*. The functional Φ_x^S is S-p.c. (as it is S-computable on $\mathcal{A}$). Let us apply the *Parameter Theorem* and move y from $\Phi_x^S(y,z)$ to the index; hence $\Phi_x^S(y,z) = \Phi_{s(x,y)}^S(z)$, for an injective computable function s (see Sect. 7.3). Now observe that the following equivalences hold: $y \in \mathcal{A} \iff \Phi_{s(x,y)}^S(s(x,y)){\downarrow} \iff s(x,y) \in \mathcal{K}^S$. So we have $y \in \mathcal{A} \iff s(x,y) \in S'$, where s is a computable function and x fixed. This means that $\mathcal{A}$ is m-reducible to S', i.e., $\mathcal{A} \leq_m S'$. Then, $\mathcal{A} \leq_T S'$ (by Theorem 11.3), and $\mathcal{A}$ is S'-decidable. $\qquad \square$

The next theorem states that every set S is S'-decidable.

Theorem 12.1. *Let S be an arbitrary set. Then $S \leq_T S'$.*

Proof. Since $S \leq_T S$ holds for every S, the set S is S-decidable and, *a fortiori*, S-c.e. Then Lemma 12.1 (with $\mathcal{A} := S$) tells us that S is S'-decidable; that is, $S \leq_T S'$.   □

The converse is not true. The next theorem states that S' is S-undecidable. However, the theorem guarantees that the S-undecidability of S' is not "excessive"; specifically, it tells us that S' is S-c.e. In short, although there is no S-TM capable of deciding the set S', there is an S-TM capable of recognizing S'.

Theorem 12.2. *Let S be an arbitrary set. Then:*

        a) S' *is S-undecidable*   *(i.e., $S' \nleq_T S$).*

        b) S' *is S-c.e.*

Proof. The proof runs along the same lines as the proof of Lemma 8.1 (see Sect. 8.2) that $\mathcal{K}$ is undecidable. The main difference is that now we will be talking of S-TMs (instead of ordinary TMs) and of the set $\mathcal{K}^S$ (instead of $\mathcal{K}$). We therefore move at a somewhat faster pace.

a) Suppose that $S' \leq_T S$. Then the set $S' \stackrel{\text{def}}{=} \{\langle T^S \rangle \mid T^S$ halts on input $\langle T^S \rangle\}$ is S-decidable, and there is an S-TM capable of deciding the question "Does T^S halt on input $\langle T^S \rangle$?" for any T^S. Let us denote this hypothetical decider by $D_{\mathcal{K}}^S$.

Now we construct a new S-TM that will use $D_{\mathcal{K}}^S$. Since it will call an S-TM, it will itself be an S-TM. So let us denote it by N^S. The input to N^S will be the code $\langle T^S \rangle$ of an arbitrary T^S. The machine N^S must operate as follows. First, it doubles the input $\langle T^S \rangle$ into $\langle T^S, T^S \rangle$, and then sends this to $D_{\mathcal{K}}^S$. The decider takes this as the question $\langle T^S, T^S \rangle \in ?\mathcal{K}^S$, eventually halts, and answers either YES or NO. If the answer is YES, then N^S calls $D_{\mathcal{K}}^S$ again with the same question; otherwise, N^S outputs its own answer YES and halts.

But there is a catch: If N^S is given as input its own code $\langle N^S \rangle$, it puts the $D_{\mathcal{K}}^S$ in trouble. Namely, if $D_{\mathcal{K}}^S$ has answered the first question $\langle N^S, N^S \rangle \in ?\mathcal{K}^S$ with YES, then N^S starts endless cycling, during which $D_{\mathcal{K}}^S$ stubbornly repeats that N^S *will* halt. If, however, $D_{\mathcal{K}}^S$ has answered $\langle N^S, N^S \rangle \in ?\mathcal{K}^S$ with NO, and so predicted that N^S would *not* halt, N^S halts in the very next step.

In short, it is *not* true that $D_{\mathcal{K}}^S$ correctly answers *any* question $\langle T^S, T^S \rangle \in ?\mathcal{K}^S$. Actually, it fails when $T^S := N^S$. This contradicts our supposition. Consequently, $S' \nleq_T S$.

b) Let $R_{\mathcal{K}}^S$ be an S-TM as follows. The input to $R_{\mathcal{K}}^S$ is the code $\langle T^S \rangle$ of an arbitrary T^S. $R_{\mathcal{K}}^S$ starts simulating T^S on $\langle T^S \rangle$, and *if* it halts (i.e., T^S would halt on $\langle T^S \rangle$), then it outputs YES. So $R_{\mathcal{K}}^S$ is a *recognizer* of the set $\{\langle T^S \rangle \mid T^S$ halts on input $\langle T^S \rangle\} = S'$. Hence, S' is S-c.e.   □

From Theorems 12.1 and 12.2a it follows that $S <_T S'$, for any set S. Therefore, S and S' are not $\equiv_T$-equivalent, but still comparable T-degrees.

Corollary 12.1. *Let S be an arbitrary set. Then: $deg(S) < deg(S')$.*

By taking $S := \mathcal{K}$ in the above corollary, we obtain $\deg(\mathcal{K}) < \deg(\mathcal{K}')$.

NB *We have discovered that there is a T-degree that is higher than $deg(\mathcal{K})$. Hence, there exist decision problems that are more difficult than the Halting Problem.*

12.3 Hierarchies of T-Degrees

Because S' is a set, we can apply the function $'$ to S' too. This leads to an even higher T-degree $\deg((S')')$. Since we can repeat this as many times as we wish, it follows that there are higher and higher $<$-comparable T-degrees—and this never ends. We conclude that there is a hierarchy of at least $\aleph_0$ T-degrees and that there is no highest T-degree. Let us now see the details.

In the same fashion as we constructed the Turing jump of the set S, we can construct the Turing jump of the set S': We take the oracle for $S'(= \mathcal{K}^S)$, make it available to each o-TM T^*, define for the obtained S-TM $T^{S'}(= T^{\mathcal{K}^S})$ the corresponding halting problem, and finally obtain the associated language (set) $\mathcal{K}^{S'}$. Then, we can rewrite $\mathcal{K}^{S'}$ as follows:

$$\mathcal{K}^{S'} = \mathcal{K}^{\mathcal{K}^S} = (\mathcal{K}^S)' = (S')'.$$

We call this set the *second Turing jump of S* and denote it simply by S'' or $S^{(2)}$.

In the same manner we define the sets $S^{(3)}, S^{(4)}, \dots$ In general, we construct $S^{(i+1)}$ by "elevating" $S^{(i)}$ to the oracle set, making it available to all o-TMs, and collecting in $S^{(i+1)}$ the codes $\langle T^{S^{(i)}} \rangle$ of those $S^{(i)}$-TMs that halt on their own codes:

$$S^{(i+1)} = \{ \langle T^{S^{(i)}} \rangle \,|\, T^{S^{(i)}} \text{ halts on input } \langle T^{S^{(i)}} \rangle \}$$
$$= \{ x \,|\, T_x^{S^{(i)}} \text{ halts on input } x \}$$
$$= \{ x \,|\, \Phi_x^{S^{(i)}}(x){\downarrow} \}.$$

Definition 12.2. (*n*th Turing Jump) The *n*th **Turing jump of the set** S is the set $S^{(n)}$, which is inductively defined as follows:

$$S^{(n)} \stackrel{\text{def}}{=} \begin{cases} S & \text{if } n = 0 \\ (S^{(n-1)})' & \text{if } n \geqslant 1 \end{cases}$$

The relation between $S^{(i)}$ and $S^{(i+1)}$ is described by the following theorem. The proof of the theorem would run in the same way as the proofs of Lemma 12.1, Theorems 12.1 and 12.2, and Corollary 12.1. We therefore leave it as an exercise to the reader.

Theorem 12.3. *Let S be an arbitrary set. Then:*

a) $S^{(n)} <_T S^{(n+1)}$
b) $S^{(n+1)}$ *is $S^{(n)}$-c.e.*
c) $\deg(S^{(n)}) < \deg(S^{(n+1)})$

We see that each set $\mathcal{S}$ is the origin of an infinite hierarchy of sets:

$$\mathcal{S}^{(0)} <_T \mathcal{S}^{(1)} <_T \mathcal{S}^{(2)} <_T \ldots <_T \mathcal{S}^{(i)} <_T \mathcal{S}^{(i+1)} <_T \ldots$$

Associated with this hierarchy is the infinite hierarchy of T-degrees:

$$\deg\big(\mathcal{S}^{(0)}\big) < \deg\big(\mathcal{S}^{(1)}\big) < \deg\big(\mathcal{S}^{(2)}\big) < \ldots < \deg\big(\mathcal{S}^{(i)}\big) < \deg\big(\mathcal{S}^{(i+1)}\big) < \ldots$$

There are at least $\aleph_0$ T-degrees that are comparable with the relation $<$. Why have we said *at least*? The only reason is because we are cautious: At this point, nothing excludes the possibility of the existence of T-degrees *between* $\deg\big(\mathcal{S}^{(i)}\big)$ and $\deg\big(\mathcal{S}^{(i+1)}\big)$, for some i. Such T-degrees would by *passed over* by the T-jump and hence not constructible by it.

We have thus discovered one more surprising fact:

NB *For every degree of unsolvability there is a higher degree of unsolvability. For every decision problem, even an undecidable one, there is a more difficult decision problem. There is no most difficult decision problem.*

12.3.1 The Jump Hierarchy

Until now, $\mathcal{S}$ denoted an arbitrary subset of $\mathbb{N}$. In this subsection we will choose for $\mathcal{S}$ a particular set.

To do this, we intuitively reason as follows. The power of the oracle for $\mathcal{S}$ will depend on the (un)decidability of the chosen set. If we choose for $\mathcal{S}$ a *decidable* set, then we expect that the oracle for $\mathcal{S}$ will help o-TMs *as little as possible*. Actually, an o-TM with such an oracle will be of equivalent power to an ordinary TM (see Sect 11.1). We therefore expect that the difference between two successive Turing jumps of the set $\mathcal{S}$, i.e., the sets $\mathcal{S}^{(i)}$ and $\mathcal{S}^{(i+1)}$, will be as small as possible. This should result in a denser hierarchy $\deg\big(\mathcal{S}^{(i)}\big)$, $i = 0, 1, 2, \ldots$, that would, hopefully, reveal *all*[2] T-degrees and hence more of their properties.

So, for $\mathcal{S}$ we will pick a decidable set. As decidable sets are T-equivalent, and the set $\emptyset$ is decidable, we will take $\mathcal{S} := \emptyset$. We obtain the following *jump hierarchy of sets*

$$\emptyset^{(0)} <_T \emptyset^{(1)} <_T \emptyset^{(2)} <_T \ldots <_T \emptyset^{(i)} <_T \emptyset^{(i+1)} <_T \ldots$$

and the associated *jump hierarchy of T-degrees*

$$\deg\big(\emptyset^{(0)}\big) < \deg\big(\emptyset^{(1)}\big) < \deg\big(\emptyset^{(2)}\big) < \ldots < \deg\big(\emptyset^{(i)}\big) < \deg\big(\emptyset^{(i+1)}\big) < \ldots$$

---

[2] However, we will learn in Chap. 13 that this reasoning is too optimistic. It will prove again that intuition can be misleading.

Example 12.1. (Degrees of Some Sets) The undecidable sets $\mathcal{K}_0, \mathcal{K}_1, \mathcal{F}in, \mathcal{C}of, \mathcal{T}ot, \mathcal{E}xt$, which we defined in Sects. 8.2 and 8.3.5, belong to the initial T-degrees of the jump hierarchy. In particular,

$$\mathcal{K}, \mathcal{K}_0, \mathcal{K}_1 \in \deg(\emptyset^{(1)})$$
$$\mathcal{F}in, \mathcal{T}ot \in \deg(\emptyset^{(2)})$$
$$\mathcal{C}of, \mathcal{E}xt \in \deg(\emptyset^{(3)})$$

This tells us that

- the following three incomputable problems are equally difficult:

$$\mathcal{D}_H = \text{"Does } T \text{ halt on input } \langle T \rangle?\text{"}$$
$$\mathcal{D}_{Halt} = \text{"Does } T \text{ halt on input } w?\text{"}$$
$$\mathcal{D}_{\mathcal{K}_1} = \text{"Is dom}(\varphi) \text{ empty?"}$$

- the next two decision problems are equally difficult, yet more difficult than the above three:

$$\mathcal{D}_{\mathcal{F}in} = \text{"Is dom}(\varphi) \text{ finite?"}$$
$$\mathcal{D}_{\mathcal{T}ot} = \text{"Is } \varphi \text{ total?"}$$

- the next two decision problems are equally difficult, but more difficult than the above five:

$$\mathcal{D}_{\mathcal{C}of} = \text{"Is } \varphi \text{ undefined on finitely many elements?"}$$
$$\mathcal{D}_{\mathcal{E}xt} = \text{"Can } \varphi \text{ be extended to a (total) computable function?"}$$

Written succinctly:

$$\mathcal{D}_H \equiv_T \mathcal{D}_{Halt} \equiv_T \mathcal{D}_{\mathcal{K}_1} <_T \mathcal{D}_{\mathcal{F}in} \equiv_T \mathcal{D}_{\mathcal{T}ot} <_T \mathcal{D}_{\mathcal{C}of} \equiv_T \mathcal{D}_{\mathcal{E}xt}$$

This situation is depicted in Fig. 12.2.

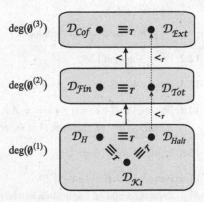

Fig. 12.2 The undecidability of some decision problems. The problems in the same T-degree are equally difficult, but more difficult than the problems in lower T-degrees

If, for instance, $\mathcal{D}_{\mathcal{T}ot}$ were decidable, $\mathcal{D}_{\mathcal{F}in}$ and $\mathcal{D}_H$, $\mathcal{D}_{Halt}$, $\mathcal{D}_{\mathcal{K}_1}$ would also be decidable; however, $\mathcal{D}_{\mathcal{C}of}$ and $\mathcal{D}_{\mathcal{E}xt}$ would remain undecidable. □

12.4 Chapter Summary

We compare the difficulty of computational problems by relating the problems using the Turing reduction.

Using the Turing jump operator, we can construct, for an arbitrary set, a set that is at a higher T-degree than the original set.

Correspondingly, for every decision problem, even an undecidable one, there is a more difficult decision problem.

This enables us to construct the jump hierarchy of T-degrees. The hierarchy starts with the T-degree $\deg(\emptyset)$ representing decidable decision problems, and continues with infinitely many T-degrees. There is no highest T-degree in the jump hierarchy. At this point nothing suggests that there might exist T-degrees other than those that are members of the jump hierarchy, i.e., constructible by the Turing jump operator.

This means that the class of decision problems does not divide into just two subclasses, $\deg(\emptyset)$ and $\deg(\mathcal{K})$, one consisting of decidable and the other of undecidable problems, which are equally difficult as the *Halting Problem*. Instead, the subclass of undecidable decision problems is further partitioned into infinitely many subclasses consisting of more and more difficult problems.

All of this is surprising and proves once again that human experience and intuition can be deceptive.

Problems

12.1. Prove:

    (a) $\emptyset' = \mathcal{K}$;

    (b) $\emptyset'' = \mathcal{K}^{\mathcal{K}}$;

    (c) $\emptyset''' = \mathcal{K}^{\mathcal{K}^{\mathcal{K}}}$.

    [*Hint.* Use $\mathcal{S} = \emptyset$ in Definitions 12.1 and 12.2.]

12.2. Prove Theorem 12.3.

12.3. Prove: $\mathcal{A}$ is $\mathcal{B}$-c.e. $\Longleftrightarrow \mathcal{A} \leq_1 \mathcal{B}'$.

12.4. Prove: $\mathcal{A}$ is $\mathcal{B}$-c.e. $\wedge \mathcal{B} \leq_T \mathcal{C} \Longrightarrow \mathcal{A}$ is $\mathcal{C}$-c.e.

    Remark. Thus, in Theorem 11.6, the conditions for the transitivity of $\leq_T$ can be relaxed.

12.5. Prove: $\mathcal{A} \leq_T \mathcal{B} \Longleftrightarrow \mathcal{A}' \leq_1 \mathcal{B}'$.

12.6. Prove: $\mathcal{A} \equiv_T \mathcal{B} \Longrightarrow \mathcal{A}' \equiv_1 \mathcal{B}'$.

    Remark. A similar implication will be proved in the next chapter (see Theorem 13.1, p. 274).

12.7. Prove: $\mathcal{A}$ is $\mathcal{B}$-c.e. $\Longleftrightarrow \mathcal{A}$ is $\overline{\mathcal{B}}$-c.e.

Bibliographic Notes

- The *jump* operator was first informally used in Post [187] when he, as well as Kleene, struggled to formally define the abstract notion of completeness of c.e. sets. Post used the Turing jump S' of a set S but did not formally define the operator $'$. So, the first published proof that for any set S there exists a set S' which is complete for S appeared in Kleene [124]. But, the jump operator was finally defined in Kleene and Post [127].
- The first results about the jump hierarchy, consisting of degrees yielded by the Turing jump operator, appeared in Kleene [121, 123, 124], Mostowski [163, 164], Post [187], and Davis [52].
- For more on degrees of unsolvability, see Lerman [139], Soare [241], Odifreddi [173], Cooper [43], Ambos-Spies and Fejer [11], and Soare [246]. See also Sacks [209], Cutland [51], and Enderton [67].

Chapter 13
The Class $\mathcal{D}$ of Degrees of Unsolvability

A structure is something that consists of parts connected together in an ordered way.

Abstract We now know that there are 1) infinitely many T-degrees; 2) the relation $<$ defined between these T-degrees; and 3) the operator Turing jump that constructs from a given set a new set at a higher T-degree. In this chapter, T-degrees will become the main object of our research. It will be useful to view T-degrees as members of a certain class, $\mathcal{D}$. We will define this class as a mathematical structure endowed with a relation and a function that we will found on $<$ and $'$, respectively. This view will simplify our expression and the investigation of the properties of the structure.

13.1 The Structure $(\mathcal{D}, \leq, ')$

Recall that a mathematical structure is a class endowed with certain relations and functions defined on the class (see the footnote on p. 40). In what follows, our intention is to define a structure whose class will contain all T-degrees, while the relation and function on this class will be based on the relation $<$ and the Turing jump operator $'$, respectively.

First, observe that instead of viewing a T-degree as a $\equiv_T$-equivalence class of subsets of $\mathbb{N}$, we can view it as a *member* of some other set. Of course, this is the *quotient set of* $2^{\mathbb{N}}$ *relative to* $\equiv_T$, i.e., the class $2^{\mathbb{N}}/\equiv_T$ of all $\equiv_T$-equivalence classes of $2^{\mathbb{N}}$. We will denote this class by $\mathcal{D}$.

Definition 13.1. (Class $\mathcal{D}$) The **class** $\mathcal{D}$ of all T-degrees is $\mathcal{D} \overset{\text{def}}{=} 2^{\mathbb{N}}/\equiv_T$.

$\mathcal{D}$ is not empty; we have seen in Chap. 12 that it has infinitely many members.

Remarks. (1) We can interpret $\mathcal{D}$ as the class of all degrees of unsolvability of decision problems. (2) It is customary to denote the members of $\mathcal{D}$ by boldface characters, e.g., a, b, c, d, or 0. This notation indicates no representatives of T-degrees. When we will need a representative of a T-degree, we will indicate it explicitly, e.g., $S \in d$ or $d = \deg(S)$, saying that S is of degree d.

© The Author(s), under exclusive license to Springer-Verlag GmbH, DE,
part of Springer Nature 2020
R. Robič, *The Foundations of Computability Theory*, https://doi.org/10.1007/978-3-662-62421-0_13

Second, we already have a relation defined on $\mathcal{D}$; this is the relation $<$, which we introduced to compare T-degrees $\deg(\emptyset)$ and $\deg(\mathcal{K})$. It turned out that there are infinitely many T-degrees $\deg(\mathcal{S}^{(i)})$, $i \in \mathbb{N}$, and that they are linearly ordered by $<$. Now recall that linear order is just a special case of partial order (see Appendix A). We want to be able to consider other partial orders on $\mathcal{D}$, *if any*. For this reason we will "relax" the relation $<$ by replacing it with its reflexive closure $\leq$, defined as usual by $a \leq b \overset{\text{def}}{\Longleftrightarrow} a < b \lor a = b$.

Finally, we also have a function $'$, the Turing jump. But we must be careful before we apply it to the members of $\mathcal{D}$. Why? We have defined $'$ on *sets* (Definition 12.1, p. 265), and *not* on T-degrees, so it is not so obvious that $\mathcal{D}$ can inherit $'$ as it inherited the relation $\leq$. What we must clear up is the following question: If sets $\mathcal{A}$ and $\mathcal{B}$ are in the same T-degree, can $\mathcal{A}'$ and $\mathcal{B}'$ be in different T-degrees? If this could happen, then $'$ would not be well defined on $\mathcal{D}$. Luckily, the answer to the question is *no*. Thus the following theorem.

Theorem 13.1. $\mathcal{A} \equiv_T \mathcal{B} \Longrightarrow \mathcal{A}' \equiv_T \mathcal{B}'$

Proof. First, we prove that $\mathcal{A} \leq_T \mathcal{B} \Longrightarrow \mathcal{A}' \leq_T \mathcal{B}'$. So let $\mathcal{A} \leq_T \mathcal{B}$. It follows that $\mathcal{A}'$ is $\mathcal{A}$-c.e. (by Theorem 12.2 b). But then $\mathcal{A}'$ is $\mathcal{B}$-c.e. Hence, $\mathcal{A}' \leq_T \mathcal{B}'$ (by Lemma 12.1). We have thus proved $\mathcal{A} \leq_T \mathcal{B} \Longrightarrow \mathcal{A}' \leq_T \mathcal{B}'$. Second, we prove that $\mathcal{B} \leq_T \mathcal{A} \Longrightarrow \mathcal{B}' \leq_T \mathcal{A}'$. This is easy: We only have to swap $\mathcal{A}$ and $\mathcal{B}$ in the previous proof. Thus, $\mathcal{B} \leq_T \mathcal{A} \Longrightarrow \mathcal{B}' \leq_T \mathcal{A}'$. Finally, the two proved relations together result in $\mathcal{A} \equiv_T \mathcal{B} \Longrightarrow \mathcal{A}' \equiv_T \mathcal{B}'$. □

Informally, this means that T-jumps of sets of the same T-degree are sets of the same T-degree. Consequently, we can extend the definition of $'$ to be a function that maps a T-degree into a (single) T-degree. Thus we can talk about the Turing jump of a *whole* T-degree. In short, $'$ is a well-defined function on the class $\mathcal{D}$. Here is the definition.

Definition 13.2. (T-jump of a T-degree) The **Turing jump of a T-degree** $d \in \mathcal{D}$ is the T-degree $d' \overset{\text{def}}{=} \deg(\mathcal{S}')$, where $\mathcal{S}$ is an arbitrary member of d.

The nth T-jump $d^{(n)}$ of a T-degree d is defined in the same fashion as for sets (see Definition 12.2, p. 267), so we omit the formal definition. Writing $\mathbf{0}^{(i)}$ instead of $\deg(\emptyset^{(i)})$, the jump hierarchy of T-degrees is now

$$\mathbf{0}^{(0)} \leq \mathbf{0}^{(1)} \leq \mathbf{0}^{(2)} \leq \cdots \leq \mathbf{0}^{(i)} \leq \mathbf{0}^{(i+1)} \leq \cdots$$

The initial T-degrees we usually denote by $\mathbf{0} = \mathbf{0}^{(0)}$, $\mathbf{0}' = \mathbf{0}^{(1)}$, $\mathbf{0}'' = \mathbf{0}^{(2)}$, $\mathbf{0}''' = \mathbf{0}^{(3)}$.

To summarize, the class $\mathcal{D}$ has been endowed with the relation $\leq$ and the function $'$. In the following sections we will list some properties of the structure $(\mathcal{D}, \leq, ')$.

13.2 Some Basic Properties of $(\mathcal{D}, \leq, ')$

In this section we will list some of the properties that concern $(\mathcal{D}, \leq, ')$ as a whole. Specifically, we will explore facts about the cardinality and structure of $(\mathcal{D}, \leq, ')$.

13.2.1 Cardinality of Degrees and of the Class $\mathcal{D}$

Concerning the cardinality, two questions are of interest:

1. *Cardinality of T-degrees.* Given any T-degree, how many sets are there in it?
2. *Cardinality of the class* $\mathcal{D}$. How many T-degrees are there in $\mathcal{D}$?

Remark. Restating the above questions in view of decision problems we obtain:
1) How many decision problems share a given degree of unsolvability?
2) How many degrees of unsolvability are there?

The first question is answered by the following theorem.

Theorem 13.2. *Every T-degree is countable.*

Recall that a set is countable if it is equinumerous to a subset of $\mathbb{N}$. Thus, the cardinality of a countable set is either a natural number or $\aleph_0$.

Proof. Let $\mathcal{B} \subseteq \mathbb{N}$ be an arbitrary set. Define $\mathrm{lcone}(\mathcal{B})$ to be the set of all sets T-reducible to $\mathcal{B}$; that is, $\mathrm{lcone}(\mathcal{B}) \stackrel{\text{def}}{=} \{\mathcal{A} \mid \mathcal{A} \leq_T \mathcal{B}\}$. The plan of the proof is this: We will prove that $\mathrm{lcone}(\mathcal{B})$ is countable; since $\deg(\mathcal{B}) \subseteq \mathrm{lcone}(\mathcal{B})$, it will then follow that $\deg(\mathcal{B})$ is countable too.

To prove that $\mathrm{lcone}(\mathcal{B})$ is countable, we must construct an injection $f : \mathrm{lcone}(\mathcal{B}) \to \mathbb{N}$. How can we do that? Since $\mathcal{A} \leq_T \mathcal{B}$, there is a $\mathcal{B}$-TM $T^{\mathcal{B}}$ capable of computing $\chi_{\mathcal{A}}(a)$ for any $a \in \mathcal{A}$. So there are countably infinitely many $\mathcal{B}$-TMs equivalent to $T^{\mathcal{B}}$, in the sense that each of them computes $\chi_{\mathcal{A}}$ (see Sect. 10.1.4). The set $\mathrm{ind}^{\mathcal{B}}(\mathcal{A})$ of all indexes of such machines is countably infinite (see Sect. 10.2.1). We now define the function $f : \mathrm{lcone}(\mathcal{B}) \to \mathbb{N}$ by letting f assign to $\mathcal{A}$ the smallest index in $\mathrm{ind}^{\mathcal{B}}(\mathcal{A})$; that is, $f(\mathcal{A}) \stackrel{\text{def}}{=} \min\{x \mid T_x^{\mathcal{B}} \text{ computes } \chi_{\mathcal{A}}\}$. (Such an index exists, as $\mathrm{ind}^{\mathcal{B}}(\mathcal{A}) \neq \emptyset$.) The function f is injective. (Otherwise, there would exist in $\mathrm{lcone}(\mathcal{B})$ two *different* sets $\mathcal{A}_1$, $\mathcal{A}_2$ that would be decided by the *same* $\mathcal{B}$-TM $T_{f(\mathcal{A}_1)}^{\mathcal{B}} = T_{f(\mathcal{A}_2)}^{\mathcal{B}}$. This would imply that $\chi_{\mathcal{A}_1} = \chi_{\mathcal{A}_2}$, which is impossible because $\mathcal{A}_1 \neq \mathcal{A}_2$.) Consequently, $\mathrm{lcone}(\mathcal{B})$ is countable. As $\deg(\mathcal{B}) \subseteq \mathrm{lcone}(\mathcal{B})$, so is $\deg(\mathcal{B})$. $\square$

But we can do more: Kleene and Post proved that each T-degree has cardinality $\aleph_0$.

Proposition 13.1. *Every T-degree is countably infinite.*

Proof. We omit the proof. See the Bibliographic Notes to this chapter. $\square$

We now turn to the second question. How many T-degrees exist in $\mathcal{D}$? We have seen that the jump hierarchy $\mathbf{0}^{(0)} < \mathbf{0}^{(1)} < \mathbf{0}^{(2)} < \ldots < \mathbf{0}^{(i)} < \mathbf{0}^{(i+1)} < \ldots$ contains $\aleph_0$ T-degrees. So, $\mathcal{D}$ contains at least as many T-degrees. What about other hierarchies starting in decidable sets $\mathcal{S}$ besides $\emptyset$? Since all decidable sets are in $\mathbf{0}$, all their Turing jumps are in $\mathbf{0}'$ (by Theorem 13.1). Thus, it seems that there can be no other hierarchy besides the jump hierarchy. On the other hand, nothing says that the Turing jump is the only way to discover new T-degrees. If other ways exist, also other degrees of unsolvability may exist. So the question is whether or not the total number of degrees of unsolvability is larger than $\aleph_0$. In other words: Is the class $\mathcal{D}$ countable or uncountable? Here is the answer.

Theorem 13.3. *The class $\mathcal{D}$ is uncountable; its cardinality is $2^{\aleph_0}$.*

Proof. There are $2^{\aleph_0}$ subsets of $\mathbb{N}$. Since each is of a certain degree of unsolvability, there can be *at most* $2^{\aleph_0}$ T-degrees in $\mathcal{D}$. Now, if there were only countably many (i.e., at most $\aleph_0$) T-degrees in $\mathcal{D}$, then—knowing (by Theorem 13.2) that each contains countably many sets—there would be countably many sets contained in *all* T-degrees. (Note that "countably many $\times$ countably many = countably many"; see Appendix A.) We conclude that there must be $2^{\aleph_0}$ T-degrees in $\mathcal{D}$. $\square$

Remarks. Let us interpret the above results in the world of decision problems. There are $2^{\aleph_0}$ degrees of unsolvability (Theorem 13.3). (This is as many as there are real numbers, assuming the *Continuum Hypothesis*, p. 16.) For each degree of unsolvability there are $\aleph_0$ decision problems of that degree of unsolvability (Theorem 13.2). For instance, $\aleph_0$ decision problems are as difficult as the *Halting Problem* (see Fig. 12.2). There are $\aleph_0$ decision problems as difficult as the problem $\mathcal{D}_{\mathcal{T}ot} = $ "Is a p.c. function φ total?" A similar situation occurs for problems that are as difficult as the problem $\mathcal{D}_{\mathcal{E}xt} = $ "Can a p.c. function φ be extended to a computable one?"

Now we see that our cautiousness in Sect. 12.3 was well grounded: Besides the $\aleph_0$ T-degrees $\mathbf{0}^{(0)} < \mathbf{0}^{(1)} < \mathbf{0}^{(2)} < \ldots < \mathbf{0}^{(i)} < \mathbf{0}^{(i+1)} < \ldots$ there are many more T-degrees. But where are they? Intuitively, we can identify two possibilities:

- there exist T-degrees that are not within the jump hierarchy;
- there exist intermediate T-degrees between a T-degree and its T-jump.

In the following, we will see that both are true.

13.2.2 The Class $\mathcal{D}$ as a Mathematical Structure

The class $\mathcal{D}$ is endowed with the relation $\leq$. Does this relation reveal any particular, distinguished order in the class $\mathcal{D}$? It is easy to prove the following theorem.

Theorem 13.4. $(\mathcal{D}, \leq)$ is partially ordered.

Proof. We leave it to the reader to check that $\leq$ is reflexive ($a \leq a$, for any $a \in \mathcal{D}$), transitive ($a \leq b \wedge b \leq c \Rightarrow a \leq c$, for any $a, b, c \in \mathcal{D}$), and anti-symmetric ($a \leq b \wedge b \leq a \Rightarrow a = b$, for any $a, b \in \mathcal{D}$). □

Incomparable *T*-Degrees

Is it, perhaps, that $(\mathcal{D}, \leq)$ is even linearly ordered, as is, for example, $(\mathbb{N}, \leqslant)$? Since $(\mathcal{D}, \leq)$ is partially ordered, to answer this question we must find out whether or not every two members of $\mathcal{D}$ are $\leq$-comparable, i.e., whether $a \leq b \vee b \leq a$, for every $a, b \in \mathcal{D}$. Surprisingly, the answer is *no*. In 1954, Kleene and Post proved that there exist sets $\mathcal{A}, \mathcal{B}$, both *T*-reducible to the set $\emptyset'$, such that $\mathcal{A} \not\leq_T \mathcal{B} \wedge \mathcal{B} \not\leq_T \mathcal{A}$. Since $\mathcal{A}$ and $\mathcal{B}$ are $\leq_T$-*incomparable*, so are $\leq$-*incomparable* $\deg(\mathcal{A})$ and $\deg(\mathcal{B})$. (Note that $\deg(\mathcal{A}) \leq \mathbf{0'}$ and $\deg(\mathcal{B}) \leq \mathbf{0'}$.) We write $a|b$ when a, b are $\leq$-incomparable. The proof of the following Kleene-Post theorem is instructive because its idea will soon be developed further and used in *Post's Problem* (see Chap. 14).

Theorem 13.5. *There exist T-degrees a, b such that $0 \leq a, b$ and $a, b \leq 0'$ and $a|b$.*

Proof. To prove the theorem, we must show that there exist two sets $\mathcal{A}$ and $\mathcal{B}$ such that $\mathcal{A} \leq_T \emptyset'$, $\mathcal{B} \leq_T \emptyset'$, and $\mathcal{A} \not\leq_T \mathcal{B} \wedge \mathcal{B} \not\leq_T \mathcal{A}$. Since we expect that proving the existence of $\mathcal{A}$ and $\mathcal{B}$ in one fell swoop might be too difficult (if not impossible), we will take a different approach: We will design a set of guidelines by which $\mathcal{A}$ and $\mathcal{B}$ can be constructed, at least in principle, in a systematic, algorithmic way. There are several ingredients in this method.

First, observe that the condition $\mathcal{A} \not\leq_T \mathcal{B}$ means that *no* $\mathcal{B}$-TM can decide the set $\mathcal{A}$. Hence the condition can be replaced by a conjunction $R_0 \wedge R_1 \wedge R_2 \wedge \ldots$ of countably many simpler requirements, where R_e requires that $\mathcal{A}$ cannot be decided by $T_e^\mathcal{B}$, the *e*th $\mathcal{B}$-TM. Equivalently, R_e demands that $\chi_\mathcal{A}$, the characteristic function of $\mathcal{A}$, not be the proper functional $\Phi_e^\mathcal{B}$:

$$R_e : \chi_\mathcal{A} \neq \Phi_e^\mathcal{B}.$$

In the same fashion we replace the condition $\mathcal{B} \not\leq_T \mathcal{A}$ with the sequence $S_0 \wedge S_1 \wedge S_2 \wedge \ldots$, where

$$S_e : \chi_\mathcal{B} \neq \Phi_e^\mathcal{A}.$$

Consequently, to prove the theorem we must show how to construct $\mathcal{A}$ and $\mathcal{B}$ such that R_e and S_e will be fulfilled for every e.

The *plan* is this: We will construct $\chi_\mathcal{A}$ and $\chi_\mathcal{B}$ in an infinite sequence of *stages*; at any stage s, only the current *approximations* f_s and g_s to $\chi_\mathcal{A}$ and $\chi_\mathcal{B}$, respectively, will exist; we will ensure that the current f_s and g_s fulfill $R_0 \wedge \ldots \wedge R_\ell$ and $S_0 \wedge \ldots \wedge S_\ell$, respectively, for some $\ell = \ell(s)$; and we will ensure that the length $\ell(s)$ of the fulfilled conjunctions will increase monotonically from stage to stage. As a consequence, in the limit, the condition $\mathcal{A} \not\leq_T \mathcal{B} \wedge \mathcal{B} \not\leq_T \mathcal{A}$ will be fulfilled.

At any stage s, the to-be-constructed sets $\mathcal{A}$ and $\mathcal{B}$ will be approximated by sets denoted by $\mathcal{A}_s$ and $\mathcal{B}_s$, respectively. Correspondingly, $\chi_\mathcal{A}$ and $\chi_\mathcal{B}$ will be approximated by f_s and g_s, respectively. So, in the limit, we expect $\mathcal{A}_\omega = \mathcal{A}$, $\mathcal{B}_\omega = \mathcal{B}$ and $f_\omega = \chi_\mathcal{A}$, $g_\omega = \chi_\mathcal{B}$. The construction of $\chi_\mathcal{A}$

and $\chi_{\mathcal{B}}$ will start at stage $s = 0$ with f_0 and g_0 representing $\mathcal{A}_0 = \mathcal{B}_0 = \emptyset$. At each next stage, we will try to fulfill first the requirement R_e, and then the requirement S_e, for some e. (Thus, at $s = 1$ we will fulfill R_0 and S_0; at $s = 2$, R_1 and S_1; and so on.) But we will try to do this in such a way that *once a requirement has been fulfilled, it will remain fulfilled forever*. (We also say that no requirement will be *injured*.) If we succeed in this plan, the requirements will become fulfilled in the order $R_0, S_0, R_1, S_1, R_2, S_2, \ldots$, which will satisfy the condition $\mathcal{A} \not\leq_T \mathcal{B} \wedge \mathcal{B} \not\leq_T \mathcal{A}$. (What about the other two conditions, $\mathcal{A} \leq_T \emptyset'$, $\mathcal{B} \leq_T \emptyset'$, that are set by the theorem? The reasons for having these will become apparent soon.)

We have come to the question about how to fulfill R_e and S_e, given that all previous Rs and Ss have been fulfilled. The answer will be given in an induction-like way by outlining what the next stage $s + 1$ should do in order to preserve the situation similar to that at stage s. So, *assume that, at stage s*, we have constructed f_s and g_s in such way that, for some n, both are defined everywhere on the initial segment $\{0, 1, \ldots, n\}$ of $\mathbb{N}$, and all the requirements $R_0, S_0, \ldots, R_{e-1}, S_{e-1}$ are fulfilled. We can represent the function f_s by a sequence of its values on the segment, that is, by the word $a_s = f_s(0)f_s(1) \ldots f_s(n) \in \{0, 1\}^*$. Of course, $f_s(i)$ tells us whether or not $i \in \mathcal{A}_s$ (when $0 \leq i \leq n$), while $f_s(i) \uparrow$ for $i > n$, as the *status* (membership) of these numbers in $\mathcal{A}$ is still open. (The same holds for g_s, which is represented by a word $b_s \in \{0, 1\}^*$.) *Then, in the next stage $s + 1$*, we will try to define f_{s+1} and g_{s+1} in such way that

- $a_s \subset a_{s+1}$ and $b_s \subset b_{s+1}$ will hold. That is, a_{s+1} is to be a *proper extension* of the word a_s, or, equivalently, a_s is to be a proper prefix of the word a_{s+1}. (Similarly for b_{s+1}.) This means that f_{s+1} is to be defined everywhere on $\{0, 1, \ldots, n, \ldots, m\}$, for some $m > n$. (Similarly for g_{s+1}.)
- R_e and S_e will be fulfilled and none of $R_0, S_0, \ldots, R_{e-1}, S_{e-1}$ will be injured.

If we attain the two objectives, the limit functions $f_\omega, g_\omega : \mathbb{N} \to \{0, 1\}$ will be *total*, and hence *characteristic functions* of the sets $\mathcal{A}_\omega = \mathcal{A}$ and $\mathcal{B}_\omega = \mathcal{B}$. These sets will fulfill all requirements R and S.

Clearly, attaining the above objectives is the crux of the proof. We will prove the following lemma.

Lemma. *Let e be an arbitrary natural number. Given a_s, b_s, there exist extensions a_{s+1}, b_{s+1} such that for any a, b that extend a_{s+1}, b_{s+1} and represent $\chi_{\mathcal{A}}$, $\chi_{\mathcal{B}}$, the $\mathcal{B}$-TM $T_e^{\mathcal{B}}$ does not decide $\mathcal{A}$.*

Proof. Let x be an arbitrary natural number for which f_s is *not* defined: $f_s(x) \uparrow$. (So x is a number whose membership in $\mathcal{A}_s$, and hence in $\mathcal{A}$, has not been determined.) The crucial question is:

    Is there a set $\mathcal{B}$ such that $T_e^{\mathcal{B}}$ would halt on input x and return either YES *or* NO*?*   (*)

(i) If such a set $\mathcal{B}$ *does not* exist, then we can take the trivial extensions $a_{s+1} := a_s$ and $b_{s+1} := b_s$. (ii) If, however, such a $\mathcal{B}$ *exists*, there is more work to do in order to construct a_{s+1} and b_{s+1}. First, we run $T_e^{\mathcal{B}}$ on x. Before halting, the machine asks the oracle finitely often whether or not a number is in $\mathcal{B}$. Let y be the largest of these numbers. Then, let b_{s+1} be the shortest extension of b_s that covers y. Now it remains to construct a_{s+1}. Note that $T_e^{\mathcal{B}}$, if run on x, would return the same answer as $T_e^{\mathcal{B}_{s+1}}$. So, to ensure that $T_e^{\mathcal{B}_{s+1}}$ (and hence $T_e^{\mathcal{B}}$) will *fail* to answer correctly the question $x \in ?\mathcal{A}_{s+1}$ (and hence the question $x \in ?\mathcal{A}$), we add x into either $\mathcal{A}_{s+1}$ or its complement $\overline{\mathcal{A}}_{s+1}$, depending on whether $T_e^{\mathcal{B}_{s+1}}$'s answer is NO or YES, respectively. As a consequence, the limit machine $T_e^{\mathcal{B}}$ will fail to decide the limit set $\mathcal{A}$.

It should be obvious that, once it has been determined which of the possibilities (i, ii) holds, the construction of a_{s+1} and b_{s+1} is computable in the ordinary sense. But how can we answer the question (*)? A systematic search for $\mathcal{B}$ is out of question. But we can obtain the answer by focusing on the program $\widetilde{\delta}_e$ of the o-TM T_e^*. Each instruction of $\widetilde{\delta}_e$ branches in two directions. This results in the *computation tree of $\widetilde{\delta}_e$*, a tree representing all possible executions of $\widetilde{\delta}_e$. Given a particular oracle set $\mathcal{B}$, the actual execution of $\widetilde{\delta}_e$ on x is represented by a branch in this tree. Halting executions are represented by finite branches, and non-halting executions by infinite branches. Answers (YES, NO) are in the leaves of the tree (i.e., at the end of finite branches). We now see: There exists a set $\mathcal{B}$ for which $T_e^{\mathcal{B}}$ halts on x and returns a YES or NO *iff* the computation tree of $\widetilde{\delta}_e$ has a leaf bearing the answer YES or NO. So we can apply an ordinary TM T that systematically checks the leaves of the computation tree of $\widetilde{\delta}_e$ and halts with an answer *found* as soon as a leaf with an answer YES or NO has been found. In this case we know that a set $\mathcal{B}$ with the above properties exists. *But* what if T never halts? So, we must find out whether or not T halts. Recall that this is the *Halting Problem*, which can be solved by an o-TM with the oracle set $\mathcal{K}_0$. Consequently, the question (∗) is $\mathcal{K}_0$-decidable, and the construction of a_{s+1} and b_{s+1} is $\mathcal{K}_0$-computable. (This is where the condition $\mathcal{B} \leq_T \emptyset'$ of the theorem comes from.) The lemma is proven.

Observe that we can exchange the roles of $\mathcal{A}$ and $\mathcal{B}$ in the lemma and prove that a_s, b_s can be extended so that ultimately the $\mathcal{A}$-TM $T_e^{\mathcal{A}}$ will not decide $\mathcal{B}$. Hence, applying the lemma twice, we can fulfill both requirements R_e and S_e. Now we can finally see the whole process of the construction of $\mathcal{A}$ and $\mathcal{B}$. We consider oracle Turing machines in succession, $T_0^*, T_1^*, T_2^*, \ldots$, and, based on the above lemma, ensure that none of them decides $\mathcal{A}$. In doing so, we alternate, for each o-TM T_e^*, the roles of the sets $\mathcal{A}$ and $\mathcal{B}$ and in this way ensure that the construction of both $\mathcal{A}$ and $\mathcal{B}$ proceeds. □

So, there are incomparable T-degrees between $\mathbf{0}$ and $\mathbf{0}'$. Is this situation specific to the pair $\mathbf{0}$, $\mathbf{0}'$? Far from it: $\leq$-incomparable T-degrees exist between $\boldsymbol{d}$ and $\boldsymbol{d}'$, for *any* T-degree $\boldsymbol{d}$. Here is a generalization of the previous theorem. (See Fig. 13.1.)

Theorem 13.6. *For any T-degree $\boldsymbol{d}$ there exist T-degrees $\boldsymbol{a}, \boldsymbol{b}$ such that $\boldsymbol{d} \leq \boldsymbol{a}, \boldsymbol{b}$ and $\boldsymbol{a}, \boldsymbol{b} \leq \boldsymbol{d}'$ and $\boldsymbol{a} | \boldsymbol{b}$.*

Proof. The proof is a relativization of the above proof. See the Bibliographic Notes to this chapter. □

Fig. 13.1 For any T-degree $\boldsymbol{d}$ there exist $\leq$-incomparable T-degrees $\boldsymbol{a}, \boldsymbol{b}$

Theorems 13.3 and 13.6 suggest that there might be uncountably many $\leq$-incomparable pairs of T-degrees. That this is indeed so is stated in the next theorem.

Theorem 13.7. *There are $2^{\aleph_0}$ mutually $\leq$-incomparable T-degrees.*

Proof. See the Bibliographic Notes to this chapter. □

Distinguished T-Degrees

Since there are $\leq$-incomparable elements in $\mathcal{D}$, the relation $\leq$ does not linearly order $\mathcal{D}$. This gives rise to a series of new questions about the existence of certain distinguished elements:

1. Are there $\leq$-minimal, $\leq$-least, $\leq$-greatest, or $\leq$-maximal elements in $\mathcal{D}$?
2. Do every two members of $\mathcal{D}$ have a $\leq$-upper bound or even a $\leq$-lub?[1]
3. Do every two members of $\mathcal{D}$ have a $\leq$-lower bound or even a $\leq$-glb?[2]
4. Is $(\mathcal{D}, \leq)$ a lattice, and if so, of what kind?

Let us answer these questions.

There can be no $\leq$-*greatest* and no $\leq$-*maximal* element, because the Turing jump constructs, for arbitrary T-degree d, a higher T-degree d'. But, there is a $\leq$-*least* element in $(\mathcal{D}, \leq)$.

Theorem 13.8. *There is a $\leq$-least T-degree in $(\mathcal{D}, \leq)$; this is 0.*

Proof. Since $\emptyset$ is decidable, $\emptyset \leq_T \mathcal{A}$ for every set $\mathcal{A}$. Hence, $0 \leq a$ for every $a \in \mathcal{D}$. □

This is not surprising. The decidability of a decidable set is insensitive to the decidability of other sets. Thus, the degrees of true unsolvability are in $\mathcal{D} - \{0\}$.

Is $(\mathcal{D}, \leq)$ a lattice? Well, to be a lattice, any two T-degrees a, b must have a $\leq$-lub and a $\leq$-glb. The first requirement is satisfied, as the following theorem states.

Theorem 13.9. *Any two T-degrees have a $\leq$-least upper bound.*

Proof. Let $a = \deg(\mathcal{A})$ and $b = \deg(\mathcal{B})$ be arbitrary T-degrees. Consider the *join* $\mathcal{A} \oplus \mathcal{B}$ of $\mathcal{A}$ and $\mathcal{B}$, that is, the set $\mathcal{A} \oplus \mathcal{B} \stackrel{\text{def}}{=} \{2x \,|\, x \in \mathcal{A}\} \cup \{2y+1 \,|\, y \in \mathcal{B}\}$. The members of $\mathcal{A}$ and $\mathcal{B}$ are injectively mapped into even and odd members of the join, respectively. Informally, $\mathcal{A} \oplus \mathcal{B}$ remembers the origin of each of its members. $\mathcal{A} \oplus \mathcal{B}$ is the $\leq_T$-lub of $\mathcal{A}$ and $\mathcal{B}$. To prove this, we must check that 1) $\mathcal{A} \leq_T \mathcal{A} \oplus \mathcal{B}$ and $\mathcal{B} \leq_T \mathcal{A} \oplus \mathcal{B}$, and 2) $\mathcal{A} \oplus \mathcal{B} \leq_T \mathcal{C}$, for any $\leq_T$-upper bound $\mathcal{C}$ of $\mathcal{A}, \mathcal{B}$. (This we leave as an exercise.) It follows that $\deg(\mathcal{A} \oplus \mathcal{B})$ is the $\leq$-lub of $a = \deg(\mathcal{A})$ and $b = \deg(\mathcal{B})$. □

We denote the $\leq$-least upper bound of $a, b \in \mathcal{D}$ by $a \vee b$. In particular, if $a = \deg(\mathcal{A})$ and $b = \deg(\mathcal{B})$, then $a \vee b = \deg(\mathcal{A} \oplus \mathcal{B})$. Informally, the set $\mathcal{A} \oplus \mathcal{B}$ remembers every member of $\mathcal{A}$ and every member of $\mathcal{B}$ by keeping track of the origin of each of its members. Thus, each representative of $a \vee b$ bears more information about its contents than any representative of a or b. Finally, of course, any finite set of T-degrees has a $\leq$-least upper bound.

In contrast to the above theorem, the $\leq$-glb of two degrees need not always exist. This was first proved by Kleene and Post in 1954.

---

[1] least upper bound
[2] greatest lower bound

Theorem 13.10. *There is a pair of T-degrees that have no $\le$-greatest lower bound.*

Proof. We omit the proof. See the Bibliographic Notes to this chapter. □

We must conclude that $(\mathcal{D}, \le)$ is not a lattice. However, because of the existence of least upper bounds, we say that $(\mathcal{D}, \le)$ is an *upper semi-lattice*.

Remarks. Let us look at the above results in the world of decision problems. There are pairs of decision problems such that neither is more difficult than the other (Theorem 13.6). For any finite set of undecidable decision problems there is a "superproblem" (i.e., a decision problem whose solution would make all problems in the set decidable) which is the easiest among all "superproblems" of the set (Theorem 13.9).

13.2.3 Intermediate T-Degrees

Can there be *intermediate degrees* between $\mathbf{0}^{(i)}$ and $\mathbf{0}^{(i+1)}$, that is, degrees that are passed over when T-jump takes $\mathbf{0}^{(i)}$ to $\mathbf{0}^{(i+1)}$? The answer to this question has already been obtained by Theorems 13.5 and 13.6. But we can ask further: How many T-degress can be passed over by a T-jump? The following theorem, which was also proved by Kleene and Post in 1954, tells us that for each degree $\mathbf{d}$ there are *infinitely many* pairwise $\le$-incomparable degrees between $\mathbf{d}$ and $\mathbf{d}'$. (See Fig. 13.2.)

Theorem 13.11. *For any T-degree $\mathbf{d}$ and $n \ge 1$, there are pairwise $\le$-incomparable T-degrees $\mathbf{c}_1, \ldots, \mathbf{c}_n$ such that $\mathbf{d} < \mathbf{c}_k < \mathbf{d}'$, for $k = 1, \ldots, n$.*

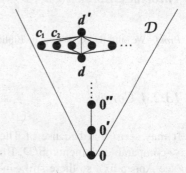

Fig. 13.2 For any $\mathbf{d}$, there are infinitely many pairwise $\le$-incomparable T-degrees $\mathbf{c}_k$ that are passed over by T-jump from $\mathbf{d}$ to $\mathbf{d}'$

Proof. See the Bibliographic Notes to this chapter. □

The above theorem guarantees the existence of $\leq$-*incomparable* T-degrees between d and d'. Can it happen that, for some d, there are T-degrees between d and d' that are comparable or even *linearly* ordered by $\leq$? If so, can such a sequence of intermediate T-degrees have infinitely many members? The answer to both questions is yes. (See Fig. 13.3.)

Theorem 13.12. *For any T-degree d and $n \geqslant 1$, there are T-degrees $c_1, \ldots, c_n$ such that $d < c_1 < \ldots < c_n < d'$.*

Fig. 13.3 For any d, there are infinitely many linearly ordered T-degrees c_i that are passed over by T-jump from d to d'

Proof. We omit the proof. See the Bibliographic Notes to this chapter.   □

Moreover, Kleene and Post proved that linearly ordered intermediate T-degrees are *dense*, in the sense that if a and b are any such T-degrees, then there is a T-degree c between them. This is stated in the next theorem.

Theorem 13.13. *If $d < a < b < d'$, then there is a T-degree c such that $a < c < b$.*

Proof. We omit the proof. See the Bibliographic Notes to this chapter.   □

13.2.4 Cones

It may seem that because of Theorems 13.3 and 13.7 there is not much left for $\leq$-comparable elements in $\mathcal{D}$. But infinite sets contain equipollent proper subsets (see Appendix), so there still can be many elements of $\mathcal{D}$ that are $\leq$-comparable. This gives rise to the question: How many members of $\mathcal{D}$ are $\leq$-comparable to a given $d \in \mathcal{D}$?

Let us pick an arbitrary $d \in \mathcal{D}$ and consider all the elements of $\mathcal{D}$ that are $\leq$-comparable to d. We divide these elements into two sets, the set of elements that are higher than (or equal to) d, and the set of elements that are lower than (or equal to) d. These sets we call the *upper cone* and *lower cone* of d, respectively. See Fig. 13.4. Here is the definition.

Definition 13.3. (Upper and Lower Cone) The **upper cone** of a T-degree d is the set $\mathrm{ucone}(d) \overset{\text{def}}{=} \{x \in \mathcal{D} \mid d \leq x\}$, and the **lower cone** of d is the set $\mathrm{lcone}(d) \overset{\text{def}}{=} \{x \in \mathcal{D} \mid x \leq d\}$.

Fig. 13.4 The upper and lower cone of a T-degree d. Any T-degree that is $\leq$-comparable to d is in d's upper or lower cone

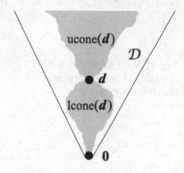

The next two theorems tell us what the population is of each of the two cones.

Theorem 13.14. *The upper cone $\mathrm{ucone}(d)$ is uncountable, for any T-degree d.*

Proof. We omit the proof. See the Bibliographic Notes to this chapter. □

Since we can construct with the Turing jump only $\aleph_0$ T-degrees, Theorem 13.14 tells us that the Turing jump cannot uncover all T-degrees above a given T-degree. We have seen one method for constructing T-degrees not reachable by the Turing jump in the proof of Theorem 13.5, and another will be described in the next chapter.

Theorem 13.15. *The lower cone $\mathrm{lcone}(d)$ is countable, for any T-degree d.*

Proof. We omit the proof. See the Bibliographic Notes to this chapter. □

13.2.5 Minimal T-Degrees

By Theorem 13.8, the T-degree 0 is lower than any other T-degree d. Pick an arbitrary $d(\neq 0)$. There may exist a T-degree c which is strictly between 0 and d, that is, $0 < c < d$. (For instance, take $d = 0''$ and $c = 0'$; or, take $d = 0'$ and $c = a$ in Theorem 13.5.)

The following question immediately arises: Does there exist a T-degree $d(\neq 0)$ such that no T-degree is strictly between 0 and d? If such a d exists, it is called *minimal*. (See Fig. 13.5.) Here is the definition.

Definition 13.4. (Minimal Degree) A T-degree d is **minimal** if $d \neq 0$ and there is no T-degree c such that $0 < c < d$.

Fig. 13.5 An element $d \neq 0$ is minimal if no element other than 0 is lower than d

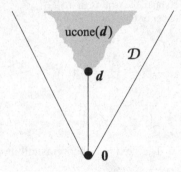

In 1956, Spector[3] proved that minimal degrees actually exist. He proved the existence of a minimal degree below $0''$.

Theorem 13.16.   *There exists a minimal T-degree d; in addition, $d \leq 0''$.*

Proof. We omit the proof. See the Bibliographic Notes to this chapter. $\square$

In 1961, Sacks[4] proved the existence of minimal degrees that are even below $0'$.

Remarks. Let us interpret the above results in the world of decision problems. For every degree of unsolvability there are $\aleph_0$ decision problems of that difficulty (Proposition 13.1). There are uncountably many degrees of unsolvability that can have a decision problem (Theorem 13.3). There exist pairs of decision problems such that neither problem is easier (or more difficult) than the

---

[3] Clifford Spector, 1931–1961, American mathematician.

[4] Gerald Enoch Sacks, b. 1933, American mathematician and logician.

other (Theorems 13.5 and 13.6). Actually, there are uncountably many pairs of such incomparable decision problems (Theorem 13.7). For any finite set of decision problems there exists a "superproblem" whose decidability would imply the decidability of every decision problem in the set. In addition, there is no easier superproblem for this set of decision problems (Theorem 13.9). For any decision problem, there are uncountably many more difficult decision problems (Theorem 13.14), and only countably many easier decision problems (Theorem 13.15). There are undecidable decision problems for which there exist no easier decision problems besides the decidable ones (Theorem 13.16).

13.3 Chapter Summary

Every T-degree is countably infinite. The class of all T-degrees is uncountable and is partially ordered by the relation $\leq$. The Turing jump operator maps a T-degree into a higher T-degree. There exist T-degrees that are incomparable with the relation $\leq$; in fact, there are uncountably many pairs of such incomparable degrees. The T-degree of decidable decision problems, $\mathbf{0}$, is the $\leq$-least T-degree in the class. Any two T-degrees have a $\leq$-least upper bound, but not necessarily a $\leq$-greatest lower bound. Between any T-degree and its T-jump there are infinitely many pairwise incomparable T-degrees and infinitely many linearly ordered T-degrees. The upper cone of any T-degree contains uncountably many T-degrees, and its lower cone contains only countably many T-degrees. There are minimal degrees, some of which are below the T-degree $\mathbf{0}''$, or even below $\mathbf{0}'$.

Problems

13.1. Prove: $(\mathcal{D}, \leq)$ is partially ordered.

13.2. Prove: There are sets $\mathcal{A}$ and $\mathcal{B}$ such that $\mathcal{A} <_T \mathcal{B}$ and $\mathcal{A}' \equiv_T \mathcal{B}'$.

Remark. Thus, the converse of the implication $\mathcal{A} \equiv_T \mathcal{B} \Longrightarrow \mathcal{A}' \equiv_T \mathcal{B}'$ (Theorem 13.1) does *not* hold; we may have $\mathcal{A} \not\equiv_T \mathcal{B}$ and still $\mathcal{A}' \equiv_T \mathcal{B}'$.

13.3. Prove: The Turing jump operator $'$ defined on the class $\mathcal{D}$ is *not* injective.

[*Hint.* See Problem 13.2.]

13.4. Complete the proof of Theorem 13.9.

Bibliographic Notes

- The abstraction from degrees of undecidability of *sets* to *degrees* as such was introduced in Kleene and Post [127]. This paper revealed many important facts about degrees of unsolvability: The T-jump operator was defined on T-*degrees* [127, §1]; the existence of infinitely many $\leq$-*incomparable* T-degrees between any two d, d' with $d < d'$ was proved [127, §2]; $(\mathcal{D}, \leq)$ was proved to be an *upper semi-lattice* [127, §1] but not a lattice [127, §4]; the class $\mathcal{D}$ was proved to be uncountable; the *cardinality* of each T-degree was proved to be $\aleph_0$; the fact that each T-jump *passes over* infinitely many degrees was proved; the density of the T-degrees was proved.

- In describing the idea of the proof of Theorem 13.5 we have leaned on Soare [241, Chap. VI] and Shen and Vereshchagin [215].

- The existence of *minimal* T-degrees was proved in Spector [247], and the existence of minimal T-degrees below $\mathbf{0}'$ in Sacks [208].

- The reader can also find proofs of Theorems 13.6, 13.7, 13.12, and 13.15 in Soare [241].

- The degrees of unsolvability are also covered by (in order of appearance) Rogers [202], Cutland [51], Odifreddi [173], Cooper [43], Ambos-Spies and Fejer [11], Enderton [67], and Weber [275].

- A *survey* of the results on $(\mathcal{D}, \leq, ')$ to the end of the 1970s is Simpson [232], to the mid-1980s Lerman [139], and from the early 1980s on Shore [219]. A survey of current knowledge of this area is Slaman [235] and Soare [246].

Chapter 14
C.E. Degrees and the Priority Method

*If something has priority over other things, it is regarded as
being more important than them and is dealt with first.
An injury is damage done to somebody's body.*

Abstract Among the Turing degrees, the so-called computably enumerable (c.e.)
degrees are all-important. This is because they stem from c.e. sets, the sets that often
spring up in practice. In this chapter we will present the basic facts of c.e. degrees.
We will then describe *Post's Problem*, a problem about c.e. degrees that was posed
by Emil Post in 1944. After a series of attempts by Post and others, the problem
was finally solved in 1956 by Muchnik and Friedberg. They simultaneously and
independently devised a method, called the *Priority Method*, and applied it to solve
the problem. We will describe *Post's Problem* and the *Priority Method*.

14.1 C.E. Turing Degrees

Undecidable c.e. sets are interesting for several reasons. The first reason is that most
of the undecidable decision problems that have arisen in practice, i.e., outside pure
Computability Theory, are represented by undecidable c.e. sets. (See Sect. 8.3 for a
number of such problems that emerged in various areas of science.) The second reason is that, roughly speaking, any such set is in a way "close" to the decidable sets,
in the sense that it is more amenable to recognition than any other, non-c.e. set. Put
differently, for any such set the membership problem is semi-decidable since there
exists an algorithm that is capable of recognizing any member of the set, though
incapable of recognizing every non-member of the set. (See Sect. 8.2.)

Remarks. An undecidable c.e. set is mirrored in the corresponding decision problem. There exists
an algorithm capable of solving any positive instance of the problem, but no algorithm is capable
of also solving any negative instance. Informally, such a decision problem is just semi-solvable.

As a consequence, much research has been and still is devoted to c.e. sets and their
degrees of unsolvability. We will describe some of this research in this chapter.

Completeness and c.e. Degrees

In 1944, Post proved that the set $\mathcal{K}$ has the following interesting property: Every c.e. set is 1-reducible to $\mathcal{K}$. (See Definition 9.2 (p. 228) and Problem 9.3 (p. 228).) Since 1-reducibility is also m-reducibility (Sect. 9.2.4), and hence T-reducibility, it followed that every c.e. set is T-reducible to $\mathcal{K}$. Post called a set such as $\mathcal{K}$ *complete* relative to T-reducibility. We now define the concept of completeness more generally, i.e., relative to an arbitrary reduction, $\leq_C$. (See discussion on $\leq_C$ in Sect. 9.2.1.)

Definition 14.1. (Complete Set) Let $\leq_C$ be an arbitrary reducibility. A c.e. set $\mathcal{S}$ is said to be $\mathcal{C}$-**complete** if $\mathcal{A} \leq_C \mathcal{S}$ for every c.e. set $\mathcal{A}$.

So, Post was aware that $\mathcal{K}$ is T-complete. As this played an important role in his further research, we state it as a theorem.

Theorem 14.1. *The set $\mathcal{K}$ is T-complete.*

Just as c.e. sets are important, so are their degrees of unsolvability. These we introduce in the next definition.

Definition 14.2. (c.e. Degree) A T-degree is **computably enumerable (c.e.)** if it contains a c.e. set.

Remark. It is customary to call a T-degree that is computably enumerable just a *c.e. degree*.

Since $\emptyset$ and $\mathcal{K}$ are c.e. sets (see Sect. 8.2.1), both $\mathbf{0}$ and $\mathbf{0}'$ are c.e. degrees.

14.2 Post's Problem

In 1944, Post asked whether there are any c.e. degrees strictly between $\mathbf{0}$ and $\mathbf{0}'$. Here is the citation from his paper of 1944:

> A primary problem in the theory of recursively enumerable sets is the problem of determining the degrees of unsolvability of the unsolvable decision problems thereof. We shall early see that for such problems there is certainly a highest degree of unsolvability. Our whole development largely centers on the single question of whether there is, among these problems, a lower degree of unsolvability than that, or whether they are all of the same degree of unsolvability.

Remarks. "Recursively enumerable sets" is the old name for c.e. sets. The problems of interest to Post are decision problems associated with undecidable c.e. sets, that is, undecidable semi-decidable decision problems. Clearly, the highest degree of unsolvability of such problems is $0'$. Post asks whether or not any such problem can be of lower degree of unsolvability than $0'$.

This question became known as *Post's Problem.*

Definition 14.3. (Post's Problem) Is there a c.e. degree c such that $0 < c < 0'$?

We already know that there are T-degrees strictly between 0 and $0'$ (Theorem 13.5). However, the T-degrees constructed in the proof of this theorem need not be c.e. This is why Post embarked on solving this problem. In the next section we will describe his approach. Although Post's approach proved to be only partially successful, it nevertheless brought many new ideas and, more importantly, motivated researchers to search for different methods that would lead to a solution to the problem.

14.2.1 Post's Attempt at a Solution to Post's Problem

To solve this problem positively, Post wanted to prove that there is a c.e. set $\mathcal{A}$ such that $\emptyset <_T \mathcal{A} <_T \mathcal{K}$. How would he do that? He had the following idea: If he managed to prove that there exists an undecidable c.e. set $\mathcal{A}$ such that $\mathcal{K} \not\leq_T \mathcal{A}$, then $\emptyset <_T \mathcal{A} <_T \mathcal{K}$ would follow. In other words, Post wanted to find an *undecidable c.e.* set that is *T-incomplete*. So he set a list of goals—called *Post's Program*—that should be attained in order to solve the problem.

Post's Program had three goals:

A. define a *property* of a c.e. set;
B. prove that any set with this property is *undecidable* and *T-incomplete*;
C. prove that there *exist* c.e. sets with this property.

He started modestly, with the m-reducibility $\leq_m$, which is more special than $\leq_T$, the Turing reducibility. Then, if he succeeded in fulfilling his program with $\leq_m$, he would try to generalize the proof to the reducibility $\leq_T$.

Suppose that $\mathcal{K} \leq_m \mathcal{A}$ for some c.e. set $\mathcal{A}$. Then, there is a computable function r such that $r(\mathcal{K}) \subseteq \mathcal{A}$ and $r(\overline{\mathcal{K}}) \subseteq \overline{\mathcal{A}}$ (see Problem 9.1b on p. 228). Post's idea was to achieve $\mathcal{K} \not\leq_m \mathcal{A}$ by defining $\mathcal{A}$ so that $\overline{\mathcal{A}}$ would be unable to contain $r(\overline{\mathcal{K}})$, for any computable function r. Thus, the complement $\overline{\mathcal{A}}$ would be too "sparse" to accommodate $r(\overline{\mathcal{K}})$, in the sense that it would lack certain objects which, in contrast, abound in $\overline{\mathcal{K}}$. Post hoped that this property of $\mathcal{A}$ would prevent the m-reduction of $\mathcal{K}$ to $\mathcal{A}$.

How should he define this property of "having a sparse complement"? Post analyzed the properties of the set $\mathcal{K}$ and discovered that the complement $\overline{\mathcal{K}}$ contains infinitely many c.e. subsets. (See the details in Box 14.1.) So, if $\overline{\mathcal{A}}$ had *no* c.e. subsets at all, it might be unable to host $r(\overline{\mathcal{K}})$, for any computable function r. This led him to the definition of *simple* sets. Informally, a c.e. set $\mathcal{A}$ is simple if $\overline{\mathcal{A}}$ is "sparse," in the sense that, although infinite, $\overline{\mathcal{A}}$ does not contain any infinite c.e. set. Here is the definition.

Definition 14.4. (Simple Set)   A set $\mathcal{A}$ is **simple** if $\mathcal{A}$ is c.e. and $\overline{\mathcal{A}}$ is infinite but $\overline{\mathcal{A}}$ contains no infinite c.e. set.

Next, in accordance with goal B of his program, it was necessary to prove that a simple set $\mathcal{A}$ would actually do the job, i.e., guarantee that $\mathcal{K} \not\leq_m \mathcal{A}$. In other words, a simple set would be *m*-incomplete. That this is indeed so, Post proved in the following theorem.

Theorem 14.2.   *If a set $\mathcal{A}$ is simple, then $\mathcal{A}$ is not m-complete.*

It remained to be proved that a simple set actually exists (goal C). Using diagonalization, Post proved the following theorem.

Theorem 14.3.   *There exists a simple set $\mathcal{A}$.*

Proof. We omit the proof. See the Bibliographic Notes to this chapter.   □

In summary, Post discovered that there exist c.e. sets that are strictly between $\emptyset$ and $\mathcal{K}$ in the $\leq_m$ ordering. So he could write the following corollary.

Corollary 14.1.   *There is a c.e. set $\mathcal{A}$ such that $\emptyset <_m \mathcal{A} <_m \mathcal{K}$.*

Yet, this was not the final solution to his problem, because the question was whether the statement of Corollary 14.1 holds for the Turing reducibility $\leq_T$.

In order to generalize the above result to the ordering $\leq_T$, Post defined a more general notion of reducibility, the *bounded truth-table reducibility* $\leq_{btt}$. (See the Bibliographic Notes to this chapter for definitions of the notions that we mention in this paragraph.) Post was able to prove that if $\mathcal{A}$ is simple, then $\mathcal{A}$ is *btt*-incomplete (i.e., $\mathcal{K} \not\leq_{btt} \mathcal{A}$). Thus, there is a c.e. set $\mathcal{A}$ such that $\emptyset <_{btt} \mathcal{A} <_{btt} \mathcal{K}$. This success led him to try with the still more general notion of *truth-table reducibility* $\leq_{tt}$ (that is, $\leq_{btt}$ with no bounds). However, it turned out that if $\mathcal{A}$ is simple, the relation

$\mathcal{K} \nleq_{tt} \mathcal{A}$ *cannot* be proved. Since this indicated that the notion of the simple set was too weak, he defined the more powerful notion of the *hyper-simple set*. He was then able to prove that (goal B) if $\mathcal{A}$ is hyper-simple, then $\mathcal{A}$ is *tt*-incomplete (i.e., $\mathcal{K} \nleq_{tt} \mathcal{A}$), and (goal C) a hyper-simple set actually exists. It followed that there is a c.e. set $\mathcal{A}$ such that $\emptyset <_{tt} \mathcal{A} <_{tt} \mathcal{K}$.

The reducibilities $\leq_1, \leq_m, \leq_{btt}$, and $\leq_{tt}$ are called *strong*. This is because they are obtained by imposing additional conditions on the Turing reducibility $\leq_T$. We now see that Post aimed to gradually relax the strength of the additional conditions and eventually obtain the Turing reducibility; during this he would define, for each new weaker condition $\mathcal{C}$, a new kind of c.e. sets (goal A), prove that such sets meet the condition $\emptyset <_{\mathcal{C}} \mathcal{A} <_{\mathcal{C}} \mathcal{K}$ (goal B), and prove the existence of such sets (goal C).

Unfortunately, his progress stalled after success with the truth-table reducibility ($\mathcal{C} = tt$), and this was still far from the Turing reducibility. To continue, he defined *hyperhyper-simple* sets, but he couldn't prove their existence.

Box 14.1 (Creative Sets).

When Post analyzed the set $\mathcal{K}$ and its complement $\overline{\mathcal{K}}$, he discovered that $\mathcal{K}$ possesses an interesting property, which he then called *creativity*. Here is the definition of a creative set.

Definition 14.5. (Creative Set) A set $\mathcal{C}$ is **creative** if $\mathcal{C}$ is c.e. and there is a p.c. function φ such that $(\forall x)[\mathcal{W}_x \subseteq \overline{\mathcal{C}} \implies \varphi(x)\downarrow \wedge \varphi(x) \in \overline{\mathcal{C}} - \mathcal{W}_x]$.

What does that mean? First, observe that $\mathcal{W}_x$ denotes the xth c.e. set. (To see that, combine Definition 6.4 (p. 135) and Problem 6.7 (p. 151).) The function φ produces, for any $\mathcal{W}_x$, an element $\varphi(x)$ *witnessing* that $\overline{\mathcal{C}} \neq \mathcal{W}_x$ (as $\varphi(x) \in \overline{\mathcal{C}} - \mathcal{W}_x$). We call φ the *production function* of the creative set $\mathcal{C}$. Since we can effectively produce, for any $\mathcal{W}_x$, a counterexample for the assertion $\overline{\mathcal{C}} = \mathcal{W}_x$, we say that $\overline{\mathcal{C}}$ is *effectively* non-c.e. Hence, $\mathcal{C}$ is an effectively undecidable set.

The set $\overline{\mathcal{C}}$ contains an infinite c.e. set; moreover, it contains infinitely many infinite c.e. sets (see Problem 14.2).

Since the set $\mathcal{K}$ is creative (see Problem 14.3), $\overline{\mathcal{K}}$ is effectively non-c.e. and contains infinitely many infinite c.e. sets. Now Post's next step was obvious: For the searched-for set $\mathcal{A}$, take a *non-creative* set. *Simple* sets are suitable non-creative sets (see Definition 14.4).

In summary, the ultimate goal of *Post's Program* for solving *Post's Problem* was to define an appropriate *structural* property of c.e. sets that would guarantee the existence, undecidability, and incompleteness of c.e. sets having this property. As we have seen, the structural property that attracted Post was "sparseness of the complement." Could Post have succeeded in attaining his program had he continued his investigation in this direction? The answer is no. In 1965, Yates[1] proved that the structural property of having a "sparse complement", and the new hyperhyper-simple sets could not lead to a positive solution to *Post's Problem*.

---

[1] C. E. M. Yates, British mathematician, logician, and computer scientist.

14.3 The Priority Method and Priority Arguments

Nevertheless, today we know that the answer to *Post's Problem* is positive: There is a c.e. set $\mathcal{A}$ such that $\emptyset <_T \mathcal{A} <_T \mathcal{K}$, and there is a c.e. degree a such that $0 < a < 0'$. How was the problem resolved?

Post posed his problem and described his attempt at a solution in 1944. Ten years later, in 1954, when Kleene and Post published their seminal paper containing many other discoveries and ideas (see Chap. 13), the problem was still open. In the paper, Kleene and Post also introduced the *method of finite extensions*, which they used to prove the existence of $\leq$-incomparable T-degrees (see Theorem 13.5). This attracted the attention of several researchers. In 1956, two of them, Friedberg[2] and Muchnik[3], simultaneously and independently upgraded the finite extensions method into a subtler one, the *Finite-Injury Priority Method*, as it is called today. By applying it, Friedberg and Muchnik obtained a positive answer to *Post's Problem*.

14.3.1 The Priority Method in General

We will now describe the *Priority Method* in general. Let P be a property sensible for sets. Is there a c.e. set with the property P? To answer the question affirmatively, we can embark on the construction of such a set. We can try to construct a c.e. set S with the property P by adhering to the guidelines described in the following.

1. The set S will be constructed step by step, in an infinite sequence of *stages*. Each stage, say i, will construct a finite set S_i, which will be an approximation of the set S. Informally, S_i will be obtained by adding new elements into S_{i-1} and/or banning certain elements from entering S_i. Intuitively, we want the sets S_i to monotonically grow as i increases, and eventually (in the limit) develop into the set S. The plan is to define the stages in such a way that two objectives will be achieved:

Objective 1: $S_{i-1} \subseteq S_i$, for every $i \geqslant 1$;

Objective 2: $\bigcup_{i=1}^{\infty} S_i = S$.

That is, each S_i should be a better approximation of S than S_{i-1}, so that $\lim_{i \to \omega} S_i = S$. But, isn't our plan unrealistic? How can we approximate an unknown object such as S? If S exists, it will become fully known only after infinitely many stages of the construction; and yet, we intend to obtain, at each stage i, a *better* approximation S_i of S. How will we evaluate the quality of S_i? Will S_i correctly extend S_{i-1}? This information will be needed to avoid missing the objectives.

---

[2] Richard Michael Friedberg, b. 1935, American theoretical physicist.

[3] Albert Abramovič Mučnik, 1934–2019, Russian mathematician.

2. The above situation is resolved by making a radical change in our view of what is approximated during the construction. Instead of approximating the unknown set S, we approximate the known property P. The idea is that the properties of each finite set S_i should approximate the property P; that is, they should satisfy, *at least partially*, the property P. To implement this idea, we must somehow *atomize P*, i.e., break it into a set of primitive properties. Only then will a set S_i be capable of fulfilling a part of P. The primitive properties are called the *requirements*. Clearly, it is sensible to atomize P in such a way that any set fulfilling *all* the requirements will fulfill the whole of P. (Such would be the set S.) To ensure this, we must break P into a *conjunction* of requirements. So we have developed the first two guidelines:

G1: Write P as a conjunction $R_0 \wedge R_1 \wedge R_2 \dots$ of countably many requirements R_i.
G2: Do G1 in such a way that a set has the property $P \Longleftrightarrow$ the set fulfills every R_i.

3. What is a requirement and how is it fulfilled? In order to keep it as simple as possible, a requirement is only allowed to specify that certain numbers must be added (i.e., *enumerated*) into the set S_i and/or that certain other numbers must be kept out of this set (for some i). So, a requirement will be fulfilled by carrying out finitely many instructions of the form

$$x \in! S_i \quad \text{(add } x \text{ into } S_i)$$
or
$$x \notin! S_i \quad \text{(ban } x \text{ from } S_i).$$

Note that $x \notin! S_i$ does not mean that x is deleted from S_i; it means that x is *kept out of S_i*. A requirement R is *initiated* (tried to be fulfilled) at the stage when it *receives attention*. At that stage, say i, R can be fulfilled simply by ensuring the presence or absence of a particular number in the set S_i, i.e., by imposing either $x \in! S_i$ or $x \notin! S_i$, where x is a *candidate*—a number whose *status* (potential membership in S_j) has not been considered, for any $j < i$. In general, there are infinitely many candidates and we must choose one of them. This is summarized in the following guideline:

G3: At stage i, a requirement R is fulfilled by deciding, for a candidate x, whether $x \in! S_i$ or $x \notin! S_i$.

Remark. Fulfilling R uncovers just a small part of the information about S, namely, the status of the candidate number. The current information about the contents of S, which has been gathered by fulfilling requirements up to the end of stage i, is represented by S_i. Since S_i fulfills more requirements than S_{i-1}, it meets P better than S_{i-1}, and hence better approximates S.

4. It is now obvious that we must arrange the construction in such way that each and every requirement will eventually be considered. If we succeed in finding such an arrangement, then each and every requirement will get a chance to be fulfilled, so that *all* the requirements *may* eventually become fulfilled. In that case, they will be fulfilled by the limit set $\lim_{i \to \omega} S_i$. According to $G2$, this set will have the property P, so it will be the searched-for set S.

An intuitively appealing guideline for considering the requirements would be:
Consider R_0, then consider R_1, then consider R_2, and so on. Actually, this arrangement worked in the proof of Theorem 13.5 (p. 277), and the reason for this is easy to
find: *A requirement, once fulfilled, always remained fulfilled.* The proving method
where this holds is called the *Finite Extension Method*, for short the *FEM*. But the
FEM is not considered to be a true *Priority Method*, because it cannot deal with the
more general situation, which we describe in the following.

5. Unfortunately, considering the requirements in a simple succession may not
suffice. Why? By any stage only finitely many requirements can be fulfilled, which
leaves infinitely many Rs to be considered and fulfilled in the rest of the construction. But, the future requirements may not be independent of those already fulfilled;
they may interact with each other. So, we are faced with a permanent lack of information about future Rs—and this may have significant consequences.

In particular, it may turn out, at any stage i, that it is impossible to fulfill the
current requirement R. How can that happen? There can be two reasons. First, at
some previous stage $j < i$, a requirement R' was fulfilled in a *wrong* way; that is,
R' was fulfilled by banning a candidate y from S_j, i.e., setting $y \notin! S_j$, and this
decision now, at stage i, *prevents* us from fulfilling R. In short, we banned from S_j a
candidate that we shouldn't have—and we could not anticipate that at stage j. The
second reason can be that R' and R are contradictory requirements and cannot both
be fulfilled. (Isn't there also a third possibility? Cannot $y \in! S_j$ be a wrong decision?
Since the set S is to be c.e., we assume that the decision $y \in! S_j \subset S$ is well grounded
and can be made effectively, so there will be no need to revoke it.)

We see that there should be a possibility of returning to a previously fulfilled requirement and fulfilling it in some other way that would allow us to proceed with the
construction. But, things are more complicated than that: Changing a decision about
S_j (e.g., from $y \notin! S_j$ to $y \in! S_j$) may affect the sets $S_{j+1}, \ldots, S_{i-1}$ (which were constructed from S_j), so also the requirements fulfilled at stages $j + 1, \ldots, i - 1$ may
have to be reconsidered and fulfilled in some other way.

6. How can the bewildering situation described in **5** be controlled so that all the
changes will be systematically enforced? Here, Friedberg and Muchnik introduced
a new ingredient in the method: They assumed that different requirements have
different *priorities*. Informally, R's priority is used to represent the *importance* of R.
Here is the new guideline.

G4: With different requirements associate different priorities.

It proves to be useful to index the Rs in such a way that their priorities decrease as
the index increases; that is, R_k is of higher priority than R_{k+1}, for any $k \geqslant 0$. So, a
requirement with a lower index has higher priority and is therefore more important.
From now on we will assume that such an indexing has been done.

7. The construction is arranged so that, at each stage, the next initiated requirement is the one that has the highest priority among the requirements that require attention at that stage. We say that such a requirement *receives attention*.

G5: At any stage initialize the highest-priority R that requires attention at that stage.

8. How did Friedberg and Muchnik apply the concept of priorities to resolve the situation described in paragraph **5**? Recall the situation: To fulfill the current R, a previously fulfilled R' should be reconsidered and fulfilled in some other way. Whether or not this is allowed will depend on the priorities of R and R' in the following way:

- R *has higher priority than* R'. If R' was fulfilled by wrongly choosing $y \notin! \, S_j$, we are allowed to return to R', revoke that choice, and change it to $y \in! \, S_j$. This will enable the fulfillment of R. However, it will also turn R' unfulfilled again. We say that R' has been *injured* by R. (We will have to cure R', that is, fulfill R' in some other way that does not affect R. We will do that by considering some other candidate number y' and choosing either $y' \in! \, S_j$ or $y' \notin! \, S_j$.)
- R *has a lower priority than* R'. In this case R is not allowed to injure R'.

This is summarized in the next guideline.

G6: A requirement can be injured only by a higher-priority requirement.

9. Friedberg and Muchnik's next key assumption was that an injured requirement R will be reconsidered *after* all the injured requirements having higher priority than R are cured. Here is the guideline.

G7: An injured R will be initiated after all higher-priority injured Rs are cured.

If R_k is injured, then there can be at most k more important injured requirements: $R_0, \ldots, R_{k-1}$. It will take finitely many steps to cure all of them, so an attempt at R_k's recovery is guaranteed to start after a finite number of stages.

10. Although only finitely many requirements can injure an R, there might still exist a requirement that would injure R infinitely many times. In such a case, R would never stop being injured and could not recover once and for all. Could the entire conjunction $R_0 \wedge R_1 \wedge R_2 \ldots$ then be fulfilled? To prevent such a situation, Friedberg and Muchnik assumed that each requirement can be injured only finitely many times. So, they introduced the following guideline.

G8: A requirement can be injured only finitely many times.

This concludes the general description of the *Priority Method* for the construction of a set S with property P. However, there are several variations of the method. For example, when the method adheres to all of the guidelines *G1, ..., G8*, it is called the *Finite-Injury Priority Method (FIPM)*. When *G8* cannot be assumed, we obtain the *Infinite-Injury Priority Method (IIPM)*.

A proof using any kind of *Priority Method* is called a *priority argument*. Priority arguments have been classified into a hierarchy based on their complexity.

14.3.2 The Friedberg-Muchnik Solution to Post's Problem

Using their *Priority Method*, Friedberg and Muchnik proved that there exist *c.e.* sets $\mathcal{A}$ and $\mathcal{B}$ such that $\mathcal{A} \not\leq_T \mathcal{B} \wedge \mathcal{B} \not\leq_T \mathcal{A}$. (As $\mathcal{A}, \mathcal{B}$ are c.e., also $\mathcal{A} \leq_T \emptyset'$ and $\mathcal{B} \leq_T \emptyset'$.) It immediately followed that $\deg(\mathcal{A})$ and $\deg(\mathcal{B})$ are $\leq$-incomparable c.e.-degrees— and *Post's Problem* was solved positively. Here is the Friedberg-Muchnik theorem.

Theorem 14.4. *There exist incomparable c.e.-degrees.*

Proof. We describe the idea of the proof; for the details see the Bibliographic Notes to this chapter. Friedberg-Muchnik's general goal was to extend the proof of Theorem 13.5 (p. 277) to the case of c.e. sets. So the requirements stated in that proof remain the same (now they are all denoted by Rs):

$$R_{2e} : \chi_{\mathcal{A}} \neq \Phi_e^{\mathcal{B}}$$
$$R_{2e+1} : \chi_{\mathcal{B}} \neq \Phi_e^{\mathcal{A}}$$

Fulfilling a Single Requirement. To fulfill R_{2e} we first associate with it a candidate number, an x whose status (membership in $\mathcal{A}$) is still open. Then we look for a stage $s+1$ such that $\Phi_e^{\mathcal{B}_s}(x){\downarrow} = 0$. If there is no such stage, then this means that $x \notin \mathcal{A}$ and $\Phi_e^{\mathcal{B}}(x){\uparrow} \vee \Phi_e^{\mathcal{B}}(x){\downarrow} \neq 0$, implying that x fulfills R_{2e}. If, however, such an $s+1$ exists, then R_{2e} will *require attention* at that stage.

When, at stage $s+1$, R_{2e} actually *receives attention*, we do the following:

- add x into $\mathcal{A}_{s+1}$ (and, hence, into $\mathcal{A}$);
- protect the construction. We do this by trying to restrain too-small numbers y from later entering $\mathcal{B}$ by any requirement of lower priority than R_{2e}. More specifically, we choose new candidates for all (lower-priority) requirements $R_k, k > 2e$, and initialize them. This ensures that only (higher-priority) requirements $R_k, k < 2e$, can later injure R_{2e} by adding some small y into $\mathcal{B}$.

This fulfills the requirement R_{2e}. If later a requirement of higher priority than R_{2e} is enumerated into $\mathcal{B}$ and injures R_{2e}, then R_{2e} is initialized and must be fulfilled again by another candidate. (To fulfill a requirement R_{2e+1} we use the same strategy as for R_{2e} but with the roles of $\mathcal{A}$ and $\mathcal{B}$ reversed.)

Fulfilling All Requirements. We saw that, occasionally, we must choose new candidate numbers for some requirements. There are two restrictions on this: First, for any R_k, we can choose another candidate for R_k only finitely often; and second, candidates for different requirements must be distinct. The latter restriction is met by choosing all candidates for a requirement R_k from the set $\mathbb{N}^{[k]} \stackrel{\text{def}}{=} \{\langle n, k \rangle \mid n \in \mathbb{N}\}$.

Construction of $\mathcal{A}$ and $\mathcal{B}$. Initially, we have $\mathcal{A}_0 = \mathcal{B}_0 = \emptyset$. At stage $s > 0$ we do the following. Let R_k be the highest-priority unfulfilled requirement and let r be the stage at which R_k was most recently initialized; of course, $r < s$. Then we determine x in the following way:

- if $k = 2e$: Let x be the least $x \in \mathbb{N}^{[k]} - \mathcal{A}_{s-1}$ such that $x > r$ and $\Phi_e^{\mathcal{B}_{s-1}}(x) = 0$;
- if $k = 2e+1$: Let x be the least $x \in \mathbb{N}^{[k]} - \mathcal{B}_{s-1}$ such that $x > r$ and $\Phi_e^{\mathcal{A}_{s-1}}(x) = 0$.

In the first case we add x into $\mathcal{A}$, and in the second case we add x into $\mathcal{B}$. (Observe that $x < s$.) We then declare R_k fulfilled and initialize all requirements of lower priority (as described above).

It can be proved that, for every k, requirement R_k receives attention at most finitely often, is injured at most $2^k - 1$ times, and is eventually fulfilled forever.   □

14.3.3 Priority Arguments

We have seen that Friedberg and Muchnik's solution to *Post's Problem* is somewhat difficult to follow. This is not a coincidence. Today, when the *Priority Method* is the main technique for establishing results about c.e. sets, it is known that priority arguments are usually very complex and sophisticated. First, the requirements and the strategy by which they are fulfilled must be carefully constructed to produce the required result—and this must be done for each problem separately. In addition, requirements can be first fulfilled and later injured, so the membership of a number in the constructed set can be first determined and later undone.

However, as the results obtained by priority arguments have been multiplying, attempts to make priority arguments more systematic and intelligible have also started appearing. The aim of these attempts is to isolate the general principles that are common to existing priority arguments (and, hopefully, to those yet to be constructed). Ultimately, the attempts should lead to a *framework* offering a uniform approach to the construction of priority arguments.

While waiting for such a systematic simplification of priority arguments, it has become desirable to prove results without priority arguments, or to see whether results proved with priority arguments can also be proved without them. For example, in 1986, Kučera[4] devised a proof of *Post's Problem* without using the priority method. Kučera's proof is involved too, but the resulting set is less artificial.

14.4 Some Properties of C.E. Degrees

When it became known that there are more than just two c.e. degrees, research into c.e. degrees started to flourish. Many of the results were (and are being) obtained by priority arguments. We now briefly list some of the first results. For further results see the Bibliographic Notes to this chapter.

1. *Every c.e. degree is $\leq \mathbf{0}'$.*
2. *Not every T-degree which is $< \mathbf{0}'$ is a c.e. degree.*
3. *Density Theorem: Between any two c.e. degrees there is a third c.e. degree.*
4. *There are two c.e. degrees with no glb in the c.e. degrees.*
5. *There is a pair of nonzero c.e. degrees whose glb is $\mathbf{0}$.*
6. *Nondiamond Theorem: There is no pair of c.e. degrees whose glb is $\mathbf{0}$ and lub is $\mathbf{0}'$.*

The above theorems were proved in the mid-1960s. In particular, 3 was proved in 1964 by Sacks; 4 and 5 were proved in 1966 by Lachlan[5] and Yates; and 6 was proved in 1966 by Lachlan. See Bibliographic Notes to this chapter for the details.

---

[4] Antonín Kučera, Czech mathematician, logician, and computer scientist.

[5] Alistair H. Lachlan, Canadian mathematician, logician, and computer scientist.

14.5 Chapter Summary

A c.e. set is said to be Turing complete (T-complete) if every c.e. set is T-reducible to it. The set $\mathcal{K}$ is T-complete. A T-degree is said to be c.e. if it contains a c.e. set. Both $\mathbf{0}$ and $\mathbf{0}'$ are c.e. degrees. *Post's Problem* asks whether or not there exists a c.e. degree c that is strictly between $\mathbf{0}$ and $\mathbf{0}'$, that is, $\mathbf{0} < c < \mathbf{0}'$.

Post attempted a solution to his problem by devising a program which is now called *Post's Program*. His aim was to define a structural property of c.e. sets that would guarantee the existence, undecidability, and Turing incompleteness of c.e. sets having this property. In trying to achieve that, Post defined various reducibilities and special kinds of c.e. sets, but did not succeed in attaining his program.

A positive solution to *Post's Problem* was obtained in 1965 by Friedberg and Muchnik, who devised and used a new method called the *Priority Method*. In this method, a conjunction of simple requirements must be fulfilled. As more and more requirements are fulfilled, a larger and larger part of the set to be constructed is uncovered. However, the requirements are interrelated, so the requirements that have already been fulfilled may become unfulfilled again, thus temporarily concealing a part of the uncovered set. Such injured requirements must be fulfilled again in some other way, and the process repeats. Although involved, the *Priority Method* is today the main technique for establishing results about c.e. sets.

Proofs that use this method are called priority arguments. There are attempts to isolate the general principles common to priority arguments and to integrate them into a framework that would offer a uniform approach to the construction of priority arguments.

At the same time, priority-free proofs are searched for because the sets constructed by them are less artificial. One such is Kučera's priority-free proof of *Post's Problem*.

Problems

14.1. Prove: If $\mathcal{C}$ is a creative set, then $\overline{\mathcal{C}}$ contains an infinite c.e. subset.

[*Hint.* Let φ be the production function of $\mathcal{C}$. Let n be an index of the empty set, i.e., $\emptyset = \mathcal{W}_n$. Consider the set $\mathcal{W} = \{x_1, x_2, \ldots\}$, where x_i is defined inductively as follows: $\mathcal{W}_{x_1} = \{\varphi(n)\}$ and $\mathcal{W}_{x_{i+1}} = \mathcal{W}_{x_i} \cup \{\varphi(x_i)\}$. So, the construction of the set $\mathcal{W}$ starts with $\mathcal{W}_n = \emptyset$ as the first current set and then, in each step, adds to the current set its witness to obtain the next current set.]

14.2. Prove: If $\mathcal{C}$ is a creative set, then $\overline{\mathcal{C}}$ contains infinitely many infinite c.e. subsets.

[*Hint.* See Problem 14.1 and use any $n \in \text{ind}(\emptyset)$.]

14.3. Prove: $\mathcal{K}$ is a creative set.

Bibliographic notes

- Post formulated what is now called *Post's Problem* in [184]. See also Post [187].
- That $\mathcal{K}$ (actually $\mathcal{K}_0$) is 1-complete and hence also m-complete was proved in Post [184].
- The existence of *simple* sets was proved in Post [184]. See also Cooper [43, Chap. 6], Weber [275, Chap. 5], and Soare [241, Chap. V].
- *Strong reducibilities* $\leq_m, \leq_{htt}, \leq_{tt}$ were introduced in Post [184]. For more on these reducibilities see Odifreddi [172].
- That *Post's Program* could not succeed with the structural property of "sparseness of the complement" was proved in Yates [280].
- The solution to *Post's Problem* appeared in Muchnik [165, 166] and Friedberg [75]. In describing the solution we followed Nies [170] and Soare [241]. See also Cooper [43].
- A wealth of information about the ideas, techniques, and applications of the *priority argument* is Cooper [43, Chap. 12]. For the *Finite Injury* and *Infinite Injury Priority Method*, see Soare [241, Chaps. VII and VIII]. A framework for developing priority arguments is described in Lerman [140].
- A solution to *Post's Problem* that does not use the priority method and therefore has no injury to the requirements, was obtained in 1986 by Kučera [134]. See Cooper [43, Chap. 15], Nies [170, Chap. 4], and Soare [241, Chap. VII] for expositions of this proof.
- Theorems 3,4,5, and 6 from Sect. 14.4 were proved by Sacks [210], Lachlan [135, 136], and Yates [281], respectively.
- For an overview of c.e. sets, see Soare [243], and for an overview of c.e. degrees, see Shore [220].

Chapter 15
The Arithmetical Hierarchy

For every sensible question there is an answer; for every answer there is a sensible question.

Abstract In this chapter we will introduce a different view of sets of natural numbers. Sometimes such a set can be defined by a property of its members, where the property is expressed by a formula of *Formal Arithmetic*. Sets defined by formulas of the same complexity constitute an *arithmetical class*. Different complexities of formulas give rise to different arithmetical classes. There is also an ordering between these classes, so they form the so-called *Arithmetical Hierarchy*. We will show that the *Arithmetical Hierarchy* is closely connected with the *Jump Hierarchy*.

15.1 Decidability of Relations

Before we delve deeper into the main subject of this chapter, we must prepare the ground by defining a few new notions and proving some basic facts about them.

A k-ary *relation* on a set S is a subset $\mathcal{R}$ of S^k. If $\mathcal{R}$ is a k-ary relation on S, it is customary to write $R(a_1, \ldots, a_k)$ to indicate that $(a_1, \ldots, a_k) \subset \mathcal{R}$. When $k = 1$ we say that R is a *property* defined on S; when $k = 2$ we call R a *binary* relation on S. Just like any other set, the set $\mathcal{R}$ can also be decidable, semi-decidable, undecidable, or can have any other property sensible for sets. In this case we say that R has (or does not have) such a property. Here are the definitions that we will need.

Definition 15.1. (Decidable Relation) A k-ary relation R on a set S is **decidable** (or **semi-decidable**, or **undecidable**) if the corresponding set $\mathcal{R} \subseteq S^k$ is decidable (or semi-decidable, or undecidable).

Remarks. 1) Instead of decidable (semi-decidable, undecidable) relation, we may say computable (c.e., incomputable) relation. 2) If R is decidable, we can decide, for any $(a_1, \ldots, a_k) \in S^k$, whether or not $R(a_1, \ldots, a_k)$. If R is undecidable but still semi-decidable, there is an algorithm guaranteed to return an answer (a YES) only for k-tuples that are in $\mathcal{R}$, and which returns NO or fails to terminate for k-tuples in $\overline{\mathcal{R}}$.

R. Robič, *The Foundations of Computability Theory*, https://doi.org/10.1007/978-3-662-62421-0_15

We will now fix the set S to $S = \mathbb{N}$. Relations on $\mathbb{N}$ are said to be *arithmetical*.

Example 15.1. (Relation R_{Halt}) Let us define an arithmetical relation $R_{Halt}(e,x,s)$ as follows:

$$R_{Halt}(e,x,s) \equiv \text{"Turing machine } T_e \text{ halts on input } x \text{ in at most } s \text{ steps and returns a result."}$$

This relation is decidable: Given any $(e,x,s) \in \mathbb{N}^3$, construct T_e (see Sect. 7.2), start T_e on x, and let it run *until* the first s steps have been completed or a result has been output. □

Let us now take an arbitrary decidable relation $R(x,y)$ on $\mathbb{N}$ and define the set $\mathcal{A}$ to be $\mathcal{A} = \{x \in \mathbb{N} \mid \exists y R(x,y)\}$. So $\mathcal{A}$ consists of those numbers that are R-related to some number. What can be said about the decidability of $\mathcal{A}$? Here is the answer.

Theorem 15.1. *A set $\mathcal{A} \subseteq \mathbb{N}$ is c.e. iff $\mathcal{A} = \{x \in \mathbb{N} \mid \exists y R(x,y)\}$ for some decidable relation R on $\mathbb{N}$.*

Proof. ($\Leftarrow$) Let R be an arbitrary decidable relation on $\mathbb{N}$. Let $D_{\mathcal{R}}$ be a decider of the corresponding set $\mathcal{R}$, and $G_{\mathbb{N}^2}$ a pair generator (see Sect. 6.3.4 and Box 6.3). Then we can construct a generator $G_{\mathcal{A}}$ of the set $\mathcal{A}$ as follows. $G_{\mathcal{A}}$ repeats the following sequence of steps: (1) it calls $G_{\mathbb{N}^2}$ to generate the next (x,y); (2) it calls $D_{\mathcal{R}}$ to see whether (x,y) is in $\mathcal{R}$; (3) if $(x,y) \in \mathcal{R}$ then it generates (outputs) x. Since $\mathcal{A}$ can be generated by a TM, it is a c.e. set. ($\Rightarrow$) If $\mathcal{A}$ is c.e., then it is the domain of a p.c. function, φ_e (Problem 6.7). So $\mathcal{A} = \{x \in \mathbb{N} \mid \exists s R_{Halt}(e,x,s)\}$, with R_{Halt} from Example 15.1. □

15.2 The Arithmetical Hierarchy

In the 1940s, Kleene and Mostowski[1] were exploring the sets of natural numbers that are defined as $\{x \in \mathbb{N} \mid F(x)\}$, where $F(x)$ is a formula of *Formal Arithmetic*, and x is a free individual variable in F (see Sect. 3.2).

Kleene was investigating how the syntactical complexity of $F(x)$ affects the decidability of the set $\{x \in \mathbb{N} \mid F(x)\}$. The obvious question was how to measure the syntactical complexity of $F(x)$. Here, Kleene leaned on the well-known fact, which was published already in 1885 by Peirce, that every formula can be transformed into *prenex normal form* (pnf), i.e., a logically equivalent formula consisting of a string of quantifiers followed by a quantifier-free formula. Kleene could therefore assume that $F(x)$ is of the form $Q_1 y_1 \ldots Q_k y_k R(x,y_1,\ldots,y_k)$, where $y_i \neq y_j$ for $i \neq j$, Q_i is $\forall$ or $\exists$, and R is a decidable arithmetical relation (predicate). In addition, adjacent quantifiers of the same kind (i.e., those having the same quantification symbol) can be contracted and replaced by a single quantifier of that kind (see Box 15.1). After all possible contractions have been performed, the resulting prenex normal form has a sequence of *alternating* quantification symbols, i.e., the formula $F(x)$ has been transformed either into the form

---

[1] Andrzej Mostowski, 1913-1975, Polish mathematician.

$$\exists y_1 \forall y_2 \exists y_3 \ldots Qy_n R(x, y_1, y_2, \ldots, y_n),$$

where Q is $\exists$ if n is odd, and $\forall$ if n is even; or into the form

$$\forall y_1 \exists y_2 \forall y_3 \ldots Qy_n R(x, y_1, y_2, \ldots, y_n),$$

where Q is $\forall$ if n is odd, and $\exists$ if n is even.

Box 15.1 (Contraction of Quantifiers in Prenex Normal Forms).

We describe the contraction of quantifiers on an example. Let $\exists y_1 \exists y_2 \forall y_3 \forall y_4 \forall y_5 P(y_1, y_2, y_3, y_4, y_5)$ be a formula. We introduce two individual variables $u = (y_1, y_2)$ and $v = (y_3, y_4, y_5)$. Recall that the projection function π_i^k returns the ith component of a k-tuple (Box 5.1, p. 82). Using u, v, π^2 and π^3 we transform the formula into an equivalent formula $\exists u \forall v P(\pi_1^2(u), \pi_2^2(u), \pi_1^3(v), \pi_2^3(v), \pi_3^3(v))$. This is the contracted pnf. Generally, given a formula $Q_1 y_1 \ldots Q_k y_k P(y_1, \ldots, y_k)$, we first partition $Q_1 y_1 \ldots Q_k y_k$ into maximal subsequences of adjacent quantifiers of the same kind; then we introduce, for each subsequence, a new individual variable (tuple); next, we replace each subsequence by the corresponding quantifier of that kind; and use projection functions in the predicate P.

So, Kleene could assume that $F(x)$ is already in the contracted prenex normal form. Any set $\{x \in \mathbb{N} \mid F(x)\}$ defined by such a distinguished $F(x)$ he called *arithmetical*.

Definition 15.2. (Arithmetical Set) A set $\mathcal{A}$ is an **arithmetical set** if $\mathcal{A} = \{x \in \mathbb{N} \mid F(x)\}$, such that, for some $n \geqslant 0$, the predicate $F(x) = \exists y_1 \forall y_2 \exists y_3 \ldots Qy_n R(x, y_1, y_2, \ldots, y_n)$ or $F(x) = \forall y_1 \exists y_2 \forall y_3 \ldots Qy_n R(x, y_1, y_2, \ldots, y_n)$, and R is a decidable arithmetical relation. In particular, when $n = 0$, we define $F(x) = R(x)$.

Then he defined the syntactical complexity of $F(x)$ as the number n of quantification symbols in $F(x)$. Depending on n and the first quantification symbol of $F(x)$, he classified the arithmetical sets $\{x \in \mathbb{N} \mid F(x)\}$ into various classes. He called these classes *arithmetical* and denoted them by Σ_n, Π_n, and Δ_n. Here is the definition.

Definition 15.3. (Arithmetical Classes) The **arithmetical classes** Σ_n, Π_n, and Δ_n are defined as follows:

$\Sigma_n = $ class of all sets $\{x \in \mathbb{N} \mid F(x)\}$, where $F(x) = \exists y_1 \forall y_2 \exists y_3 \ldots Qy_n R(x, y_1, \ldots, y_n)$
for some decidable arithmetical relation R;

$\Pi_n = $ class of all sets $\{x \in \mathbb{N} \mid F(x)\}$, where $F(x) = \forall y_1 \exists y_2 \forall y_3 \ldots Qx_n R(x, y_1, \ldots, y_n)$
for some decidable arithmetical relation R;

$\Delta_n - $ class of all sets $\{x \in \mathbb{N} \mid F(x)\}$ that are in $\Sigma_n \cap \Pi_n$.

Example 15.2. ($\mathcal{K}$ is in Σ_1) The set $\mathcal{K}$ is in Σ_1, because $\mathcal{K} \overset{\text{def}}{=} \{x \mid \varphi_x(x)\downarrow\} = \{x \mid \exists s R_{Halt}(x,x,s)\}$, where R_{Halt} is the (decidable) relation from Example 15.1. □

Example 15.3. ($\mathcal{K}_0$ is in Σ_1) This is because $\mathcal{K}_0 \overset{\text{def}}{=} \{\langle e,x\rangle \mid \varphi_e(x)\downarrow\} = \{\langle e,x\rangle \mid \exists s R_{Halt}(e,x,s)\} = \{\langle e,x\rangle \mid \exists s R_{Halt}(\langle e,x\rangle_1, \langle e,x\rangle_2, s)\}$, where $\langle e,x\rangle_1 = e$, $\langle e,x\rangle_2 = x$, and R_{Halt} is from Example 15.1. □

We now justify the title of this chapter, the *Arithmetical Hierarchy*. So, is there a hierarchy of arithmetical classes? Yes; in 1943, Kleene proved the following

Theorem 15.2. *For any $n \geqslant 0$, the following hold:*

$$\text{a) } \Sigma_n \subset \Sigma_{n+1} \qquad\qquad \text{b) } \Pi_n \subset \Pi_{n+1}$$
$$\text{c) } \Sigma_n \subset \Pi_{n+1} \qquad\qquad \text{d) } \Pi_n \subset \Sigma_{n+1}$$
$$\text{e) } \Delta_n \subset \Sigma_n$$
$$\text{f) } \Delta_n \subset \Pi_n$$

Proof idea. In the first part we prove, for each of the above relations, that the left-hand side is related to the right-hand side with the relation $\subseteq$. To do this, we introduce a *dummy* variable y_{n+1}. In the second part we prove, for each $\mathcal{X} \subseteq \mathcal{Y}$ of the relations $\Sigma_n \subseteq \Sigma_{n+1}$, $\Pi_n \subseteq \Pi_{n+1}$, $\Sigma_n \subseteq \Pi_{n+1}$, $\Pi_n \subseteq \Sigma_{n+1}$, $\Delta_n \subseteq \Sigma_n$, $\Delta_n \subseteq \Pi_n$, that $\mathcal{X} \neq \mathcal{Y}$. We use *diagonalization* to prove that there is a set that is in $\mathcal{Y}$ but not in $\mathcal{X}$. The diagonal argument is a generalization of the argument that we used in proving the existence of undecidable c.e. sets. (See Box 15.2 for further details.)

Box 15.2 (Proof of Theorem 15.2).

a) Let S be an arbitrary element of the arithmetical class Σ_n. Then $S = \{x \in \mathbb{N} \mid F(x)\}$, where $F(x) = \exists y_1 \forall y_2 \exists y_3 \ldots Q y_n R(x, y_1, \ldots, y_n)$ for some decidable arithmetical relation R. Now define a new relation $R'(x, y_1, \ldots, y_n, y_{n+1}) \overset{\text{def}}{=} R(x, y_1, \ldots, y_n) \wedge (y_{n+1} = y_{n+1})$ and a new formula $F'(x) \overset{\text{def}}{=} \exists y_1 \forall y_2 \exists y_3 \ldots Q y_n Q' y_{n+1} R'(x, y_1, \ldots, y_n, y_{n+1})$, where Q' denotes the alternate of Q. Finally, observe that $S = \{x \in \mathbb{N} \mid F'(x)\} \in \Sigma_{n+1}$.

b) The proof is similar to the proof of a), except that $F(x) = \forall y_1 \exists y_2 \forall y_3 \ldots Q y_n R(x, y_1, \ldots, y_n)$ and $F'(x) \overset{\text{def}}{=} \forall y_1 \exists y_2 \forall y_3 \ldots Q y_n Q' y_{n+1} R'(x, y_1, \ldots, y_n, y_{n+1})$.

c) Let $S \in \Sigma_n$ and $S = \{x \in \mathbb{N} \mid F(x)\}$, where $F(x) = \exists y_1 \forall y_2 \exists y_3 \ldots Q y_n R(x, y_1, \ldots, y_n)$ for some decidable arithmetical relation R. Observe that $F(x) = \exists y_1 \forall y_2 \exists y_3 \ldots Q y_n R(x, y_1, \ldots, y_n) = \forall y_{n+1} \exists y_1 \forall y_2 \exists y_3 \ldots Q y_n [R(x, y_1, \ldots, y_n) \wedge (y_{n+1} = y_{n+1})]$, where we have introduced a new variable y_{n+1}. Now define a new relation $R'(x, y_1, \ldots, y_n, y_{n+1}) \overset{\text{def}}{=} R(x, y_1, \ldots, y_n) \wedge (y_{n+1} = y_{n+1})$ and a new formula $F'(x) \overset{\text{def}}{=} \forall y_{n+1} \exists y_1 \forall y_2 \exists y_3 \ldots Q y_n R'(x, y_1, \ldots, y_n, y_{n+1})$. Then, after appropriate renaming of the variables y_i, we see that $S = \{x \in \mathbb{N} \mid F'(x)\} \in \Pi_{n+1}$.

d) The proof is similar to the proof of c), except that $F(x) = \forall y_1 \exists y_2 \forall y_3 \ldots Q y_n R(x, y_1, \ldots, y_n)$ and $F'(x) \overset{\text{def}}{=} \exists y_{n+1} \forall y_1 \exists y_2 \forall y_3 \ldots Q y_n R'(x, y_1, \ldots, y_n, y_{n+1})$.

e,f) $\Delta_n \subseteq \Sigma_n$ follows directly from the definition of Δ_n. The same holds for $\Delta_n \subseteq \Pi_n$.

We demonstrate the second part of the proof in cases e) and f). We will construct a set $\mathcal{P}$ such that $\mathcal{P} \in \Sigma_n - \Pi_n$ and $\overline{\mathcal{P}} \in \Pi_n - \Sigma_n$. Since $\Delta_n = \Sigma_n \cap \Pi_n$, parts e) and f) of the theorem will follow.

Each element of Σ_1 is a c.e. set and hence the domain of a p.c. function (see Problem 6.7). But p.c. functions can be effectively enumerated (see Proposition 6.1 and Definition 6.4). It follows that we can effectively enumerate the elements of Σ_1. We can also enumerate the elements of Π_1. This is because if $\mathcal{B} \in \Pi_1$, then $\mathcal{B} = \overline{\mathcal{A}_e}$ for some $\mathcal{A}_e \in \Sigma_1$, and we can rename $\mathcal{B}$ as $\mathcal{B}_e$. From the two enumerations we can construct enumerations of elements of Σ_2 and Π_2 and then of Σ_n and Π_n for higher n. Thus we can speak, for any $n \geqslant 1$, of the eth set of the class Σ_n or the class Π_n.

Now define a set $\mathcal{S}$ as follows: $\mathcal{S} \stackrel{\text{def}}{=} \{\langle e, x \rangle \in \mathbb{N} \mid \text{the } e\text{th set in } \Sigma_n \text{ contains } x\}$. The set $\mathcal{S}$ is in Σ_1, because we can write $\mathcal{S} = \{\langle e, x \rangle \in \mathbb{N} \mid \exists s R_{Halt}(e, x, s)\}$, where R_{Halt} is the relation from Example 15.1. Consequently, $\mathcal{S}$ is in Σ_n, for any $n \geqslant 1$. (Notice that $\mathcal{S}$ is related to the universal set (language) $\mathcal{K}_0$; see Definition 8.5 on p. 181.)

Based on $\mathcal{S}$ we finally define the set $\mathcal{P}$ as follows: $\mathcal{P} \stackrel{\text{def}}{=} \{x \in \mathbb{N} \mid \langle x, x \rangle \in \mathcal{S}\}$. The set $\mathcal{P}$ is in Σ_1, because $\mathcal{P} = \{x \in \mathbb{N} \mid \exists s R_{Halt}(x, x, s)\}$. ($\mathcal{P}$ is related to the diagonal set $\mathcal{K}$; see Definition 8.6.) Hence, $\mathcal{P} \in \Sigma_n$. However, $\mathcal{P} \notin \Pi_n$. (*Suppose the contrary:* $\mathcal{P} \in \Pi_n$. Then it would follow that $\overline{\mathcal{P}} \in \Sigma_n$, so $\overline{\mathcal{P}}$ would be the e'th set in Σ_n for some e'. Then, by definition of $\mathcal{P}$, we would have that $e' \in \overline{\mathcal{P}} \Leftrightarrow \langle e', e' \rangle \in \mathcal{S}$. But, by definition of $\mathcal{S}$, we would also have $e' \in \overline{\mathcal{P}} \Leftrightarrow e' \notin \mathcal{P} \Leftrightarrow \langle e', e' \rangle \notin \mathcal{S}$, which would be a contradiction.) So, $\mathcal{P} \in \Sigma_n - \Pi_n$. Similarly we find that $\overline{\mathcal{P}} \in \Pi_n - \Sigma_n$.

Informally, Theorem 15.2 e,f tell us that not every arithmetical set can be defined in both ways, that is, both as a member of Σ_n and as a member of Π_n. Next, for any $n \geqslant 0$, there are arithmetical sets $\{x \in \mathbb{N} \mid F(x)\}$ that cannot be defined by properties $F(x)$ having just n alternating quantifiers (Theorem 15.2 a,b,c,d).

The inclusions between the arithmetical classes are depicted in Fig. 15.1.

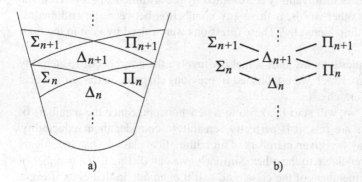

a)   b)

Fig. 15.1 a) The inclusions between the arithmetical classes at two successive levels; b) Hasse diagram representing the classes of two successive levels ordered by inclusion

To get some feeling for the arithmetical classes, let us see what kinds of sets are gathered in the classes at the lowest levels of the hierarchy. For $n = 0$, any formula $F(x)$ is just a decidable unary relation $R(x)$, so the corresponding set $\{x \in \mathbb{N} \mid F(x)\}$ is decidable. Obviously, this is also true of the members of Π_0 and Δ_0. Thus,

$$\Sigma_0 = \Pi_0 = \Delta_0 = \textit{class of all decidable sets.}$$

Now take $n = 1$. The class Σ_1 contains all the sets defined by $\{x \in \mathbb{N} \mid \exists y_1 R(x, y_1)\}$. By Theorem 15.1 any such set is c.e., so

$$\Sigma_1 = class\ of\ all\ c.e.\ sets.$$

What about the class Π_1? By definition it is the class of all the sets that are defined by $\{x \in \mathbb{N} \mid \forall y_1 R(x, y_1)\}$, or, equivalently, by $\{x \in \mathbb{N} \mid \neg\exists y_1 \neg R(x, y_1)\}$. But any such set is the complement of some c.e. set—namely, the set $\{x \in \mathbb{N} \mid \exists y_1 \neg R(x, y_1)\}$—and is therefore called *co-c.e.* Consequently,

$$\Pi_1 = class\ of\ all\ co\text{-}c.e.\ sets.$$

The class Δ_1 is, by definition, equal to $\Sigma_1 \cap \Pi_1$, so it contains all the sets that are both c.e. and co-c.e. But, by Theorem 7.3 (p. 156), such sets are decidable. It follows that

$$\Delta_1 = class\ of\ all\ decidable\ sets.$$

15.3 The Link with the Jump Hierarchy

What about the arithmetical classes $\Sigma_n, \Pi_n, \Delta_n, n = 2, 3, \ldots$, which reside on higher levels of the arithmetical hierarchy? What kinds of sets are gathered in these classes? Do all the sets in a given arithmetical class share the same degree of unsolvability, as was the case with $n = 0$ and partly with $n = 1$? If so, then is there any relationship between the degrees of unsolvability represented by the arithmetical classes and the Turing degrees? In other words, is there any connection between the arithmetical hierarchy and the jump hierarchy? These questions were raised by Post in the mid-1940s.

To answer the questions we must somehow involve the concepts of reducibility and the Turing jump, which we established in previous chapters, in the arithmetical classes and their hierarchy.

To achieve this, we will start by defining a new concept. Since the members of arithmetical classes are sets, it is perfectly sensible to consider their reducibility. For example, for any two given members of an arithmetical class we can investigate whether one is m-reducible to the other. Similarly, we can distinguish a member of a class from other members of the class, and call it *complete* in that class if every member of the class is m-reducible to it. Here is the definition.

Definition 15.4. (Σ_n-Complete Set) A set $\mathcal{A}$ is Σ_n-**complete** if $\mathcal{A} \in \Sigma_n$ and $\mathcal{X} \leq_m \mathcal{A}$ for every $\mathcal{X} \in \Sigma_n$. Π_n-**complete** and Δ_n-**complete** sets are defined similarly.

The class Σ_1 will give us some clues about what might be the relationship between the jump and arithmetical hierarchies:

a) We know that $\mathcal{K} \in \Sigma_1$ (see Example 15.2). We also know that $\mathcal{K}$ is m-complete (see Problem 9.2, p. 228). So, by Definition 15.4, the set $\mathcal{K}$ is Σ_1-complete. Now recall that $\mathcal{K} = \emptyset'$ (see Definition 12.1, p. 265). It follows that $\emptyset'$ *is* Σ_1-*complete*. Observing this we notice a possible pattern which we express in the speculation:

Is it perhaps true that for *every* $n \geqslant 0$ the set $\emptyset^{(n+1)}$ is Σ_{n+1}-complete?

b) Let $\mathcal{A} \in \Sigma_1$. Then $\mathcal{A}$ is c.e. (see previous page) and there is an ordinary TM recognizing $\mathcal{A}$. The recognizer is equivalent to an o-TM with the oracle set $\emptyset$, say $T^\emptyset$. So, $\mathcal{A}$ *is* $\emptyset$-*c.e.* Observing this we may be led to the next bold speculation:

Is it perhaps true that for *every* $n \geqslant 0$ the class Σ_{n+1} consists of $\emptyset^{(n)}$-c.e. sets?

Indeed, in 1948 Post announced that the answer to both questions is *yes*. This is stated in the following *Post's Theorem*, which relates the jump hierarchy to the arithmetical hierarchy.

Theorem 15.3. *Let $\mathcal{A} \subseteq \mathbb{N}$ and $n \geqslant 0$. Then:*

> *a)* $\emptyset^{(n+1)}$ *is* Σ_{n+1}-*complete;*
> *b)* $\mathcal{A} \in \Sigma_{n+1} \Longleftrightarrow \mathcal{A}$ *is* $\emptyset^{(n)}$-*c.e.*
> *c)* $\mathcal{A} \in \Delta_{n+1} \Longleftrightarrow \mathcal{A} \leq_T \emptyset^{(n)}$.

Proof idea. First, part *b* is proven by induction on n. This is then used to prove parts *a* and *c*. For details see the Bibliographic Notes to this chapter. □

Since the theorem is important, we invest some time in commenting on it:

- Part *a* reveals the link between the concepts of the Turing jump and the class Σ_n in terms of m-reducibility. For $n = 0$ it tells us that the set $\emptyset'(= \mathcal{K}^\emptyset = \mathcal{K})$ is in Σ_1 and that any set $\mathcal{X} \in \Sigma_1$ is m-reducible to $\mathcal{K}$. Well, we already knew that. But, for $n = 1$, it tells us that the second jump $\emptyset''(= (\emptyset')' = \mathcal{K}' = \mathcal{K}^\mathcal{K})$ is in Σ_2 and every set $\mathcal{X} \in \Sigma_2$ is m-reducible to $\mathcal{K}^\mathcal{K}$. And for $n = 2$ it says that $\emptyset'''(= \mathcal{K}^{\mathcal{K}^\mathcal{K}}) \in \Sigma_3$ and any set in Σ_3 is m-reducible to $\emptyset'''$; that is, $\emptyset'''$ is Σ_3-complete.
- Part *b* reveals the connection between any two consecutive classes Σ_n and Σ_{n+1} in terms of their relative computability. Specifically, Σ_{n+1} contains exactly the sets that would become c.e. if there existed an oracle for the set $\emptyset^{(n)}$. For example, the sets in Σ_1 are $\emptyset$-c.e. (that is, c.e.), which is nothing new. But the sets in Σ_2 are $\emptyset'$-c.e. (that is, $\mathcal{K}$-c.e.), so they can be recognized by o-TMs with the oracle set $\mathcal{K}$. Similarly, any set in Σ_3 is $\emptyset''$-c.e., so it can be recognized with an o-TM with the oracle set $\mathcal{K}^\mathcal{K}$.
- Part *c* reveals the link between the concepts of the arithmetical class Δ_{n+1} and the Σ_n-complete set in terms of Turing reducibility. Indeed, it tells us that any set of the class Δ_{n+1} can be T-reduced to $\emptyset^{(n)}$, a Σ_n-complete set.

15.4 Practical Consequences: Proving Incomputability

The link between the jump hierarchy and the arithmetical hierarchy gives rise to yet another method for proving the undecidability of sets and, consequently, of decision problems. Recall that a decision problem $\mathcal{D}$ is represented by the corresponding formal language $L(\mathcal{D})$, a subset of the standard universe Σ^* (see Sect. 8.1.2). But $L(\mathcal{D})$ is associated, via a bijection $f : \Sigma^* \to \mathbb{N}$, with the set $f(L(\mathcal{D})) \subseteq \mathbb{N}$ (see Sect. 6.3.6). Now, *if* this set is arithmetical, i.e., $f(L(\mathcal{D})) = \{x \in \mathbb{N} \,|\, F(x)\}$ with $F(x)$ complying with Definition 15.2, then finding $F(x)$ with minimal number n of quantifiers, such that $f(L(\mathcal{D}))$ is at least in one of the classes Σ_n and Π_n, uncovers the degree of undecidability of the set $L(\mathcal{D})$—and, therefore, of the decision problem $\mathcal{D}$.

So, the method is as follows. Given a decision problem $\mathcal{D}$, construct the simplest arithmetical description of the set $f(L(\mathcal{D}))$, i.e., construct the predicate $F(x)$ with minimal number n of quantifiers that complies with Definition 15.2 such that $f(L(\mathcal{D})) = \{x \in \mathbb{N} \,|\, F(x)\}$.

The steps of the method are then as follows.

Method. (Proof by Arithmetical Hierarchy) The degree of undecidability of a decision problem $\mathcal{D}$ can be found as follows:

1. Consider the set $L(\mathcal{D})$.
2. Try to prove: For some $n \geqslant 0$, the set $f(L(\mathcal{D})) = \{x \in \mathbb{N} \,|\, F(x)\}$, where $F(x)$ is either $\exists y_1 \forall y_2 \ldots Q y_n R(x, y_1, y_2, \ldots, y_n)$ or $\forall y_1 \exists y_2 \ldots Q y_n S(x, y_1, y_2, \ldots, y_n)$, and R, S are decidable relations.
3. If step 2 succeeded and the obtained n is the minimal number satisfying 2, then $f(L(\mathcal{D}))$ is at least in one of the classes Σ_n and Π_n.

We now give some examples.

Example 15.4. (*Halting Problem $\mathcal{D}_{Halt}$*) The language of $\mathcal{D}_{Halt}$ is $L(\mathcal{D}_{Halt}) = \{\langle T, w \rangle \,|\, T$ halts on $w\}$ $= \mathcal{K}_0$. We can rewrite it as $\mathcal{K}_0 = \{\langle T, w \rangle \,|\, \exists s : T$ halts on w in at most s steps and returns a result$\}$. The relation $R(T, w, s) \overset{\text{def}}{=}$ "T halts on w in at most s steps and returns a result" is decidable. So, $\mathcal{K}_0 = \{\langle T, w \rangle \,|\, \exists s R(T, w, s)\}$. The corresponding subset of $\mathbb{N}$ is $\mathcal{K}_0 = \{x \,|\, \exists s R_{Halt}(\pi_1^2(x), \pi_2^2(x), s)\}$, where π_1^2, π_2^2 are projection functions and R_{Halt} is from Example 15.1. Thus, $\mathcal{K}_0 \in \Sigma_1$.   □

Example 15.5. (*Empty Proper Set, $\mathcal{D}_{Emp}$*) The proper set of a Turing machine T is the set $L(T) \overset{\text{def}}{=} \{w \in \Sigma^* \,|\, T$ accepts $w\}$ (see Definition 6.8, p. 141), and the decision problem $\mathcal{D}_{Emp}$ asks whether or not $L(T) = \emptyset$ (see Sect. 8.3.1). The language of $\mathcal{D}_{Emp}$ is $L(\mathcal{D}_{Emp}) = Emp = \{\langle T \rangle \,|\, L(T) = \emptyset\}$. We can restate it: $Emp = \{\langle T \rangle \,|\, \forall w \forall s : T$ does not accept w in s steps$\}$. The relation $R(T, w, s) \overset{\text{def}}{=}$ "T does not accept w in s steps" is decidable (just run T on w for at most s steps). So, $Emp = \{\langle T \rangle \,|\, \forall w \forall s R(T, w, s)\}$. The corresponding subset of $\mathbb{N}$ is $Emp = \{x \,|\, \forall t \forall s R(x, t, s)\}$, where $x = f(\langle T \rangle)$ and $t = f(w)$. The prefix $\forall t \forall s$ can be contracted into $\forall y$, where $t = \pi_1^2(y)$ and $s = \pi_2^2(y)$, so $Emp = \{x \,|\, \forall y R(x, \pi_1^2(y), \pi_2^2(y))\}$. Hence, $Emp \in \Pi_1$. It remains to show that $n = 1$ is minimal.   □

Example 15.6. (*Finite Proper Set*) This is the decision problem $\mathcal{D} \equiv$ "Is $L(T)$ finite?" Its language is $L(\mathcal{D}) = \{\langle T \rangle \,|\, L(T)$ is finite$\}$. Clearly, $L(T)$ is finite if and only if there is an upper bound on the length of its words. That is, $L(T)$ is finite *iff* there exists an ℓ such that, for any $w \in L(T)$, $|w| \leqslant \ell$. Therefore, $L(\mathcal{D}) = \{\langle T \rangle \mid \exists \ell \forall w (w \in L(T) \Rightarrow |w| \leqslant \ell)\}$. Using the equivalence $(A \Rightarrow B) \Leftrightarrow (\neg A \vee B)$, we obtain $L(\mathcal{D}) = \{\langle T \rangle \mid \exists \ell \forall w (w \notin L(T) \vee |w| \leqslant \ell)\}$. A word is not in $L(T)$ *iff* T does not accept it, regardless of the number s of steps performed. Hence, $L(\mathcal{D}) = \{\langle T \rangle \mid \exists \ell \forall w \forall s ((T \text{ does not accept } w \text{ in } s \text{ steps}) \vee |w| \leqslant \ell)\}$. Now define the relation $R(T, w, s, \ell) \stackrel{\text{def}}{=}$ "(T does not accept w in s steps) $\vee$ $|w| \leqslant \ell$." The relation R is decidable (run T on w for s steps, and if w has been accepted, compare $|w|$ with ℓ). Hence, $L(\mathcal{D})$ can be expressed as $L(\mathcal{D}) = \{\langle T \rangle \mid \exists \ell \forall w \forall s R(T, w, s, \ell)\}$. After contracting $\forall w \forall s$ into $\forall t$ we obtain $L(\mathcal{D}) = \{\langle T \rangle \mid \exists \ell \forall t R(T, \pi_1^2(t), \pi_2^2(t), \ell)\}$. The corresponding subset of $\mathbb{N}$ is in Σ_2. What remains to be shown is that $n = 2$ is minimal. □

Example 15.7. (Finite Function Domain, $\mathcal{D}_{\mathcal{F}in}$) This is the problem "Is $\mathrm{dom}(\varphi)$ finite?" Its language is $\mathcal{F}in = \{e \mid \mathrm{dom}(\varphi_e(x)) \text{ is finite}\}$. Leaning on Example 15.6, we can easily prove that $\mathcal{F}in \in \Sigma_2$. What remains to be shown is that $n = 2$ is minimal. □

Example 15.8. (Function Totality, $\mathcal{D}_{\mathcal{T}ot}$) "Is a p.c. function φ total?" This is the decision problem $\mathcal{D}_{\mathcal{T}ot}$. Its language is $L(\mathcal{D}_{\mathcal{T}ot}) = \mathcal{T}ot = \{e \mid \forall x \varphi_e(x) \downarrow\}$, or rewritten, $\mathcal{T}ot = \{e \mid \forall x \exists s R_{Halt}(e, x, s)\}$; see Example 15.1. Thus, $\mathcal{T}ot \in \Pi_2$. What remains to be shown is that $n = 2$ is minimal. □

Figure 15.2 shows the initial part of the arithmetical hierarchy for $n = 1, 2, 3, 4$. The corresponding Σ_n-complete sets are $\emptyset'$, $\emptyset''$, $\emptyset'''$, $\emptyset^{(4)}$, and the Π_n-complete sets are $\overline{\emptyset'}$, $\overline{\emptyset''}$, $\overline{\emptyset'''}$, $\overline{\emptyset^{(4)}}$. We mention without proof that some of these are the sets $\mathcal{K}, \mathcal{F}in, \mathcal{I}nf, \mathcal{C}of, \mathcal{T}ot$ (and their complements) that we introduced in Sect. 8.3.5 and used in Example 12.1 (p. 269). Actually, the following holds:

$$\emptyset''' \equiv_m \mathcal{C}of \qquad \overline{\emptyset'''} \equiv_m \overline{\mathcal{C}of}$$
$$\emptyset'' \equiv_m \mathcal{F}in \qquad \overline{\emptyset''} \equiv_m \mathcal{T}ot \equiv_m \mathcal{I}nf$$
$$\emptyset' \equiv_m \mathcal{K} \qquad \overline{\emptyset'} \equiv_m \overline{\mathcal{K}}$$

Fig. 15.2 The lowest levels of the arithmetical hierarchy: $(n = 1)$ Σ_1 is the class of c.e. sets; Π_1 is the class of co-c.e. sets; Δ_1 is the class of decidable sets; $\emptyset'$ is Σ_1-complete; $\overline{\emptyset'}$ is Π_1-complete; $(n = 2)$ Σ_2 is the class of $\emptyset'$-c.e. sets; Π_2 is the class of $\emptyset'$-co-c.e. sets; Δ_2 is the class of $\emptyset'$-decidable sets; $\emptyset''$ is Σ_2-complete; $\overline{\emptyset''}$ is Π_2-complete; $(n = 3)$ Σ_3 is the class of $\emptyset''$-c.e. sets; Π_3 is the class of $\emptyset''$-co-c.e. sets; Δ_3 is the class of $\emptyset''$-decidable sets; $\emptyset'''$ is Σ_3-complete; $\overline{\emptyset'''}$ is Π_3-complete; $(n = 4)$ Σ_4 is the class of $\emptyset'''$-c.e. sets; Π_4 is the class of $\emptyset'''$-co-c.e. sets; Δ_4 is the class of $\emptyset'''$-decidable sets; $\emptyset^{(4)}$ is Σ_4-complete; and $\emptyset^{(4)}$ is Π_4-complete

15.5 Chapter Summary

A k-ary relation R is decidable if there is a Turing machine capable of deciding, for any given k-tuple of elements $(a_1, a_2, \ldots, a_k)$, whether or not $R(a_1, a_2, \ldots, a_k)$. A set $\mathcal{A}$ is arithmetical if it can be represented as $\mathcal{A} = \{x \in \mathbb{N} \mid F(x)\}$ such that, for some $n \geqslant 0$, the predicate $F(x) = \exists y_1 \forall y_2 \exists y_3 \ldots Q y_n R(x, y_1, y_2, \ldots, y_n)$ or $F(x) = \forall y_1 \exists y_2 \forall y_3 \ldots Q y_n R(x, y_1, y_2, \ldots, y_n)$, and R is a decidable relation on $\mathbb{N}$. The sets defined by the properties $F(x) = \exists y_1 \forall y_2 \exists y_3 \ldots Q y_n R(x, y_1, y_2, \ldots, y_n)$ are in the arithmetical class Σ_n, and the sets defined by $F(x) = \forall y_1 \exists y_2 \forall y_3 \ldots Q y_n R(x, y_1, y_2, \ldots, y_n)$ are in the arithmetical class Π_n. The intersection of Σ_n and Π_n is the arithmetical class Δ_n. The classes Σ_n and Π_n are proper subclasses of both Σ_{n+1} and Π_{n+1}, and Δ_n is a proper subclass of Σ_n and Π_n. A member $\mathcal{A}$ of Σ_n is Σ_n-complete if every member of Σ_n is m-reducible to $\mathcal{A}$; and a member $\mathcal{B}$ of Π_n is Π_n-complete if every member of Π_n is m-reducible to $\mathcal{B}$.

There is a connection between the arithmetical hierarchy and the jump hierarchy; for $n \geqslant 0$ the following hold: a) the set $\emptyset^{(n+1)}$ is Σ_{n+1}-complete ; b) the class Σ_{n+1} consists of exactly the $\emptyset^{(n)}$-c.e. sets; and c) any member of Δ_{n+1} is T-reducible to $\emptyset^{(n)}$. In particular, Δ_1 contains decidable sets, Σ_1 contains c.e. sets, and Π_1 contains co-c.e. sets. Thus, the jump hierarchy is interleaved with the arithmetical hierarchy.

We can use the arithmetical hierarchy in proving the undecidability of a set (or a decision problem). To do this, we must express the set (or the language of the problem) as an arithmetical set. Then, its degree of undecidability is obtained from the index of the lowest arithmetical class that contains it.

Problems

Definition 15.5. (Graph of a Function) The **graph** of a partial function φ is the relation $\mathrm{graph}(\varphi)$ defined by $(x, y) \in \mathrm{graph}(\varphi) \Longleftrightarrow \varphi(x) = y$.

15.1. Prove the following *Graph Theorem*.

Theorem 15.4. (Graph Theorem) *A partial function φ is p.c. $\Longleftrightarrow \mathrm{graph}(\varphi)$ is c.e.*

15.2. Prove:

    (a) $\Sigma_n \subseteq \Sigma_m \cap \Pi_m$, for any $m > n$.

    (b) $\Pi_n \subseteq \Sigma_m \cap \Pi_m$, for any $m > n$.

    [*Hint.* Use Theorem 15.2.]

15.3. Prove:

    (a) $\mathcal{A} \in \Sigma_n \Longrightarrow \mathcal{A} \times \mathcal{A} \in \Sigma_n$.

    (b) $\mathcal{A}, \mathcal{B} \in \Sigma_n \Longrightarrow \mathcal{A} - \mathcal{B} \in \Delta_{n+1}$.

15.4. Prove:

(a) $\mathcal{A}, \mathcal{B} \in \Sigma_n \implies \mathcal{A} \cup \mathcal{B} \in \Sigma_n$.

(b) $\mathcal{A}, \mathcal{B} \in \Sigma_n \implies \mathcal{A} \cap \mathcal{B} \in \Sigma_n$.

(c) $\mathcal{A}, \mathcal{B} \in \Pi_n \implies \mathcal{A} \cup \mathcal{B} \in \Pi_n$.

(d) $\mathcal{A}, \mathcal{B} \in \Pi_n \implies \mathcal{A} \cap \mathcal{B} \in \Pi_n$.

[*Hint.* $\forall y_1 \exists y_2 \ldots R \wedge \forall z_1 \exists z_2 \ldots S = \forall y_1 \forall z_1 \exists y_2 \exists z_2 \ldots (R \wedge S)$.]

15.5. Prove:

(a) $\mathcal{A} \leq_m \mathcal{B} \wedge \mathcal{B} \in \Sigma_n \implies \mathcal{A} \in \Sigma_n$.

(b) $\mathcal{A} \leq_m \mathcal{B} \wedge \mathcal{B} \in \Pi_n \implies \mathcal{A} \in \Pi_n$.

[*Hint.* Let $x \in \mathcal{A} \Leftrightarrow f(x) \in \mathcal{B}$ where f is a computable function. If $R(x, y_1, \ldots, y_n)$ is a decidable relation, so is $R(f(x), y_1, \ldots, y_n)$.]

Bibliographic Notes

- The *prenex normal form* (PNF) was discussed and presaged (if not demonstrated at full strength) in 1885 by Peirce [175, pp. 196–198]. For the description of the Tarski-Kuratowski transformation algorithm see Mendelson [155, Chap. 2] or Rogers [202, Chap. 14].
- The *arithmetical hierarchy* was first studied by Kleene [123] and Mostowski [163].
- The link between the jump hierarchy and the arithmetical hierarchy appeared in Post [187]. For the proof of Theorem 15.3, see Cooper [43, Chap. 10] or Soare [246, p.84]. There, the reader can also find the proofs that sets $\mathcal{K}, \mathcal{F}in, \mathcal{I}nf, \mathcal{C}of, \mathcal{T}ot$ and their complements are Σ_n- or Π_n-complete for small n.

Part IV
BACK TO THE ROOTS

The crisis in the foundations of mathematics bequeathed fundamental questions: What is an algorithm? What is computing? What is computable? In Part I, we explained how these questions led to the birth of *Computability Theory* when the informally, intuitively understood concepts of "algorithm", "computation", and "computable function" were rigorously defined by the *Computability* (*Church-Turing*) *Thesis*.

In Parts II and III we showed how this thesis opened the door to a mathematical treatment of these intuitive concepts. In particular, we explained how the thesis enabled the discovery of the *universal Turing machine*, which was soon physically realized in the form of a *general-purpose computer*. *Computability Theory* continued its development in different directions and yielded many important theoretical and practical discoveries about computation and its application. Simultaneously, general-purpose computers kept improving and evolved into powerful computing machines capable of solving complex computational problems. Taking everything into account, the *Computability* (*Church-Turing*) *Thesis* brought about consequences that tremendously changed human knowledge and civilization.

Realizing its immense importance, we will revisit the *Computability* (*Church-Turing*) *Thesis* in Part IV in much greater detail. There are good reasons for this decision: Due to the fast development of general-purpose computers and the recent rise of different proposals for new computing paradigms, some scientists have questioned the adequacy of the thesis for the suggested paradigms, whether realistic (e.g., parallel or quantum computation) or speculative ones (e.g., hypercomputation). As a result, new versions of the thesis have recently emerged, each involving a certain notion or concept in the original thesis (e.g., physical implementability of algorithms, computational complexity, or physical computability). The necessity, adequacy, and relative power of each of the proposed versions are currently intensely discussed. Since these matters are in active discussion within the academic community, a well-grounded understanding of the *Computability* (*Church-Turing*) *Thesis* is needed in order to keep up with them.

Chapter 16
Computability (Church-Turing) Thesis Revisited

> With this concept [Turing computability] one has for the first
> time succeeded in giving an absolute definition of an interesting
> epistemological notion, i.e., one not depending on the formalism
> chosen. (Gödel, 1946)

Abstract The *Computability* (*Church-Turing*) *Thesis* formalized the informal notions of computation and enabled their mathematical treatment. The thesis is by
many viewed as an unproved or even unprovable proposition, although it has been
subjected to continuous examination. Nevertheless, the consequences of this breakthrough are immense. Recent developments in computer science have brought some
similar theses. We describe in greater detail the evolution of the thesis from its beginnings to some of its modern versions, and review some of their open questions.

16.1 Introduction

Hilbert, who in 1899 gave an axiomatization of geometry, reduced the question of
the consistency of geometry to the question of the consistency of arithmetic. To
prove the latter, he proposed using the kind of reasoning, called *finitism*, that would
only use finite mathematical objects and constructive methods, at least in principle. Hilbert showed that finitism would reduce mathematical proving to mechanical
manipulation of finite strings of symbols, free of deceptive intuitive meaning.[1]

The idea of *mechanical* manipulation of finite strings stimulated interest in
"*procedures*" (or "algorithms"[2]) that could mechanically and systematically—in
other words *effectively*—carry out such manipulations.[3]

Another major part of Hilbert's program concerned the *Entscheidungsproblem*,
the problem of finding a "procedure" for deciding the validity of *any* mathematical statement. With such a decision "procedure" available, we could answer any
mathematical question in a purely mechanical manner, at least in principle.[4]

---

[1] This would bring us a kind of *mathematical paradise* in which creative thinking would no longer
be needed to pursue research. (This would have nothing to do with "Cantor's paradise", the situation in mathematics just before paradoxes were discovered in Cantor's naive set theory.)

[2] We use the terms "algorithm" and "effective procedure" interchangeably.

[3] See Sects. 2.2.4, 3.1.1, 3.1.2, and 3.1.3.

[4] See Sects. 4.1.1 and 4.1.2.

Hilbert's ideas were deeply founded on formal axiomatic systems that included *First-Order Logic* (based on *Principia Mathematica*) and *Formal Arithmetic* (based on *Peano Arithmetic*). Alas, in 1931 Gödel published his *Incompleteness Theorems* about such formal axiomatic systems. He proved that if **F** is any such (consistent) system, then (*i*) there are *undecidable* propositions in **F**, and (*ii*) the consistency of **F** cannot be proved with **F**'s own means. Although Gödel's discovery unexpectedly brought Hilbert's program to a close[5] the fall of Hilbert's program bequeathed new burning questions. These arose from the notion of a "procedure" that was used by Hilbert and Gödel.

The notion of a "procedure" was *informal*, understood intuitively as "a kind of a recipe, or plan, telling how a human can carry out a given task in a purely mechanical way." But after Gödel's discovery it turned out that a rigorous, mathematically precise, that is, *formal* definition of the concept of a "procedure" became indispensable. Since a "procedure" could be viewed as a *mapping* of its inputs to its outputs,[6] a formal characterization of the notion of a "computable" *function* became indispensable too.

NB *In this chapter we use quotation marks to refer to human informal, intuitive understanding of the notions of computing. For example, "procedure", "finite and mechanical procedure", "effective procedure", "computable" number, "computable" function, "computation", "creative thinking", "insight", "intuition", "ingenuity", "process", and "argument" were at the time only intuitively (and imprecisely) defined. The exceptions to this rule will be quoted passages from other sources.*

16.2 The Intuitive Understanding of the Notion of a "Procedure"

Generally speaking, the term *procedure* means a usual or correct way of doing something; it is a series of actions conducted in a certain prescribed order and manner to achieve something. The informal concept of a *mathematical* procedure, i.e., the "procedure" that Hilbert and Gödel were concerned with, complied with this definition, but it also bore its own characteristics. Namely, the concept of a "procedure" implicitly emerged with Euclid in ancient Greece, and gradually acquired, especially during the first two decades of the 1900s, specific additional connotations of some concepts relevant to mathematical problem solving. As a result, at the time of Hilbert and Gödel it became required that mathematical "procedures" were "effective".

The intuitive, common-sense understanding of the idea of an "effective procedure" was formulated informally, such as in the following definition.

---

[5] See Sects. 4.2.3 and 4.2.5.

[6] Accordingly, this mapping is treated *extensionally*, in the sense that it is determined by the set of all ordered pairs of its arguments and values, and not by the particular way in which it is defined.

Definition 16.1. ("Effective procedure") A "procedure" for achieving some mathematical result is said to be **"effective"** if it

1. consists of a *finite* number of *exact* instructions;

2. produces the result in a *finite* number of steps, if carried out without error;

3. can be carried out by a *human* aided only by paper and pencil;

4. demands from the human *no "creative thinking"* ("insight, intuition, ingenuity").

The definition is informal because it doesn't precisely tell us what the instructions are and what "creative thinking" is. This is not surprising: Humans were (and still are) perfectly aware that human understanding of the term "creative thinking" and its mental phenomena such as "insight", "intuition", and "ingenuity" are only founded on humans' *limited experience* and *deceptive intuition*. No mathematically precise characterization of these terms seemed (and still seems) to be within human reach.

16.3 Toward the Thesis

The main protagonists of the search for a precise mathematical characterization of the idea of an "effective procedure" were Gödel, Church, and Turing; others also contributed, most notably Herbrand, Post, Kleene, and Rosser. In the following we will describe in greater detail the events following the year 1931 through which these young[7] mathematicians and logicians endeavored to answer the seemingly undemanding questions:

> *What exactly is a mathematical "procedure"?*
> *What exactly is an "effective procedure"?*
> *When exactly is something "computable"?*
> And after all, *What exactly is "computing"?*

16.3.1 Gödel

After publishing his famous *Incompleteness Theorems* in 1931, Gödel continued his research aiming to extend the theorems to formal mathematical systems in general. In the course of this research he realized that a *precise* definition of "computable" functions and "procedures" that compute the values of such functions was *necessary* in order to mathematically secure both his general definition of a formal axiomatic system and the generalization of his incompleteness theorems.

---

[7] In 1931, Turing was 19, Kleene 22, Herbrand 23, Rosser 24, Gödel 25, Church 28, and Post 34.

Gödel at Princeton

In 1934, during the period February through May, Gödel delivered a series of lectures at the Princeton *Institute for Advanced Study*, where he explicated undecidable propositions of formal axiomatic systems in general. He started with the following definition of a formal axiomatic system:

> We require that the rules of inference, and the definitions of meaningful formulas and axioms, be constructive; that is, for each rule of inference there shall be a finite procedure for determining whether a given formula B is an immediate consequence (by that rule) of given formulas $A_1, \ldots, A_n$, and there shall be a finite procedure for determining whether a given formula A is a meaningful formula or an axiom.

In this passage he assumed the existence of two "procedures", the first one for deciding whether or not a given formula directly follows from other given formulas, and the second one for distinguishing axioms from other well-formed formulas. In addition to being *"finite"*, the "procedures" should be *"mechanical"*, in the sense he explained for the first "procedure" a year earlier:

> [The] outstanding feature of the rules of inference [is] that they are purely formal, i.e., refer only to the outward structure of the formulas, not to their meaning, so that they could be applied by someone who knew nothing about mathematics, or by a machine.

Gödel was of course well aware that an irreproachable development of his ensuing generalized incompleteness theorems would require an *exact* characterization of the concept of the "finite and mechanical procedure". As he later explained:

> When I first published my paper about undecidable propositions the result could not be pronounced in this generality, because for the notions of mechanical procedure and of formal system no mathematically satisfactory definition had been given at that time. [...] The essential point is to define what a procedure is.

But at the time of his 1934 Princeton lectures he had no such characterization. Neither did he have a precise idea of what could be "computed" by such "procedures". Indeed, when speaking of functions—before he introduced the new concept of a *general*[8] recursive function—he reviewed the *primitive*[9] recursive functions, which he introduced in 1931, and stated an *easy* observation that primitive recursive functions are "computable" (can be "computed" by "finite and mechanical procedures"):

> [Primitive] [r]ecursive functions have the important property that, for each given set of values of the argument, the value of the function can be computed by a finite procedure.

To this, however, he added a footnote stating a corresponding *hard* observation, namely that to show the *converse* (i.e., that the "computable" functions are general recursive, or recursive of the most general kind) would be quite a different matter:

---

[8] Gödel developed general recursive functions from Herbrand's ideas; see Sect. 5.2.1, p. 84.

[9] See Sect. 5.2.1, p. 81.

The converse seems to be true, if, besides [primitive] recursions [...] recursions of other forms (e.g., with respect to two variables simultaneously) are admitted. This cannot be proved, since the notion of finite computation is not defined, but it serves as a heuristic principle.

Gödel's footnote indicates that he *believed* that the class of functions obtainable by recursions *of the most general kind* (possibly the general recursive functions) was the same as the class of "computable" functions. However, as he clarified later, at the time of the Princeton lectures he was *not at all* convinced that his and Herbrand's concept of *general* recursion encompassed *all possible recursions*; in addition, at the time of his lectures the equivalence between his general recursiveness and μ-recursiveness, which was then investigated by Kleene (see below), was not quite trivial. Thus, Gödel's above footnote did not anticipate Church's Thesis (which will be described shortly).

In sum, in 1934 Gödel had the concept of general recursiveness, but he was not convinced that general recursiveness is the recursiveness of the most general kind, nor did he believe that it could be *rigorously proved* that this recursiveness captures the *informal* notion of functions "computable by finite and mechanical procedures".

16.3.2 Church

At Princeton three years earlier, Church started an attempt to develop a logical system in which the notion of a function would play a fundamental role. In the context of the system, the λ-*notation*[10] arose in a very natural way. In this notation he defined the well-formed expressions (λ-terms) and the operations (β-reductions) that transform λ-terms without altering their meanings. Church identified *positive integers* with certain λ-formulas, called *Church numerals*. Then he defined a function of positive integers to be λ-*definable* if the function is representable by a λ-term that is β-reducible to a Church numeral exactly when the numeral represents the function's value at a given number represented by the corresponding Church numeral. In 1931, he had just two λ-definable functions, the successor function $\sigma(n) = n + 1$ and the sum function $m + n$ (and, in a certain way, the functions $m - n$ and $m \times n$).

It was obvious that every λ-definable function is "computable"—or, in Church's wording, "effectively calculable"—since the "effective procedure" for obtaining the function's values is implicit in the function's λ-term. But Church was speculating that also the converse may be true, i.e., that every "computable" function is λ-definable. So he set his doctoral student Kleene to investigate the λ-definability of some particular "computable" functions. Kleene exceeded Church's expectations: By 1934, he proved that *all the usual* "computable" number-theoretic functions are λ-definable.

---

[10] See Sect. 5.2.1, p. 85.

Church's Thesis (1934)

Encouraged by this evidence and by his own intuition, Church presented to Gödel—who was at the time at Princeton giving his lectures as mentioned above—a *thesis* on "computable" functions. In the thesis he proposed that λ-*definability* of functions be *adopted* as *the* formal characterization of their "computability". In other words, Church proposed that function "computability" simply be *identified with* (or, *defined as*) function λ-definability. Here is his thesis.

> **Church's Thesis (1934)** *A function is "computable" iff it is λ-definable.*

If we denote by $\mathcal{F}$ the informal class of all "computable" functions, by $\mathcal{C}$ the class of all Church's λ-definable functions, and by $a := b$ the assignment operation "a becomes b", then Church's Thesis postulated that

$$\mathcal{F} := \mathcal{C}.$$

Remark. Kleene, who also heard of the thesis, was doubtful of it. He tried right away to *disprove* it by diagonalization.[11] However, he unexpectedly found that the diagonalization procedure *failed* to produce a contradiction. The surprising incapability of diagonalization in the case of Church's Thesis sufficed to convince Kleene and turned him into a supporter of the thesis.

Doubts about Church's Thesis (1934)

Gödel, by contrast, was *not* convinced by the available evidence, and he rejected Church's Thesis. Church described this event in a letter to Kleene:

> In regard to Gödel and the notions of recursiveness and effective calculability, the history is the following. In discussion with him the notion of lambda-definability, it developed that there was no good definition of effective calculability. My proposal that lambda-definability be taken as a definition of it he regarded as thoroughly unsatisfactory.

Church's immediate reply to Gödel was that they further test λ-definability:

> I replied that if he would propose any definition of effective calculability which seemed even partially satisfactory I would undertake to prove that it was included in lambda-definability.

However, Gödel's counter-suggestion to Church was to focus on "computability":

> His [Gödel's] only idea at the time was that it might be possible, in terms of effective calculability as an undefined notion, to state a set of axioms which would embody the generally accepted properties of this notion [effective calculability], and to do something on that basis.

---

[11] See Sect. 5.3.3, p. 100 and Sect. 9.1.

We can see that Gödel's belief was that (*i*) the notion of "computability" should be better understood by further investigations; (*ii*) if some of the discovered properties of "computability" became generally accepted, then these properties could serve as postulates of an informal theory of "computability"; and (*iii*) such a theory might reveal an appropriate formalization of the notion of "computability".

Church's Thesis (1936)

Church, however, went on to publicly announce his thesis. In 1935, he gave a talk at a meeting of the *American Mathematical Society*, saying:

> Following a suggestion of Herbrand, but modifying it in an important respect, Gödel has proposed (in a set of lectures at Princeton, N.J., 1934) a definition of the term [general] *recursive function*, in a very general sense. In this paper a definition of the [general] *recursive function of positive integers* which is essentially Gödel's is adopted. And it is maintained that the notion of an effectively calculable function of positive integers should be identified with that of a [general] recursive function, since other plausible definitions of effective calculability turn out to yield notions which are either equivalent to or weaker than recursiveness.

In 1936, he published a paper in the *American Journal of Mathematics* stating:

> [In §1:] The purpose of the present paper is to propose a definition of effective calculability which [the definition] is thought to correspond satisfactorily to the somewhat vague intuitive notion [effective calculability] in terms of which problems [...] are often stated [...]

> [In footnote 3:] As will appear, this definition of effective calculability can be stated in either of two equivalent forms, (1) that a function of positive integers shall be called effectively calculable if it is λ-definable [...], (2) that a function of positive integers shall be called effectively calculable if it is [general] recursive [...]

> [In §6:] THEOREM XVI. *Every [general] recursive function of positive integers is λ-definable*. THEOREM XVII. *Every λ-definable function of positive integers is [general] recursive.* [...]

> [In §7:] We now define the notion, already discussed, of an *effectively calculable* function of positive integers by identifying it with the notion of a [general] recursive function of positive integers (or of a λ-definable function of positive integers). This definition is thought to be justified by the considerations which follow, so far as positive justification can ever be obtained for the selection of a formal definition to correspond to an intuitive notion. [...]

The reader may have noted that in the above announcements Church formulated his thesis in terms of Gödel-Herbrand's (general) recursiveness, rather than in terms of his own λ-definability. Why did he do so? There were three reasons for this change: (*i*) according to Kleene, there were rather chilly receptions from audiences around 1933–1935 to disquisitions on λ-definability; (*ii*) Church and Kleene had each already proved that λ-definable functions are (general) recursive (see THEOREM XVI); (*iii*) in 1936, Kleene also proved that (general) recursive functions are λ-definable (THEOREM XVII). Therefore, at the time of submitting his paper, Church already knew that λ-definability and (general) recursiveness are formally equivalent. The reader should also take note that Church's "computability" as defined in the above passage refers to functions of *positive integers* only.

Accordingly, Church proposed in effect the following new version of his thesis.

Church's Thesis (1936)   *A function of positive integers is "computable" iff it is (general) recursive (or, equivalently, λ-definable).*

If we denote by $\mathcal{F}^+$ the informal class of all "computable" functions *of positive integers*, by $\mathcal{G}^+$ the class of all Gödel's (general) recursive functions *of positive integers*, and by $\mathcal{C}^+$ the class of all Church's λ-definable functions (*of positive integers*), then the second version of Church's Thesis is

$$\mathcal{F}^+ := \mathcal{G}^+ \,(= \mathcal{C}^+).$$

This version too postulates that the *vaguely* defined class $\mathcal{F}^+$ *coincides* with the *precisely* defined class $\mathcal{G}^+$ (and with $\mathcal{C}^+$ due to the proved confluence $\mathcal{G}^+ = \mathcal{C}^+$).

Doubts about Church's Thesis (1936)

We have seen that, in his second thesis, Church *defined* the class $\mathcal{F}^+$ of "computable" functions of positive integers to coincide with a particular class of functions, $\mathcal{G}^+$ (or $\mathcal{C}^+$). Post, however, was greatly opposed to speaking of Church's Thesis as a definition; he viewed the thesis as a *working hypothesis*, which needs to be *continually verified*. Namely, for any new function $f \in \mathcal{F}^+$ that one might conceive, one would still need to *prove* that $f \in \mathcal{G}^+$ (or $f \in \mathcal{C}^+$), because f's membership in $\mathcal{G}^+$ (or $\mathcal{C}^+$) *does not* logically follow by (or from) Church's Thesis, which is just an arbitrarily *postulated* definition. On account of this, Post criticized Church for masking this hypothesis in the guise of a definition:

> And to our mind such [working hypothesis] is Church's identification of effective calculability with recursiveness. [...] Actually the work already done by Church and others carries this identification considerably beyond the working hypothesis stage. But to mask this identification under a definition hides the fact that a fundamental discovery [...] has been made and blinds us to the need of its continual verification.

Gödel too remained reluctant to accept $\mathcal{G}^+ = \mathcal{C}^+$, the confluence of (general) recursiveness and λ-definability, as decisive new evidence for Church's Thesis. He was still demanding a deeper understanding of the ideas of "computable" function and "effective procedure" computing it, an understanding that would help to identify those properties of the two ideas that could be generally accepted as characteristic of them.

16.3.3 Kleene

Kleene was Church's doctoral student at Princeton who importantly contributed to the birth and development of *Computability Theory*. In 1935–1936, Kleene discovered his own formulation of the (general) recursive functions. How did he do that? First, in 1934 he found that Gödel's *primitive* recursiveness can be augmented by a new rule of function construction, which he called the µ-*operation*. The functions constructible in this way are now called the µ-*recursive functions*.[12] Then Kleene discovered his *Normal Form Theorem*, which asserts that every (general) recursive function φ can be constructed from two primitive recursive functions T and U (where T is a predicate) by a single application of the µ-operator.

Theorem 16.1. (Normal Form Theorem) *There is a primitive recursive predicate* $T(e,x,y)$ *and a primitive recursive function* $U(y)$ *such that, for any (general) recursive function* $\varphi(x)$, *a number e can be found such that* $\varphi(x) = U(\mu y\, T(e,x,y))$.

Remark. Here the predicate $T(e,x,y)$ is true *iff* y is the code of some computation of the value $\varphi(x)$ and e represents the system $\mathscr{E}(\varphi)$ of equations defining the function φ (see Sect. 5.2.1, p. 84). The function U extracts the value $\varphi(x)$ from y. Thus, the theorem states that the value $\varphi(x)$ of any (general) recursive function φ can be computed as follows: Find the shortest code y in the set of all encoded computations of the value $\varphi(x)$—here the function φ is defined by the system $\mathscr{E}(\varphi)$ of equations whose index is e—and then extract the computed value $\varphi(x)$ from y.

By the *Normal Form Theorem* every (general) recursive function is µ-recursive. Kleene also proved the converse: Every µ-recursive function is (general) recursive. So he proved the equivalence of (general) recursiveness and his µ-recursiveness, i.e.,

$$\mathcal{G}^+ = \mathcal{K}^+,$$

where we denote by $\mathcal{K}^+$ the class of all µ-recursive functions (of positive integers).

16.3.4 Rosser

Rosser was Church's doctoral student at Princeton who importantly contributed to the development of the λ-calculus. In collaboration with Kleene he proved that the original λ-calculus, which Church introduced in 1932, was inconsistent.[13] In 1936, in collaboration with Church, he proved the *Church-Rosser Theorem*,[14] which states that if a λ-term can be β-reduced to two different λ-terms, then there is a λ-term to which both λ-terms can be β-reduced. This means that if two different computations of a λ-defined function terminate, they return virtually equal results. Besides this, Rosser also made important discoveries in other fields of mathematics.[15]

---

[12] See Sect. 5.2.1, p. 81.

[13] Nevertheless, the λ-calculus later developed into a calculating system that now has strong impact on some fields of computing, e.g., functional programming languages.

[14] See Box 5.2, p. 86.

[15] Rosser proved that the requirement for ω-consistency in Gödel's *First Incompleteness Theorem* can be weakened to the requirement for usual consistency. (See Sects. 5.2.1, p. 85 and 4.2.4, p. 66.)

16.3.5 Post

At the age of twelve, Post lost his left arm below the shoulder in an accident when he reached for a lost ball under a parked car and a second car crashed into it. This changed his childhood ambitions from astronomy to mathematics. After receiving a B.S. in mathematics from City College, he was a doctoral student at Columbia University from 1917 to 1920. In this time, he became interested in modern mathematical logic; so he completed his doctoral studies with a dissertation in this area. A shortened version of the dissertation was published in 1921, with its main result being the first published proof of completeness and decidability of the propositional calculus of *Principia Mathematica* (*PM*).[16] In 1920–1921, Post held a postdoctoral fellowship at Princeton University. In the course of the fellowship, independently of Hilbert's formalist program he created his own research project. It was as part of this project that Post made astounding discoveries in logic that *anticipated* the work that was done on incompleteness and undecidability *a decade and a half later* in different ways by Gödel, Church, Turing, and others.[17] Alas, the excitement caused by his discoveries precipitated Post's first attack of bipolar disorder, a lifelong condition that heavily affected him. To avoid undue excitement, he developed a routine that restricted his time spent on research to just three hours a day (4–5 p.m. and 7–9 p.m.). Due to this regimen and illness, he failed to promptly publish his ideas; indeed, it was only after other researchers had caught up with his discoveries that he became able to publish. Thus he published—with his wife assisting him by typing his papers and letters—from 1935 to his death in 1954—some of his most remarkable papers. Since Post's point of view and ideas were substantially different from the approaches of other logicians, these papers are highly original and influential. For example, in contrast to other researchers, Post never emphasized "computable *functions*". So when he read the paper presenting *Church's Thesis* (1936), he retained the point of view adopted in his earlier work (f.a.s. expressed as rules for string manipulation). Accordingly, he chose to analyze the intuitive notion of a "computing *process*", rather than that of a "computable function". As a result, he proposed in 1936 his model of computation, called the *finite combinatory process*.[18]

---

[16] In this paper (see [180]), Post stressed that the authors of the *PM* adopted a restricted view and methods whereby they could develop the logical foundation of mathematics in a *fixed* axiomatic system. This resulted in a logical formalism that lacks universality and avoids any considerations of the system itself. To recover generality, Post adopted a metamathematical view of the *PM*. He developed a logical method (similar to formalism) by allowing mostly constructive methods (though he didn't explicitly restrict to finitary methods). Conforming to this view and methods, he then introduced the now familiar truth-tables and truth-table method. Then he proved his *Fundamental Theorem*, which states that a propositional function is assertible *iff* it is a tautology. This provided both a completeness theorem and a decision procedure for the propositional calculus of the *PM*.

[17] In 1920–1921 (see [182]), Post investigated abstractions of the classical propositional logic, which he called the *postulate systems*. He defined the notion of *reduction* of a system to another system; then he defined a series of reductions of systems through which he reduced a decision problem to a *system in normal form* (an extremely simple postulate system); next, by applying a diagonalization argument he discovered that the decision problem for systems in normal form is *unsolvable*; from this he finally deduced his *incompleteness results*.

[18] Today, the standard terminology for the model is *Post machine*.

Finite Combinatory Processes

A computational problem (called a *general* problem by Post) is a class of problem instances (called *specific* problems.) The computation of the solution to a given problem instance is carried out by what Post called the *problem solver* (or *worker*). The idea is that the problem solver has at his disposal a *workspace* and a *program*, which Post called the *symbol space* and the *set of directions*, respectively.

The **workspace** is a two-way infinite sequence of *boxes* where the work leading from a problem instance to its solution is carried out. A box can be either *empty* or *marked* (i.e., contains a single symbol, say the stroke ⏐). One box is singled out and called the *starting box* (*starting point* by Post).

The **problem solver** is assumed to be capable of moving and working in his workspace with the proviso that he can observe and act on just *one* box at a time. He can perform any of the following five *primitive acts*: (*a*) mark the observed empty box; (*b*) erase the mark in the observed marked box; (*c*) move to the box on his right; (*d*) move to the box on his left; and (*e*) determine whether the observed box is marked or empty. The problem solver does not act freely; he is assumed to strictly follow the associated program and perform the prescribed primitive acts.

The **program** is identical for all instances of the problem and consists of a heading and a body. The *heading* instructs the problem solver to observe the starting box and follow instruction 1 of the body. The *body* consists of a finite sequence of *instructions* (called *directions* by Post) numbered $1, 2, 3, \ldots, s$. The ith instruction is of one of the following three forms:

- $i : O_i\ j_i$ ... perform primitive act $O_i \in \{a, b, c, d\}$ and follow instruction j_i;
- $i : O_i\ j_i'\ j_i''$ ... perform primitive act $O_i = e$ and depending whether the answer is YES or NO follow instruction j_i' or j_i'', respectively;
- Stop.

How are Post's ideas related to Church's and Gödel's attempts at formalizing the notion of a "computable" function? Let P be a particular problem solver (with a fixed program). Post observed that P induces a function $f_P : \mathbb{N} \to \mathbb{N}$ defined as follows. If a given input $n \in \mathbb{N}$ is unary represented by a sequence of n marked boxes, starting at the starting box, and P's program halts on input n, leaving on P's workspace exactly $m \in \mathbb{N}$ marked boxes, then let $f_P(n) = m$. Next he defined: A function f for which there exists a problem solver P such that $f = f_P$ is said to be *Post-computable*.[19] Finally, he anticipated:

> The writer expects the present formulation [Post computability] to turn out to be logically equivalent to recursiveness in the sense of the Gödel-Church development.

If we denote by $\mathcal{P}^+$ the class of Post-computable functions (of positive integers), then Post foresaw the equivalence of general recursiveness and his computability, that is,

$$\mathcal{G}^+ = \mathcal{P}^+.$$

---

[19] In fact Post called such a function the *1-function*, a term based on the notions in his 1936 paper.

16.3.6 Turing

In England, independently and unaware of the related research being done at the time at Princeton, Turing characterized the informal notions of human "computation" in terms of an *abstract computing machine*. In short, his principal idea was to show that any "process" that a human could carry out during his "computation" can be analyzed into a succession of simple operations of a particular abstract computing machine. Turing's approach to such a characterization of "computation" was to (*i*) start with an *ideal human*[20] carrying out a "computation" of a number, (*ii*) eliminate all the irrelevant details of the relevant observable "processes" making up such human "computations", and (*iii*) carry out the obtained simplified "processes" by his *machine*. In this way Turing gave arguments that his machines could compute any "computable" number and, consequently, the value of any "computable" function.

The machine which Turing conceived for simulating human "computation" he called the *a-machine* (for automatic machine).[21] We will call it the *Turing machine*. In 1936, Turing published his ideas and their highly influential applications[22] in the *Proceedings of the London Mathematical Society*.

We now comment on some of the sections of this paper in which Turing developed his characterization of the basic notions of computing.

NB *From now on and until p. 333 the subject will be "computation" of numbers.*

Turing's Thesis (numbers)

Turing formulated two theses. In the first, which we call *Turing's Thesis (numbers)*, he compared the numbers computed by his machine with the numbers "computed" by a human; in the second thesis, called *Turing's Thesis (operations)*, he compared the operations that make up machine computations of numbers with the "operations" which make up human "computations" of numbers. From the two theses the usual *Turing's Thesis (functions)* can easily be synthesized. This thesis links the functions whose values can be computed by a Turing machine with those whose values can be "computed" by a human. We now start with the first thesis.

---

[20] A human is *ideal* because his computation is not bounded by practical limits in time, space, or resources. An ideal human is error-free, immortal, free of boredom and fatigue, and not troubled by insufficiency of paper, pencils, or any other simple tool needed in his computation.

[21] Later, after some Post's practical adaptations, the *a*-machine was renamed the *Turing machine*.

[22] Using his ideas, Turing discovered the *universal computing machine* [§§6,7] (see Sect. 6.2); proved the undecidability of the *Halting Problem* [§8] (see Sect. 8.2) and the unsolvability of the *Entscheidungsproblem* [§11] (see Sect. 9.2.4); and outlined a proof that, over *positive integers*, his computability and Church's λ-definability are *equivalent* [Appendix]. Regarding the Appendix, Kleene later wrote:

> Turing learned of the work at Princeton on λ-definability and general recursiveness just as he was ready to send off his manuscript, to which he then added an appendix outlining a proof of the equivalence of his computability to λ-definability.

Turing-computable numbers. Initially, Turing defined a new notion, that of a *computable number*, which we will right away rename a *Turing-computable number*. Here is the definition:

> According to my definition, a number is [Turing-] computable if its decimal [representation] can be written down [digit by digit] by a machine.

So Turing-computable numbers are those whose decimal representations can be generated progressively, digit by digit, by a machine. At this point, Turing said nothing about the machine that would be capable of writing down the numbers being computed by it. This he would do shortly and there it would become clear that some Turing-computable numbers can be *real*, not just positive integers (as was the case with Church's Thesis).

Next, Turing involved in his consideration the (intuitively understood) "computable" numbers. Actually, he called them numbers "which would naturally be regarded as computable." (Clearly, these numbers coincide with Church's "effectively calculable" numbers and also with Gödel's numbers "computable by finite and mechanical procedures.")

Finally, Turing declared his intention to show that the set of all Turing-computable numbers, $\mathcal{T}$, *contains* the set $\mathcal{N}$ of all "computable" numbers.[23] Here is his claim:

> In [later sections] I give some arguments with the intention of showing that the [Turing-] computable numbers [set $\mathcal{T}$] include all numbers which would naturally be regarded as [humanly] computable [set $\mathcal{N}$].

This claim is today sometimes called **Turing's Thesis** (in terms of *numbers*).

Turing's Thesis (numbers) *The set $\mathcal{T}$ of Turing-computable numbers contains the set $\mathcal{N}$ of "computable" numbers; that is, $\mathcal{N} \subseteq \mathcal{T}$.*

Turing's Thesis (operations)

Turing machine. In the next step, Turing defined an abstract mechanical *machine* that would be capable of writing down progressively, digit by digit, decimal representations of the numbers being computed by it. Here is the passage of his paper describing the workings of the machine:[24]

---

[23] So $\mathcal{N}$ is the set of "computable" *numbers* and $\mathcal{F}$ is the class of "computable" *functions*

[24] Instead of Turing's original terms "m-configuration", "square", and "configuration" we now use the terms "state", "cell", and "the pair (current state, currently scanned symbol)", respectively. In the passage, we indicate the present-day terms in square brackets.

[(*i*)] We may compare a man in the process of computing a real number to a machine which is   [(*ii*)] only capable of a finite [not infinite] number of conditions [...] which will be called "*m*-configurations" [states]. The machine is supplied with a "tape" (the analogue of paper) running through it, and divided into sections (called "squares") [cells] each capable of bearing a "symbol".   [(*iii*)] At any moment there is just one [bounded number] square [cell] [...] bearing the symbol [...] which is "in the machine" [currently scanned]. [(*iv*)] The "scanned symbol" is the only one of which the machine is [...] "directly aware." However, by altering its *m*-configuration [state] the machine can effectively remember some [just finitely many] of the symbols which it has "seen" (scanned) previously.   [(*v*)] The possible behaviour of the machine at any moment is determined by the *m*-configuration [state] [...] and the scanned symbol. [...] This pair [the pair (current state, currently scanned symbol)] [...] will be called the "configuration". In some of the configurations in which the scanned symbol is blank [...] the machine writes down a new symbol on the scanned square [cell]: in other configurations it erases the scanned symbol. The machine may also change the square [cell] which is being scanned, but only by shifting the tape one [bounded number] place to the right or left. In addition to any of these operations the *m*-configuration [state] may be changed.   [(*vi*)] Some of the symbols written down will form the sequence [...] which is the decimal of the real number [not just positive integer] which is being computed.   [(*vii*)] The others are just rough notes [e.g., intermediate results] to "assist the memory". It will only be these rough notes which will be liable to erasure.

In the following comment on the quoted passage, we will include in several places Turing's *tacit* assumptions about humans and their "computations". These assumptions do not explicitly appear in the quoted passage. Turing explicated them later, so that the motivation for his particular definition of the machine clears up after reading the rest of the paper.

So, in the quoted passage, Turing revealed several issues: (*i*) In defining this machine, Turing had in mind a *human* equipped with a pen and paper in the process of "computing".   (*ii*) He attributed to the machine a *finite* number of memory states (*tacitly* assuming that human memory is *limited*) and a *one*-dimensional tape divided into cells each capable of containing a symbol (*tacitly* assuming that the usual two-dimensional paper used by a human could easily be replaced, without loss of generality, by one-dimensional paper tape).   (*iii*) Turing defined that, at any step of the computation of his machine, there is just *one* cell on the tape whose symbol is currently scanned by, and hence known to, the machine (*tacitly* assuming that humans too can observe, and be aware of, only finitely many symbols at a time).   (*iv*) Turing pointed out that finitely many previously scanned symbols could be memorized by encoding them in the machine's states.[25]   (*v*) Furthermore, the action (i.e., local behavior[26]) of the machine must, at any step of its computation, be *deterministically* determined by the *pair* consisting of the current state of the machine and the symbol which the machine currently scans. Based on this, the machine makes three *simple operations*: (*a*) erases the scanned symbol or writes a new one to the scanned cell; (*b*) shifts the tape to one of the two *adjacent* cells; and (*c*) changes its state. (Here Turing *tacitly* took into account that a human too decides, depending on his current "state of mind" and the symbols he currently observes, how to deal with these symbols; which *not too distant* symbols on the paper

---

[25] See Sect. 6.1.2, p. 117 and Sect. 6.1.3, p. 119.

[26] See Sect. 7.2.

to observe next; and how to prepare his "state of mind" for dealing with them.)   (*vi*) If the computation of the machine terminates, then the number computed by the machine is left on the tape as a sequence of symbols. Note that the computed number can be *real*, not just a positive integer.   (*vii*) During the computation, the machine can also use its tape for writing down temporary results and other notes. In this way the machine can avoid memorizing (finitely many) tape symbols by encoding them in its states.   (*viii*) Finally, the defined machine is *finite*, in the sense that it uses, at any step of its computation, only finite means: working components of finite size, finitely many states, finitely many cells, and finitely many symbols; a fixed bounded number of currently scanned symbols (actually just one); and a fixed bound on the maximal magnitude of tape shifts (actually by just one cell).[27] Moreover, the machine operates in a *purely mechanical* way, in the sense that its local behavior is determined exclusively by its program consisting of a finite set of indivisible instructions whose execution is purely mechanical (involving no "creative thinking").

Machine operations. Next, Turing claimed that the operations used by the Turing machine *encompass* all the "operations" that a *human* uses in the "computation" of a number:

> It is my contention that these operations [of the Turing machine] include all those which are used [by a human] in the computation of a number.

This claim is today sometimes called **Turing's Thesis** (in terms of *operations*).

Turing's Thesis (operations)   *Operations of the Turing machine include all the "operations" that a human uses in the "computation" of a number.*

Here Turing reached the point where the two theses should be proved.[28] We describe how he overcame the difficulties he was faced with in his search for a proof.

Justification of Turing's Thesis (numbers)

Could Turing *rigorously* prove that Turing-computable numbers make up a set $\mathcal{T}$ that *contains* the set $\mathcal{N}$ of all "computable" numbers? The usual way of proving the relation $\mathcal{N} \subseteq \mathcal{T}$ would be to prove that $x \in \mathcal{N}$ implies $x \in \mathcal{T}$, for every x. In Turing's case, he should prove that any "computable" number $x \in \mathcal{N}$ is also Turing-computable, $x \in \mathcal{T}$. This, however, wouldn't work in Turing's favor. The reason was that the informal notion of a "computable" number might implicitly involve vaguely understood, informal ingredients of human mental "processes", such as "creative thinking" ("insight", "intuition", and "ingenuity"). But what *exactly* is "creative

---

[27] Turing's and Post's ideas bear strong resemblance but they pursued their research independently.

[28] In fact, the theses were stated in §§0,1, clarified in §§2,3,4,5 and justified in §§9,10 of the paper.

thinking" or "insight", "intuition", and "ingenuity"? How can we know when exactly *none* of them is present in a human "computation"? That is, when *exactly* does a human "compute" a number x in a *purely mechanical* way, so that $x \in \mathcal{N}$ beyond any doubt? Turing should have known precise answers to these questions in order to rigorously prove the relation $x \in \mathcal{N} \Rightarrow x \in \mathcal{T}$, because the machine he defined has no "insight", "intuition", "ingenuity" or any other kind of creative thinking. But, of course, he was well aware that answers to these questions would necessarily be intuitive and imprecise, and so too would be the proof of $x \in \mathcal{N} \Rightarrow x \in \mathcal{T}$. In his words:[29]

> No attempt has been made to show that the "computable" numbers [set $\mathcal{T}$] include all numbers which would naturally be regarded as computable [set $\mathcal{N}$]. All arguments which can be given are bound to be, fundamentally, appeals to intuition, and for this reason rather unsatisfactory mathematically.

Justification of Turing's Thesis (operations)

What about *Turing's Thesis (operations)*? We can demonstrate that this thesis implies *Turing's Thesis (numbers)*. (If a Turing machine were able to imitate every "operation" used by a "computing" human, then it could compute anything the human can "compute"; since this could be a number, $\mathcal{N} \subseteq \mathcal{T}$ would follow.) So Turing focused on his second thesis. To *prove* it rigorously, he should prove that every "process" a human carries out in his "computation" of a number can be *reduced to* (or *simulated by*) a process carried out by the Turing machine. But to do the latter, Turing should better understand the "processes" involved in human "computation". As he said:

> The real question at issue is "What are the possible processes which can be carried out [by a human] in computing a number?"

However, human "process" is an informal idea lacking a precise characterization. Because of this, finding a *rigorous* proof that a Turing machine can simulate every "process" of a "computing" human seemed to be rather difficult, if not impossible.

Human "operations". In this situation, Turing reflected on how the "processes" carried out by a human in his "computation" could be convincingly analyzed to an extent that would allow a Turing machine to plausibly imitate them. So he proceeded with an informal analysis of the relevant "processes" carried out by a "computing" human. (He called such a human the *computer*.) Turing assumed that the human computer is on the one hand *ideal* (in the sense that he is not bounded by any practical limits on the time, space, or other resources required for the "computation"), yet still *limited* by nature (in his memory capacity and sensory capability) on the other. Here is Turing's informal analysis:

---

[29] *Warning*: In this passage "computable" means Turing-computable, not informally computable.

[(*i*)] [Human] [c]omputing is normally done by writing certain symbols on paper. We may suppose that this paper is divided into squares like a child's arithmetic book. [...] [T]he two-dimensional character of paper is no essential of computation. I assume then that the [human] computation is carried out on one-dimensional paper, i.e. on a tape divided into squares.   [(*ii*)] I shall also suppose that the number of symbols which may be printed is finite. If we were to allow an infinity of symbols, then there would be symbols differing to an arbitrarily small extent. The effect of this restriction of the number of symbols is not very serious. It is always possible to use sequences of symbols in the place of single symbols. Thus an Arabic numeral 17 or 999999999999999 is normally treated as a single symbol. Similarly in any European language words are treated as single symbols. (Chinese, however, attempts to have an enumerable infinity of symbols.) The differences from our point of view between the single and compound symbols is that the compound symbols, if they are too lengthy, cannot be observed [by a human] at a glance. This is in accordance with experience. We cannot tell at a glance whether 9999999999999999 and 999999999999999 are the same.   [(*iii*)] The behaviour of the [human] computer at any moment is determined by the symbols which he is observing, and his "state of mind" at that moment.   [(*iv*)] We may suppose that there is a bound B to the number of symbols or squares which the [human] computer can observe at one moment. If he wishes to observe more, he must use successive observations.   [(*v*)] We will also suppose that the number of states of mind which need to be taken into account is finite. The reasons for this are of the same character as those which restrict the number of symbols. If we admitted an infinity of states of mind, some of them will be "arbitrarily close" and will be confused. Again, the restriction is not one which seriously affects [human] computation since use of more complicated states of mind can be avoided by writing more symbols on the tape.   [(*vi*)] Let us imagine the operations performed by the [human] computer to be split up into "simple operations" which are so elementary that it is not easy to imagine them further divided.   [(*vii*)] Every such operation consists of some change of the physical system consisting of the [human] computer and his tape [paper]. We know the state of the system if we know the sequence of symbols on the tape, which of these are observed by the [human] computer [...], and the state of mind of the [human] computer.   [(*viii*)] We may suppose that in a simple operation not more than one symbol is altered. Any other changes can be split up into simple changes of this kind. The situation in regard to the squares whose symbols may be altered in this way is the same as in regard to the observed squares. We may, therefore, without loss of generality, assume that the squares whose symbols are changed are always "observed" squares. Besides these changes of symbols, the simple operations must include changes of distribution of observed squares. The new observed squares must be immediately recognizable by the [human] computer. I think it is reasonable to suppose that they can only be squares whose distance from the closest of the immediately previously observed squares does not exceed a certain fixed amount. Let us say that each of the new observed squares is within L squares of an immediately previously observed square. [...]

This analysis isolated several issues concerning the "processes" and "operations" that compose human "computing": (*i*) A human "computes" a number by using a pen and paper of finite size divided into equally sized squares to contain symbols. This is his working space. Of course, when necessary, he can add an additional finite amount of paper. Turing noted that the usual two-dimensional paper can be replaced by *one*-dimensional *paper tape*, which is also divided into equally sized squares and potentially infinite: When necessary, additional paper tape with finitely many squares can be added to the right or left end of the paper tape. The transition to a one-dimensional working space is not an essential limitation since two- or

higher-dimensional arrays of squares can be represented by one-dimensional ones.[30]
Note also that a jump to a nonadjacent square in a two- or higher-dimensional
paper can be simulated by a finite sequence of moves to adjacent squares in the
one-dimensional paper tape. Thus, at any moment of his "computation", the working space (paper tape) of a human is a finite sequence of squares that can contain certain symbols which he has previously written, at most one symbol to a
square.   (ii) Turing then justified that the human needs only *finitely* many symbols to carry out a "computation". For if there were uncountably many symbols,
then—because of the finite size of the square—some of them would be too similar to be distinguishable; even if there were only countably infinitely many, say
$a_1, a_2, a_3, \ldots$, then of course they could be encoded by just two symbols, say s and
$'$, so that $a_1 = s', a_2 = s'', a_3 = s''', \ldots$ but the problem of distinguishing between
them would remain.   (iii) At any step of the "computation", the human action is
determined by (a) the symbols in the currently observed squares, and (b) his current
"state of mind".[31]   (iv) Due to his limited sensory capability, there must be a *fixed
bound B* on the number of squares (and their symbols) that the human can observe
at one moment. If he wishes to observe more, he must use successive observations.
(v)   Due to his limited memory capacity, the human can only use *finitely* many
"states of mind" to carry out the "computation"; otherwise, some of the "states of
mind" would become indistinguishable.[32]   (vi) Turing recognized that the "processes" that the human carries out in the "computation" can be split up into *simple*,
indivisible elementary "operations".   (vii) Generally speaking, each "simple operation" changes the state of the system consisting of the human and paper tape,
where the state of the system consists of (a) the sequence of all symbols written on
the paper tape; (b) the symbols in the currently observed sequence of squares of the
paper tape; and (c) the current "state of mind" of the human.   (viii) More precisely,
each "simple operation" can (a) alter not more than *one* symbol, which, in addition, must be currently observed and have been recognized by the human; (b) alter
the currently observed sequence of squares, provided that the distance between any
newly and any immediately previously observed square is less than a *fixed bound L*,
and that the symbols in the newly observed squares are *immediately recognizable*
by the human; and (c) alter the current "state of mind" of the human.

Then Turing summed up the isolated "simple operations" of human "computation":

> The most general single operation [of a human in computation] must therefore be taken to
> be one of the following:
> (A) A possible change of symbol together with a possible change of state of mind.
> (B) A possible change of observed squares, together with a possible change of state of mind.
> The operation actually performed is determined [...] by the state of mind of the [human]
> computer and the observed symbols. In particular, they determine the state of mind of the
> [human] computer after the operation is carried out.

---

[30] See Sect. 6.1.2, p. 118 and Sect. 6.1.3, p. 121.

[31] The "state of mind" of a human is an informal notion which we only comprehend intuitively.

[32] This is the weakest of Turing's assumptions for lack of our understanding of human "states of
mind".

Reduction of "human" operations to machine operations. Now Turing returned to his machine to show how the machine can simulate any "simple operation" of a computing human and hence any human "computation" of a number. In his words:

> We may now construct a machine [Turing machine] to do the work of this [human] computer. To each state of mind of the [human] computer corresponds an "*m*-configuration" [state] of the machine. The machine scans *B* squares [cells] corresponding to the *B* squares observed by the [human] computer. In any move the machine can change a symbol on a scanned square [cell] or can change any one of the scanned squares [cells] to another square [cell] distant not more than *L* squares [cells] from one of the other scanned squares [cells]. The move which is done, and the succeeding configuration [pair (current state, scanned symbol)], are determined by the scanned symbol and the *m*-configuration [state].

It should now be evident that Turing defined his machine (*i*) according to the results of his analysis of human "computation" and (*ii*) according to his intention to simulate "computation' by the machine.

Summary of Turing's ideas

Turing's construction of the Turing machine clearly shows that he conceived his machine with the *intention* of providing a *finite and mechanical description* of the "computations" performed by a *human* whose mental and physical capabilities are limited (and hence realistic), while the time and space available for human "computation" are not limited (thus allowing the "computation" carried out by one human to be continued by another human). The method that Turing followed was to (*i*) analyze the informal idea of a "computation" of a number carried out by a human, and remove, one after the other, successive layers of all the irrelevant details; (*ii*) distill the observable relevant "processes" that compose such "computations"; (*iii*) isolate the elementary, indivisible "simple operations" that make up the "processes"; (*iv*) identify the key properties of the "simple operations"; (*v*) define an abstract computing machine, the *Turing machine*; and, finally, (*vi*) demonstrate that the "simple operations" of a human can be simulated by the Turing machine.

In his paper, Turing presented his work in a different order, which enabled him to describe the implications of his discovery before giving justifications of it.

The Usual Turing's Thesis (functions)

What about "computable" *functions*? While Gödel, Church, and others searched for a characterization of "computable" functions, Turing investigated "computable" *numbers*. Why did he limit his attention to numbers? As said, Turing learned of the research pursued at Princeton just as he was ready to send off his manuscript. Nevertheless, at the beginning of his paper he did explain his approach:

> Although the subject of this paper is ostensibly the computable *numbers*, it is almost equally easy to define and investigate computable functions of an integral variable or a real or computable variable, computable predicates, and so forth. The fundamental problems involved are, however, the same in each case, and I have chosen the computable numbers for explicit treatment as involving the least cumbrous technique.

For instance, we can say that the function $g : \mathbb{N} \to \{0, \dots, 9\}$ defined by $g(n) \stackrel{\text{def}}{=} x_n$ is Turing-computable *iff* the number $x = 0 . x_1 x_2 \dots \in (0, 1)$ is Turing-computable.

So how can we extend Turing's discoveries to computation of function values? Turing justified that any humanly "computable" number is also Turing-machine computable. What if a number to be "computed" is the value of a function at a given argument?

Let $f : \mathbb{R} \to \mathbb{R}$ be an arbitrary function, $x \in \mathbb{R}$ an arbitrary argument, and suppose that $f(x)$ is defined, i.e., $f(x)\!\downarrow$. Human "computation" of the number $f(x)$ can be viewed as a finite sequence of numbers written on the paper tape such that (*i*) the last number of the sequence is $f(x)$, and (*ii*) each number of the sequence is either the argument x or has been directly "computed" from some of the preceding numbers of the sequence by one of the "simple operations" which—according to Turing's analysis—a human can carry out in his "computation" of a number. In the case that such a sequence exists, we informally say that f is "computable" *in x*. Then we informally say that the *function f is "computable" iff* it is "computable" in every x for which it is defined. In other words, a function is "computable" *iff* its value can be "computed" by a human whenever the value is defined.[33]

Now suppose that the function $f : \mathbb{R} \to \mathbb{R}$ is "computable," that is,

$$f \in \mathcal{F}.$$

Let x be an arbitrary number for which $f(x)$ is defined. Since f is "computable" *in x*, there is a sequence corresponding to the "computation" of the number $f(x)$. But Turing has shown that his machine can simulate this "computation" by (*a*) simulating each "simple operation" of the human, and (*b*) writing down on its tape the intermediate results (numbers) as notes assisting the machine's memory. Hence, $f(x)$ is a Turing-computable number; for this reason we say that f is Turing-computable *in x*. We have justified that if f is "computable" in x, it is also Turing-computable in x. Since we supposed that f is "computable" (i.e., "computable" in every x for which it is defined), it follows that f is Turing-computable in every x for which it is defined. Such a function f is said to be a *Turing-computable function*. If we denote by $\mathcal{T}$ the class of all *Turing-computable functions*, we conclude that

$$f \in \mathcal{T}.$$

We have justified that $f \in \mathcal{F} \Rightarrow f \in \mathcal{T}$, for any function $f : \mathbb{R} \to \mathbb{R}$. This means that

$$\mathcal{F} \stackrel{\text{just}}{\subseteq} \mathcal{T},$$

where $\stackrel{\text{just}}{\subseteq}$ denotes a justified (informally proved) inclusion. This was the so-called hard half of Turing's Thesis (functions).

---

[33] Note the similarity between this definition of a "computation" and the definition of a *deduction* (see Sect. 2.1.1, p. 10). This will become important in Kripke's argument (see Sect. 16.4.3, p. 340).

The easy half is to justify the converse:

$$\mathcal{F} \overset{\text{just}}{\supseteq} \mathcal{T}.$$

Here is a proof: A human can imitate the operations of the Turing machine as these do not violate any of the limitations of human memory or human sensory capability.

In sum, we have justified that

$$\mathcal{F} \overset{\text{just}}{=} \mathcal{T},$$

where $\overset{\text{just}}{=}$ denotes a justified (informally proved) equality. This is a concise form of the usual *Turing's Thesis*.

Turing's thesis. *A function is "computable" iff it is Turing-computable.*

Reactions to Turing's Thesis

Gödel approved Turing's analysis and Turing's Thesis as satisfactorily convincing. In addition to that, Turing's method partly complied with Gödel's suggestions to Church in 1934 (see p. 320).

Church endorsed Turing's characterization of "computability" in his review in 1937:

> [C]omputability by a Turing machine [...] has the advantage of making the identification with effectiveness in the ordinary (not explicitly defined) sense evident immediately.

Kleene wrote:

> For rendering the identification with effective calculability the most plausible—indeed, I believe compelling—Turing computability has the advantage of aiming directly at the goal.

Post, whose paper from 1936 contains the same idea as Turing's, accepted Turing's Thesis and soon proposed some improvements to the definition of a Turing machine.

Turing's work also convinced Gödel of Church's Thesis. Here is what Kleene wrote:

> According to a November 29, 1935, letter from Church to me, Gödel "regarded as thoroughly unsatisfactory" Church's proposal to use λ-definability as a definition of effective calculability. [...] It seems that only after Turing's formulation appeared did Gödel accept Church's thesis.

But there were more reasons to endorse Turing's work. Namely, Tarski and Gödel considered Turing computability and the equivalent (general) recursiveness of great importance for *metamathematical* reasons, because the two proposed concepts *absolutely* defined the epistemological notion of computability. Why?

In 1936, Gödel already proved that, given a sequence of increasingly strong formal axiomatic systems (f.a.s.) F_1, F_2, ..., passing from F_i to a stronger F_{i+1} may enable us to prove certain propositions that were not provable in F_i. So Gödel defined the concept of a function (of positive integers) *computable in an f.a.s.* and focused on the *provability of propositions* about such functions in different f.a.s. He discovered that once we have a *sufficiently strong* f.a.s. F_k—which is in fact *Formal Arithmetic*[34] A—there is no need to consider stronger f.a.s. $F_{k+1}, \ldots,$ because in $F_k (= A)$ can be proved anything about such functions that could be proved in a stronger f.a.s.

In the particular case of the provability of *function computability* this means that a function of positive integers is computable in any f.a.s. F containing A *iff* it is computable in A. It also means that the notion of computability of functions of positive integers is epistemologically invariant (stable) in all f.a.s. that contain A. In this sense, the definition of function computability in terms of Turing computability or (general) recursiveness is *absolute* (i.e., stable from $F_k = A$ on). In 1946, Gödel briefly and clearly described this fortunate situation:[35]

> Tarski has stressed in his lecture (and I think justly) the great importance of the concept of general recursiveness (or Turing's computability). It seems to me that this importance is largely due to the fact that with this concept one has for the first time succeeded in giving an absolute definition of an interesting epistemological notion, i.e., one not depending on the formalism chosen. In all other cases treated previously, such as demonstrability or definability, one has been able to define them only relative to a given language, and for each individual language it is clear that the one thus obtained is not the one looked for. For the concept of computability however, although it is merely a special kind of demonstrability or decidability the situation is different. By a kind of miracle it is not necessary to distinguish orders, and the diagonal procedure does not lead outside the defined notion.

In 1951, Gödel summarized the results important to mathematics and philosophy of the last few decades and emphasized Turing's role in this:

> Research in the foundations of mathematics during the past few decades has produced some results of interest, not only in themselves, but also in regard to their implications for the traditional philosophical problems about the nature of mathematics. [...] The greatest improvement was made possible through the precise definition of the concept of the finite procedure ["equivalent to the concept of a 'computable function of integers'"], which plays a decisive role in these results. There are several different ways of arriving at such a definition, which, however, all lead to exactly the same concept. The most satisfactory way, in my opinion, is that of reducing the concept of a finite procedure to that of a machine with a finite number of parts, as has been done by the British mathematician Turing.

---

[34] See Sect. 3.2, p. 45.

[35] At the time it was already believed that the concept of "computability" is *formalism-independent*, in the sense that all formal characterizations of the concept pick out the *same* class of functions.

16.4 Church-Turing Thesis

We have seen that, in 1936, the quest for the formalization of informal notions of computing brought to the surface two theses which concerned computability of functions, Church's Thesis and Turing's Thesis. Some researchers supported the first thesis and others approved the second one. Naturally, they started comparing the two theses and eventually spotted several differences, which we sum up in the following.

16.4.1 Differences Between Church's and Turing's Theses

The main differences between Church's Thesis and Turing's Thesis are:

1. Church simply and boldly *postulated* (defined) that

$$\mathcal{F}^+ := \mathcal{G}^+ (= \mathcal{C}^+),$$

   with the only indication for this being the rigorously proved confluence $\mathcal{G}^+ = \mathcal{C}^+$, while Turing *justified* that $\mathcal{F} \overset{\text{just}}{\subseteq} \mathcal{T}$ and, along with the easy part $\mathcal{F} \overset{\text{just}}{\supseteq} \mathcal{T}$, that

$$\mathcal{F} \overset{\text{just}}{=} \mathcal{T}.$$

2. Church provided in support of his thesis *no* reference to any means that could mechanically carry out computation of (general) recursive or λ-definable functions, while Turing established and justified his thesis by introducing an abstract computing *machine* and linking it to the intuitive idea of "computability". The ability of this machine, the Turing machine, to encapsulate the fundamental logical principles of computation had a profound significance for the emerging science of automatic computation and the realization of the stored-program general-purpose digital computer.

3. Church focused on functions of *positive integers*, while Turing considered functions of *positive integers* or *Turing-computable real* numbers. In fact, Turing's concerns were even more general than Church's: they also included computable predicates and functions of real variables.

4. While the *intention* of both Church and Turing was to refer to functions "computable" by an idealized human, Turing's additional intention was to provide an idealized description of numerical "computation" as a particular human activity. Because of this difference between their intentions, we say that Church's and Turing's Theses are *not intensionally equivalent*.

16.4.2 The Church-Turing Thesis

In 1952, Kleene observed that by restricting attention to functions of positive integers Turing's Thesis reduces to

$$\mathcal{F}^+ \overset{\text{just}}{=} \mathcal{T}^+,$$

where $\mathcal{F}^+$ and $\mathcal{T}^+$ are the class of "computable" functions of positive integers and the class of Turing-computable functions of positive integers, respectively. Since Turing outlined a proof of $\mathcal{T}^+ = \mathcal{C}^+$ (in the Appendix to his paper of 1936), and gave a full proof of $\mathcal{T}^+ = \mathcal{C}^+$ and $\mathcal{T}^+ = \mathcal{G}^+$ (in his paper of 1937), the following was known about the classes of functions of positive integers:

Church's Thesis: $\mathcal{F}^+ := \mathcal{G}^+ (= \mathcal{C}^+)$

$\\\quad //$

Turing's Thesis (for positive integers): $\mathcal{F}^+ \overset{\text{just}}{=} \mathcal{T}^+$

So, after restricting to functions of positive integers, both theses characterized the same class, $\mathcal{F}^+$, in spite of the fact that their original intentions were distinct. This is why we say that Church's and Turing's Theses are *extensionally equivalent*.

Now note that we can combine the two theses into one. Start with Turing's Thesis (for functions of positive integers) and equate its right-hand side ($\mathcal{T}^+$) with the right-hand side of Church's Thesis, *bypassing* Church's postulated definition (:=). The resulting thesis is today called the *Church-Turing Thesis* (see below the line):

Church's Thesis: $\mathcal{F}^+ := \mathcal{G}^+ (= \mathcal{C}^+)$

$\\\quad //$

Turing's Thesis (for positive integers): $\mathcal{F}^+ \overset{\text{just}}{=} \mathcal{T}^+$

_____

Church-Turing Thesis: $\mathcal{F}^+ \overset{\text{just}}{=} \mathcal{T}^+ = \mathcal{G}^+ = \mathcal{C}^+$

The new thesis is often shortened to

$$\mathcal{F}^+ \overset{\text{just}}{=} \mathcal{T}^+ (= \mathcal{C}^+),$$

but still called the *Church-Turing Thesis*.

Church-Turing Thesis. *A function of positive integers is "computable" iff it is Turing-computable (or λ-definable).*

16.4.3 Justifications of the Church-Turing Thesis

Since the inception of the Church-Turing Thesis, several types of arguments have been given to justify it. In the following we classify them according to Kleene and Shoenfield:[36]

1. **The argument from non-refutation.**
   In spite of sustained and ongoing attempts to find a counter example (e.g., a "computable" function that is *not* Turing-computable), the Church-Turing Thesis has never been refuted. All the "computable" functions that researchers have constructed have been shown to be Turing-computable. Moreover, all the known *techniques* for constructing new "computable" functions from old ones have been shown to lead from Turing-computable functions to Turing-computable functions. Thus, there is no idea how to produce a "computable" but not Turing-computable function.

2. **The argument from replacement.**
   As Shoenfield pointed out, we become convinced of the Church-Turing Thesis after detailed study of *Computability Theory*, because all the results of the theory become quite reasonable or even obvious when the term Turing-computable is replaced by the term "computable".

3. **The argument from confluence.**
   The first proposals for the characterization of function "computability", that is, λ-definability, Turing computability, μ-recursiveness, and Post's finite combinatory processes, quickly proved to be extensionally equivalent, i.e., they defined the *same* subclass $\mathcal{T}^+ = \mathcal{G}^+ = \mathcal{C}^+ = \mathcal{K}^+ = \mathcal{P}^+$ of the class $\mathcal{F}^+$ of all "computable" functions of positive integers. Subsequently, many other—in terms of the approach and formal details—quite different proposals for the characterization of $\mathcal{F}^+$ were given, such as Markov algorithms,[37] RAM,[38] and cellular automata. But again, each of them turned out to encompass the *very same* subclass $\mathcal{T}^+ (= \mathcal{G}^+ = \mathcal{C}^+ = \mathcal{K}^+ = \mathcal{P}^+)$ of $\mathcal{F}^+$. So the class $\mathcal{T}^+$ turned out to be significantly *formalism-independent*, in the sense that *all known*, superficially diverse formalisms pick out exactly this subclass of $\mathcal{F}^+$. This striking confluence of many *fundamentally different* proposals for the characterization of $\mathcal{F}^+$ suggests that the distinguished class $\mathcal{T}^+ = \mathcal{G}^+ = \mathcal{C}^+ = \mathcal{K}^+ = \mathcal{P}^+ = \cdots$ is a very natural class. It also gives us grounds for the belief that the confluence of ideas will continue for the characterizations of $\mathcal{F}^+$ that may be proposed in future. All of this strengthens our confidence that $\mathcal{T}^+$ actually characterizes the informally defined target class $\mathcal{F}^+$.

---

[36] Joseph Robert Shoenfield, 1927–2000, American mathematical logician.

[37] See Sect. 5.2.3, p. 93.

[38] See Sect. 6.2.7, p. 132.

4. **The argument from Turing's analysis (Argument I).**
   Turing's analysis (which we presented in Sect. 16.3.6) reveals five general and
   intuitive constraints satisfied by a *"computing" human*: (*i*) Human behavior is
   at any moment determined by his current "state of mind" and the currently
   observed symbols. (*ii*) There is a bound on the number of squares which a human
   can observe at one moment. (*iii*) There is a bound on the distance between
   the newly observed squares and the immediately previously observed squares.
   (*iv*) In a simple operation, a human alters at most one symbol. (*v*) Only finitely
   many human "states of mind" need to be taken into account. Then Turing's
   analysis demonstrates that *if* a human "computation" is subject to these five reasonable and natural constraints, *then* it can be simulated by a *Turing machine*,
   a finite machine that operates in a purely mechanical manner. Under these constraints, any humanly "computable" function is also Turing-computable. This
   argument is often called *Argument I*.

5. **The argument from first-order logic (Argument II).**
   In addition to Argument I, Turing gave an argument which he titled "A proof of
   the equivalence of two definitions" and which today is often called Argument II.
   This argument has mostly been ignored in discussions of Turing's paper, perhaps
   because its presentation is somewhat demanding, while Argument I is accessible and is well accepted by most of the research community. In Argument II,
   Turing stated that he had constructed a Turing machine that can generate all
   the provable formulas of the First-Order Logic **L** (he used the term Hilbert
   restricted functional calculus). Conversely, after showing that some sequences of
   symbols can be defined by formulas of **L**, he proved that sequences definable in
   L are Turing-computable. The latter statement is today often called the Turing's
   Provability Theorem.

   Theorem 16.2. (Turing's Provability Theorem) *Every formula provable in the
   First-Order Logic **L** can be proved by the universal Turing machine.*

   As said, Turing expressed a belief that his thesis is not susceptible to rigorous
   proof because it is not a mathematically precise statement (see p. 330). Accordingly, his Argument I is an informal justification of the thesis. Turing's intentions
   with Argument II were presumably similar, so he may have presented the above
   theorem as part of an intended justification. This is the basis of the next argument.

6. **Kripke's Logical Orientation.**
   In 2013, Kripke[39] revived Argument II. He took Turing's *Provability Theorem*
   as the main point of Argument II, and building on this, he advocated a *logical* orientation to the Church-Turing Thesis. In particular, Kripke proposed that
   derivability from a finite set of instructions expressible in a first-order language
   be accepted as a basic concept of computability. He explained the rationale behind the proposition as follows. Kripke assumed that the underlying motivation

---

[38] See Appendix A, p. 364.

[39] Saul Kripke, b. 1949, American philosopher and logician.

of Argument II was Turing's belief that his characterization of "computability" in terms of his machines is *equivalent* to the characterization of "computability" in terms of deducibility in a standard formal axiomatic system, e.g., *First-Order Logic with equality*,[40] **L**. In Kripke's opinion, therefore, Turing's intention was to argue that his *Provability Theorem* could serve in proving this equivalence.

Building on this, Kripke presented the following reconstruction of Argument II in which Turing's *Provability Theorem* actually plays an important role:

a) *Premise: "Computation" is a special form of "deduction".*

This is *Kripke's Claim.* In contrast to the Church-Turing Thesis, the claim relates *two informal* notions. We do not expect that the claim can be rigorously proved, but a justification for it can be given as follows. Suppose that one is given a list (i.e., recipe or "effective procedure") of *finitely* many instructions, and perhaps some well-known and not explicitly stated mathematical premises, which can be used during "computation", such as the Peano axioms or premises about the values of some input and/or "computed" data. The "computation" is the "execution" of the given list of instructions. Every step in the "computation" is supposed to follow *deductively* from the "execution" of the current instruction and from the currently relevant implicit premises and/or explicit premises (i.e., facts) that have been established during the "computation". The "computation" can therefore be viewed as a very specialized "deduction," i.e., deductive "argument," whose conclusion is a sentence precisely describing the result of the "computation".[41] It is this sense that Kripke advocated a *logical* orientation to the characterization of the notion of "computation". The following example illustrates Kripke's claim.

Example 16.1. ("Computation is a special form of "deduction".) First a short reminder. We say that a sentence p *logically entails* a sentence q (written $p \vDash q$) iff every truth assignment that satisfies p also satisfies q. Generally, a set P of sentences logically entails a sentence q (i.e., $P \vDash q$) iff every truth assignment that satisfies all of the sentences in P also satisfies q. Now, the "computation" describing the usual "execution" of the following simple list

$$I_1: \quad a \leftarrow 3 \qquad \text{// instruction } I_1 \text{ stores 3 in a}$$
$$I_2: \quad a \leftarrow a+4 \qquad \text{// instruction } I_2 \text{ adds 4 to a}$$

can be expressed as

$$I_1; I_2$$

which means that the execution of I_1 is immediately followed by the execution of I_2. The "deduction" corresponding to the "computation" $I_1; I_2$ can be expressed as

1. $I_1: \quad a \leftarrow 3$ // execution of I_1 with no premises
2. $a = 3$ // follows from (logically entailed by) 1.
3. $I_2: \quad a \leftarrow a+4$ // execution of I_2 with explicit premise 2.
4. $a = 7$ // follows from (logically entailed by) 3.

Thus $a = 7$ is the conclusion of the "deduction" and 7 is the result of the "computation". □

---

[40] See Sect. 3.2, p. 44.

[41] In his 1936 paper, Turing developed a detailed logical notation for expressing such "arguments". See the footnote about the similarity between definitions of "computation" and deduction on p. 334.

b) *Premise: Every "deduction" can be expressed as a valid deduction in* **L**.

This statement Kripke called *Hilbert's Thesis*. It claims that any "deduction" (i.e., any finite sequence of content-dependent inferences where each conclusion undoubtedly follows from the meaning of its premises) can be restated as a *valid deduction* (preserving truth from premises to the final conclusion) in a language based on *First-Order Logic with equality*, **L**. Since the thesis identifies the intuitive notion of a "deduction" with the precise notion of a valid derivation in a standard formal axiomatic system, it is believed, in accordance with the prevailing opinion, to be justifiable only by appeal to intuition (but see a critique of this opinion in Sect. 16.4.4.).

c) *Every "computation" can be expressed as a valid deduction in* **L**.

This follows directly from *a*) and *b*).

d) *Every valid deduction in* **L** *is provable in* **L**.

This follows directly from Gödel's *Completeness Theorem*[42], which states that **L** is semantically complete (a formula has a valid deduction *iff* it is provable).

e) *Every "computation" is a provable deduction in* **L**.

This follows directly from *c*) and *d*). In other words, every human "computation" can be proved in **L**, in the sense that each step of the formalized computation in **L** can be derived from the current instruction and possibly some well-defined premises.

f) *Every provable deduction in* **L** *is provable by the universal Turing machine.*

This is Turing's *Provability Theorem* (see above).

g) *Conclusion: Every "computation" can be carried out by Turing machine.*

This follows from *e*) and *f*). So, every "computation" can be carried out by the universal Turing machine.

In summary, under the two premises (Kripke's and Hilbert's), the Church-Turing Thesis follows from Gödel's *Completeness Theorem* and Turing's *Provability Theorem*.

Fig. 16.1 The structure of Kripke's argument

---

[42] See Sect. 4.1.1, p. 58.

Thus, Kripke demonstrated that the Church-Turing Thesis is deducible from his own claim and Hilbert's Thesis. Since the latter is not susceptible to rigorous proof, Kripke's argument doesn't prove the Church-Turing Thesis, but reduces it to another thesis that also resists rigorous proof.

Nevertheless, this reduction allows one to adopt as a basic concept of computability either the machine-oriented Church-Turing thesis *or* the perhaps more intuitively appealing logically oriented Hilbert's thesis (assuming that Kripke's claim holds). This may have various advantages. One of them Kripke described as follows:

> Another relatively minor advantage of the approach of deducibility in a first-order language [...] is simply pedagogical. Ever since Post's famous paper (1944) [see Post [184]], the advantage of an intuitive presentation of computability arguments has been evident, rather than the rival approach using a formal definition of computability in the proofs. Although the experienced computability theorist will know how to convert such arguments into proofs using a formal definition, and cannot be said to be relying on an unproved thesis, this is hardly true for a beginner. However, such a beginner, if he has already studied elementary logic, will readily accept that the steps of an argument can be stated in a first-order language, even if they are given verbally. The Gödel completeness theorem guarantees that if the steps really do follow (using any implicit axioms in addition to the actually stated steps), the argument can be formalized in one of the usual systems. Granted that the proof predicate [the predicate stating that a formula *is provable* from another formula] of such a system is recursive/computable by one of the usual definitions, that one really has a technically valid proof will be readily accepted.

16.4.4 Provability of the Church-Turing Thesis

Until recently, most logicians, mathematicians, and computer scientists shared the opinion that the Church-Turing Thesis is not amenable to rigorous proof. However, in 1990, Mendelson[43] gave a powerful critique of this general opinion. His critique consists of the following three points:

1. **Mathematics and logic already tolerate vagueness.** *The concepts and assumptions supporting the notion of Turing-computable function are essentially no less vague and imprecise than the notion of "computable" function.*

For example, the concept of a function, which is used in the definition of a Turing-computable function, is rigorously defined using the notion of a set: a *function* $f : \mathcal{A} \to \mathcal{B}$ is a set f of ordered pairs (a, b), where $a \in \mathcal{A}$, $b = f(a) \in \mathcal{B}$, and where $(a, b) \in f$ and $(a, c) \in f$ implies $b = c$. The notion of the *ordered pair* (x, y) too is rigorously defined by the notion of the set: $(x, y) = \{\{x\}, \{x, y\}\}$. In contrast, the notion of a *set*, being a basic notion, is *not* rigorously defined.[44]

---

[43] Elliott Mendelson, b. 1931, American logician.

[44] See Sects. 2.1.2, 2.1.3, and 3.2, p. 47.

Consequently, the concept of a function ultimately rests on an intuitively understood and imprecisely defined notion of a set! But there is more to it than that. In
contrast with Turing-computable and "computable" functions, which are linked
by the Church-Turing Thesis, there is no thesis in mathematics claiming that the
above (seemingly) rigorous definition of a function $f : \mathcal{A} \rightarrow \mathcal{B}$ satisfactorily characterizes the intuitive concept of a "function" (which understands f to be a "rule"
for "calculating" or "constructively assigning" $b = f(a) \in \mathcal{B}$ to any $a \in \mathcal{A}$). The
concept of a function is so familiar and so well-tested in logic and mathematics
that such a thesis was never an issue. Mendelson pointed out that there are many
other intuitive notions (such as "a sentence being true in a structure", "limit",
"measure", "dimension") whose rigorous definitions ultimately rest on the intuitive notion of a set, and the adequacy of these (seemingly) rigorous definitions
is never questioned.

2. **The intuitive and rigorous can be linked.** *The general assumption that a proof
   linking vague and precise mathematical notions is* impossible *is false.*

   For example, the easy half of the Church-Turing Thesis, stating that all Turing-
   computable functions are "computable", is widely acknowledged to be obvious
   and is therefore readily accepted. This is because there is a straightforward argument for it, which, due to its simplicity, is acknowledged to be a proof, *in spite*
   of the fact that it involves the intuitive notion of "computability".[45] So, the fact
   that the argument is not in **ZF** or some other formal axiomatic system is no draw-
   back (but, as Mendelson stressed, shows that there is more to mathematics than
   appears in **ZF**).

3. **Proving is not the only way to ascertain the truth.** *The usual opinion that* proof
   is the only *way to ascertain the truth of the Church-Turing Thesis is false.*

   In mathematics and logic, proof is *not* always the only way in which a statement
   comes to be accepted as true. Equivalences between intuitive notions and apparently more precise mathematical notions often are simply "*seen*" to be true *without* proof, or are based on arguments that are a mixture of such (non-empirical)
   intuitive perceptions and standard logical and mathematical reasoning. Some
   such intuitive notions were mentioned above in the first item.

   In summary, today it appears that the opinion that the Church-Turing Thesis *is*
provable is gaining acceptance. Although several proofs of the Church-Turing Thesis have recently been proposed, the matter remains the subject of active discussion
within the academic community.

---

[45] We describe the argument for μ-recursive functions. The initial functions are "computable"
(the "procedures" to "compute" their values are trivial). Composition, primitive recursion, and
μ-operation produce "computable" functions from "computable" functions (in each case we can
easily describe "procedures" that will "compute" the new "computable" function).

16.5 Résumé and Warnings

Besides the original Church-Turing Thesis, the reader may find in various contexts similar-looking statements that are, for different reasons, *mislabeled* as the Church-Turing Thesis. Below we will address some of these look-alikes.

Moreover, due to the fast development of general-purpose computers and the recent rise of different proposals for a *new computing paradigm*, several scientists have questioned the adequacy of the Church-Turing Thesis for the suggested paradigms, whether realistic (such as *parallel* and potentially *quantum computation*) or speculative ones (such as *hypercomputation*). Consequently, *new versions* of the Church-Turing Thesis have emerged, some of them only recently, each involving or emphasizing a certain notion implicit in the original Church-Turing Thesis. These notions include implementability of algorithms, computational complexity, and physical computability. Not surprisingly, questions about the original Church-Turing Thesis and the proposed versions have arisen. Thus, the necessity, adequacy, and relative power of each of them are currently actively discussed within the academic community.

To keep up with this discussion, understanding the subtle differences between the original Church-Turing Thesis on the one side and its look-alikes or the emerging new versions on the other side is important for anyone interested in the foundations of Computability Theory.

Résumé

In 2017, Copeland[46] extracted the essence of Turing's work in the following passage:

> *Turing proved that his **universal [Turing] machine can compute any function that any Turing machine can compute;** and he put forward, and advanced philosophical arguments in support of, the thesis that **effective methods [procedures] are to be identified with methods [procedures] that the universal Turing machine is able to carry out.***

Copeland then summed up:

> *Essentially, the Church-Turing Thesis says that **no human computer, or machine that mimics a human computer, can out-compute the universal Turing machine.***

---

[46] B. Jack Copeland, b. 1950, English philosopher and historian of computing, and mathematical and philosophical logic.

Some Warnings

In relation to this, here are some potentially misleading claims that have appeared in various sources. The claims are only conditionally true because they go far beyond Turing's discoveries:

- Turing showed that the universal Turing machine can specify the steps required for the solution of any problem that can be solved by <u>instructions, explicitly stated rules, or procedures</u>.

  The claim is true *only if* the general terms "instruction", "explicitly stated rule", and "procedure" are restricted so that they refer only to what can be done by "effective procedures".

- Turing had proven that his universal Turing machine can compute any function that <u>any computer, with any architecture</u>, can compute.

  The claim is true *only if* the term "computer" refers to the Turing machine or a machine equivalent to it. Turing proved that the universal Turing machine is universal with respect to other Turing machines (not to any machine capable of computing).

- Every task for which there is a <u>clear recipe composed of simple steps</u> can be performed by a very simple computer, the universal Turing machine.

  The claim is true *only if* the term "clear recipe composed of simple steps" refers to an "effective procedure".

- Turing's results entail that a standard digital computer can compute any <u>rule-governed input-output function</u>.

  The claim is true *only if* the term "rule-governed input-output function" refers to a Turing-computable function.

- A function is computable <u>by means of a machine</u> (i.e., mechanically computable), *iff* it is Turing-computable.

  The Church-Turing Thesis does not address computability of functions *by means of machines*, so it does not imply the above claim. Conversely, *if* we accept that a human is a type of machine, *then* the claim implies the Church-Turing Thesis. Thus, the claim and the Church-Turing Thesis are not equivalent.

16.6 New Questions About the Church-Turing Thesis

In 2019, Copeland and Shagrir[47] addressed the open questions that concern the fundamental place of computing in the physical universe. After giving an overview of the original Church-Turing Thesis and its equivalent versions, they described and discussed today's more powerful versions of this thesis. In the rest of this section, we follow their exposition (and abbreviate the Church-Turing Thesis to CTT).

16.6.1 Original CTT

First, of course, there is the original Church-Turing Thesis:

(**CTT**) *Any function that is intuitively computable is computable by some Turing machine.*

Copeland and Shagrir put it as follows:

(**Copeland-Shagrir (CTT-O)**) *Every function that can be computed by the idealized human computer (i.e., can be effectively computed), is Turing-computable.*

Now we describe *algorithmic, complexity-theoretic,* and *physical* versions of CTT.

16.6.2 Algorithmic Versions of CTT

These versions of the original Church-Turing Thesis involve the *intuitive notion of "algorithm"*[48] by asking *what can be computed by "algorithms" in general.* Here is an algorithmic version of CTT stated in 1981 by Lewis[49] and Papadimitriou[50]:

(**Lewis-Papadimitriou**) *[W]e take the Turing machine to be a precise formal equivalent of the intuitive notion of "algorithm": nothing will be considered an algorithm if it cannot be rendered as a Turing machine.*

But computing machines have been developing; for example, they are now able—in contrast to Turing machines—to change several dislocated parts of data at each step of computation. Accordingly, the concept of algorithm has been evolving, so much so that certain steps of some kinds of algorithms cannot be directly carried out by the basic steps of Turing machines. For such an algorithm, a Turing machine with an essentially different Turing program must be designed if one wants to show that the problem under consideration is still Turing computable (and not just computable by the particular kind of algorithm). But how much can algorithms evolve and still remain reasonable relative to the Turing machine? That is, what is the upper bound on the concept of an algorithm so that algorithmic computability entails Turing computability? Again we are faced with the question *What is an algorithm?*

---

[47] Oron Shagrir, b. 1961, Israeli philosopher of computing and cognitive/brain sciences.

[48] We use terms "algorithm" and "effective procedure" interchangeably.

[49] Harry Roy Lewis, b. 1947, American computer scientist and mathematician.

[50] Christos Harilaos Papadimitriou, b. 1949, Greek theoretical computer scientist.

Besides various views of what kind of *abstract entities* are algorithms, it is also debated to what extent algorithms should be *implementable*, that is, reducible to restricted forms that are based on the *familiar models of computation* (e.g., Turing machine and RAM). Since implementable algorithms are expressible in terms of familiar models of computation, do *non*-implementable algorithms make any sense? *Theoretically*, they do. To provide a development of the theory of algorithms within axiomatic set theory, Moschovakis[51] adopted an *abstract* notion of algorithm (with recursion as a primitive operation) that is so wide as to admit even *non*-implementable algorithms. In contrast, focusing on the *practical* aspects of algorithms, Harel[52] advocates that an algorithm only has to be *expressible* in a "reasonable" programming language. Thus, Harel suggested the following algorithmic version of CTT:

> (**Harel**) *[A]ny algorithmic problem for which we can find an algorithm that can be programmed in some programming language,* any *language, running on some computer,* any *computer, even one that has not been built yet but* can *be built, and even one that will require unbounded amounts of time and memory space for even-larger inputs, is also solvable by a Turing machine.*

Building on this and setting aside the issue of algorithm implementability, Copeland and Shagrir formulated in 2019 the following algorithmic version of CTT:

> (**Copeland-Shagrir (CTT-A)**) *Every algorithm can be expressed by means of a program in some (not necessarily currently existing) Turing-equivalent programming language.*

16.6.3 Complexity-Theoretic Versions of CTT

These are versions of the original Church-Turing Thesis involving *issues related to Computational Complexity Theory*, such as the relation between time complexities of computational problems in different (reasonable general) models of computation. That such models are connected in a particularly appealing way is stated by the Cobham[53]-Edmonds[54] thesis, which has become part of computer science folklore.

> (**Cobham-Edmonds**) *If a computational problem's time complexity is t in some (general and reasonable) model, then its time complexity is assumed to be poly(t) in the single-tape Turing machine model.*

There are of course different views about which models count as *reasonable*, one being that a model is reasonable iff it is physically realizable (at least in principle).

But why is this thesis important? Take an arbitrary NP-complete computational problem. Assuming the Cobham-Edmonds thesis, it follows that if the problem is not solvable in polynomial time on a single-tape Turing machine, then neither is it solvable in polynomial time in *any other* reasonable general model of computation. This means that the notion of polynomial (un)solvability of computational problems is *robust*, insensitive to the model of computation. And so is the question $P \overset{?}{=} NP$.

---

[51] Yiannis N. Moschovakis, b. 1938, Greek-American logician and theoretical computer theorist.

[52] David Harel, b. 1950, Israeli computer scientist.

[53] Alan Belmont Cobham, 1927-2011, American mathematician and theoretical computer scientist.

[54] Jack R. Edmonds, b. 1934, American computer scientist.

Some forms of the Cobham-Edmonds thesis use a *probabilistic* Turing Machine[55] instead of a single-tape Turing machine. For example, Bernstein[56] and Vazirani[57] pointed out that *Computational Complexity Theory* rests upon the following thesis:

(Bernstein-Vazirani) *Any reasonable model of computation can be efficiently simulated on a probabilistic Turing machine.*

In 2013, Aharonov[58] and Vazirani reformulated the Bernstein-Vazirani thesis and stated the following *Extended Church-Turing Thesis* (CTT-E):

(Aharonov-Vazirani (CTT-E)) *[A]ny reasonable computational model can be simulated efficiently by the standard model of classical computation, i.e., a probabilistic Turing machine.*

Here, "classical" refers to any computation that is not quantum computation. This is because CTT-E is also used in quantum-computation research. And it is this research that brought about a potential counterexample to CTT-E. Namely, the fastest known classical algorithm solves the PRIME FACTORIZATION[59] problem on any classical computer in *exponential time*, while a *quantum computer* can solve it in *polynomial* time by Shor's quantum algorithm. Does this falsify CTT-E? Not quite, as some still hold that a quantum computer is *not* a physically reasonable model of computation.

16.6.4 Physical Versions of CTT

These versions of the original Church-Turing Thesis involve *physical reality* by asking *What can be computed by physical systems in general?* For example, in 1985, Wolfram[60] suggested the following physical version of CTT:

(Wolfram) *[U]niversal computers are as powerful in their computational capacities as any physically realizable system can be, so that they can simulate any physical system.*

Also in 1985, and independently of Wolfram, Deutsch proposed a similar thesis, which is now named the *Church-Turing-Deutsch-Wolfram Thesis* (CTDW):

(Church-Turing-Deutsch-Wolfram (CTDW)) *Every finite physical system can be simulated to any specified degree of accuracy by a universal Turing machine.*

Here, "simulation" is not a perfect simulation in the sense of achieving absolute accuracy; otherwise, CDTW would be falsified by classical *continuous* physical systems since these involve *incomputable* real numbers, proved Deutsch.

---

[55] A probabilistic TM picks one from a set of alternative transitions according to some probability distribution.

[56] Ethan Bernstein was a student of U. Vazirani.

[57] Umesh Virkumar Vazirani, Indian-American computer scientist.

[58] Dorit Aharonov, b.1970, Israeli computer scientist.

[59] PRIME FACTORIZATION is the problem of decomposing a composite number into a product of prime numbers. Currently, no classical algorithm is known that can solve the problem in polynomial time. Neither the existence nor non-existence of such an algorithm has been proved (it is believed that it does not exist and that the problem is not in class P). The problem is in class NP. But it has not been proved to be or not be NP-complete (it is believed not to be NP-complete).

[60] Stephen Wolfram, b. 1959, British-American computer scientist, physicist, and businessman.

Since simulation of a system is just computation of a particular aspect of the system's behavior, namely the system's time evolution from an initial state to a final state, Copeland and Shagrir suggested a much stronger version of the above thesis, calling it the *Total Physical Computability Thesis* (CTT-P):

> (**Copeland-Shagrir (CTT-P)**) *Every physical aspect of the behavior of any physical system can be calculated (to any specified degree of accuracy) by a universal Turing machine.*

Here, "physical system" means any actual, non-actual, or idealized system whose behavior is in accordance with the actual laws of physics.

Is CTT-P true? Copeland and Shagrir listed several *potential* counterexamples to CTT-P that have been presented by researchers. Here we describe two of them that have recently emerged from *quantum mechanics*:

- In 2012, Eisert, Müller, and Gogolin proved that OUTCOME SEQUENCES[61], a decision problem about the results of repeated quantum measurements, is *undecidable*. Remarkably, if we require that the measurements are classical instead of quantum, the problem becomes *decidable*.

- In 2015, Cubitt, Perez-Garcia, and Wolf proved that the SPECTRAL GAP[62] problem is *undecidable* (the proof is by reduction of the problem HALTING OF TURING MACHINES[63]). So there cannot exist an algorithm or a computable criterion that solves the SPECTRAL GAP problem in general. But SPECTRAL GAP is one of the most important physical properties of quantum many-body systems and, consequently, an important determinant of a material's properties. Thus, a *major physics problem* is undecidable.

  How is this result connected to CTT-P? Cubitt et al. proved undecidability of the SPECTRAL GAP problem for an *infinite* system of two-dimensional lattices of atoms. But the proof also applies to *finite* systems whose size *increases*. Such systems are physically relevant and, as the proof revealed, there exists no effective method for computing their future behavior from complete descriptions of their current and past states. So these finite systems of increasing size offer a *counterexample to* CTT-P.

  However, there is the open question of whether the SPECTRAL GAP problem becomes decidable if all involved mathematical structures (e.g., Hilbert spaces) are bound to realistically low dimensions as imposed by the actual physical world.

Although also other *theoretical* counterexamples in which CTT-P is false have been found, there is currently still no evidence that CTT-P is false in the *actual* universe. But, of course, this does not imply that it is true.

---

[61] OUTCOME SEQUENCES is to determine whether or not a given sequence can occur as outcome sequence in repeated quantum measurements.

[62] A *spectral gap* is the energy difference between the ground state and first excited state of a quantum many-body system. Such a system is *gapless* if it has a continuous spectrum above the ground state in the thermodynamic limit, and is *gapped* if it has a unique ground state and a constant lower bound on the spectral gap. The SPECTRAL GAP problem is to determine, given the matrices describing the local interactions of a many-body system, whether the system is gapless or gapped.

[63] See Sect. 8.3.1, p 187.

In addition to the above mentioned, there are several other physical versions of CTT in the literature. By extracting their shared essences, Piccinini[64] classified them into two types of theses, calling them the *bold* and the *modest* physical CTT:

(**Piccinini (BoldCTT-P)**) *Any physical process—anything doable by a physical system—is computable by a Turing machine.*

(**Piccinini (ModestCTT-P)**) *Any function that is computable by a physical system is computable by a Turing machine.*

So, anything a physical system can *do* can be simulated by a TM (BoldCTT-P); and anything a physical system can *compute* can be computed by a TM (ModestCTT-P).

But what is the difference between the two theses? Isn't physical *computation* in a physical system something the system actually *does*? True, but the converse is not: In BoldCTT-P, a physical process *may not* be physical computation in the sense of yielding the *value* of some dynamic physical variable, but it can be simply the time evolution of the physical system from its initial state at a given instant to a final state at some later instant. Moreover, even if a physical process is physical computation of the value of some dynamic physical variable, the process may be *unusable* in the sense that the resulting value cannot be obtained by an *observer* whose behavior is supposed to be appropriately influenced by the result of the physical process. Specifically, the observer, who can be a human being or a functionally organized system, may want to use the physical process in the physical system to obtain the value of a given *function* whose argument has been encoded in the system's initial state.

Thus, BoldCTT-P *may not* be about what an observer can discover about the values of a given function. Because of this, Piccinini argued that BoldCTT-P is *irrelevant* to the epistemological concerns motivating the original Church-Turing Thesis.

Consequently, Piccinini stated his *Usability Constraint*, which distinguishes between physical processes and physical computations as follows:

A physical process should not count as a computation unless a finite[65] observer can use it to generate the desired values of a given function.

Then Piccinini proceeded to his ModerateCTT-P, grounding it on a notion of physical computation that now complies with the *Usability Constraint*. How did he do that? We have seen that the *Usability Constraint* filters out all physical processes that are irrelevant because their results cannot be obtained and used by finite observers. The remaining physical processes are in this sense usable, but it is still unclear whether or not these processes must pass any additional constraints in order to qualify as genuine physical computations. Facing the question of *what a genuine physical computation is*, Piccinini put forward the following open-ended list of constraints and their sub-constraints imposed on any physical process P and any physical system S if P is to be counted as a *physical computation* on S:

---

[64] Gualtiero Piccinini, b. 1970, Italian-American philosopher.

[65] A finite observer is an observer of *bounded capacities*.

- *Executability*: P can be set in motion by a finite observer to generate the values of a given function until it generates a readable result: The sub-constraints are:

  - *Readable Inputs and Outputs*: The inputs and outputs of P must be readable, i.e., they can be specified and measured to the desired degree of accuracy.

  - *Process-Independent Rule*: The function computed by P must be definable independently of P (and similarly for the problem solved by P).

  - *Repeatability*: P can be repeated by any competent finite observer.

  - *Settability*: S can be reset to its initial state.

  - *Physical Constructibility*: S can be constructed by arranging the relevant physical materials.

- *Automaticity*: P must run with no intuitions, ingenuity, invention, or guesses.

- *Uniformity*: P does not need to be redesigned or modified for different inputs.

- *Reliability*: P generates results at least some of the time and the results are correct. The sub-constraints are:

  - S's components must not break too often.

  - S is designed so that noise and external disturbances don't interfere with results.

According to Piccinini's classification and constraints, CTT-P is neither bold nor modest; indeed, it can be labeled "super-bold". Therefore, Copeland and Shagrir weakened their CTT-P and stated the following modest thesis:

(**Copeland-Shagrir (CTT-P-C)**) *Every function computed by any physical computing system is Turing-computable.*

By CTT-P-C, if a function's values can be computed by the physical processes of some physical system, then it can also be computed by some Turing machine. Equivalently, if a function is *not* Turing-computable, neither is it computable by any physical computing system. Here is an example. Since the *Halting Problem* is undecidable,[66] its characteristic function

$$\chi_{\mathcal{K}_0}(\langle m,n \rangle) = \begin{cases} 1 & \text{if TM } T_m \text{ halts on } n; \\ 0 & \text{if TM } T_m \text{ does not halt on } n \end{cases}$$

is not Turing-computable; then—if CTT-P-C is true—neither is it computable by any physical computing system.

To sum up: *If* CTT-P-C is true, then Turing computability is the *upper bound* on what CTT-P-C deems physically computable. In other words, if CTT-P-C is true, then Turing computability is a computational barrier that cannot be broken, and is sometimes called the *Church-Turing barrier*.

But is CTT-P-C true?

---

[66] See Sect. 8.2, p. 180.

16.6.5 Hypercomputing?

While many researchers believe that CTT-P-C is true, several researchers have been trying to conceive a physical computing system that would *breach* the Church-Turing barrier. Recently, several such physical computing systems have been proposed. They are based on very different concepts, but their common intention is to squeeze infinitely many computational steps into a finite span of time. For example, *accelerating machines* carry out the next operation in half the time of the previous one, and *relativistic machines* make use of relativistic gravitational time dilation.

Physical computing systems that would be capable of "computing" beyond the Church-Turing barrier are nowadays collectively called *hypercomputers*, and their modes of "computing" are collectively called *hypercomputing*. If physically realistic, any hypercomputer would falsify CTT-P-C. However, to this day, no proposed hypercomputer has been proved to be entirely physically realistic; for now, hypercomputers remain *notional* computing systems.

We describe Németi's[67] *relativistic machine* from 2006, a hypercomputer conceived to compute the values of the halting function $\chi_{\mathcal{K}_0}$ by using relativistic phenomena.

Relativistic Machine. Let $T_\oplus$ be a universal Turing machine and $T_\bullet$ an ordinary Turing machine that can communicate with each other. (We will explain the meaning of indexes $\oplus$ and $\bullet$ shortly.) The pair $(T_\oplus, T_\bullet)$ is a physical system intended to compute the value $\chi_{\mathcal{K}_0}(\langle m,n \rangle)$ of the halting function $\chi_{\mathcal{K}_0}$ for any pair $(m,n) \in \mathbb{N}^2$. Here is how the system operates.

Fig. 16.2 The Relativistic Machine. Turing machine $T_\bullet$ writes provisional result $\chi_{\mathcal{K}_0}(\langle m,n \rangle) = 0$ and commands the universal Turing machine $T_\oplus$ to simulate T_m on input n. If $T_\oplus$ terminates in time $t_\oplus$, it signals to $T_\bullet$, which then replaces the provisional result with the final result $\chi_{\mathcal{K}_0}(\langle m,n \rangle) = 1$. But $T_\bullet$ operates in a stronger gravitational field than $T_\oplus$, so $T_\bullet$'s proper time is passing more slowly than $T_\oplus$'s proper time. In effect, $T_\bullet$ can wait reasonably long even for an unreasonably long simulation

---

[67] István Németi, b. 1942, Hungarian mathematician and theoretical computer scientist.

There is nothing spectacular in this scenario: $T_\bullet$ writes the provisional result $\chi_{\mathcal{K}_0}(\langle m,n\rangle) = 0$ on its tape and hands over the computation of $\chi_{\mathcal{K}_0}(\langle m,n\rangle)$ to $T_\oplus$, which, in turn, starts simulating T_m on n for at most $t_\oplus$ seconds. (The simulation may terminate sooner.) Meanwhile, $T_\bullet$ waits at most $t_\bullet \geqslant t_\oplus$ seconds for the result of the simulation. Thus, $T_\bullet$ learns in at most $t_\bullet$ seconds whether or not $T_\oplus$ terminated the simulation in at most $t_\oplus$ seconds. If it did, then, obviously, T_m halted on n, so $T_\bullet$ replaces the provisional result on its tape with the final result $\chi_{\mathcal{K}_0}(\langle m,n\rangle) = 1$ and halts; *otherwise*, nothing can be said about the final value $\chi_{\mathcal{K}_0}(\langle m,n\rangle)$, as the simulation might *either* terminate later (for a larger $t_\oplus$) *or* never terminate (even for infinite $t_\oplus$). (If the simulation does not terminate in $t_\oplus$ seconds, the system $(T_\oplus, T_\bullet)$ could extend deadlines $t_\bullet$ and $t_\oplus$. But this would only pay off if in truth T_m halts on n; otherwise, $(T_\oplus, T_\bullet)$ would keep extending $t_\bullet$ and $t_\oplus$ indefinitely and never halt.)

We tacitly assumed that both $T_\bullet$ and $T_\oplus$ reside on Earth and run under the same physical conditions. *What if* they were on *two very different* astronomical objects?

Box 16.1 (Black Holes).

The *General Theory of Relativity* predicts that a sufficiently compact mass can deform spacetime into an astronomical object called a *black hole*, a spheroidally bounded region of spacetime exhibiting gravitational "pull" so strong that nothing can escape from it. The boundary of the region is called the *event horizon* and is the defining feature of the black hole. It is a bubble-like hypersurface in spacetime surrounding the region's interior such that matter and radiation can pass only inward, towards the gravitational center of the black hole, and nothing can escape outward from the region. The exhibited gravitational "pull" tends to infinity as the distance to the event horizon decreases. Why? The black hole's mass warps spacetime so that escaping paths bend back to the horizon.

Any object that crosses the event horizon is irreversibly consumed by the black hole. There are no warnings; the object cannot detect any difference between the gravitational field of a black hole and that of any other spheroidal object of the same mass; neither can it detect any characteristic local observables. When it reaches and passes through the event horizon, the object feels nothing peculiar; e.g., the time it measures is perfectly smooth. (Of course, as the object approaches the event horizon, the gravitational field becomes very strong, and the big difference between the gravity forces exerted on the upper and lower parts of the object stretches, or "spaghettifies", the object and can even tear it up. But such *tidal forces* act in every non-homogeneous gravitational field, not just near black holes.) If an event occurs inside the event horizon, *no information* about it can reach an outside observer, making it impossible to determine anything about the event.

A black hole can form when, at the end of its life cycle, a massive star collapses under its own gravitational pull. Once across a certain radius (r^*), gravitational collapse to a *singularity* (where size is zero and density is infinite) is inevitable. The newborn black hole can continue to grow by absorbing mass from the surrounding stars or merging with other black holes. So, black holes range from *micro* ($M \leqslant M_L$) via *stellar* ($M \leqslant 10 M_\odot$) and *intermediate-mass* ($M \leqslant 10^3 M_\odot$) to *supermassive* ($M \leqslant 10^{10} M_\odot$) black holes, where M, M_L, and $M_\odot$ are the black hole, lunar, and solar mass, respectively. The *General Theory of Relativity* predicts four kinds of black holes: While every black hole has a mass M, some also have angular momentum J or electric charge Q. The simplest, non-rotating ($J = 0$) and non-charged ($Q = 0$) kind is called a *Schwarzschild* black hole. Its mass M is so compressed that its radius r is less than the radius $r^* = 2 G_N M / c^2$ of its event horizon. The other three kinds are called *Kerr* ($J > 0, Q = 0$), *Reissner-Nordström* ($J = 0, Q > 0$), and *Kerr-Newman* ($J > 0, Q > 0$) black holes. Of these, Kerr black holes are the most relevant to realistic astrophysical situations. A Kerr black hole, due to its rotation, has the singularity spread out over a ring. This is surrounded by an *inner* event horizon (r_-) that is surrounded by an *outer* event horizon (r_+) and that by a *stationary limit surface* (r_e). The region between r_+ and r_e is the *ergosphere*, where a stationary object can approach or move away from the event horizon.

Suppose that $T_\bullet$ is near a Schwarzschild[68] black hole and $T_\oplus$ is on Earth. (We denote the black hole by $\bullet$ and Earth by $\oplus$.) Putting aside the distance between $\bullet$ and $\oplus$, which brings communication delay due to the finite speed of electromagnetic waves carrying signals, there is a difference in the strengths of their gravitational fields. Now comes to the fore Einstein's *General Theory of Relativity*, which tells us, informally, that *a clock in a stronger gravitational field runs more slowly*. This phenomenon is called *gravitational time dilation*. Here is how Cheng[69] describes it:

> The time unit itself changes in the presence of gravity. The clocks run at different rates when situated at different gravitational field points: there is a **gravitational time dilation** effect. The clock at the higher gravitational potential point [where the field is weaker] will run faster. Here we are saying that two clocks, even at rest with respect to each other, run at different rates if the gravitational fields at their respective locations are different. The observer at a higher gravitational potential point [where the field is weaker] sees the lower clock [which is in a stronger field] run slow, and the lower observer sees the higher clock run fast.

The gravitational time dilation is not just a theoretical result; it can be detected, measured, and used in reality, explains Cheng:

> The gravitational time dilation effects have been tested directly by comparing the times kept by two cesium atomic clocks: one flown in an airplane at high altitude h (about 10 km) in a holding pattern, for a long time τ, over the ground station where the other clock sits. [...T]he high altitude clock was found to gain over the ground clock by a time interval of $\Delta\tau = (gh/c^2)\tau$ in agreement with the expectation [g is the gravitational field, c is the speed of light].

In particular, near the black hole $\bullet$, time passes more slowly than on Earth $\oplus$. We must therefore distinguish between the time measured by the clock $C_\bullet$ at $T_\bullet$ (called the *proper time* of $T_\bullet$) and the time measured by the clock $C_\oplus$ at $T_\oplus$ (proper time of $T_\oplus$). Whenever $C_\bullet$ ticks one second ($T_\bullet$ waits one second of *its* proper time), $C_\oplus$ ticks more than one second ($T_\oplus$ simulates more than one second of *its* proper time). So, from $T_\bullet$'s perspective, $T_\oplus$ *carries out in its proper time $t_\oplus$ more of the simulation if $T_\bullet$ waits $t_\bullet$ of its proper time near the black hole (instead of on Earth).*

Can $T_\bullet$ use this phenomenon to find the value $\chi_{\mathcal{K}_0}(\langle m, n\rangle)$? Let $T_\bullet$ be slowly approaching the black hole's event horizon. Simultaneously, the gravitational field is getting stronger, $T_\bullet$'s proper time is slowing down relative to $T_\oplus$'s proper time, $C_\oplus$'s ticking is speeding up relative to $C_\bullet$'s ticking, and the part of the simulation carried out in $T_\bullet$'s proper time $t_\bullet$ is growing with no upper bound.

If in truth T_m halts on n, then there is a finite time $t_\oplus$ in which $T_\oplus$ will terminate the simulation and signal HALTED to $T_\bullet$. (Even if $t_\oplus$ is colossal, $T_\bullet$ will wait for the signal only a reasonable proper time $t_\bullet$ if $T_\bullet$ has sufficiently approached the event horizon.) Then $T_\bullet$ turns back, thus making the final result $\chi_{\mathcal{K}_0}(\langle m, n\rangle) = 1$ on its tape obtainable by an outside observer. In sum, if in truth $\chi_{\mathcal{K}_0}(\langle m, n\rangle) = 1$, then $T_\bullet$ can learn this in a reasonable proper time $t_\bullet$. The observer learns this with a finite time delay.

---

[68] Karl Schwarzschild, 1873–1916, German physicist and astronomer.

[69] Ta-Pei Cheng, b. 1941, Chinese-American particle physics theorist and author of books on Einstein's physics, relativity, and cosmology as well as particle physics.

What about *if in truth T_m does not halt on n*? Can $T_\bullet$ decide that the simulation never terminates? Since the simulation doesn't terminate, $T_\bullet$ receives in its proper time $t_\bullet$ no signal from $T_\oplus$ and continues approaching the event horizon with the provisional result $\chi_{\mathcal{K}_0}(\langle m, n \rangle) = 0$ on its tape. But $T_\bullet$ cannot detect the exact moments of reaching and crossing the event horizon since there are no observables characterizing the two events (although $T_\bullet$'s proper time runs smoothly and $T_\bullet$ can measure it). After $T_\bullet$ has crossed the event horizon, the provisional result $\chi_{\mathcal{K}_0}(\langle m, n \rangle) = 0$ on its tape tacitly but correctly becomes final (although $T_\bullet$ cannot detect when that happens). In sum, if in truth $\chi_{\mathcal{K}_0}(\langle m, n \rangle) = 0$, then $T_\bullet$ can *obtain* it in a reasonable proper time $t_\bullet$. But how could $T_\bullet$ *decide* that the provisional result $\chi_{\mathcal{K}_0}(\langle m, n \rangle) = 0$ is in fact final? This problem is addressed by introducing a *time bound*, b, on $T_\bullet$'s proper time $t_\bullet$. When $T_\bullet$ reaches the event horizon, the gravitational field becomes so strong that, from $T_\oplus$'s perspective, $T_\bullet$'s time stops (though it continues from $T_\bullet$'s perspective). Thus, from $T_\oplus$'s perspective, $T_\bullet$'s crossing of the event horizon takes infinitely long. Suppose that during the simulation $T_\oplus$ rarely but periodically signals SIMULATING to $T_\bullet$. Since the whole infinite time span of $T_\oplus$'s simulation matches some finite time span of $T_\bullet$'s waiting, $T_\bullet$ will receive all these signals in this finite time span. If $T_\bullet$ picked a large enough finite b, then—when $T_\bullet$'s proper time reaches b—$T_\bullet$ will determine that the simulation must be over and the result $\chi_{\mathcal{K}_0}(\langle m, n \rangle) = 0$ is final.

Let us now focus on the physical realizability of the physical system $(T_\oplus, T_\bullet)$. How can $T_\bullet$ approach the event horizon and survive spaghettification? The answer is offered by *Kerr black holes*.[70] These black holes rotate (in contrast to Schwarzschild black holes), so the centrifugal forces acting on the synchronously rotating $T_\bullet$ reduce the destructive effect of tidal forces on it. In addition, the Kerr black hole must be large enough, because larger black holes induce weaker tidal forces (due to their large radii). Then $T_\bullet$ can peacefully travel towards the event horizon, free from danger of being torn up. Németi further elaborated a possible use for the structure of Kerr black holes by taking into account the two event horizons and the ergosphere between the outer horizon and the stationary limit surface. (The reader can find further details in the Bibliographic Notes to this chapter.)

We conclude this description of Németi's relativistic machine with a remark. When $T_\bullet$ reaches the event horizon, the black hole irrevocably consumes $T_\bullet$, together with its final result $\chi_{\mathcal{K}_0}(\langle m, n \rangle) = 0$. After that, *nothing*, even electromagnetic waves carrying this result, can escape through the event horizon. To an external observer the final result $\chi_{\mathcal{K}_0}(\langle m, n \rangle) = 0$ (even if known to $T_\bullet$) is irretrievable. It seems that the described relativistic hypercomputation of the value $\chi_{\mathcal{K}_0}(\langle m, n \rangle)$ does not completely fulfill Piccinini's constraints about physical computation.

Nevertheless, Németi's construction demonstrates that there is a fascinating interplay between *Computability Theory* and the *General Theory of Relativity* that might lead to new discoveries about computing.

---

[70] Roy Kerr, b. 1934, New Zealand physicist who in 1963 found the metric for a rotating black hole.

Bibliographic Notes

The following sources were consulted for the subjects covered in this chapter:

- Notes of Gödel's Princeton lectures in 1934 were taken by Kleene and Rosser and published in Davis [57]. Gödel's work on the analysis of mechanical procedures is summarized in Sieg [223]. Gödel's speech on the situation in the foundations of mathematics in 1933 can be found in [86].
- Church published his thesis in [37]. Post expressed his reservations about Church's Thesis in [181]. That Gödel accepted Church's Thesis only after Turing's formulation is from Kleene [126].
- Kleene proved his *Normal Form Theorem* and the equivalence $\mathcal{G}^+ = \mathcal{K}^+$ of (general) recursiveness and μ-recursiveness in [121]; see also his [124, §58]. Kleene's survey of the investigations in 1931–1933 in which he played a vital role is [126]. Shoenfeld [218] is an overview of Kleene's mathematical work from the origins of *Computability Theory* to objects of higher types.
- In describing Post's contributions to the *Computability (Church-Turing) Thesis* and *Computability Theory* we strongly leaned on the excellent exposition of Post's biography and work by Urquhart [265]. A shortened version of Post's doctoral dissertation is [180]. In 1941, Post submitted for publication a survey [182] of his work of 1920–1921, but the paper was rejected. (It was published two and a half decades later.) Post characterized computation in terms of *Finite Combinatory Processes* in [181].
- Turing described his ideas and their consequences in [262]. The term "Turing machine" appeared in 1937 in Church [38]. Turing published the full proof that the class $\mathcal{T}^+$ of Turing-computable functions of positive integers is equal to the class $\mathcal{C}^+$ of λ-definable functions of positive integers and to the class $\mathcal{G}^+$ of (general) recursive functions of positive integers in [263]. Church endorsed Turing's approach to formalization of computability in [38]. (But see Hodges [108] and Sieg [225] for a critical analysis of Church's observations on Turing's work.) Post commented on the Turing machine and proposed some improvements in its definition in [186]. Kleene's comment on Turing's characterization of "computability" is from [126]. The metamathematical importance of Turing's results is stressed in Gödel [83] and the importance of Turing's work for mathematics and philosophy in Gödel [85]. The archive of Turing's papers is http://www.alanturing.net/turing_archive/archive/index/archiveindex.html.
- That Church's Thesis (1936) and Turing's Thesis are equivalent when attention is restricted to functions of positive integers was first expressed and proved in Kleene [124, p. 376]. The term Church-Turing Thesis was coined in Kleene [124, p. 382]. The classification of arguments supporting the Church-Turing Thesis was given by Kleene [124, pp. 319–323] and Shoenfield [216, pp. 26–28]. There were also attempts to refute the Church-Turing Thesis. In 1959, Kalmár [119] published one which was four years later refuted by Mendelson [154].
- Mendelson [156] renounced the standard view that the Church-Turing Thesis cannot be proved. (See also recently Mendelson [157].) Thus, Sieg [222, 225] argued that the Church-Turing Thesis is amenable to proof and offered an axiomatized version of Turing's Argument I. Similarly, Dershowitz and Gurevich [60] offered a proof of Church's Thesis, adding that a similar approach could prove Turing's Thesis. See also a detailed analysis of this proof in Sieg [228].
- Kripke revived Turing's Argument II and presented his logical orientation to the problem of formalization of the notion of "computability" in [131]. A detailed logical notation for expressing computations as special deductions was also developed in Turing [262, §§3,4,5].
- Copeland's contribution [45] to the Stanford Encyclopedia of Philosophy is a rich source for the Church-Turing Thesis.
- The importance of understanding the differences between Church's Thesis, Turing's Thesis, and other similar-looking statements was emphasized in Ben-Amram [15], Hodges [108], and Copeland [45]. Davis [56] critically discusses various claims made by researchers of hyper-computation.

- An excellent and approachable description of Turing's work, its relation to the work of Hilbert, Gödel, Church, Kleene, Post, and others, and its consequences, is Herken [99]. An excellent overview of the ideas presented in this chapter is Davis [55]. Copeland and Shagrir [48] present fresh interpretations of the positions of Turing and Gödel on computability and the mind. Copeland, Posy, and Shagrir [46] is an excellent recent collection of chapters on various aspects of computability written by renowned present-day logicians, mathematicians, computer scientists, and philosophers.
- A recent brief overview of various theses about computability ranging from the original Church-Turing Thesis to the present-day versions is Copeland and Shagrir [49]. The formulation CTT-O of the original Church-Turing Thesis is from [49, p. 68].
- Lewis and Papadimitriou's *algorithmic version* of the Church-Turing Thesis is from Lewis and Papadimitriou [142, p. 223]. Harel's algorithmic version of the Church-Turing Thesis is from Harel and Feldman [96, p. 228]. The development of the theory of algorithms within axiomatic set theory is described in Moschovakis [160, 161], and the relation between an implementable algorithm and its implementations is given in Moschovakis and Paschalis [162, pp. 87–118]. The algorithmic version CTT-A is stated in [49, p. 68].
- The Bernstein-Vazirani *complexity-theoretic version* of the Church-Turing Thesis was given special prominence in [19]. The Extended Church-Turing Thesis CTT-E is stated and discussed in Aharonov and Vazirani [7].
- The *physical version* of Church-Turing Thesis was formulated in Wolfram [278] and a similar thesis in Deutsch [61]. The *Total Physical Computability Thesis* and its modest version CTT-P-C appeared in Copeland and Shagrir [49]. Undecidability of the problem OUTCOME SEQUENCES was proved in Eisert et al. [66] and that the SPECTRAL GAP problem is undecidable was proved in Cubitt et al. [50]. The classification of physical versions of the Church-Turing Thesis into the *bold* and *modest* ones is described in Piccinini [178].
- The *accelerating Turing machine* and its operation was described in Copeland and Shagrir [47]. The *relativistic machine* was devised in Németi and Dávid [169]. For the background in *General Theory of Relativity*, in particular gravitational time dilation and black hole spacetime geometry, we consulted the following excellent sources: Carroll [31], Cheng [34], Guidry [92], Hartle [97], Lambourne [137], and Schutz [212]. For a critical account of *hypercomputing* see Davis [56].
- Sieg [226] is an in-depth treatise on the historical developments of computability theory that are deeply intertwined with metamathematical work in the foundations of mathematics. Sieg [229] discusses the philosophical challenge to answer whether there exists a rigorous argument from Gödel's incompleteness theorems to the claim that machines can never replace mathematicians (or, more generally, that the human mind infinitely surpasses any finite machine).
- A nice background on the contemporary varieties and aspects of computation is given in Shagrir [213]. Sieg [224] addresses the question of whether there are strictly broader notions of effectiveness in view of mathematical reasoning that transcends mechanical computing.

Chapter 17
Further Reading

Greg, is that all?

Oh no, Becky, that was just the beginning!

Euclid
Boole
Frege
Peano
al Khwarizmi
Pascal
Leibniz
Babbage
Cantor
Whitehead
Russell
Hilbert
Brouwer
Zermelo
Fraenkel
von Neumann
Bernays
Gödel
Hilbert?
Gödel!
?
Gödel
Kleene
Herbrand
Gödel
Church
Turing
Post
Turing & Church
Post
Cantor
Rice
Turing
Post
Friedberg
Kleene
Mučnik

This text has been designed so that it can serve as a stepping stone to a more advanced study of *Computability Theory*, or as an introduction to *Computational Complexity Theory*. Here are some suggestions for further reading.

Robert I. Soare, *Turing Computability: Theory and Applications*, Theory and Applications of Computability, Springer (2016).

Soare's recent monograph emphasizes three important concepts: computability (or effective calculability), Turing (or classical) computability, and the art (or mathematical aesthetic) of computability. The monograph is a survey of the results in *Computability Theory* up to the mid-2010s. Through the parts on the foundations of computability theory, computably open and closed sets of reals, minimal Turing degrees, and games in computability theory it will bring you closer to the frontiers of the current research in this theory. In the last, fifth part you will find a short history of computability theory.

R. Robič, *The Foundations of Computability Theory*, https://doi.org/10.1007/978-3-662-62421-0_17

Robert I. Soare, *Recursively Enumerable Sets and Degrees*, Springer (1987).

Soare's monograph is a concise survey of *Computability Theory* during the periods 1931–1943, 1944–1960, and 1961–1987. It will deepen and broaden the fundamental concepts that you are now acquainted with, and bring you deep into *Post's Problem*, oracle constructions, and finitary and infinitary methods for constructing c.e. sets and degrees. There are many exercises and you will find a lot of information there. Proofs will often demand additional work.

S. Barry Cooper, *Computability Theory*, Chapman & Hall/CRC Mathematics (2004).

Cooper's monograph consists of three parts. Being now acquainted with the fundamental concepts of *Computability Theory*, you should have no problems with the first part. But carefully reading it, you will complement your current knowledge and view many issues from different perspectives. The second part starts with oracle computations, which should be easy for you, and proceeds to topics fundamental to *Computational Complexity Theory*. The third part brings in advanced topics about degree structures, forcing, determinacy, and applications to mathematics and science. There are many examples and exercises.

Rebecca Weber, *Computability Theory*, Student Mathematical Library, vol. 62, American Mathematical Society (2012).

Weber's monograph, similarly to Cooper's, will complement your current knowledge and present it from other perspectives. The second half of the monograph will give you additional explanation of methods, tools, and the arithmetical hierarchy, and the last chapter will give you a taste of various areas of *Computability Theory* where research is currently active.

Hartley Rogers, *Theory of Recursive Functions and Effective Computability*, 2nd ed., MIT Press (1987).

Rogers's monograph is a complete and concise presentation of classical *Computability Theory* up to 1970 or so. Its central concerns are the theories of computably enumerable (c.e.) sets, of degrees of unsolvability, and of Turing degrees in particular. The subjects are developed in a succinct, mathematically mature manner free from the details of any particular formalism. Because of the clear exposition at a level accessible to the reader with little training in logic or other special mathematics subjects, the monograph became one of the most influential textbooks on *Computability Theory* in the 1970s and 1980s. There are many exercises and a lot of additional information that will augment your current knowledge.

Rodney G. Downey and Denis R. Hirschfeldt, *Algorithmic Randomness and Complexity*, Theory and Applications of Computability, Springer (2010).

Downey-Hirschfeldt's monograph will introduce you to a research area that has been flourishing since the late 1990s. It will explain to you how relative computability, information content, and randomness interact.

André Nies, *Computability and Randomness*, Oxford Logic Guides, Oxford University Press (2009).

Nies's monograph will explain to you how *Computability Theory* is used in the study of randomness of sets of natural numbers; conversely, it will show you how ideas originating from randomness are used to enrich *Computability Theory*. You will find many advanced topics that will extend your current knowledge.

Piergiorgio Odifreddi, *Classical Recursion Theory: The Theory of Functions and Sets of Natural Numbers*, volume I and II, 2nd ed., Elsevier (1999).

Odifreddi's two-volume monograph contains a wealth of information. The author has opted for breadth rather than depth so the book provides rudiments of many branches of classical *Computability Theory*. It is a good reference for those with a moderate background.

Appendix A
Mathematical Background

In this appendix, we review the basic notions, concepts, and facts of logic, set theory, algebra, analysis, and formal-language theory that are used throughout this book. For further details see, for example, [106, 155, 196] for logic, [88, 95, 133] for set theory, [64, 115, 179] for algebra, [79, 205, 231] for formal languages, and [118, 206, 248] for mathematical analysis.

Propositional Calculus P

Syntax

- An *expression* of **P** is a finite sequence of symbols. Each symbol denotes either an individual constant or an individual variable, or it is a logic connective or a parenthesis. *Individual constants* are denoted by $a, b, c, \ldots$ (possibly indexed). *Individual variables* are denoted by $x, y, z, \ldots$ (possibly indexed). *Logic connectives* are $\vee, \wedge, \Rightarrow, \Leftrightarrow$, and $\neg$; they are called disjunction, conjunction, implication, equivalence, and negation, respectively. *Punctuation marks* are parentheses.
- Not every expression of **P** is well formed. An expression of **P** is *well formed* if it is either 1) an individual-constant or individual-variable symbol, or 2) one of the expressions $F \vee G$, $F \wedge G$, $F \Rightarrow G$, $F \Leftrightarrow G$, and $\neg F$, where F and G are well-formed expressions of **P**. A well-formed expression of **P** is called a *sentence*.

Axioms and Rules of Inference

- If F, G, H are arbitrary sentences, the following are axioms of **P**:

  - $F \Rightarrow (G \Rightarrow F)$
  - $(F \Rightarrow (G \Rightarrow H)) \Rightarrow ((F \Rightarrow G) \Rightarrow (F \Rightarrow H))$
  - $(\neg G \Rightarrow \neg F) \Rightarrow ((\neg G \Rightarrow F) \Rightarrow G)$

- The only rule of inference is *Modus Ponens*: G is a direct consequence of F and $F \Rightarrow G$.

Semantics

- The standard meanings of the logic connectives are: "or" ($\vee$), "and" ($\wedge$), "implies" ($\Rightarrow$), "if and only if" ($\Leftrightarrow$), "not" ($\neg$).
- Let $\{\top, \bot\}$ be a set. The elements $\top$ and $\bot$ are called *logic values* and stand for "true" and "false", respectively. Often, 1 and 0 are used instead of $\top$ and $\bot$, respectively.

- Any sentence has either the truth-value $\top$ or $\bot$. A sentence is said to be *true* if its truth-value is $\top$, and *false* if its truth-value is $\bot$. Individual constants and individual variables obtain their truth-values by assignment. When logic connectives combine sentences into new sentences, the truth-value of the new sentence is determined by the truth-values of its component sentences. Specifically, let E and F be sentences. Then:

  - $\neg$E is true if E is false, and $\neg$E is false if E is true.
  - E $\vee$ F is false if both E and F are false; otherwise E $\vee$ F is true.
  - E $\wedge$ F is true if both E and F are true; otherwise E $\wedge$ F is false.
  - E $\Rightarrow$ F is false if E is true and F is false; otherwise E $\Rightarrow$ F is true.
  - E $\Leftrightarrow$ F is true if E and F are either both true or both false; otherwise, E $\Leftrightarrow$ F is false.

- The following hold: $\neg(E \vee F) \Leftrightarrow (\neg E) \wedge (\neg F)$ and $\neg(E \wedge F) \Leftrightarrow (\neg E) \vee (\neg F)$.

First-Order Logic L

Syntax

- An *expression* of **L** is a finite sequence of symbols, where each symbol is an individual-constant symbol (e.g., a, b, c), an individual-variable symbol (e.g., x, y, z), a logic connective ($\vee, \wedge, \Rightarrow, \Leftrightarrow, \neg$), a function symbol (e.g., f, g, h), a predicate symbol (e.g., P, Q, R), a quantification symbol ($\forall, \exists$), or a punctuation mark (e.g., colon, parenthesis). (Predicates are also called *relations*.)
- We are only interested in the well-formed expressions of **L**. To define these, we need two definitions. First, a *term* is either 1) an individual-constant or individual-variable symbol, or 2) a function symbol applied to terms (e.g., $f(a, x)$). Second, an *atomic formula* is a predicate symbol applied to terms (e.g., $P(y, f(a, x))$). Finally, we say that an expression of **L** is *well formed* if it is either 1) an atomic formula, or 2) one of the expressions F $\vee$ G, F $\wedge$ G, F $\Rightarrow$ G, F $\Leftrightarrow$ G, $\neg$F, $\forall \tau$F, and $\exists \tau$F, where F and G are well-formed expressions of **L** and τ is an individual-variable symbol. A logic expression of **L** that is well formed is called a *formula*.

Axioms and Rules of Inference

- If F, G, H are arbitrary formulas, the following are axioms of **L**:

  - $F \Rightarrow (G \Rightarrow F)$
  - $(F \Rightarrow (G \Rightarrow H)) \Rightarrow ((F \Rightarrow G) \Rightarrow (F \Rightarrow H))$
  - $(\neg G \Rightarrow \neg F) \Rightarrow ((\neg G \Rightarrow F) \Rightarrow G)$
  - $\forall x F(x) \Rightarrow F(t)$ if t is a term free for x in $F(x)$
  - $\forall x (F \Rightarrow G) \Rightarrow (F \Rightarrow \forall x G)$ if x is not free in F

- The rules of inference are:

  - Modus Ponens: G follows from F and $F \Rightarrow G$.
  - *Generalization:* $\forall x$F follows from F

Semantics

- The standard meanings of the quantification symbols are: "for all" ($\forall$), "exists" ($\exists$). For the meanings of logic connectives, see *Propositional Calculus* **P** above.
- The truth-value of a formula is determined as follows. Let E and F be formulas. Then:

  - $\forall \tau$F is true if F is true for every possible assignment of a value to τ.
  - $\exists \tau$F is true if F is true for at least one possible assignment of a value to τ.
  - For the truth-values of F $\vee$ G, F $\wedge$ G, F $\Rightarrow$ G, F $\Leftrightarrow$ G, $\neg$F, see *Propositional Calculus* **P** above.

Sets

Basics

- Given any objects $a_1, \ldots, a_n$, the *set* containing $a_1, \ldots, a_n$ as its only elements is denoted by $\{a_1, \ldots, a_n\}$. More generally, given a property P, the set of those elements having the property P is written as $\{x \mid P(x)\}$. If an element x is in a set $\mathcal{A}$, we say that x is a *member* of $\mathcal{A}$ and write $x \in \mathcal{A}$; otherwise, we write $x \notin \mathcal{A}$ and say that x is not a member of $\mathcal{A}$. The set with no members is called the *empty set* and denoted by $\emptyset$.
- For sets $\mathcal{A}$ and $\mathcal{B}$, we say that $\mathcal{A}$ is a *subset* of $\mathcal{B}$, written $\mathcal{A} \subseteq \mathcal{B}$, if each member of $\mathcal{A}$ is also a member of $\mathcal{B}$. A set $\mathcal{A}$ is a *proper subset* of $\mathcal{B}$, written $\mathcal{A} \subsetneq \mathcal{B}$, if $\mathcal{A} \subseteq \mathcal{B}$ but there is a member of $\mathcal{B}$ not in $\mathcal{A}$. Instead of $\subsetneq$ we also write $\subset$.
- Sets $\mathcal{A}$ and $\mathcal{B}$ are *equal*, written $\mathcal{A} = \mathcal{B}$, if $\mathcal{A} \subseteq \mathcal{B}$ and $\mathcal{B} \subseteq \mathcal{A}$.
- Given a set $\mathcal{A} = \{x_\iota \mid \iota \in \mathcal{I}\}$, the set $\mathcal{I}$ is called the *index set* of $\mathcal{A}$.
- By $(a_1, \ldots, a_n)$, or also by $\langle a_1, \ldots, a_n \rangle$, we denote the *ordered n-tuple* of objects $a_1, \ldots, a_n$. When $n = 2$, the n-tuple is called an ordered *pair*. Two ordered n-tuples $(a_1, \ldots, a_n)$ and $(b_1, \ldots, b_n)$ are *equal*, denoted by $(a_1, \ldots, a_n) = (b_1, \ldots, b_n)$, if $a_i = b_i$ for $i = 1, \ldots, n$.

Operations on Sets

- The *union* of sets $\mathcal{A}$ and $\mathcal{B}$, written as $\mathcal{A} \cup \mathcal{B}$, is the set of elements that are members of at least one of $\mathcal{A}$ and $\mathcal{B}$.
- The *intersection* of sets $\mathcal{A}$ and $\mathcal{B}$, written as $\mathcal{A} \cap \mathcal{B}$, is the set of elements that are members of both $\mathcal{A}$ and $\mathcal{B}$. We say that $\mathcal{A}$ and $\mathcal{B}$ are *disjoint* if $\mathcal{A} \cap \mathcal{B} = \emptyset$.
- The *difference* of sets $\mathcal{A}$ and $\mathcal{B}$, written as $\mathcal{A} - \mathcal{B}$, is the set of those members of $\mathcal{A}$ that are not in $\mathcal{B}$.
- If $\mathcal{A} \subseteq \mathcal{B}$, then the *complement of $\mathcal{A}$ with respect to $\mathcal{B}$* is the set $\mathcal{B} - \mathcal{A}$.
- The *power set* of a set $\mathcal{A}$ is the set of all subsets of $\mathcal{A}$ and is denoted by $2^{\mathcal{A}}$.
- The *Cartesian product* of a finite sequence of sets $\mathcal{A}_1, \ldots, \mathcal{A}_n$ is the set of all ordered n-tuples $(a_1, \ldots, a_n)$, where $a_i \in \mathcal{A}_i$ for each i. In this case it is denoted by $\mathcal{A}_1 \times \ldots \times \mathcal{A}_n$. If $\mathcal{A}_1 = \ldots = \mathcal{A}_n = \mathcal{A}$, the Cartesian product is denoted by $\mathcal{A}^n$. By convention, $\mathcal{A}^1$ stands for $\mathcal{A}$.

Relations

Basics

- An n-ary *relation* on a set $\mathcal{A}$ is a subset of $\mathcal{A}^n$. When $n = 2$ we say that the relation is *binary*, or for short, a *relation*. If R is a relation, we write xRy to indicate that $(x, y) \in R$. A 1-ary relation on $\mathcal{A}$ is a subset of $\mathcal{A}$, and is called a *property* on $\mathcal{A}$.
- A relation R on $\mathcal{A}$ is:

  - *reflexive* if xRx for each $x \in \mathcal{A}$.
  - *irreflexive* if xRx for no $x \in \mathcal{A}$.
  - *symmetric* if xRy implies yRx, for arbitrary $x, y \in \mathcal{A}$.
  - *asymmetric* if xRy implies that not yRx, for arbitrary $x, y \in \mathcal{A}$.
  - *anti-symmetric* if xRy and yRx imply $x = y$, for arbitrary $x, y \in \mathcal{A}$.
  - *transitive* if xRy and yRz imply xRz, for arbitrary $x, y, z \in \mathcal{A}$.

Ordered Sets

- A *preordered set* is a pair $(\mathcal{A}, R)$ where $\mathcal{A}$ is a set and R a binary relation on $\mathcal{A}$ such that (i) R is reflexive, and (ii) R is transitive. In this case, we say that R is a *preorder* on $\mathcal{A}$. Two elements $x, y \in A$ are *incomparable* by R (for short, *R-incomparable*) if neither xRy nor yRx.

- A *partially ordered* set is a pair $(\mathcal{A}, R)$ where $\mathcal{A}$ is a set and R a binary relation on $\mathcal{A}$ such that (i) R is reflexive, (ii) R is transitive, and (iii) R is anti-symmetric. In this case, we say that R is a *partial order* on $\mathcal{A}$. A partial order is often denoted by $\leq, \leqslant, \preceq, \preccurlyeq$ or any other symbol indicating the properties of this order.
- Let $(\mathcal{A}, \preccurlyeq)$ be a partially ordered set. The relation $\prec$ on $\mathcal{A}$ is the *strict partial order* corresponding to $\preccurlyeq$ if $a \prec b \Leftrightarrow a \preccurlyeq b \wedge a \neq b$, for arbitrary $a, b \in \mathcal{A}$. We say that $\prec$ is the *irreflexive reduction* of $\preccurlyeq$. Conversely, $\preccurlyeq$ is the *reflexive closure* of $\prec$, since $a \preccurlyeq b \Leftrightarrow a \prec b \vee a = b$.
- Let $(\mathcal{A}, \preccurlyeq)$ be a partially ordered set and $a, b \in \mathcal{A}$. When $a \preccurlyeq b$, we say that a is *smaller than or equal to* (or *lower than or equal to*) b. Correspondingly, we say that b is *larger than or equal to* (or *higher than or equal to*) a. When $a \prec b$, we say that a is *smaller than* (or *lower than*, or *below*) b. Correspondingly, we say that b is *larger than* (or *higher than*, or *above*) a.
- Let $(\mathcal{A}, \preccurlyeq)$ be a partially ordered set and $a, b, c, d \in \mathcal{A}$. Then we say:

  - a is $\preccurlyeq$-*minimal* if $x \preccurlyeq a$ implies $x = a$ for all $x \in \mathcal{A}$ (nothing in $\mathcal{A}$ is smaller than a).
  - b is $\preccurlyeq$-*least* if $b \preccurlyeq x$ for all $x \in \mathcal{A}$ (b is smaller than any other in $\mathcal{A}$).
  - c is $\preccurlyeq$-*maximal* if $c \preccurlyeq x$ implies $x = c$ for all $x \in \mathcal{A}$ (nothing in $\mathcal{A}$ is greater than c).
  - d is $\preccurlyeq$-*greatest* if $x \preccurlyeq d$ for all $x \in \mathcal{A}$ (d is greater than any other in $\mathcal{A}$).

  When the relation $\preccurlyeq$ is understood, we can drop the prefix "$\preccurlyeq$-". The least and greatest elements are called the *zero* (0) and *unit* (1) element, respectively.
- Let $(\mathcal{A}, \preccurlyeq)$ be a partially ordered set, $\mathcal{B} \subseteq \mathcal{A}$, and $u, v, w, z \in \mathcal{A}$. Then we say:

  - u is a $\preccurlyeq$-*upper bound* of $\mathcal{B}$ if $x \preccurlyeq u$ for all $x \in \mathcal{B}$.
  - v is a $\preccurlyeq$-*least upper bound* (or $\preccurlyeq$-*lub*) of $\mathcal{B}$ if v is a $\preccurlyeq$-upper bound of $\mathcal{B}$ and $v \preccurlyeq u$ for every $\preccurlyeq$-upper bound u of $\mathcal{B}$.
  - w is a $\preccurlyeq$-*lower bound* of $\mathcal{B}$ if $w \preccurlyeq x$ for all $x \in \mathcal{B}$.
  - z is a $\preccurlyeq$-*greatest lower bound* (or $\preccurlyeq$-*glb*) of $\mathcal{B}$ if z is a $\preccurlyeq$-lower bound of $\mathcal{B}$ and $w \preccurlyeq z$ for every $\preccurlyeq$-lower bound w of $\mathcal{B}$.

  When the relation $\preccurlyeq$ is understood, we can drop the prefix "$\preccurlyeq$-".
- A *lattice* is a partially ordered set $(\mathcal{A}, \preccurlyeq)$ in which any two elements have an lub and a glb. The lub of $a, b \in \mathcal{A}$ is denoted by $a \vee b$, and the glb by $a \wedge b$. An *upper semi-lattice* is a partially ordered set $(\mathcal{A}, \preccurlyeq)$ in which any two elements have an lub (but not necessarily a glb).
- A *linearly* (or *totally*) *ordered* set is a partially ordered set $(\mathcal{A}, \preccurlyeq)$ such that for all $x, y \in \mathcal{A}$ either $x \preccurlyeq y$ or $y \preccurlyeq x$. In this case we say that $\preccurlyeq$ is a *linear order* on $\mathcal{A}$.
- A *well-ordered* set is a linearly ordered set $(\mathcal{A}, \preccurlyeq)$ such that every nonempty subset of $\mathcal{A}$ has a $\preccurlyeq$-least element. We say that such a $\preccurlyeq$ is a *well-order* on $\mathcal{A}$.
- Associated with every well-ordered set $(\mathcal{A}, \preccurlyeq)$ is the corresponding *Principle of Complete Mathematical Induction*: If P is a property such that, for any $b \in \mathcal{A}$, $P(b)$, whenever $P(a)$ for all $a \in \mathcal{A}$ such that $a \prec b$, then $P(x)$ for all $x \in \mathcal{A}$. When $\mathcal{A}$ is infinite, a proof using this principle is called a proof by *transfinite induction*.

Equivalence Relations

- A relation R on $\mathcal{A}$ is an *equivalence relation* if (i) R is reflexive, (ii) R is symmetric, and (iii) R is transitive. In this case, the *R-equivalence class* of $a \in \mathcal{A}$ is the set $\{x \in \mathcal{A} \mid xRa\}$. Elements of the R-equivalence class of a are said to be *R-equivalent* to a. If $\mathcal{C}$ is an equivalence class, any element of $\mathcal{C}$ is called a *representative* of the class $\mathcal{C}$.
- A *partition* of $\mathcal{A}$ is any collection $\{\mathcal{A}_i \mid i \in \mathcal{I}\}$ of nonempty subsets of $\mathcal{A}$ such that (i) $\mathcal{A} = \bigcup_{i \in \mathcal{I}} \mathcal{A}_i$, and (ii) $\mathcal{A}_i \cap \mathcal{A}_j = \emptyset$, for all $i, j \in \mathcal{I}$ with $i \neq j$. So, $\mathcal{A}$ is the disjoint union of the sets in the partition.
- Any equivalence relation on $\mathcal{A}$ is associated with a partition of $\mathcal{A}$, and vice versa. If R is an equivalence relation on $\mathcal{A}$, then the associated partition of $\mathcal{A}$ is called the *quotient set of $\mathcal{A}$ relative to R* and is denoted by $\mathcal{A}/R$. The members of $\mathcal{A}/R$ are the R-equivalence classes of $\mathcal{A}$. The function $f : \mathcal{A} \to \mathcal{A}/R$ that associates with each element $a \in \mathcal{A}$ the R-equivalence class of a is called the *natural map* of $\mathcal{A}$ relative to R.
- An equivalence relation is often denoted by $\sim, \simeq, \equiv$, or any other symbol indicating the properties of this relation.

Functions

Basics

- A total *function f from $\mathcal{A}$ into $\mathcal{B}$* is a triple $(\mathcal{A}, \mathcal{B}, f)$ where $\mathcal{A}$ and $\mathcal{B}$ are nonempty sets and for every $x \in \mathcal{A}$ there is a unique member, denoted by $f(x)$, of $\mathcal{B}$. We call $\mathcal{A}$ the *domain* of f and denote it by $\text{dom}(f)$. The set $\mathcal{B}$ we call the *co-domain* of f and denote it by $\text{codom}(f)$. We usually write $f : \mathcal{A} \to \mathcal{B}$ instead of $(\mathcal{A}, \mathcal{B}, f)$. A function is also called a *mapping*.
- In specifying a definition of $f : \mathcal{A} \to \mathcal{B}$ we say that f is *well defined* if we are assured that f is single-valued, i.e., with each member of $\mathcal{A}$, f associates a *unique* member of $\mathcal{B}$.
- When the domain of a function consists of ordered n-tuples, the function is said to be *of n arguments*. A (total) *function of n arguments on a set $\mathcal{S}$* is a function f whose domain is $\mathcal{S}^n$. We write $f(a_1, \ldots, a_n)$ instead of $f(\langle a_1, \ldots, a_n \rangle)$.
- Let $f : \mathcal{A} \to \mathcal{B}$ and $\mathcal{C} \subseteq \mathcal{A}$. The *image* of $\mathcal{C}$ under f is a set denoted by $f(\mathcal{C})$ and defined by $f(\mathcal{C}) = \{f(x) \mid x \in \mathcal{C}\}$. In particular, $f(\mathcal{A})$ is called the *range* of f and denoted by $\text{rng}(f)$.
- A function $f : \mathcal{A} \to \mathcal{B}$ is:

    - *injective* if $f(x) \neq f(y)$ whenever $x \neq y$; we also say that such an f is an *injection*.
    - *surjective* if $f(\mathcal{A}) = \mathcal{B}$; we also say that such an f is a *surjection*.
    - *bijective* if it is injective and surjective; we also say that such an f is a *bijection*.

- An element $a \in \mathcal{A}$ is called a *fixed point* of a function $f : \mathcal{A} \to \mathcal{A}$ if $f(a) = a$.
- Let $f : \mathcal{A} \to \mathcal{B}$ and $\mathcal{C} \subseteq \mathcal{A}$. Then a function $g : \mathcal{C} \to \mathcal{B}$ is the *restriction* of f to $\mathcal{C}$ if $g(x) = f(x)$ for each $x \in \mathcal{C}$. The restriction of f to $\mathcal{C}$ is denoted by $f|_{\mathcal{C}}$. In that case f is an *extension* of g to $\mathcal{A}$.
- Let $f : \mathcal{A} \to \mathcal{B}$ and $g : \mathcal{C} \to \mathcal{D}$ be functions and $f(\mathcal{A}) \subseteq \mathcal{C}$. Then the *composite function* (or *composition*) of f and g, denoted by $g \circ f$, is the function $g \circ f : \mathcal{A} \to \mathcal{D}$ defined by $(g \circ f)(x) = g(f(x))$, for each $x \in \mathcal{A}$.
- Let $\mathcal{A}$ and $\mathcal{U}$ be sets, and let $\mathcal{A} \subseteq \mathcal{U}$. The *characteristic function* of $\mathcal{A}$ is the function $\chi_{\mathcal{A}} : \mathcal{U} \to \{0, 1\}$ such that $\chi_{\mathcal{A}}(u) = 1$ if $u \in \mathcal{A}$ and $\chi_{\mathcal{A}}(u) = 0$ if $u \notin \mathcal{A}$.
- Let $\mathcal{A}$ and $\mathcal{B}$ be nonempty sets. Then the set of all functions having the domain $\mathcal{A}$ and co-domain $\mathcal{B}$ is denoted by $\mathcal{B}^{\mathcal{A}}$.

Cardinality

- Two sets $\mathcal{A}$ and $\mathcal{B}$ are said to be *equinumerous* (or *equipollent* or of the same *power*), which is denoted by $\mathcal{A} \simeq \mathcal{B}$, if there exists a bijection $f : \mathcal{A} \to \mathcal{B}$. In that case we say that $\mathcal{A}$ and $\mathcal{B}$ have the *same cardinal number*. The cardinal number of a set $\mathcal{A}$ is denoted by $|\mathcal{A}|$. The relation $\simeq$ is an equivalence relation.
- A cardinal number $|\mathcal{A}|$ is *smaller* than a cardinal number $|\mathcal{B}|$, written $|\mathcal{A}| < |\mathcal{B}|$, if there is an injection $f : \mathcal{A} \to \mathcal{B}$, but $\mathcal{A}$ and $\mathcal{B}$ are not equinumerous.
- *Cantor's Theorem* states that $|\mathcal{A}| < |2^{\mathcal{A}}|$, for any set $\mathcal{A}$.
- A set $\mathcal{A}$ is

    - *finite* if either $\mathcal{A} = \emptyset$ or $\mathcal{A} \simeq \{1, 2, \ldots, n\}$ for some natural n;
    - *infinite* if it is not finite;
    - *countable* (or *enumerable*, or *denumerable*) if $\mathcal{A} \simeq \mathcal{B}$ for some $\mathcal{B} \subseteq \mathbb{N}$; when $\mathcal{B} = \mathbb{N}$, the set $\mathcal{A}$ is said to be *countably infinite*;
    - *uncountable* if it is not countable.

- If a set $\mathcal{A}$ is infinite, then there is $\mathcal{B} \subsetneq \mathcal{A}$ such that $\mathcal{B} \simeq \mathcal{A}$.
- Any subset of a countable set is countable. The union of countably many countable sets is countable. The Cartesian product of two countable sets is countable.
- Let n be a natural number, $\aleph_0 = |\mathbb{N}|$, and $\mathbf{c} = |\mathbb{R}|$ the cardinality of continuum. Then: $\aleph_0 + n = \aleph_0$, $\aleph_0 + \aleph_0 = \aleph_0$, $n \cdot \aleph_0 = \aleph_0$, $\aleph_0^n = \aleph_0$, $\mathbf{c} + \aleph_0 = \mathbf{c}$, and $\aleph_0 \cdot \mathbf{c} = \mathbf{c}$.

- A *sequence* is a function f defined on $\mathbb{N}$, the set of natural numbers. If we write $f(n) = x_n$, for $n \in \mathbb{N}$, we also denote the sequence f by $\{x_n\}$, or by $x_0, x_1, x_2, \ldots$ When $x_n \in \mathcal{A}$ for all $n \in \mathbb{N}$, we say that $\{x_n\}$ is a *sequence of elements of* $\mathcal{A}$. The elements of any at most countable set can be arranged in a sequence.
- The cardinality of $\mathcal{B}^{\mathcal{A}}$, the set of all functions mapping $\mathcal{A}$ into $\mathcal{B}$, is $|\mathcal{B}|^{|\mathcal{A}|}$.

Operations and Algebraic Structures

- An *n-ary operation* on a set $\mathcal{A}$ is a function $\star : \mathcal{A}^n \to \mathcal{A}$. When $n = 2$, we say that the operation is *binary*. In this case we write $a \star b$ instead of $\star(a, b)$ When $n = 1$, the operation is said to be *unary* and we write $a^\star$ instead of $\star(a)$.
- A binary operation on a set $\mathcal{A}$ is:

  - *associative* if $a \star (b \star c) = (a \star b) \star c$, for all $a, b, c \in \mathcal{A}$.
  - *commutative* if $a \star b = b \star a$, for all $a, b \in \mathcal{A}$.

- A *semigroup* is a pair $(\mathcal{A}, \star)$, where $\star$ is an associative binary operation on $\mathcal{A}$.
- A *group* is a semigroup $(\mathcal{A}, \star)$ satisfying the following requirements:

  - there exists an element $e \in \mathcal{A}$ such that $a \star e = e \star a = a$, for all $a \in \mathcal{A}$ (e is called the *identity* of $\mathcal{A}$);
  - for each $a \in \mathcal{A}$ there exists an element $a^{-1} \in \mathcal{A}$ such that $a \star a^{-1} = a^{-1} \star a = e$ (a^{-1} is called the *inverse* of a).

Natural Numbers

- Natural numbers are $0, 1, 2, \ldots$. The set of all natural numbers is denoted by $\mathbb{N}$. The cardinal number of $\mathbb{N}$ is denoted by $\aleph_0$ (aleph zero).
- A *prime* is a natural number greater than 1 that has no positive divisors other than 1 and itself. There are infinitely many primes. A natural number greater than 1 that is not a prime is called a *composite*.
- The *Fundamental Theorem of Arithmetic* states: Any positive integer ($\neq 1$) can be expressed as a product of primes; this expression is unique except for the order in which the primes occur. Thus, any positive integer $n(\neq 1)$ can be written as $p_1^{\alpha_1} p_2^{\alpha_2} \ldots p_r^{\alpha_r}$, where $p_1, p_2, \ldots, p_r$ are primes satisfying $p_1 < p_2 < \ldots < p_r$, and $\alpha_1, \alpha_2, \ldots, \alpha_r$ are positive integers.
- The *Principle of Mathematical Induction* is: Any subset of $\mathbb{N}$ that contains 0 and, for every natural k, contains $k + 1$ whenever it contains k, is equal to $\mathbb{N}$.
- The set $(\mathbb{N}, <)$ is well-ordered. It is also denoted by ω.
- The *Principle of Complete Mathematical Induction*: Any subset of ω that, for every natural k, contains k whenever it contains every natural $i < k$ is equal to ω.
- The set of all subsets of $\mathbb{N}$, i.e., the set $2^{\mathbb{N}}$, is uncountable. Its cardinality is $2^{\aleph_0}$. This is equal to $\mathsf{c} = |\mathbb{R}|$, the cardinality of continuum.
- Functions $f : \mathbb{N} \to \mathbb{N}$, $k \geq 1$, are called *numerical*.
- The set $\mathbb{N}^{\mathbb{N}}$ of all functions $f : \mathbb{N} \to \mathbb{N}$ is uncountable: $|\mathbb{N}^{\mathbb{N}}| = 2^{\aleph_0} = \mathsf{c}$. In particular, the set $\{0, 1\}^{\mathbb{N}}$ of all characteristic functions $\chi : \mathbb{N} \to \{0, 1\}$ is equinumerous to the set $2^{\mathbb{N}}$. Since each χ is identified with an infinite sequence of 0s and 1s, the set of all infinite binary sequences is also uncountable.
- The *join* of two sets $\mathcal{A}, \mathcal{B} \subseteq \mathbb{N}$ is the set denoted by $\mathcal{A} \oplus \mathcal{B}$ and defined by $\mathcal{A} \oplus \mathcal{B} \overset{\text{def}}{=} \{2x \mid x \in \mathcal{A}\} \cup \{2y + 1 \mid y \in \mathcal{B}\}$. Informally, $\mathcal{A} \oplus \mathcal{B}$ "remembers" every member of $\mathcal{A}$ and every member of $\mathcal{B}$.

Formal Languages

Basics

- An *alphabet* Σ is a finite nonempty set of abstract *symbols*.
- A *word* of length $k \geqslant 0$ *over the alphabet* Σ is a finite sequence $x_1, \ldots, x_k$ of symbols in Σ. A word $x_1, \ldots, x_k$ is usually written without commas, i.e., as $x_1 \ldots x_k$.
- The length of a word w is denoted by $|w|$. The word of length zero is called the *empty word* and denoted by ε.
- If $w = x_1 \ldots x_k$ is a word, then the word $w^R = x_k \ldots x_1$ is called the *reversal of w*.
- Two words $x_1 \ldots x_r$ and $y_1 \ldots y_s$ over the alphabet Σ are *equal*, written $x_1 \ldots x_r = y_1 \ldots y_s$, if $r = s$ and $x_i = y_i$ for each i.
- Let x and y be words over the alphabet Σ. The word x is a *subword* of y if $y = uxv$ for some words u and v. The word x is a *proper* subword of y if x is a subword of y, but $x \neq y$.
- Let x and y be words over the alphabet Σ. The word x is a *prefix* of y, written $x \subseteq y$, if $y = xv$ for some word v. The word x is a *proper* prefix of y, written $x \subset y$, if x is a prefix of y, but $x \neq y$.
- The *set of all words, including* ε, *over the alphabet* Σ is denoted by Σ^*.
- The set Σ^* is countably infinite.
- Each subset $\mathcal{L} \subseteq \Sigma^*$ is called a *formal language* (or *language* for short).

Operations on Languages

- If $x = x_1 \ldots x_r$ and $y_1 \ldots y_s$ are words, then xy, called the *concatenation* of x and y, is the word $x_1 \ldots x_r y_1 \ldots y_s$.
- For languages $\mathcal{L}_1$ and $\mathcal{L}_2$, the *concatenation* (or *product*) of $\mathcal{L}_1$ and $\mathcal{L}_2$ is a language denoted by $\mathcal{L}_1 \mathcal{L}_2$ and defined by $\mathcal{L}_1 \mathcal{L}_2 = \{xy \mid x \in \mathcal{L}_1 \wedge y \in \mathcal{L}_2\}$.
- For a language $\mathcal{L}$ let $\mathcal{L}^0 = \{\varepsilon\}$ and, for each $n \geqslant 1$, let $\mathcal{L}^n = \mathcal{L}^{n-1} \mathcal{L}$. The *Kleene star of* $\mathcal{L}$ is the language denoted by $\mathcal{L}^*$ and defined by $\mathcal{L}^* = \bigcup_{i=0}^{\infty} \mathcal{L}^i$. Similarly, the *Kleene plus of* $\mathcal{L}$ is the language denoted by $\mathcal{L}^+$ and defined by $\mathcal{L}^+ = \bigcup_{i=1}^{\infty} \mathcal{L}^i$. In particular, for the alphabet Σ, the language Σ^n contains all words of length n over Σ, and Σ^* contains all words over Σ.

Orders on Languages

- Let $\leq$ be a linear order on the alphabet Σ. A *lexicographic order* $\leq_{\text{lex}}$ on Σ^n, induced by $\leq$, is the order in which $x_1 \ldots x_n <_{\text{lex}} y_1 \ldots y_n$ if there is a $j, 1 \leq j \leq n$, such that $x_i = y_i$ for each $i = 1, \ldots, j-1$, but $x_j < y_j$.
- A *shortlex order* on a language $\mathcal{L} \subseteq \Sigma^*$ is the order in which words of $\mathcal{L}$ are primarily ordered by increasing length, and words of the same length are then lexicographically ordered. The shortlex order is a well-order on Σ^* and, consequently, on $\mathcal{L}$.

Appendix B
Notation Index

Chapter 4

Chapter 5

Glossary

A

abstract computing machine an instance of a model of computation, e.g., a particular Turing machine, or a particular Post machine, or a particular Markov grammar

abstraction (in λ-calculus) an operation in λ-calculus that binds a free variable in a λ-term and thus constructs a new λ-term that denotes a function of that variable

accelerating machine a hypercomputer that carries out each next basic operation (called by the program) in half the time taken to carry out the previous one

Ackermann function a very rapidly growing function that is defined inductively on pairs of natural numbers and is μ-recursive but not primitive recursive

algorithm (intuitively) a finite list of precisely described unambiguous instructions that is supposed to be applied and mechanically followed through to a conclusion in order to accomplish a specified computation or other task

algorithm (formally) a Turing program (or a construction of a μ-recursive function; or a system $\mathscr{E}(f)$ of equations; or a λ-term; or a Post program; or a Markov grammar)

alphabet a finite, nonempty set whose elements are referred to as symbols

arithmetical class (of sets) a class of arithmetical sets characterized by predicates with n alternating quantification symbols and the same first quantification symbol; if the first quantification symbol is $\exists\,(\forall)$, the class is $\Sigma_n\,(\Pi_n)$; in addition, $\Delta_n \overset{\text{def}}{=} \Sigma_n \cap \Pi_n$

arithmetical hierarchy (of arithmetical classes) a hierarchy consisting of the arithmetical classes Σ_n, Π_n, and Δ_n, for $n = 0, 1, 2, \ldots$, and inclusions between them

arithmetical relation a relation defined on the set of natural numbers; that is, a subset of $\mathbb{N}^k$, for some k

arithmetical set a set of natural numbers x characterized by a predicate of the form $\exists y_1 \forall y_2 \exists y_3 \ldots Qy_n R(x, y_1, y_2, \ldots, y_n)$ or $\forall y_1 \exists y_2 \forall y_3 \ldots Qy_n R(x, y_1, y_2, \ldots, y_n)$, where $n \geqslant 0$ and R is a decidable arithmetical relation

arithmetization (of a theory) the treatment of the theory by methods involving only the fundamental concepts and operations of arithmetic, such as natural and prime numbers and their sums and products

axiom (of a theory) a statement whose truth is either to be taken as self-evident or to be assumed

axiomatic method a method of developing a theory by first choosing a set of basic notions and a set of axioms about these notions, and then discovering their consequences (by deducing new theorems only from axioms or previously deduced theorems, and defining new notions only using the basic or previously defined notions)

axiomatic set theory a theory of sets developed by the axiomatic method from some basic notions that include those of the "set" and the "membership" relation, and some axioms that are consistent and in accord with intuitive beliefs about sets

axiomatic system a collection of basic notions and axioms about the basic notions

axiomatizable theory a theory for which there is an algorithm that can decide, for any formula, whether or not the formula is an axiom of the theory

B

behavior (of a mechanism) mechanisms may differ in their local behavior (when their moves are governed by different mechanical rules) but they may still be equal in their global behavior (i.e., they produce equal results)

Burali-Forti's paradox the unacceptable conclusion derived in Cantor's set theory stating that there exists a well-ordered set (namely the set of *all* ordinal numbers) whose ordinal number is larger than itself

busy beaver (n-state) a Turing machine with n states that writes the largest number of 1s that any n-state Turing machine can write and still halt

busy beaver function the function whose value at n is the number of 1s written by an n-state busy beaver

busy beaver problem the problem of deciding, for an arbitrary Turing machine, whether or not the machine is an n-state busy beaver for some n

C

canonical system a generator consisting of a start symbol S, an alphabet Σ, and a finite set of productions that can transform S through various sequences of production applications into various words in Σ^* (and thus generate a set of words over Σ)

Cantor's paradox the unacceptable conclusion derived in Cantor's set theory stating that there exists a set (namely the set of *all* sets) whose cardinality both is and is not strictly less than the cardinality of its power set

Cantor's theorem the theorem stating that the cardinality of a set is strictly less than the cardinality of its power set

c.e. degree *see* computably enumerable degree

c.e. set *see* computably enumerable set

characteristic function (of a set) a function whose value is 1 for every member of the set, and 0 otherwise

characterization (of the notion of algorithm) the process of searching for (or the state after finding) a property that is shared by all algorithms and algorithms only

Church-Turing barrier the asserted fundamental logical limit on what can be computed no matter how far and in what multitude of ways computers develop; synonymous with Turing-computability

Church-Turing thesis *see* computability thesis

class (generally) a generalization of the notion of set specifying that every set is a class, but some classes, called proper classes, are not sets

class $\mathcal{D}$ (of all degrees of unsolvability) the class of all Turing degrees

coding function (for a problem) a function that transforms every instance of a computational problem, which is to be solved on a machine, into a form understandable by the machine (namely into a word over the input alphabet of the machine)

compactness theorem the theorem stating that a first-order theory has a model *if* every finite part of the theory does

completeness requirement (in defining "computable" functions) the requirement that any formalization of the informal notion of the intuitively computable function must include *all* such functions and *nothing else*

complexity (of an algorithm) an asymptotic estimation of the amount of computational resources needed for a complete execution of the algorithm; usually expressed in terms of the size of the input to the algorithm

computability theory the theory that classifies computational problems according to whether or not they can be algorithmically solved at least in principle, i.e., their solving has at its disposal unrestricted computational resources, e.g., time and space

computability thesis the assertion that the intuitive notion of what is computable is adequately formalized by the notion of computable by the Turing machine (or, equivalently, by any model of computation equivalent to the Turing machine)

computable function a total function $\varphi : \mathcal{A} \to \mathcal{B}$ for which there exists a Turing machine that can compute the value $\varphi(x)$ for any argument $x \in \mathcal{A}$

computable problem a computational problem for which there exists a Turing machine capable of computing the solution to any instance of the problem, *if* the solution is defined

computable set *see* decidable set

computably enumerable (or **c.e.**) **degree** a Turing degree containing some c.e. set

computably enumerable (or **c.e.**) **set** a set whose members can be listed (enumerated) by a Turing machine (which may run forever, if necessary)

computation intuitively, a sequence of steps that a human or a device makes while following a finite list of instructions that tell how to solve a problem; formally, a sequence of steps that a Turing machine makes while following its Turing program

computational complexity theory a theory that classifies computational problems according to whether or not they can be solved with appropriately restricted computational resources, e.g., within appropriately bounded time or space

computational problem a problem whose instances require certain computations to yield their solutions; it can be a decision, search, counting, or generating problem

configuration (internal, of an abstract computing machine) the collection of statuses of the relevant components of the machine at a particular step of computation; informally, a snapshot of the machine at a particular point of its computation

consistency problem the problem asking whether or not a given theory is consistent

consistent (theory) a theory that does not contain a contradiction, i.e., there is no formula such that both the formula and its negation are derivable in the theory

continuum hypothesis the assertion that there exists no set whose cardinality is strictly between that of the integers and the real numbers

counting problem a computational problem whose each instance asks for the number of the members of a given set that have a given property

creative set a c.e. set C for which there is a p.c. function φ such that, for any c.e. set $\mathcal{W}_x$, the value $\varphi(x)$ witnesses that $\overline{C} \neq \mathcal{W}_x$; informally, C is effectively undecidable

D

decidability problem the problem asking whether or not a given theory is decidable

decidable problem a decision problem for which there is a Turing machine that can compute, for any instance of the problem, whether the answer to the instance is YES or NO

decidable property (of p.c. functions) an intrinsic property of functions for which there is a Turing machine that can decide, for any p.c. function, whether or not the function has the property

decidable property (of c.e. sets) an intrinsic property of sets for which there is a Turing machine that can decide, for any c.e. set, whether or not the set has the property

decidable relation (k-ary, on a set $\mathcal{S}$) a subset $\mathcal{R}$ of $\mathcal{S}^k$ for which there is a Turing machine that can decide, for any $(a_1, \dots, a_k) \in \mathcal{S}^k$, whether or not $(a_1, \dots, a_k) \in \mathcal{R}$

decidable set a set for which there is a Turing machine that can decide, for any element, whether or not the element is a member of the set

decidable theory a consistent and syntactically complete theory for which there exists a Turing machine capable of answering—in finite time, and for any formula F of the theory—the question "Is F derivable in the theory?"

decision problem a computational problem whose each instance asks for an answer that is either YES or NO

decider (of a set) a Turing machine capable of deciding, for any element, whether or not the element is a member of the set

deduction the process of reaching a conclusion about something because of other things, called premises, that we know or assume to be true

degree of unsolvability (of an unsolvable problem) an informal notion that represents our intuitive understanding of the hardness of the unsolvable problem; for decision problems it is formalized by the concept of Turing degree

density theorem the theorem stating that between any two c.e. degrees there exists a third c.e. degree

derivation (of a word v from a word u) a finite sequence of substitutions that transform u into v while complying with a given set of rules (productions)

derivation (of a theorem F in a theory) a finite sequence of formulas whose last formula is F and each formula in the sequence is either an axiom of the theory or directly follows from some of the preceding formulas by one of the rules of inference

diagonalization a technique for proving that a set $\mathcal{S}$ cannot be exhibited by listing the elements of $\mathcal{T} \subseteq \mathcal{S}$, in which $\mathcal{T}$ is represented by a table whose rows represent members of $\mathcal{T}$ and whose appropriately changed diagonal would represent a member of $\mathcal{S}$ yet differ from every row, and hence represent none of the members of $\mathcal{T}$

diagonal language the language $\mathcal{K}$ of the problem of deciding, for an arbitrary Turing machine T, whether or not T halts on its code, i.e., $\mathcal{K} = \{\langle T, T \rangle \mid T \text{ halts on } \langle T \rangle\}$

Δ_n-complete set a Δ_n-set such that every Δ_n-set is m-reducible to it

Δ_n-set a set of natural numbers that is both a Σ_n-set and a Π_n-set

domain (of interpretation) a structure in which a theory developed in a formal axiomatic system is interpreted (given meaning); a field of interest in which formulas of a formal theory become true or false statements about the objects of the field

dovetailing a technique that avoids getting trapped in any of countably many potentially non-halting computations by executing them simultaneously, i.e., by periodically making small progress in each of the computations (or in each of a growing subset of them)

E

effective procedure (for calculating a function's values) a finite set of instructions in any language that, given any input in the domain of the function, completes in a finite number of mechanically executed steps and returns the function's value

effectively calculable (function) a function whose values can be computed by an effective procedure; that is, a "computable" (intuitively computable) function

effectiveness requirement (in defining "computable" functions) the requirement that any formalization of the informal notion of the intuitively computable function must exhibit, for each such function, an effective procedure for calculating its values

elementary function a function constructed from finitely many exponentials $e^{(\cdot)}$, logarithms $\log(\cdot)$, roots $\sqrt[(\cdot)]{(\cdot)}$, real constants, and the variable x, by using finitely many function compositions and operations $+$, $-$, $\times$, and $\div$

Entscheidungsproblem the decidability problem of mathematics, i.e., the question asking whether or not there is an algorithm that can decide, for any formula of a theory formalizing mathematics, whether or not the formula is derivable in the theory

enumeration function (of a set $\mathcal{S}$) a computable function $g : \mathbb{N} \to \mathcal{S}$ such that $\{g(i) \,|\, i \in \mathbb{N}\} = \mathcal{S}$; if such a g exists, then $\mathcal{S}$ is enumerated by g, in the sense that an $x \in \mathcal{S}$ is taken to be nth in $\mathcal{S}$ *iff* n is the smallest $i \in \mathbb{N}$ for which $x = g(i)$

enumeration problem *see* generation problem

equivalent models (of computation) models of computation that have the same global behavior, e.g., accept the same language

equivalent grammars grammars that generate the same language

extensional definition a definition that gives the meaning of a term by listing everything that falls under that term

F

finitism the kind of mathematical reasoning which—in order to avoid deceptive intuition and doubts arising from the use of the notion of infinity—preferably uses finite objects and finite methods that are constructive, at least in principle

finite-injury priority method a priority method in which a requirement can only be injured finitely many times

first incompleteness theorem (Gödel's) the theorem stating that if the *Formal Arithmetic* is consistent, then it is semantically incomplete, i.e., incapable of proving all *Truths* about the natural numbers

first incompleteness theorem (generalized) the theorem stating that any consistent extension of the set of axioms of *Formal Arithmetic* induces a semantically incomplete theory, i.e., a theory incapable of proving all *Truths* about the natural numbers

fixed-point theorem *see* recursion theorem

formal arithmetic the formal axiomatic system and the induced theory that formalize the arithmetic of natural numbers

formal axiomatic system a system consisting of a symbolic language (for writing formulas), a set of axioms (selected formulas), and a set of rules of inference (specifying the conditions under which formulas can be derived from other formulas)

formalism the treatment of mathematics that replaces contentual[1] mathematics by mechanical manipulation of meaningless symbols in accordance with accepted string manipulation rules

formalization of computation a definition that formally defines, in terms of a model of computation, the basic intuitive notions of algorithmic computation, i.e., the algorithm, the required environment, and the execution of the algorithm in it

formal theory a set of formulas in the symbolic language of a formal axiomatic system such that any formula derivable from the axioms of the system by the rules of inference of the system is called a theorem of the theory

foundations of mathematics the study of the philosophical and logical basis of mathematics; in a broader sense, the mathematical investigation of what underlies the philosophical theories concerning the nature of mathematics

function application (in the λ-calculus) an operation in the λ-calculus that substitutes every bound occurrence of a variable in one λ-term (denoting a function of that variable) with another λ-term (denoting the argument of the function)

G

(general) recursive function a function that can be well defined by a system of equations in standard form, and whose values can be computed from the system by using only the rules of substitution and replacement

generalized padding lemma the lemma stating that an $\mathcal{O}$-p.c. function has countably infinitely many indexes, and given one of them, countably infinitely many others can be generated

---

[1] relating to content

generator (of a set) a model that formally defines the intuitive concept of the algorithmic generation of a set, i.e., algorithmic listing of all the elements of a set

generation problem a computational problem whose each instance asks for a list (i.e., enumeration) of the members of a given set that have a given property; a listed member is labeled with n *iff* the first occurrence of the member is nth in order

global behavior (of an abstract computing machine) the results that the machine can or cannot produce (regardless of its workings, i.e., local behavior)

Gödel numbering (of a formal theory) a computable injective function that maps the syntactic objects of the theory (symbols, formulas, finite sequences of formulas) to the natural numbers, such that there is an algorithm that, for any natural n, decides whether n is the image of some syntactic object, and if so, identifies the object

grammar a quadruple consisting of a set of symbols called nonterminals, a disjoint set of symbols called terminals, a designated nonterminal called the start symbol, and a finite set of rules called productions for transforming words into other words

H

halting problem the problem of deciding, for an arbitrary Turing machine T and an arbitrary input x, whether or not T halts on x

Hilbert's tenth problem the problem of deciding, for an arbitrary multivariate polynomial equation $p(x_1, \ldots, x_n) = 0$, whether or not the equation is solvable in the integers

Hilbert's program (for mathematics) a formalistic attempt that would use formal axiomatic systems to eliminate all known and unknown paradoxes from mathematics

hypercomputer a notional computing machine able to compute beyond the Church-Turing barrier, i.e., beyond Turing-computability

hypercomputation any mode of computation that goes beyond what is permitted by the Church-Turing barrier, i.e., is not bounded by Turing-computability

I

incomputable function a function for which there is no Turing machine capable of computing the function's values everywhere the function is defined

incomputable problem a computational problem for which there exists no Turing machine capable of computing the solution to every instance of the problem where the solution is defined

incomputable set *see* undecidable set

index (of a Turing machine) a natural number that encodes a Turing machine; informally, a number that represents an algorithm

index (of a p.c. function) the index of a Turing machine that can compute the values of the function

index set (of a p.c. function) the set of indexes of all Turing machines capable of computing the values of the function (anywhere the function is defined); informally, the set of all the (encoded) algorithms that compute the function's values

index set (of a semi-decidable set) the set of indexes of all Turing machines that recognize the set; informally, the set of all the (encoded) recognizers of the set

index set (of an $\mathcal{O}$-p.c. function) the set of indexes of all $\mathcal{O}$-TMs that can compute the values of the function (whenever defined); informally, the set of all the (encoded) algorithms that compute the function's values with the help of the oracle for the set $\mathcal{O}$

injury the change of the status of a satisfied requirement R back to unsatisfied, because a later action A' (taken to satisfy some unsatisfied requirement R') conflicted with the preservation of the previous action A (taken to satisfy the requirement R)

intended model (of a theory) a particular structure (i.e., field of interest) for the investigation of which a formal axiomatic system is established and theory developed

intensional definition a definition that gives the meaning of a term by specifying necessary and sufficient conditions for when the term should be used. In the case of nouns, this is equivalent to specifying the properties that an object needs to have in order to be counted as a referent of the term

internal configuration *see* configuration

interpretation (of a formal theory) a mapping that gives the formal theory meaning in a field of interest, i.e., defines, for every closed formula of the theory, how the formula is to be understood as a statement about the objects of the field

intrinsic property (of p.c. functions) an essential property of functions that is insensitive to the algorithm, machine, and program that compute the functions' values

intrinsic property (of c.e. sets) an essential property of sets that is insensitive to the algorithm, machine, and program solving the membership problem for the sets

intuitionism a mathematical school that argued for greater mathematical rigor in the process of proving; it advocated a (non-Platonic) view that the existence of a mathematical object is closely connected to the existence of its mental construction

J

jump *see* Turing jump operator

jump hierarchy (of sets) the hierarchy $\emptyset^{(0)} <_T \ldots <_T \emptyset^{(i)} <_T \emptyset^{(i+1)} <_T \ldots$, where the set $\emptyset^{(i)}$ is the ith Turing jump of the decidable set $\emptyset$

jump hierarchy (of Turing degrees) the hierarchy $\mathbf{0}^{(0)} < \ldots < \mathbf{0}^{(i)} < \mathbf{0}^{(i+1)} < \ldots$, where Turing degree $\mathbf{0}^{(i)}$ is the ith Turing jump of $\mathbf{0}^{(0)}$, the Turing degree of the decidable sets

K

L

λ-calculus a model of computation that transforms, via a sequence of reductions, a given initial λ-term, which represents a function and its arguments, into a final λ-term, which represents the corresponding value of the function

λ-definable function a function f of one positive integer for which there exists a λ-term F such that if $f(m) = n$ and M and N are λ-terms denoting m and n, respectively, then λ-term FM can be transformed into N with a sequence of reductions

λ-term a well-formed expression in λ-calculus built from variables and other well-formed expressions by finitely many abstractions and function applications

language (formal) a set of finite words consisting of symbols from an alphabet

language (accepted by a Turing machine) *see* proper set

language (generated by a grammar G) the set of all words that consist of G's terminals only and can be derived from G's start symbol by using G's productions

language (of a decision problem) the set of the codes of all the positive instances of the problem

liar paradox the sentence *"This sentence is false"*, which complies with syntactic and semantic rules yet cannot consistently be assigned a truth-value because either of the premises—the sentence is *true*; the sentence is *false*—implies its own negation

local behavior (of an abstract computing machine) the way in which the machine operates when its basic instructions are performed

logical axiom an axiom that epitomizes a principle of pure logical reflection and is therefore present in every axiomatic system

logically valid formula a formula of a theory that is valid under every interpretation of the theory

logicism a school of mathematics that aimed to found mathematics on pure logic; as a side-effect it developed a symbolic language of mathematics, concisely formulated its rules of inference, and thus gave mathematics concise and precise expression

Löwenheim-Skolem theorem the theorem stating that if a theory has a model, then it has a countable model

M

Markov algorithm a finite sequence $\alpha_1 \to \beta_1, \ldots, \alpha_n \to \beta_n$ of productions that transform a given input word via a sequence of intermediate words into some output word by always applying the first applicable production to the last intermediate word

Markov-computable function a function for which there is a Markov algorithm that can compute the values of the function anywhere the function is defined

m-**complete** (set) a c.e. set such that every c.e. set is *m*-reducible to it

mechanism a device with predictable local behavior, in the sense that each move of the device is governed by some mechanical rule

membership problem (for a set) the problem of deciding, for an arbitrary element, whether or not the element is in the set

metalanguage a language used to describe or analyze another language (called the object language), that is, to make statements about statements of the object language

metamathematics the study of mathematics itself using mathematical methods, i.e., the field of study that deals with the structure and formal properties of mathematics

metatheory a theory whose subject matter is some theory; in mathematics, for example, a metatheory is a mathematical theory about some other mathematical theory

model (of a theory) an interpretation of the theory under which *all* the axioms of the theory are valid; intuitively, a field of interest that the theory sensibly formalizes

model (of computation) a definition that formally describes and characterizes the basic notions of algorithmic computation: what the algorithm is; what the environment capable of executing algorithms is; and what computation is

μ-operation *see* unbounded minimization

mortal matrix problem the undecidable decision problem asking whether or not the matrices of a given finite set of square matrices can be multiplied in some order, possibly with repetitions, so that the product is the zero matrix

μ-recursive function a (partial) function that can be constructed from the functions $\zeta(n) = 0$ (zero), $\sigma(n) = n + 1$ (successor), and $\pi_i^k(x_1, \ldots, x_k) = x_i$ (projection) using the operations of composition, primitive recursion, and μ-operation

m-**reduction** (between decision problems) a transformation of a decision problem $\mathcal{P}$ into a decision problem $\mathcal{Q}$ such that the positive (resp. negative) instances of $\mathcal{P}$ transform into the positive (resp. negative) instances of $\mathcal{Q}$

m-**reduction** (between sets) a mapping of a set $\mathcal{A}$ to a set $\mathcal{B}$ such that the members (resp. non-members) of $\mathcal{A}$ are mapped into the members (resp. non-members) of $\mathcal{B}$

N

negative instance (of a decision problem) an instance of the decision problem whose answer is NO

nondeterministic Turing machine a Turing machine whose program defines, for each pair (state, symbol), a finite set of alternative moves out of which the machine guesses the one leading to a successful termination of the computation, if such exists

nondiamond theorem the theorem stating that there exists no pair of c.e. degrees with greatest lower bound **0** (the Turing degree of decidable problems) and least upper bound **0′** (the Turing degree of the complete semi-decidable problems)

non-logical axiom *see* proper axiom

non-proper class a class that is also a set (being a set, such a class is a member of its power set, but this is a class by definition)

nonterminal (in a grammar) a symbol of the grammar that can be replaced by the right-hand side of some production of the grammar

normal system a canonical system whose productions are simplified in a particular way while retaining their generating power

O

object language a language which is the object of study, i.e., whose statements are described and analyzed in some metalanguage

$\mathcal{O}$**-computable function** a total function $\varphi : \mathcal{A} \to \mathcal{B}$ for which there exists an oracle Turing machine with oracle set $\mathcal{O}$ that can compute the value $\varphi(x)$ for any $x \in \mathcal{A}$

$\mathcal{O}$**-decidable set** a set whose characteristic function is $\mathcal{O}$-computable

$\mathcal{O}$**-incomputable function** a function for which there is no $\mathcal{O}$-TM that can compute the function's values everywhere the function is defined

$\mathcal{O}$**-p.c. function** *see* partial $\mathcal{O}$-computable function

oracle (for a set) a miraculous and unspecified means that can immediately decide, for any element, whether or not the element is in the set (the so-called oracle set)

oracle set a set whose membership problem is assumed to be decidable by an oracle

oracle Turing machine a Turing machine with specified oracle Turing program and unspecified oracle set (e.g., when the oracle set is yet to be specified or changed)

oracle Turing program a program in the Turing machine's control unit that determines the next move of the machine, based on the machine's current state, the last symbol read from the tape, *and* the oracle's answer to the program's current question

$\mathcal{O}$-**semi-decidable set** a set for which there is an $\mathcal{O}$-TM that can determine, for any element in the set, that the element *is* in the set; if in truth the element is not in the set, the machine determines that the element *is not* in the set—or *never halts*

$\mathcal{O}$-**TM** an oracle Turing machine with oracle set $\mathcal{O}$

$\mathcal{O}$-**undecidable set** a set for which there is no $\mathcal{O}$-TM capable of deciding, for every element, whether or not the element is in the set

P

padding lemma the lemma stating that a p.c. function has countably infinitely many indexes; and if one of them is given, then countably infinitely many others can be generated

pairing function a computable bijective function $f : \mathbb{N}^2 \to \mathbb{N}$ whose inverse functions g and h, defined by $f(g(n), h(n)) = n$, are computable

paradox an unacceptable conclusion derived by apparently acceptable reasoning from apparently acceptable premises

p.c. function *see* partial computable function

parameter theorem a theorem stating that the variables with fixed values (parameters) of a multi-variable p.c. function φ can always be incorporated into φ's program to obtain the program for the induced function φ' on the rest of the variables

partial function a function $\varphi : \mathcal{A} \to \mathcal{B}$ whose value $\varphi(x)$ may be undefined for $x \in \mathcal{A}$

partial computable (or **p.c.**) **function** a partial function for which there is a Turing machine that can compute the function's values anywhere the function is defined

partial $\mathcal{O}$-computable (or **$\mathcal{O}$-p.c.**) **function** a partial function for which there is an $\mathcal{O}$-TM that can compute the function's values anywhere the function is defined

Π_n-**complete set** a Π_n-set such that every Π_n-set is m-reducible to it

Π_n-**set** a set of exactly those natural numbers x for which the predicate of the form $\forall y_1 \exists y_2 \forall y_3 \ldots Q y_n R(x, y_1, y_2, \ldots, y_n)$ is true, where Q is $\forall$ ($\exists$) if n is odd (even), and R is a decidable arithmetical relation

Platonism (or **Platonic view**) the view that there are abstract mathematical objects (e.g., numbers, sets) that exist independently of us (our thought, language, practices); our statements about such objects are made true or false by the objects themselves; therefore, mathematical truths are discovered, not invented

positive instance (of a decision problem) an instance of the decision problem whose answer is YES

Post-computable function a function for which there exists a Post machine that can compute the function's values anywhere the function is defined

Post's correspondence problem the undecidable decision problem asking, given two finite lists $u_1, \ldots, u_n$ and $v_1, \ldots, v_n$ of words over an alphabet Σ, whether or not there is a sequence $i_1, \ldots, i_k$ of indexes such that $u_{i_1} \ldots u_{i_k} = v_{i_1} \ldots v_{i_k}$

Post machine a model of computation that has a control unit; a queue (for symbols) connected to the control unit; a read-only tape with cells containing symbols, a window movable to the right and connected to the control unit; and a Post program

Post program a directed graph in the control unit of the Post machine that directs the operation of the machine, i.e., each arc of the graph can trigger the instruction in the destination vertex, and instructions change the contents of the machine's queue

Post's problem the question whether there is a c.e. degree strictly between the Turing degrees $\mathbf{0}$ and $\mathbf{0}'$, i.e., whether there is a decision problem whose difficulty is strictly between that of the decidable and the complete undecidable problems

Post's program (for solving Post's problem) an attempt to solve the problem by using a structural property (namely sparseness of the complement) of c.e. sets that would guarantee the existence, undecidability, and incompleteness of such c.e. sets

Post's theorem the theorem stating that if a set and its complement are both semi-decidable then they are both decidable

Post's thesis the assertion that the intuitive notion of what (set) is generable is adequately formalized by the notion of generable by a Post normal system (or, equivalently, by any model of generation equivalent to the normal system)

prenex normal form (of a formula in first-order logic) a logically equivalent for-mula consisting of a string of quantifiers followed by a quantifier-free formula

primitive recursion a construction of a function f from two given functions and f itself in a restricted way; e.g., $f(n,0) \overset{\text{def}}{=} g(n)$ and $f(n,m+1) \overset{\text{def}}{=} h(n,m,f(n,m)), m \geqslant 0$

primitive recursive function a function that can be constructed from the zero function $\zeta(n) \overset{\text{def}}{=} 0$, the successor function $\sigma(n) \overset{\text{def}}{=} n+1$, and the projection func-tions $\pi_i^k(x_1, \ldots, x_k) \overset{\text{def}}{=} x_i$, using only function composition and primitive recursion

priority method a method for constructing a set that satisfies all requirements in an infinite list of requirements; the method organizes actions so that although previ-ously satisfied requirements can be injured, in the limit, all requirements are satisfied

production (in a grammar) a transformation rule of the form $\alpha \to \beta$ describing the conditions under which, and the manner in which, subwords of a word can be used to transform the word into a new word

projection function (k-place) the function $\pi_i^k(x_1, \ldots, x_k) = x_i$, which maps k-tuples to their ith components; informally, π_i^k extracts and returns the ith argument

proper axiom an axiom that condenses some fact about specific basic notion(s) typical of the current field of interest

proper class a class that is not a member of any class (and is, therefore, not a set; otherwise it would be a member of its power set, which is a class by definition)

proper function (of a Turing machine T) a partial function induced by T as follows: The function maps a word u to a word v *iff* T halts on u and leaves only v on the tape; otherwise the function is undefined for u

proper set (of a Turing machine T) a set induced by T as follows: A word is in the set *iff* T accepts the word, i.e., after reading it, T eventually halts in a final state

Q

quantum algorithm an algorithm that uses some essential feature of quantum computation and can run on an abstract computing machine that is an instance of a model of quantum computation (e.g., quantum Turing machine, quantum circuit)

quantum computing the computing that uses computational methods inspired by quantum-mechanical principles and phenomena, such as probabilistic universes, interference, superposition, and entanglement

quantum Turing machine a model of quantum computation that is based on the (ordinary) Turing machine and captures all of the power of quantum computation; any quantum algorithm can be expressed as a particular quantum Turing machine

R

random access machine (RAM) a computation model with several arithmetic registers and a potentially infinite number of memory registers

recognizer (of a set) a Turing machine that can determine, for any element in the set, that the element *is* in the set; if in truth the element is not in the set, the machine determines that the element *is not* in the set—or *never halts*

recursion (self-reference) the process of defining or expressing something (e.g., a function, procedure, language construct, or solution to a problem) in terms of itself

recursion (or **fixed-point**) **theorem** informally, the theorem stating that if a transformation modifies every Turing program, then some Turing program is transformed into an equivalent Turing program, i.e., a program with the same global behavior

recursive function *see* (general) recursive function

reduction (in λ-calculus) the operation that transforms a λ-term by applying one of its functions to the function's arguments and replacing the λ-term representing the function and its arguments with the λ-term representing the value of the function

reduction (of a computational problem) a strategy for solving a problem that transforms its input into the input for another problem, solves that problem on the transformed input, and transforms the solution into the solution to the original problem

relativistic machine a hypercomputer consisting of a universal Turing machine (stationed on Earth) and a Turing machine (approaching a black hole) that uses gravitational time dilation to compute the values of an incomputable function

Rice's theorem (for p.c. functions) the theorem stating that intrinsic properties of p.c. functions are decidable *iff* they are trivial

Rice's theorem (for c.e. sets) the theorem stating that intrinsic properties of c.e. sets are decidable *iff* they are trivial

Russell's paradox the unacceptable conclusion derived in Cantor's set theory stating that there exists a set that both is and is not a member of itself

S

search problem a computational problem whose each instance asks for the members of a given set that have a given property

second incompleteness theorem (Gödel's) the theorem stating that if the *Formal Arithmetic* **A** is consistent, then this cannot be proved with means available in **A**

semantically complete (theory) a consistent theory **F** such that a formula F is derivable in **F** *iff* F represents a *Truth* in **F**, i.e., F is valid in every model of **F**; informally, in such an **F**, *Truth* and nothing but the *Truth* can be derived

semantic completeness problem the problem of deciding whether or not a given theory is semantically complete

semi-decidable problem a decision problem for which there is a Turing machine that can determine, for any *positive* instance of the problem, that the answer to the instance is YES; if the instance is negative, the machine answers NO—or *never halts*

semi-decidable set a set for which there is a Turing machine that can determine, for any element in the set, that the element *is* in the set; if in truth the element is not in the set, the machine determines that the element *is not* in the set—or *never halts*

semi-Thue system (over an alphabet Σ) a finite set of rules (productions) of the form $x \rightarrow y$ (where x and y are words over Σ) used for investigating whether and how a word over Σ can be transformed, using only the given rules, into another word over Σ

simple set informally, a c.e. set whose complement is "sparse", in the sense that the complement, although infinite, does not contain any infinite c.e. set

s-m-n theorem *see* parameter theorem

Σ_n**-complete set** a Σ_n-set such that every Σ_n-set is m-reducible to it

Σ_n**-set** a set of exactly those natural numbers x for which the predicate of the form $\exists y_1 \forall y_2 \exists y_3 \ldots Q y_n R(x, y_1, y_2, \ldots, y_n)$ is true, where Q is $\exists$ ($\forall$) if n is odd (even), and R is a decidable arithmetical relation

solvable problem a problem, not necessarily a computational one, for which there exists a single procedure that can construct a solution to any instance of the problem

sound theory a theory in which every theorem is valid in every model of the theory; informally, a theory in which we cannot deduce something that is not a *Truth*

space complexity (of an algorithm) a function whose argument is the size of the algorithm's input and whose value represents the amount of computational space needed to execute the algorithm

start symbol (in a grammar) a designated nonterminal from which every word of the language (of the grammar) is generated

standard model (of a theory) a particular structure (field of interest) in which the theory is usually interpreted

syntactically complete (theory) a theory such that, for any formula F of the theory, at least one of the formulas F and $\neg F$ is derivable in the theory; so in a consistent and syntactically complete theory every formula is either provable or refutable

syntactic completeness problem the problem of deciding whether or not a given theory is syntactically complete

T

T-complete set *see* Turing-complete set

terminal (in a grammar) a symbol that cannot be replaced by other symbols

theorem (of a theory) a formula that can be derived in the theory; informally, a statement in mathematics or logic that can be proved by reasoning

theory a formal idea or set of ideas that is intended to explain something

thesis an idea that is expressed as a statement and is discussed in a logical way

Thue system a semi-Thue system where each rule $x \rightarrow y$ has the reverse rule $y \rightarrow x$

time complexity (of an algorithm) a function whose argument is the size of the algorithm's input and whose value represents the number of operations or the running time needed to execute the algorithm

total function a function $\varphi : \mathcal{A} \rightarrow \mathcal{B}$ whose value $\varphi(x)$ is defined for every $x \in \mathcal{A}$

transfinite induction a generalization of mathematical induction stating that in a well-ordered set $(\mathcal{S}, \preccurlyeq)$ a predicate P holds for every element of $\mathcal{S}$ *if* the following condition is met: P holds for $y \in \mathcal{S}$ if P holds for every $x \in \mathcal{S}$ such that $x \preccurlyeq y$

transition function (of a Turing machine) a partial function $\delta : (q, a) \mapsto (p, b, D)$ specifying the machine's local behavior: In each step, if the machine reads in state q symbol a, it writes b, moves the window one cell in direction D, and enters state p

trivial property (of p.c. functions) an intrinsic property of p.c. functions that is shared by either every or no p.c. function

trivial property (of c.e. sets) an intrinsic property of c.e. sets that is shared by either every or no c.e. set

Turing-complete (or T-**complete**) **set** a c.e. set such that every c.e. set is Turing-reducible to it

Turing-computable function a function for which there exists a Turing machine that can compute the value of the function anywhere the function is defined

Turing degree (of a set) an equivalence class consisting of all sets that are Turing-equivalent to the set

Turing equivalence (of sets) a relation between two sets stating that if either of the sets were decidable then also the other would be decidable

Turing jump (operator) an operator that maps sets to sets in such a way that the Turing degree of the image set is higher than that of the original set

Turing jump (of a set) the set resulting from a single application of the Turing jump operator to the given set

Turing jump (of a Turing degree) the Turing degree of the set which is the Turing jump of an arbitrary set in the given Turing degree

Turing machine a model of computation that has a control unit, which is always in some state; an infinite tape with cells, which may contain symbols; a movable window over the tape, which is connected to the control unit; and a Turing program

Turing program a partial function in the control unit of a Turing machine, which directs the operation of the machine, i.e., determines each next move of the machine based on the current state of the control unit and the symbol under the window

Turing reduction (between problems) a relation $\leq_T$ on computational problems such that $P \leq_T Q$ *iff* the existence of an algorithm A_Q for Q would imply the existence of an algorithm A_P for P (with A_P allowed to make finitely many calls to A_Q)

Turing reduction (between sets) a relation $\leq_T$ on sets such that $\mathcal{A} \leq_T \mathcal{B}$ *iff* the existence of a decider $D_\mathcal{B}$ for $\mathcal{B}$ would imply the existence of a decider $D_\mathcal{A}$ for $\mathcal{A}$ (with $D_\mathcal{A}$ allowed to make finitely many calls to $D_\mathcal{B}$)

U

unbounded minimization a construction of a function f from a given function g using the μ-operation; e.g., $f(n) \stackrel{\text{def}}{=} \mu x.g(n,x)$ defines $f(n)$ to be the smallest m such that $g(n,m) = 0$, if there is one; otherwise $f(n)$ is undefined

undecidable problem a decision problem for which there is no Turing machine capable of deciding, for every instance of the problem, whether the answer to the instance is YES or NO

undecidable set a set for which there is no Turing machine capable of deciding, for every element, whether or not the element is in the set

universal language the language of the *Halting Problem*; that is, the set $\mathcal{K}_0$ of the codes of all positive instances of the *Halting Problem*, i.e., $\mathcal{K}_0 = \{\langle T, x\rangle \mid T \text{ halts on } x\}$

universal Turing machine a Turing machine that can simulate execution of any other Turing machine on any input

V

valid formula (under an interpretation) a formula of a theory which, when interpreted in a field of interest, is a true statement about the elements of the field, for every assignment of elements of the field to the free variable symbols of the formula

valid formula (in a theory) a formula of the theory that is valid in each model of the theory; informally, such a formula represents a certain mathematical *Truth* expressible in the theory

W

word (over an alphabet) a finite sequence of symbols from the alphabet

word problem (for semi-groups) the undecidable decision problem asking, for any Thue system and any words u and v, whether or not u can be transformed into v in the system

word problem (for groups) the undecidable problem asking, for any Thue system such that every symbol a has an annihilating symbol b (i.e., $ba \to \varepsilon$ and $\varepsilon \to ba$ are rules) and any words u and v, whether or not u can be transformed into v

X

Y

Yes/No problem *see* decision problem

Z

References

1. Aaronson, S.: Quantum Computing Since Democritus. Cambridge University Press (2013)
2. Ackermann, W.: On Hilbert's Construction of the Real Numbers. Mathematische Annalen **99**, 118–133 (1928). (Reprint in: van Heijenoort, 1999)
3. Adams, R.: An Early History of Recursive Functions and Computability: From Gödel to Turing. Docent Press (2011)
4. Adleman, L.M.: Molecular Computation of Solutions to Combinatorial Problems. Science **266**(5187), 1021–1024 (1994)
5. Adleman, L.M.: Computing with DNA. Scientific American **279**(2), 54–61 (1998)
6. Adler, A.: Some Recursively Unsolvable Problems in Analysis. Proceedings of the American Mathematical Society **22**, 523–526 (1969)
7. Aharonov, D., Vazirani, U.V.: Is Quantum Mechanics Falsifiable? A Computational Perspective on the Foundations of Quantum Mechanics. In: B.J. Copeland, C.J. Posy, O. Shagrir (eds.) Computability: Turing, Gödel, Church, and Beyond, pp. 329–349. The MIT Press (2013)
8. Aho, A.V., Hopcroft, J.E., Ullman, J.D.: The Design and Analysis of Computer Algorithms, 3rd edn. Addison-Wesley (1974)
9. Aho, A.V., Lam, M.S., Sethi, R., Ullman, J.: Compilers: Principles, Techniques, and Tools, 2nd edn. Addison-Wesley (2006)
10. Ambos-Spies, K.: Polynomial Time Reducibilities and Degrees. In: E.R. Griffor (ed.) Handbook of Computability Theory, pp. 683–706. Elsevier/North-Holland (1999)
11. Ambos-Spies, K., Fejer, P.F.: Degrees of Unsolvability (2006). Unpublished paper. http://www.cs.umb.edu/~fejer/articles/History_of_Degrees.pdf
12. Baaz, M., Papadimitriou, C.H., Putnam, H.W., Scott, D.S., Harper, C.L. (eds.): Kurt Gödel and the Foundations of Mathematics: Horizons of Truth. Cambridge University Press (2011)
13. Barendregt, H.: The Lambda Calculus. Its Syntax and Semantics. College Publications (2012)
14. Barwise, J.: Handbook of Mathematical Logic, *Studies in Logic and the Foundations of Mathematics*, vol. 90. Elsevier/North-Holland (1977)
15. Ben-Amram, A.M.: The Church-Turing Thesis and its Look-Alikes. ACM SIGACT News **36**(3), 113–114 (2005). (There is an extended version from 2006.)
16. Benacerraf, P., Putnam, H. (eds.): Philosophy of Mathematics, 2nd edn. Cambridge University Press (1984)
17. Berger, R.: The Undecidability of the Domino Problem. Memoirs of the American Mathematical Society **66** (1966)
18. Berlekamp, E.R., Conway, J.H., Guy, R.K.: Winning Ways for Your Mathematical Plays, vol. 4, 2nd edn. Routledge (2004)
19. Bernstein, E., Vazirani, U.: Quantum Complexity Theory. SIAM Journal on Computing **26**(5) (1997)

R. Robič, *The Foundations of Computability Theory*, https://doi.org/10.1007/978-3-662-62421-0

20. Boole, G.: An Investigation of the Laws of Thought: On Which are Founded the Mathematical Theories of Logic and Probabilities. Macmillan (1854). (Cambridge University Press, reissue edn. 2009)

21. Boolos, G.S., Burgess, J.P., Jeffrey, R.C.: Computability and Logic, 4th edn. Cambridge University Press (2002)

22. Boolos, G.S., Jeffrey, R.C.: Computability and Logic, 3rd edn. Cambridge University Press (1989)

23. Boone, W.W.: The Word Problem. Annals of Mathematics **70**(2), 207–265 (1959)

24. Boothby, W.M.: An Introduction to Differentiable Manifolds and Riemannian Geometry, Revised 2nd edn. Academic Press (2003)

25. Bratley, P., Millo, J.: Computer Recreations: Self-Reproducing Programs. Software Practice and Experience **2**, 397–400 (1972)

26. Brouwer, L.E.J.: On the Significance of the Principle of Excluded Middle in Mathematics, Especially in Function Theory (1923). (Reprint in: van Heijenoort, 1999)

27. Burali-Forti, C.: Una questione sui numeri transfiniti *and* Sulle classi ben ordinate. Rendiconti Circ. Mat. Palermo **11**, 154–164, 260 (1897). (A Question on Transfinite Numbers *and* On Well-Ordered Classes. Reprint in: van Heijenoort, 1999)

28. Burger, J., Brill, D., Machi, F.: Self-Reproducing Programs. Byte **5**, 72–74 (1980)

29. Burgess, A.G., Burgess, J.P.: Truth. Princeton Foundations of Contemporary Philosophy. Princeton University Press (2011)

30. Cantor, G.: Über eine Eigenschaft des Inbegriffes aller reellen algebraischen Zahlen. J. Math. **77**, 258–262 (1874)

31. Carroll, S.: Spacetime and Geometry: An Introduction to General Relativity. Pearson (2014)

32. Chaitin, G.: Meta Maths: The Quest for Omega. Atlantic Books (2006)

33. Chang, C.C., Keisler, H.J.: Model Theory, 3rd edn. Dover Books on Mathematics. Dover Publications (2012)

34. Cheng, T.P.: Relativity, Gravitation and Cosmology: A Basic Introduction, 2nd edn. Oxford Master Series in Particle Physics, Astrophysics, and Cosmology. Oxford University Press (2010)

35. Church, A.: A Set of Postulates for the Foundation of Logic (second paper). Annals of Mathematics **34**, 839–864 (1933)

36. Church, A.: A Note on the Entscheidungsproblem. Journal of Symbolic Logic **1**, 40–41 (1936). (Reprint in: Davis, 2004)

37. Church, A.: An Unsolvable Problem of Elementary Number Theory. American Journal of Mathematics **58**, 345–363 (1936). (Reprint in: Davis, 2004)

38. Church, A.: Review of Turing [1936]. Journal of Symbolic Logic **2**(1), 42–43 (1937)

39. Cohen, P.: Set Theory and the Continuum Hypothesis. W.A. Benjamin (1966)

40. Cook, M.: Universality in Elementary Cellular Automata. Complex Systems **15**(1), 1–40 (2004)

41. Cook, S.A., Reckhow, R.A.: Time Bounded Random Access Machines. Journal of Computer and System Sciences **7**(4), 354–375 (1973)

42. Cooper, K., Torczon, L.: Engineering a Compiler, 2nd edn. Morgan Kaufmann (2011)

43. Cooper, S.B.: Computability Theory. Chapman & Hall/CRC Mathematics (2004)

44. Cooper, S.B., van Leeuwen, J. (eds.): Alan Turing: His Work and Impact. Elsevier (2013)

45. Copeland, B.J.: The Church-Turing Thesis. In: E.N. Zalta (ed.) The Stanford Encyclopedia of Philosophy, spring 2019 edn. Metaphysics Research Lab, Stanford University (2019). https://plato.stanford.edu/archives/spr2019/entries/church-turing/

46. Copeland, B.J., Posy, C.J., Shagrir, O. (eds.): Computability: Turing, Gödel, Church, and Beyond. The MIT Press (2013)

47. Copeland, B.J., Shagrir, O.: Do Accelerating Turing Machines Compute the Uncomputable? Minds and Machines **21**(2), 221–239 (2011)

48. Copeland, B.J., Shagrir, O.: Turing versus Gödel on Computability and the Mind. In: B.J. Copeland, C.J. Posy, O. Shagrir (eds.) Computability: Turing, Gödel, Church, and Beyond, pp. 1–33. The MIT Press (2013)

49. Copeland, B.J., Shagrir, O.: The Church-Turing Thesis: Logical Limit or Breachable Barrier? Communications of the ACM **62**(1), 66–74 (2019)
50. Cubitt, T.S., Perez-Garcia, D., Wolf, M.M.: Undecidability of the Spectral Gap. Nature **528**(7581), 207–211 (2015)
51. Cutland, N.J.: Computability: An Introduction to Recursive Function Theory. Cambridge University Press (1980)
52. Davis, M.: On the Theory of Recursive Undecidability. Ph.D. thesis, Princeton University (1950)
53. Davis, M.: Hilbert's Tenth Problem Is Unsolvable. The American Mathematical Monthly **80**(3), 233–269 (1973)
54. Davis, M.: Computability and Unsolvability. Dover (1982). (Originally: McGraw-Hill, 1958)
55. Davis, M.: Why Gödel Didn't Have Church's Thesis. Information and Control **54**, 3–24 (1982)
56. Davis, M.: The Myth of Hypercomputation. In: C. Teuscher (ed.) Alan Turing: Life and Legacy of a Great Thinker. Springer (2004)
57. Davis, M.: The Undecidable. Dover (2004)
58. Davis, M.: Turing Reducibility. Notices of the AMS **53**(10), 1218–1219 (2006)
59. Davis, M., Sigal, R., Weyuker, E.J.: Computability, Complexity, and Languages, Fundamentals of Theoretical Computer Science, 2nd edn. Morgan Kaufmann (1994)
60. Dershowitz, N., Gurevich, Y.: A Natural Axiomatization of Computability and Proof of Church's Thesis. The Bulletin of Symbolic Logic **14**(3), 299–350 (2008)
61. Deutsch, D.: Quantum Theory, the Church-Turing Principle and the Universal Quantum Computer. Proceedings of the Royal Society of London A: Mathematical, Physical and Engineering Sciences **400**(1818), 97–117 (1985)
62. Doner, J., Hodges, W.: Alfred Tarski and Decidable Theories. The Journal of Symbolic Logic **53**(1), 20–30 (1988)
63. Downey, R.G., Hirschfeldt, D.R.: Algorithmic Randomness and Complexity. Theory and Applications of Computability. Springer (2010)
64. Dummit, D.S., Foote, R.M.: Abstract Algebra. Prentice Hall (1991)
65. Ebbinghaus, H.D., Flum, J., Thomas, W.: Mathematical Logic, 2nd edn. Undergraduate Texts in Mathematics. Springer (1994)
66. Eisert, J., Müller, M.P., Gogolin, C.: Quantum Measurement Occurrence is Undecidable. Physical Review Letters **108**(26), 1–5 (2012)
67. Enderton, H.B.: Computability Theory: An Introduction to Recursion Theory. Academic Press (2011)
68. Epstein, R.L., Carnielli, W.A.: Computability: Computable Functions, Logic, and the Foundations of Mathematics, 3rd edn. Advanced Reasoning Forum (2008)
69. Etzion-Petruschka, Y., Harel, D., Myers, D.: On the Solvability of Domino Snake Problems. Theoret. Comput. Sci. **131**, 243–269 (1994)
70. Fernández, M.: Models of Computation: An Introduction to Computability Theory. Undergraduate Topics in Computer Science. Springer (2009)
71. Feynman, R.P.: Feynman Lectures on Computation. Penguin Books (1999)
72. Fitting, M.: Notes on the Mathematical Aspects of Kripke's Theory of Truth. Notre Dame Journal of Formal Logic **27**(1), 75–88 (1986)
73. Floyd, R.W., Beigel, R.: The Language of Machines: An Introduction to Computability and Formal Languages. Computer Science Press (1994)
74. Frege, G.: Begriffsschrift, eine der arithmetischen nachgebildete Formelsprache des reinen Denkens. Halle A/S, Louis Nebert, Halle (1879). (Begriffsschrift, a Formula Language, Modeled upon that of Arithmetic, for Pure Thought. Reprint in: van Heijenoort, 1999)
75. Friedberg, R.M.: Two Recursively Enumerable Sets of Incomparable Degrees of Unsolvability. In: Proc. Nat'l Acad. Sci. USA, vol. 43, pp. 236–238 (1957)
76. Gandy, R.: The Confluence of Ideas in 1936. In: R. Herken (ed.) The Universal Turing Machine: A Half-Century Survey, pp. 55–111. Oxford University Press (1988)
77. Gardner, M.: Mathematical Games – The Fantastic Combinations of John Conway's New Solitaire Game Life. Scientific American **223**(4), 120–123 (1970)

78. George, A., Velleman, D.J.: Philosophies of Mathematics. Wiley-Blackwell (2001)
79. Ginsburg, S.: Algebraic and Automata-Theoretic Properties of Formal Languages, *Fundamental Studies in Computer Science*, vol. 2. North-Holland (1975)
80. Gödel, K.: Die Vollständigkeit der Axiome des logischen Funktionenkalküls. Monatsh. Math. Physik **37**, 349–360 (1930). (The Completeness of the Axioms of the Functional Calculus of Logic. Reprint in: van Heijenoort, 1999)
81. Gödel, K.: Über formal unentscheidbare Sätze der Principia Mathematica und verwandter Systeme I. Monatsh. Math. Physik **38**, 173–198 (1931). (On Formally Undecidable Propositions of Principia Mathematica and Related Systems I. Reprint in: van Heijenoort, 1999)
82. Gödel, K.: On Undecidable Propositions of Formal Mathematical Systems (1934). (Lectures at Institute for Advanced Study, Princeton. Reprint in: Davis, 2004)
83. Gödel, K.: Remarks Before the Princeton Bicentennial Conference on Problems in Mathematics (1946). (Reprint in: Davis, 2004)
84. Gödel, K.: On Formally Undecidable Propositions of Principia Mathematica and Related Systems. Dover Publications (1992). (Republication with an introduction by R. B. Braithwaite)
85. Gödel, K.: Some Basic Theorems on the Foundations of Mathematics and their Philosophical Implications (1951). In: S. Feferman, J.W. Dawson, Jr., W. Goldfarb, C. Parsons, R.M. Solovay (eds.) Kurt Gödel, Collected Works, vol. III: Unpublished Essays and Lectures, pp. 304–323. Oxford University Press (1995)
86. Gödel, K.: The Present Situation in the Foundations of Mathematics (1933). In: S. Feferman, J.W. Dawson, Jr., W. Goldfarb, C. Parsons, R.M. Solovay (eds.) Kurt Gödel, Collected Works, vol. III: Unpublished Essays and Lectures, pp. 45–53. Oxford University Press (1995)
87. Gödel, K.: Undecidable Diophantine Propositions (193?, unpublished). In: S. Feferman, J.W. Dawson, Jr., W. Goldfarb, C. Parsons, R.M. Solovay (eds.) Kurt Gödel, Collected Works, vol. III: Unpublished Essays and Lectures, pp. 164–175. Oxford University Press (1995)
88. Goldrei, D.: Classic Set Theory. Chapman & Hall/CRC Mathematics (1996)
89. Goldreich, O.: Computational Complexity: A Conceptual Perspective. Cambridge University Press (2008)
90. Goldreich, O.: P, NP, and NP-Completeness: The Basics of Computational Complexity. Cambridge University Press (2010)
91. Griffor, E.R. (ed.): Handbook of Computability Theory, *Studies in Logic and the Foundations of Mathematics*, vol. 140. Elsevier/North-Holland (1999)
92. Guidry, M.: Modern General Relativity: Black Holes, Gravitational Waves, and Cosmology. Cambridge University Press (2019)
93. Halava, V., Harju, T., Hirvensalo, M.: Undecidability Bounds for Integer Matrices Using Claus Instances. International Journal of Foundations of Computer Science **18**(5), 931–948 (2007)
94. Halbach, V.: Axiomatic Theories of Truth. Cambridge University Press (2011)
95. Halmos, P.R.: Naive Set Theory. Van Nostrand (1960)
96. Harel, D., Feldman, Y.: Algorithmics: The Spirit of Computing, 3rd edn. Pearson (2004)
97. Hartle, J.B.: Gravity: An Introduction to Einstein's General Relativity. Addison-Wesley (2003)
98. Herbrand, J.: Sur la non-contradiction de l'arithmetique. Journal für die reine und angewandte Mathematik **166**, 1–8 (1932)
99. Herken, R. (ed.): The Universal Turing Machine: A Half-Century Survey, 2nd edn. Springer (1995)
100. Hermes, H.: Enumerability, Decidability, Computability: An Introduction to the Theory of Recursive Functions, 2nd edn. Springer (1969)
101. Heyting, A.: Intuitionism: An Introduction, 3rd edn. *Studies in Logic and the Foundations of Mathematics*. Elsevier/North-Holland (1971)
102. Hilbert, D.: Über die Grundlagen der Logik und der Arithmetik. Verhandlungen des Dritten Internationalen Mathematiker-Kongress in Heidelberg, 8.-13. August 1904 pp. 174–185 (1905). (On the Foundations of Logic and Arithmetic. Reprint in: van Heijenoort, 1999)

103. Hilbert, D.: Grundlagen der Geometrie, 7th edn. Teubner-Verlag, Leipzig, Berlin (1930)
104. Hilbert, D., Ackermann, W.: Grundzüge der theoretischen Logik. Springer (1928). (*Principles of Mathematical Logic*, Chelsea, 1950)
105. Hindley, J.R., Seldin, J.P.: Lambda-Calculus and Combinatory Logic: An Introduction, 2nd edn. Cambridge University Press (2008)
106. Hinman, P.G.: Fundamentals of Mathematical Logic. A K Peters (2005)
107. Hodges, A.: Alan Turing: The Enigma. Vintage Books (1992)
108. Hodges, A.: Did Church and Turing Have a Thesis about Machines? In: A. Olszewski, J. Woleński, R. Janusz (eds.) Church's Thesis After 70 Years, Ontos Mathematical Logic (Book 1), pp. 242–252. De Gruyter (2006)
109. Hodges, W.: A Shorter Model Theory. Cambridge University Press (1997)
110. Hodges, W.: Model Theory. Cambridge University Press (2008)
111. Holmes, M.R., Forster, T., Libert, T.: Alternative Set Theories. In: D.M. Gabbay, A. Kanamori, J. Woods (eds.) Handbook of the History of Logic, vol. 6. Sets and Extensions in the Twentieth Century, pp. 559–632. Elsevier (2012)
112. Homer, S., Selman, A.L.: Computability and Complexity Theory, 2nd edn. Springer (2011)
113. Hopcroft, J.E., Ullman, J.D.: Introduction to Automata Theory, Languages, and Computation. Addison-Wesley (1979)
114. Irvine, A.D.: Bertrand Russell's Logic. In: D.M. Gabbay, J. Woods (eds.) Handbook of the History of Logic, vol. 5. Logic from Russell to Church, pp. 1–28. Elsevier (2009)
115. Jacobson, N.: Basic Algebra I, 2nd edn. Dover Publications (2009)
116. Jech, T.J.: The Axiom of Choice, *Studies in Logic and the Foundations of Mathematics*, vol. 75. Elsevier/North-Holland (1973)
117. Jech, T.J.: Set Theory, The Third Millennium, revised and expanded edn. Springer Monographs in Mathematics. Springer (2011)
118. Johnsonbaugh, R., Pfaffenberger, W.E.: Foundations of Mathematical Analysis. Dover Books on Mathematics. Dover Publications (2010)
119. Kalmár, L.: An Argument Against the Plausibility of Church's Thesis. In: A. Heyting (ed.) Constructivity in Mathematics, pp. 72–80. North-Holland (1959)
120. Kanamori, A.: Set Theory from Cantor to Cohen. In: D.M. Gabbay, A. Kanamori, J. Woods (eds.) Handbook of the History of Logic, vol. 6. Sets and Extensions in the Twentieth Century, pp. 1–72. Elsevier (2012)
121. Kleene, S.C.: General Recursive Functions of Natural Numbers. Mathematische Annalen **112**(5), 727–742 (1936). (Reprint in: Davis, 2004)
122. Kleene, S.C.: On Notations for Ordinal Numbers. Journal of Symbolic Logic **3**, 150–155 (1938)
123. Kleene, S.C.: Recursive Predicates and Quantifiers. Transactions of the American Mathematical Society **53**, 41–73 (1943)
124. Kleene, S.C.: Introduction to Metamathematics. Elsevier/North-Holland (1952)
125. Kleene, S.C.: Mathematical Logic. Wiley (1967)
126. Kleene, S.C.: Origins of Recursive Function Theory. Annals of the History of Computing **3**(1), 52–67 (1981)
127. Kleene, S.C., Post, E.: The Upper Semi-lattice of Degrees of Recursive Unsolvability. Annals of Mathematics **59**(3), 379–407 (1954)
128. Kozen, D.C.: Automata and Computability. Springer (1997)
129. Kozen, D.C.: Theory of Computation. Springer (2006)
130. Kripke, S.: Outline of a Theory of Truth. The Journal of Philosophy **72**, 690–716 (1975)
131. Kripke, S.A.: The Church-Turing "Thesis" as a Special Corollary of Gödel's Completeness Theorem. In: B.J. Copeland, C.J. Posy, O. Shagrir (eds.) Computability: Turing, Gödel, Church, and Beyond, pp. 77–104. The MIT Press (2013)
132. Kunen, K.: Set Theory, An Introduction to Independence Proofs, *Studies in Logic and the Foundations of Mathematics*, vol. 102. Elsevier (1980)
133. Kunen, K.: The Foundations of Mathematics. Studies in Logic. College Publications (2009)

134. Kučera, A.: An Alternative, Priority-Free, Solution to Post's Problem. In: J. Gruska, B. Rovan, J. Wiedermann (eds.) Proc. Mathematical Foundations of Computer Science (Bratislava, 1986), *Lecture Notes in Computer Science*, vol. 233, pp. 493–500. Springer (1986)

135. Lachlan, A.H.: The Impossibility of Finding Relative Complements for Recursively Enumerable Degrees. The Journal of Symbolic Logic **31**(3), 434–454 (1966)

136. Lachlan, A.H.: Lower Bounds for Pairs of Recursively Enumerable Degrees. Proceedings of the London Mathematical Society **16**(3), 537–569 (1966)

137. Lambourne, R.J.A.: Relativity, Gravitation and Cosmology. Cambridge University Press (2010)

138. Lee, J.M.: Introduction to Topological Manifolds, 2nd edn. Graduate Texts in Mathematics. Springer (2011)

139. Lerman, M.: Degrees of Unsolvability: Local and Global Theory. Perspectives in Mathematical Logic. Springer (1983)

140. Lerman, M.: A Framework for Priority Arguments, *Lecture Notes in Logic*, vol. 34. Cambridge University Press (2010)

141. Levy, A.: Basic Set Theory. Dover Publications (2002)

142. Lewis, H.R., Papadimitriou, C.H.: Elements of the Theory of Computation. Prentice Hall (1981)

143. Machover, M.: Set Theory, Logic and Their Limitations. Cambridge University Press (1996)

144. Machtey, M., Young, P.: An Introduction to the General Theory of Algorithms. Elsevier/North-Holland (1978)

145. Manin, Y.I.: A Course in Mathematical Logic. Graduate Texts in Mathematics. Springer (1977)

146. Manna, Z.: Mathematical Theory of Computation. McGraw-Hill (1974)

147. Markov, A.A.: The Theory of Algorithms. Trudy Math. Inst. Steklova **38**, 176–189 (1951). (A.M.S. transl. 15:1–14,1960)

148. Markov, A.A.: The Theory of Algorithms. Trudy Math. Inst. Steklova **42**, 3–375 (1954)

149. Markov, A.A.: Unsolvability of the Problem of Homeomorphy. Proc. Int. Cong. Math. pp. 300–306 (1958). (Russian)

150. Massey, W.S.: Algebraic Topology: An Introduction. Graduate Texts in Mathematics. Springer (1990)

151. Matiyasevič, Y.V.: Enumerable Sets are Diophantine. Dokl. Akad. Nauk SSSR **191**, 279–282 (1970). (Russian. English translation in: Soviet Math. Doklady, 11:354–357, 1970)

152. Matiyasevič, Y.V.: Diophantine Representation of Enumerable Predicates. Izv. Akad. Nauk SSSR **Ser. Mat. 35**, 3–30 (1971). (Russian)

153. McNulty, G.F.: Alfred Tarski and Undecidable Theories. The Journal of Symbolic Logic **51**(4), 890–898 (1986)

154. Mendelson, E.: On Some Recent Criticism of Church's Thesis. Notre Dame Journal of Formal Logic **IV**(3), 201–205 (1963)

155. Mendelson, E.: Introduction to Mathematical Logic. The University Series in Undergraduate Mathematics. Van Nostrand (1964). (6th edn. Chapman & Hall/CRC Press (2015))

156. Mendelson, E.: Second Thoughts about Church's Thesis and Mathematical Proofs. The Journal of Philosophy **87**(5), 225–233 (1990)

157. Mendelson, E.: On the Impossibility of Proving the "Hard-Half" of Church's Thesis. In: A. Olszewski, J. Woleński, R. Janusz (eds.) Church's Thesis After 70 Years, Ontos Mathematical Logic (Book 1), pp. 304–309. De Gruyter (2006)

158. Moore, G.H.: Zermelo's Axiom of Choice: Its Origins, Development & Influence. Dover Publications (2013)

159. Moschovakis, J.R.: The Logic of Brouwer and Heyting. In: D.M. Gabbay, J. Woods (eds.) Handbook of the History of Logic, vol. 5. Logic from Russell to Church, pp. 77–126. Elsevier (2009)

160. Moschovakis, Y.N.: On Founding the Theory of Algorithms. In: H.G. Dales, G. Oliveri (eds.) Truth in Mathematics, pp. 71–104. Clarendon Press (1998)

161. Moschovakis, Y.N.: What is an Algorithm? In: B. Engquist, W. Schmid (eds.) Mathematics Unlimited – 2001 and Beyond, pp. 919–936. Springer (2001)

162. Moschovakis, Y.N., Paschalis, V.: Elementary Algorithms and Their Implementations. In: S.B. Cooper, B. Löwe, A. Sorbi (eds.) New Computational Paradigms: Changing Conceptions of What is Computable. Springer (2008)

163. Mostowski, A.: On Definable Sets of Positive Integers. Fundamenta Mathematicae **34**, 81–112 (1947)

164. Mostowski, A.: A Classification of Logical Systems. Studia Philosophica **4**, 237–274 (1951)

165. Muchnik, A.A.: On the Unsolvability of the Problem of Reducibility in the Theory of Algorithms. Dokl. Akad. Nauk SSSR **108**, 194–197 (1956). (Russian)

166. Muchnik, A.A.: Solution of Post's Reduction Problem and of Certain Other Problems in the Theory of Algorithms. Trudy Moskov. Mat. Obšč. **7**, 391–405 (1958)

167. Murawski, R.: E.L. Post and the Development of Mathematical Logic and Recursion Theory. Studies in Logic, Grammar and Rhetoric **2**(15), 17–30 (1998)

168. Nagel, E., Newman, J.R.: Gödel's Proof. Routledge (1958)

169. Németi, I., Dávid, G.: Relativistic Computers and the Turing Barrier. Journal of Applied Mathematics and Computation **178**(1), 118–142 (2006)

170. Nies, A.: Computability and Randomness. Oxford Logic Guides. Oxford University Press (2009)

171. Novikov, P.S.: On the Algorithmic Unsolvability of the Word Problem in Group Theory. Izvestiya Akademii Nauk **18**, 485–524 (1954). (Translated in: A.M.S. transl 9:1–124,1958)

172. Odifreddi, P.: Reducibilities. In: E.R. Griffor (ed.) Handbook of Computability Theory, pp. 89–120. Elsevier/North-Holland (1999)

173. Odifreddi, P.G.: Classical Recursion Theory, *Studies in Logic and the Foundations of Mathematics*, vol. 125. Elsevier/North-Holland (1989)

174. Peano, G.: Arithmetices Principia, Novo Methodo Exposita. Fratres Bocca, Torino (1889). (The Principles of Arithmetic, Presented by a New Method. Reprint in: van Heijenoort, 1999)

175. Peirce, C.S.: On the Algebra of Logic: A Contribution to the Philosophy of Notation. American Journal of Mathematics **7**, 180–196, 197–202 (1885)

176. Péter, R.: Recursive Functions, 3rd edn. Academic Press (1967)

177. Pettorossi, A.: Elements of Computability, Decidability and Complexity, 3rd edn. Aracne (2009)

178. Piccinini, G.: The Physical Church-Turing Thesis: Modest or Bold? The British Journal for the Philosophy of Science **62**(4), 733–769 (2011)

179. Pinter, C.C.: A Book of Abstract Algebra, 2nd edn. Dover Publications (2010)

180. Post, E.: Introduction to a General Theory of Elementary Propositions. American Journal of Mathematics **43**, 163–185 (1921). (Reprint in: van Heijenoort, 1999)

181. Post, E.: Finite Combinatory Processes – Formulation I. Journal of Symbolic Logic **1**, 103–105 (1936). (Reprint in: Davis, 2004)

182. Post, E.: Absolutely Unsolvable Problems and Relatively Undecidable Propositions: Account of an Anticipation (1941). (Submitted for publ. and rejected. Reprint in: Davis, 2004)

183. Post, E.: Formal Reductions of the General Combinatorial Decision Problem. American Journal of Mathematics **65**, 197–215 (1943)

184. Post, E.: Recursively Enumerable Sets of Positive Integers and Their Decision Problems. Bulletin of the American Mathematical Society **50**, 284–316 (1944). (Reprint in: Davis, 2004)

185. Post, E.: A Variant of a Recursively Unsolvable Problem. Bulletin of the American Mathematical Society **52**, 264–268 (1946)

186. Post, E.: Recursive Unsolvability of a Problem of Thue. Journal of Symbolic Logic **12**, 1–11 (1947). (Reprint in: Davis, 2004)

187. Post, E.: Degrees of Recursive Unsolvability. (Preliminary report). Bulletin of the American Mathematical Society **54**, 641–642 (1948). Abstract

188. Potter, M.: Set Theory and its Philosophy. Oxford University Press (2004)

189. Presburger, M.: Über die Vollständigkeit eines gewissen Systems der Arithmetik ganzer Zahlen, in welchem die Addition als einzige Operation hervortritt. Comptes Rendus du Premier Congrès des Mathématiciens des Pays Slaves pp. 92–101 (1929). Warsaw, Poland

190. Priest, G.: The Structure of the Paradoxes of Self-Reference. Mind **103**, 25–34 (1994)

191. Priest, G.: Paraconsistent Logic. In: D.M. Gabbay, F. Guenthner (eds.) Handbook of Philosophical Logic, vol. 6., 2nd edn., pp. 287–393. Kluwer Academic Publishers (2002)

192. Prijatelj, N.: Matematične strukture I., II., III. del. DMFA, Ljubljana (1985). (*Mathematical Structures, vol. I, II, III.* In Slovenian)

193. Prijatelj, N.: Osnove matematične logike, 1., 2., 3. del. DMFA, Ljubljana (1994). (*Fundamentals of Mathematical Logic, vol. 1, 2, 3.* In Slovenian)

194. Putnam, H.: Craig's Theorem. The Journal of Philosophy **62**(10), 251–260 (1965)

195. Radó, T.: On Non-Computable Functions. The Bell System Technical Journal **41**(3), 877–884 (1962)

196. Rautenberg, W.: A Concise Introduction to Mathematical Logic, 3rd edn. Universitext. Springer (2009)

197. Rice, H.G.: Classes of Recursively Enumerable Sets and Their Decision Problems. Transactions of the American Mathematical Society **74**(2), 358–366 (1953)

198. Rich, E.: Automata, Computability, and Complexity: Theory and Applications. Pearson Prentice Hall (2008)

199. Richardson, D.: Some Undecidable Problems Involving Elementary Functions of a Real Variable. The Journal of Symbolic Logic **33**(4), 514–520 (1968)

200. Robbin, J.W.: Mathematical Logic: A First Course. W.A. Benjamin (1969)

201. Robinson, J.: Definability and Decision Problems in Arithmetic. The Journal of Symbolic Logic **14**(2), 98–114 (1949)

202. Rogers, H.: Theory of Recursive Functions and Effective Computability. McGraw-Hill (1967)

203. Rogozhin, Y.: Small Universal Turing Machines. Theoretical Computer Science **168**(2), 215–240 (1996)

204. Rosenberg, A.L.: The Pillars of Computation Theory: State, Encoding, Nondeterminism. Universitext. Springer (2009)

205. Rozenberg, G., Salomaa, A. (eds.): Handbook of Formal Languages, vol. 1 Word, Language, Grammar. Springer (2012)

206. Rudin, W.: Principles of Mathematical Analysis, 3rd edn. McGraw-Hill (1976)

207. Russell, B.: Letter to Frege (1902). (Reprint in: van Heijenoort, 1999)

208. Sacks, G.E.: A Minimal Degree Less than $0'$. Bulletin of the American Mathematical Society **67**, 416–419 (1961)

209. Sacks, G.E.: Degrees of Unsolvability, *Ann. Math. Stud.*, vol. 55, 2nd edn. Princeton University Press (1963)

210. Sacks, G.E.: The Recursively Enumerable Degrees Are Dense. Annals of Mathematics **80**(2), 300–312 (1964)

211. Sainsbury, R.M.: Paradoxes, 3rd edn. Cambridge University Press (2009)

212. Schutz, B.: Gravity from the Ground Up: An Introductory Guide to Gravity and General Relativity. Cambridge University Press (2013)

213. Shagrir, O.: Advertisement for the Philosophy of the Computational Sciences. In: P. Humphreys (ed.) The Oxford Handbook of Philosophy of Science, pp. 15–42. Oxford University Press (2016)

214. Shannon, C.E.: A Universal Turing Machine with Two Internal States. In: C.E. Shannon, J. McCarthy (eds.) Automata Studies, *Ann. Math. Stud.*, vol. 34, pp. 157–165. Princeton University Press (1956)

215. Shen, A., Vereshchagin, N.K.: Computable Functions, *Student Mathematical Library*, vol. 19. American Mathematical Society (2003)

216. Shoenfield, J.R.: Mathematical Logic. Addison-Wesley (1967)

217. Shoenfield, J.R.: Recursion Theory, *Lecture Notes in Logic*, vol. 1. CRC Press (1993)

218. Shoenfield, J.R.: The Mathematical Work of S. C. Kleene. The Bulletin of Symbolic Logic **1**(1), 9–43 (1995)

219. Shore, R.A.: The Structure of the Degrees of Unsolvability. In: A. Nerode, R.A. Shore (eds.) Recursion Theory, pp. 33–51. American Mathematical Society (1985). (*Proceedings of Symposia in Pure Mathematics*, vol. 42)

220. Shore, R.A.: The Recursively Enumerable Degrees. In: E.R. Griffor (ed.) Handbook of Computability Theory, pp. 169–197. Elsevier/North-Holland (1999)

221. Sieg, W.: Mechanical Procedures and Mathematical Experience. In: A. George (ed.) Mathematics and Mind, pp. 71–117. Oxford University Press (1994)

222. Sieg, W.: Calculations by Man and Machine: Conceptual Analysis. In: W. Sieg, R. Sommer, C. Talcott (eds.) Reflections on the Foundations of Mathematics. Essays in Honor of Solomon Feferman, *Lecture Notes in Logic*, vol. 15, pp. 396–415. Association for Symbolic Logic (2002)

223. Sieg, W.: Gödel on Computing. Philosophia Mathematica **14**(2), 189–207 (2006)

224. Sieg, W.: On Mind and Turing's Machines. Natural Computing **6**(2), 187–205 (2007)

225. Sieg, W.: Church Without Dogma: Axioms for Computability. In: S.B. Cooper, B. Löwe, A. Sorbi (eds.) New Computational Paradigms, pp. 139–152. Springer (2008)

226. Sieg, W.: On Computability. In: D.M. Gabbay, P. Thagard, J. Woods, A. Irvine (eds.) Handbook of the Philosophy of Science, vol. Philosophy of Mathematics, pp. 525–621. Elsevier (2008)

227. Sieg, W.: Hilbert's Proof Theory. In: D.M. Gabbay, J. Woods (eds.) Handbook of the History of Logic, vol. 5. Logic from Russell to Church, pp. 321–384. Elsevier (2009)

228. Sieg, W.: Axioms for Computability: Do They Allow a Proof of Church's Thesis? In: H. Zenil (ed.) A Computable Universe: Understanding and Exploring Nature as Computation, pp. 99–123. World Scientific (2012)

229. Sieg, W.: Gödel's Philosophical Challenge (to Turing). In: B.J. Copeland, C.J. Posy, O. Shagrir (eds.) Computability: Turing, Gödel, Church, and Beyond, pp. 183–202. The MIT Press (2013)

230. Sieg, W.: Hilbert's Programs and Beyond. Oxford University Press (2013)

231. Simovici, D.A., Tenney, R.L.: Theory of Formal Languages with Applications. World Scientific (1999)

232. Simpson, S.G.: Degrees of Unsolvability: A Survey of Results. In: J. Barwise (ed.) Handbook of Mathematical Logic, *Studies in Logic and the Foundations of Mathematics*, vol. 90, pp. 631–652. Elsevier/North-Holland (1977)

233. Sipper, M.: Fifty Years of Research on Self-Replication: An Overview. Artificial Life **4**(3), 237–257 (1998)

234. Sipser, M.: Introduction to the Theory of Computation, 2nd edn. Thompson Course Technology (2006)

235. Slaman, T.A.: The Global Structure of the Turing Degrees. In: E.R. Griffor (ed.) Handbook of Computability Theory, pp. 155–168. Elsevier/North-Holland (1999)

236. Smith, P.: An Introduction to Gödel's Theorems. Cambridge Introductions to Philosophy. Cambridge University Press (2007)

237. Smullyan, R.M.: Gödel's Incompleteness Theorems. Oxford Logic Guides. Oxford University Press (1992)

238. Smullyan, R.M.: First-Order Logic. Dover Books on Mathematics. Dover Publications (1995)

239. Smullyan, R.M.: A Beginner's Guide to Mathematical Logic. Dover Books on Mathematics. Dover Publications (2014)

240. Smullyan, R.M., Fitting, M.: Set Theory and the Continuum Hypothesis. Dover Books on Mathematics. Dover Publications (2010)

241. Soare, R.I.: Recursively Enumerable Sets and Degrees. Perspectives in Mathematical Logic. Springer (1987)

242. Soare, R.I.: The History and Concept of Computability. In: E.R. Griffor (ed.) Handbook of Computability Theory, pp. 3–36. Elsevier/North-Holland (1999)

243. Soare, R.I.: An Overview of the Computably Enumerable Sets. In: E.R. Griffor (ed.) Handbook of Computability Theory, pp. 199–248. Elsevier/North-Holland (1999)

244. Soare, R.I.: Turing Oracle Machines, Online Computing, and Three Displacements in Computabiliy Theory. Annals of Pure and Applied Logic **160**, 368–399 (2009)
245. Soare, R.I.: Interactive Computing and Relativized Computability. In: B.J. Copeland, C.J. Posy, O. Shagrir (eds.) Computability: Turing, Gödel, Church, and Beyond, pp. 203–260. The MIT Press (2013)
246. Soare, R.I.: Turing Computability: Theory and Applications. Theory and Applications of Computability. Springer (2016)
247. Spector, C.: On Degrees of Recursive Unsolvability. Annals of Mathematics **64**(2), 581–592 (1956)
248. Spivak, M.: Calculus, 4th edn. Publish or Perish (2008)
249. Steprāns, J.: History of the Continuum in the 20th Century. In: D.M. Gabbay, A. Kanamori, J. Woods (eds.) Handbook of the History of Logic, vol. 6. Sets and Extensions in the Twentieth Century, pp. 73–144. Elsevier (2012)
250. Stewart, I.: Concepts of Modern Mathematics. Dover Publications (1995)
251. Stillwell, J.: Classical Topology and Combinatorial Group Theory, 2nd edn. Graduate Texts in Mathematics. Springer (1993)
252. Stoll, R.R.: Set Theory and Logic. Dover Publications (1979)
253. Suppes, P.: Axiomatic Set Theory. Dover Publications (1972)
254. Tarski, A.: Der Wahrheitsbegriff in die formalisierten Sprachen. Studia Philosophica **1**, 261–405 (1935). (The Concept of Truth in Formalized Languages. Translation in: Tarski, 1983, pp. 152–278)
255. Tarski, A.: A Decision Method for Elementary Algebra and Geometry. University of California Press (1951)
256. Tarski, A.: Truth and Proof. Scientific American **220**(6), 63–70, 75–77 (1969)
257. Tarski, A.: Logic, Semantics, Metamathematics: Papers from 1923 to 1938, 2nd edn. Hackett (1983)
258. Tarski, A.: Introduction to Logic and the Methodology of Deductive Sciences. Dover (1995)
259. Tarski, A., Mostowski, A., Robinson, R.M.: Undecidable Theories. North-Holland (1953)
260. Tent, K., Ziegler, M.: A Course in Model Theory, *Lecture Notes in Logic*, vol. 40. Cambridge University Press (2012)
261. Tu, L.W.: An Introduction to Manifolds, 2nd edn. Universitext. Springer (2010)
262. Turing, A.M.: On Computable Numbers, with an Application to the Entscheidungsproblem. Proceedings of the London Mathematical Society **42**(2), 230–265 (1936). (A.M. Turing, A Correction, ibid. 43:544–546, 1937) (Reprint in: Davis, 2004)
263. Turing, A.M.: Computability and λ-Definability. Journal of Symbolic Logic **2**, 153–163 (1937)
264. Turing, A.M.: Systems of Logic based on Ordinals. Proceedings of the London Mathematical Society **45**(2), 161–228 (1939). (Reprint in: Davis, 2004)
265. Urquhart, A.: Emil Post. In: D.M. Gabbay, J. Woods (eds.) Handbook of the History of Logic, vol. 5. Logic from Russell to Church, pp. 617–666. Elsevier (2009)
266. van Atten, M., Kennedy, J.: Gödel's Logic. In: D.M. Gabbay, J. Woods (eds.) Handbook of the History of Logic, vol. 5. Logic from Russell to Church, pp. 449–510. Elsevier (2009)
267. van den Dries, L.: Alfred Tarski's Elimination Theory for Real Closed Fields. The Journal of Symbolic Logic **53**(1), 7–19 (1988)
268. van Heijenoort, J. (ed.): From Frege to Gödel: A Source Book in Mathematical Logic, 1879-1931. Harvard University Press (1999)
269. Vaught, R.L.: Alfred Tarski's Work in Model Theory. The Journal of Symbolic Logic **51**(4), 869–882 (1986)
270. von Neumann, J.: First Draft of a Report on the EDVAC. (1945) (A modern interpretation in: Godfrey, M.D, Hendry, D.F.: The Computer as von Neumann Planned It, IEEE Annals of the History of Computing, **14**(3):1993)
271. von Neumann, J.: The General and Logical Theory of Automata. In: L.A. Jeffress (ed.) Cerebral Mechanisms in Behavior. The Hixon Symposium, New York, pp. 1–41. Wiley (1951)
272. von Neumann, J.: Theory of Self-Reproducing Automata. University of Illinois Press (1966). (Edited and completed by A. W. Burks.)

273. Wang, H.: Proving Theorems by Pattern Recognition, II. Bell System Technical Journal **40**, 1–41 (1961)
274. Wang, P.S.: The Undecidability of the Existence of Zeros of Real Elementary Functions. J. Assoc. Comput. Mach. **21**(4), 586–589 (1974)
275. Weber, R.: Computability Theory, *Student Mathematical Library*, vol. 62. American Mathematical Society (2012)
276. Whitehead, A.N., Russell, B.: Principia Mathematica. Cambridge University Press (1910-13). (2nd edn., 1925–27)
277. Winfried, J., Weese, M.: Discovering Modern Set Theory, vol. I. The Basics. American Mathematical Society (1996)
278. Wolfram, S.: Undecidability and Intractability in Theoretical Physics. Physical Review Letters **54**(8), 735–738 (1985)
279. Wolfram, S.: A New Kind of Science, 1st edn. Wolfram Media (2002)
280. Yates, C.E.M.: Three Theorems on the Degrees of Recursively Enumerable Sets. Duke Mathematical Journal **32**(3), 461–468 (1965)
281. Yates, C.E.M.: A Minimal Pair of Recursively Enumerable Degrees. The Journal of Symbolic Logic **31**(2), 159–168 (1966)

Index

λ-calculus, 85–87
λ-definable function, 86, 96, 319–323, 337, 338
λ-term, 85–87
  α-conversion, 85–87
  β-contraction, 85–87
  β-normal form (β-nf), 87
  β-redex, 87
  β-reduction, 87
  abstraction, 85–87
  application, 85–87
  Church numeral, 87, 319
  defining a numerical function, 87
  final, 85–87
  initial, 85–87
μ-Recursive function, 81–83, 99, 100, 161, 165, 207
  Ackermann, 81, 109
  initial, 81–83
    identity (=projection) π, 81–83
    projection (=identity) π, 81–83
    successor σ, 81–83
    zero ζ, 81–83
  non-total, 82
  primitive, 81, 161
  relative to a set, 245
  rule of construction
    μ-operation (Unbounded Search), 82–83, 99
    composition, 81–83
    primitive recursion, 81–83, 165
μ-operation, *see* μ-Recursive function

Abstract computing machine, 95, 219, 223
  basic instruction, 95
    finite effect, 95
  behavior

    global, 223
    local, 223
  capability, 95
  code $\langle \cdot \rangle$, 95
  components, 95
  internal configuration, 95, 114
  resources, xi, 95
  unrestricted, xi, 95
Ackermann, 26, 81, 109
  function, 81, 109
Adleman, 98
Aharonov, 349
Al-Khwarizmi, 6
Aleph zero $\aleph_0$, 15, 65, 186
Algorithm, 3, 57, 77
  characterization of the concept, 79
    Church's, 86
    convincing, 97
    Gödel-Kleene's, 82
    Herbrand-Gödel's, 85
    Markov's, 94
    Post's, 93
    Turing's, 90
  description, 79
  environment, 79
  execution, 78, 79
  existence, 79
  formally, 99
  intuitively, 4, 78, 98
  non-existence, 79
  offline, 247
  online, 247
  problems about, 189
  property, 208–209
  quantum, 78
  shrewd, 209
Alphabet, 27, 32, 40

Printed in the United States
by Baker & Taylor Publisher Services

Printed in the United States
by Baker & Taylor Publisher Services